Potential Theory – ICPT 94

Potential Theory – ICPT 94

Proceedings of the International Conference
on Potential Theory
held in Kouty, Czech Republic, August 13–20, 1994

Editors

Josef Král
Jaroslav Lukeš
Ivan Netuka
Jiří Veselý

Walter de Gruyter · Berlin · New York 1996

Editors

J. Král
Math. Institute
Czech Academy of Sciences
11567 Prague 1
Czech Republic

J. Lukeš
Dept. of Math. Analysis
Charles University
18600 Prague 8
Czech Republic

I. Netuka
Math. Institute
Charles University
18600 Prague 8
Czech Republic

J. Veselý
Math. Institute
Charles University
18600 Prague 8
Czech Republic

1991 Mathematics Subject Classification: 31−06

Printed on acid-free paper which falls within the guidelines of the ANSI to ensure permanence and durability.

Library of Congress Cataloging-in-Publication-Data

International Conference on Potential Theory (1994 ; Kouty, Czech Republic)
Potential theory − ICPT 94 : proceedings of the International Conference on Potential Theory, held in Kouty, Czech Republic, August 13−20, 1994 / editors, Josef Král ... [et al.].
p. cm.
ISBN 3-11-014654-1 (alk. paper)
1. Potential theory (Mathematics) − Congresses.
I. Král, Josef, DrSc. II. Title.
QA404.7.I57 1994
515'.9−dc20 96-11744
CIP

Die Deutsche Bibliothek − Cataloging-in-Publication-Data

Potential theory : proceedings of the International Conference on Potential Theory, held in Kouty, Czech Republic, August 13−20, 1994 / ICPT 94. Ed. Josef Král ... − Berlin : New York : de Gruyter, 1996
ISBN 3-11-014654-1
NE: Král, Josef [Hrsg.]; International Conference on Potential Theory <1994, Kouty>

Printed in Germany. Typeset using the authors' TEX files: I. Zimmermann, Freiburg.
Printing: Ratzlow-Druck, Berlin. Binding: Lüderitz & Bauer, Berlin.
Cover design: Thomas Bonnie, Hamburg.

Preface

International meetings of specialists in potential theory have a well-established tradition. On August 13–20, 1994 an International Conference on Potential Theory was held in Kouty, Czech Republic. (It was the second Conference in this series organized in Bohemia; the first was held in Prague in 1987.) Participants remembered the main organizer of the previous International Conference on Potential Theory held in Amersfoort in 1991, Professor Emil Bertin, who died in March 1994. We were also grieved to learn that Professor Myron Goldstein, a participant of the Conference in Kouty, died unexpectedly shortly after the Conference. Both of them represent distinguished personalities whose contribution to the field will not be forgotten.

The 1994 Conference was organized by the Faculty of Mathematics and Physics of Charles University; generous financial supports of Konsolidační banka, Praha, and Hewlett-Packard, Praha, are gratefully acknowledged. Support coming from the Czech Grant Agency (grant No. 201/93/2174), the Grant Agency of Charles University (grant No. 354/1993) and the Grant Agency of the Czech Academy of Sciences (grant No. 11957) enabled several mathematicians to take part. Thanks are due to members of the Prague Seminar on Mathematical Analysis as well as colleagues, students and staff from the Faculty who selflessly helped to overcome all the difficulties connected with the Conference.

The conference was attended by 73 participants and 13 accompanying persons from 18 countries.

The present volume includes 11 invited survey lectures, 23 contributions and formulations of 18 problems communicated by the participants. The editors are grateful to the referees who kindly revised the submitted manuscripts.

Topics included in the proceedings represent a sample of present research in potential theory which is a vast field intertwining with many modern branches of mathematics. They concern basic properties of harmonic functions and their boundary behaviour, as well as approximation properties, both in the classical context and in an abstract setting. Connections with analytic functions and the theory of linear and nonlinear differential operators and pseudodifferential operators are studied, semigroups of operators, probabilistic aspects and potentials on graphs and fractals are treated.

We hope that this volume will be of interest to mathematicians and graduate students specializing in Potential Theory and related branches of Analysis and Probability.

Praha, August 1995

Josef Král
Ivan Netuka
Jaroslav Lukeš
Jiří Veselý

Contents

Contributed Papers

Recollections, reflections, results

Introductory thoughts to the ICPT 1994

Heinz Bauer

Abstract. A survey lecture with two main directions: 1. The role of potential theory in connection with the existence of positive harmonic functions on a complete Riemannian manifold with pinched negative sectional curvature (based on the work of M. T. Anderson and R. Schoen). 2. Potential theory of the nonlinear equation $\Delta u = u^{\alpha}$ appearing in a paper by C. Loewner and L. Nirenberg [LN74] and its connection with superprocesses.

1991 Mathematics Subject Classification: 31C35, 35J60, 53C21, 60J85

The author had the previlege to give the closing address at the 1987 *Prague Conference on Potential Theory.* The last words of this address, spoken on August 24, 1987, were: "I declare the *First* Prague Conference on Potential Theory closed." He had inserted the word "First" in order to express the predominant wish of the participants to return to Prague for another such congress in a not too distant future. However, in those days the outlook for another ICPT in Czechoslovakia was not at all positive.

After these *Recollections* the author expressed his and the participants' enthousiasm and happiness that – thanks to the powerful initiative of the "Harmonic Group" of colleagues from Prague – it became possible to have another ICPT in Bohemia. This time, after breathtaking political changes in Eastern Europe, in a country deliberated from the pressure to look only eastward, a country which finally could find its way back to the center of Europe in freedom.

The author's *Reflections* centered around the widely accepted fact that mathematics becomes particularly interesting and promising whenever different fields of mathematics interfere or approach each other with the chance to amalgamate.

In the main part of the lecture, it was then the author's aim to point out *Results* in connection with two such amalgamation processes: amalgamation of potential theory with Riemannian geometry as well as amalgamation of nonlinear aspects of potential theory with the probabilistic theory of superprocesses.

In what follows the main ideas and results of this main part of the lecture will be presented in a condensed form with complete references.

1. Negatively curved manifolds, boundaries and Liouville property

We shall consider a smooth Riemannian manifold $M = (M, g)$ of dimension $n \geq 2$ (C^∞, connected, with countable base) with tangent bundle $p \mapsto T_pM$, $p \in M$. As usual $g(p) = \langle \cdot, \cdot \rangle_p$ denotes the given inner product on T_pM varying smoothly with p. In local coordinates g is described by the fundamental metric tensor (g_{ij}), i.e., by the quadratic form

$$ds^2 = g_{ij} dx^i dx^j$$

(with the usual convention about summation). The quadratic form leads to the length $L(\gamma)$ of a parametrized curve $\gamma : [0,1] \to M$ and to the internal metric

$$d(p,q) := \inf\{L(\gamma) : \ \gamma \text{ connecting } p \text{ with } q\}$$

on M. We assume that M is *complete* in this metric. According to the Hopf–Rinow-Theorem we then know that, for every $p \in M$, the exponential map $\exp_p$ is a diffeomorphism of T_pM onto M, that each two points of M can be joined by a (minimal) geodesic, and that geodesics can be extended to geodesic rays (or lines).

By means of exp we obtain the *Laplace–Beltrami* operator Δ_M of M in the form

$$\Delta_M f(p) := \Delta(f \circ \exp_p)(0),$$

where 0 is the origine of T_pM. The smooth solutions of $\Delta_M h = 0$ are the *harmonic functions*. M is said to have the *Liouville property* if all bounded harmonic functions, globaly defined on M, are constant.

The usual *sectional curvature* of M will be denoted by K_M. For $p \in M$, $K_M(p)$ is defined on the Grassmannian manifold of the 2-dimensional linear subspaces of T_pM. E. B. Dynkin, P. Malliavin and S. T. Yau have asked (around 1970) whether the Liouville property *fails* if M, besides being complete, is *simply connected* and has *pinched sectional curvature*, i.e., if K_M satisfies

$$-\infty < -b^2 \leq K_M \leq -a^2 < 0.$$

The motivation for this problem comes from the special case when K_M is a real constant K. Up to isometries, the manifold M then equals the well-known *space forms* $H^n(K)$:

For $K = 0$, $H^n(0) = \mathbb{R}^n$ with euclidean metric,
for $K > 0$, $H^n(K) = S^n(1/\sqrt{K})$ with induced euclidean metric,
for $K < 0$, $H^n(K) = B^n(\sqrt{-4/K})$ with Poincaré metric

$$g_{ij}(x) = \frac{\delta_{ij}}{1 + K/4 \sum_{i=1}^n x_i^2}$$

in standard coordinates.

Here $S^n(r)$, $B^n(r)$ denote the n-dimensional sphere or ball of radius r and with center at the origin, respectively. Obviously, the Liouville property fails for these space forms if and only if $K < 0$.

The problem of Dynkin, Malliavin and Yau has found positive solutions by Anderson [An83] and Sullivan [Su83]. An earlier solution under somewhat stronger assumptions was given by Kifer [Ki76]. The most elegant proof is due to Anderson, Schoen [AS85] with important improvements by Ancona [An87]. The main ideas of the proof of Anderson, Schoen was presented in the lecture with particular emphasis on the underlying potential theory. Here we give some of the details:

Geodesic rays lead in a natural way to a compactification

$$\widetilde{M} := M \cup S_\infty$$

by adding to M a set S_∞ of "points at infinity" which turns out to be homeomorphic to an $(n-1)$-sphere. However, this "sphere at infinity" is not smoothy attached to M. Potential theory enters into the picture via the function

$$y \mapsto \exp(-\delta d(x,y)) \qquad (x \in M),$$

which, for $\delta > 0$ small (resp., large), is Δ_M-superharmonic (resp., Δ_M-subharmonic). This leads to a good potential theory on M:

(1) M is a Brelot harmonic space in the strong sense. S_∞ is a regular boundary for M, i.e., every $\varphi \in \mathcal{C}(S_\infty)$ extends continuously to a harmonic function on M. Positivity is respected hereby. Consequently, there exists an abundance of bounded harmonic functions on M.
This shows that the Liouville property fails.

(2) For every $x \in M$ there exists a Green function $G_x = G(x, \cdot)$ with pole at x which extends continuously to $\widetilde{M}$ with zero boundary values on S_∞.

(3) S_∞ turns out to be the Martin boundary of the harmonic space M: For fixed $x_0 \in M$ and for arbitrary $\zeta \in S_\infty$, there exists a unique harmonic function $h_\zeta \geq 0$ on M with the properties:

$$h_\zeta(x_0) = 1 \text{ and } h_\zeta = \mathrm{O}(G_{x_0}) \text{ as } x \to \zeta' \in S_\infty \setminus \{\zeta\}.$$

In particular, h_ζ is minimal and $\{h_\zeta : \zeta \in S_\infty\}$ is the set of all minimal positive harmonic functions on M, normalized at x_0.

(4) The kernel function $K(x,\zeta) := h_\zeta(x)$ is continuous on $M \times S_\infty$. This Martin kernel leads to the integral representation of harmonic functions $u \geq 0$ on M: There is exactly one Borel measure $\mu \geq 0$ on S_∞ such that

$$u(x) = \int K(x,\zeta)\,\mu(d\zeta).$$

Conversely, every Borel measure $\mu \geq 0$ on S_∞ leads to a harmonic function $u_\mu \geq 0$ on M.

(5) Boundary behavior of Fatou–Doob-type: For $\zeta \in S_\infty$, denote by γ_ζ the geodesic ray connecting x_0 with ζ, and put

$$A_{\zeta,r} := \bigcup_{x \in \gamma_\zeta} B(x,r).$$

For every choice of Borel measures μ, ν on S_∞ with $\nu \neq 0$ the limit

$$\lim_{\substack{x \to \zeta \\ x \in A_{\zeta,r}}} \frac{u_\mu(x)}{u_\nu(x)}$$

exists at ν-almost all $\zeta \in S_\infty$ and equals $d\mu/d\nu$ ν-almost everywhere.

Remarks. 1) S_∞ remains the Martin boundary of M if the Laplace–Beltrami operator Δ_M is replaced by certain elliptic differential operators $\mathcal{L}$ on M. For such a result it is crucial that $\mathcal{L}$ has a nice behavior uniformly on the balls $B(x,1)$ in M with radius equal to 1. Furthermore, the existence of a strictly positive superharmonic function with respect to $\mathcal{L} + \varepsilon I$ for some $\varepsilon > 0$ is needed. The existence of a Green function, however, is not sufficient. Even on $H_n(-1)$ the Martin boundary may reduce to a singleton. (Cf. Ancona [An87].)

2) There is a recent interesting "boundary free" version of a Fatou–Doob-type theorem appearing in connection with the Liouville property. It is a result in a simple probabilistic setting, where the sets $A_{\zeta,r}$ from (5) are replaced by a family of open neighborhoods of bounded (Harnack-)diameter along the trajectories of an M-valued diffusion process. This result of Leeb [Le93] seems to be an interesting object for further studies.

2. Nonlinearity and superprocesses

This second part of the *Results* was dedicated to the memory of Karl Löwner (later: Charles Loewner). He was born on May 29, 1893 in Lány (German: Lana), a small Bohemian town 35 km east of Prague (the summer residence of the President of the Czech Republic). His last mathematical paper (Loewner, Nirenberg [LN74]) was completed by Nirenberg. The main result of this paper were obtained shortly before Loewner's death on January 8, 1968 in Stanford.

The Loewner–Nirenberg paper refers to a domain D in $\mathbb{R}^d$, $d \geq 3$, and the nonlinear differential equation

$$\Delta u = u^\alpha, \qquad (*)$$

where

$$\alpha := \frac{d+2}{d-2}.$$

D will be called *nice* if its boundary ∂D is a compact hypersurface with at most finitely many components.

The main results of the paper are the following:

(i) There exists a maximal C^∞-solution v_D (necessarily unique), i.e., $v_D \geq v$ whenever $v \geq 0$ is a C^∞-solution of $(*)$.
(ii) If D is nice there exists a unique C^∞-solution $v_\infty > 0$ such that $v(x) \to +\infty$ as $x \to \partial D$. (Observe that $v_\infty \leq v_D$ and uniqueness yield $v_\infty = v_D$.)
(iii) On $\mathbb{R}^d \setminus \{0\}$ no positive C^∞-solution with boundary data 0 at origin and $+\infty$ at infinity exists.
(iv) For D nice, $\Phi \in C^\infty(\partial D)$, $\Phi > 0$, there exists a unique function $u \in C^\infty_+(\overline{D})$ such that $u|_{\partial D} = \Phi$. Furthermore $0 < u(x) < \max \Phi$ for all $x \in D$.

We shall now present the intuitive ideas behind the notion of a superprocess.

Consider a diffusion process (ξ_t, P^x) on $\mathbb{R}^d$ with a nice elliptic operator L as generator. The process describes a *randomly moving particle.*

Imagine a branching particle system with birth and death (a "population") underlying a movement in space governed by the original process (ξ_t, P^x). After a suitable limiting procedure, one ends up with a *randomly moving cloud,* i.e., a new Markov process (X_t, P^μ) derived from the original one. The state space is no more $\mathbb{R}^d$ (or a locally compact space with countable base) but the space $\mathcal{M}$ of all bounded positive (Borel-)measures on $\mathbb{R}^d$: Hence, X_t is a random measure (describing the "cloud"), and P^μ is the steering probability with initial distribution $\mu \in \mathcal{M}$.

The new process is called a superprocess derived from the original process (ξ_t, P^x) by means of a certain "branching parameter" ψ.

In our case the original diffusion is Brownian motion associated to the operator $L = \frac{1}{2}\Delta$ on $\mathbb{R}^d$. The branching parameter is

$$\psi(x) = \frac{1}{2}|x|^{\alpha-1}x \qquad \text{on } \mathbb{R}$$

or

$$\psi(x) = \frac{1}{2}x^\alpha \qquad \text{on } \mathbb{R}_+,$$

where the condition

$$1 < \alpha \leq 2$$

is satisfied. For the probabilistic treatment this condition is essential. Hence, we study the superprocess (X_t, P^μ), in fact a super-Brownian motion, in connection with the nonlinear equation

$$\Delta u = u^\alpha \qquad (1 < \alpha \leq 2). \tag{NL}$$

In contrast to this, is the linear case

$$\Delta u = 0 \tag{L}$$

(governed by Brownian motion).

The main contributors to the field of superprocesses are D. A. Dawson, E. B. Dynkin, J. F. Le Gall and E. A. Perkins. For all technical details, references and, in particular, for the following results about the probabilistic treatment of equation $(*)$ and the discussion of the Loewner–Nirenberg results, the reader is referred to Dynkin [Dy91], [Dy93] and, in addition, to [Da93], [Le91], [Pe90].

We shall outline results about the treatment of (NL) and their classical counterpart for the equation (L) and, hence, their relation to classical potential theory.

Consider a regular bounded domain D in $\mathbb{R}^d$ and a continuous real-valued boundary function f. The unique solution of the corresponding boundary value problem is given in probabilistic terms for the equation (L) by

$$u_f(x) = E^x(f \circ \xi_\tau) \qquad (x \in D) \tag{L$_1$}$$

where $\tau = \tau_D$ is the first exit time from D.

In the nonlinear case (NL) it is possible to define X_τ (an $\mathcal{M}$-valued random variable) with $\tau = \tau_D$ for good sets and, among them, open ones. The nonlinear counterpart to (L$_1$) is then

$$u_f(x) = -\log E^{\delta_x}\left[e^{-\langle f, X_\tau\rangle}\right] \tag{NL$_1$}$$

for $f \in C_+(\partial D)$. Here δ_x is the Dirac measure at x and, by definition, $\langle f, X_\tau\rangle = \int f\, dX_\tau$. Obviously, $u_f \geq 0$.

Again in the nonlinear case for a regular bounded domain D, one has the following result:

$$u(x) = -\log P^{\delta_x}\{X_\tau = 0\} \tag{NL$_2$}$$

is the only solution $u \geq 0$ of (NL) with the boundary behavior $u(x) \to +\infty$ as $x \to z$ for arbitrary $z \in \partial D$. Consequently, $u = v_\infty$ in the situation (ii) considered by Loewner–Nirenberg.

For the study of the maximal positive solution of (NL), one has to introduce the *range* $\mathcal{R}$ of the superprocess X as the smallest closed subset of $\mathbb{R}^d$ containing all clouds, i.e.,

$$\mathcal{R} = \overline{\bigcup_{t \geq 0} \operatorname{supp} X_t}.$$

Let us underline that $\mathcal{R}$ is a set-valued random variable. Whenever $\mu \in \mathcal{M}$ has compact support, then $\mathcal{R}$ is P^μ-almost surely compact.

The following result then is the probabilistic version of the Loewner–Nirenberg result (i):

For D open $\subset \mathbb{R}^d$

$$v_D(x) = -\log P^{\delta_x}\{\mathcal{R} \subset D\} \tag{NL$_3$}$$

is the (unique) maximal solution of (NL) among all solution $v \geq 0$.

In classical potential theory, polarity of a (Borel or analytic) set $A \subset \mathbb{R}^d$ is characterized by the property

$$P^x\{\xi_t \in A \text{ for some } t > 0\} = 0 \qquad \text{for all } x \in \mathbb{R}^d. \tag{L$_4$}$$

The corresponding nonlinear notion is S-polarity: a Borel set $A \subset \mathbb{R}^d$ is called *S-polar* if

$$P^{\delta_x}\{\mathcal{R} \cap A \neq \emptyset\} = 0 \qquad \text{for all } x \in \mathbb{R}^d \setminus A. \tag{NL$_4$}$$

The connection with the maximal solution v_D then is as follows:

Let F be a closed set $\subset \mathbb{R}^d$ and D its complement. Then F is S-polar if and only if one of the following two equivalent statements hold: (NL$_5$)

$$\text{(a) } v_D = 0; \qquad\qquad \text{(b) } v_D \text{ bounded.}$$

In the linear case (L) a set A is polar if and only if $\text{cap}A = 0$ for the Newtonian capacity or – equivalently – if the Newtonian potential N^μ is unbounded for every measure $\mu \neq 0$ from $\mathcal{M}$ supported by A.

In the nonlinear case (NL) one has to replace the Newtonian kernel (for dimension $d \geq 3$)

$$N(x,y) = \int_0^{+\infty} p_t(x-y)\,dt \qquad (x, y \in \mathbb{R}^d)$$

by the *Bessel kernel*

$$B(x,y) = \int_0^{+\infty} e^{-t/2} p_t(x-y)\,dt \qquad (x, y \in \mathbb{R}^d)$$

where

$$p_t(z) = \left(\frac{1}{2\pi t}\right)^{d/2} \exp\left(-\frac{|z|^2}{2t}\right) \qquad (z \in \mathbb{R}^d).$$

Then cap_α is defined as follows:

$$\text{cap}_\alpha A = \sup\left\{\|\mu\| : \mu \in \mathcal{M} \text{ supported by } A,\ \|B^\mu\|_{L^\alpha} \leq 1\right\},$$

and one has the equivalence

$$\text{cap}_\alpha A = 0 \iff B^\mu \notin L^\alpha(\mathbb{R}^d) \text{ for all } \mu \neq 0 \text{ from } \mathcal{M}, \text{ supported by } A.$$

It then follows for Borel sets $A \subset \mathbb{R}^d$:

$$A \text{ is } S\text{-polar} \iff \text{cap}_\alpha A = 0. \qquad \text{(NL}_6\text{)}$$

In particular, a result of Brezis and Véron [BV80] states that singletons in $\mathbb{R}^d$ are S-polar if and only if $d \geq \frac{2\alpha}{\alpha-1}$ for the given α from (NL).

Finally, we again emphasize that the above mentioned probabilistic results concerning (NL) are only available for $1 < \alpha \leq 2$. In the situation of Loewner, Nirenberg [LN74] α equals $\frac{d+2}{d-2}$ for $d \geq 3$. Consequently, the probabilistic interpretation is only available for dimension $d \geq 6$.

We close by pointing out that in Constantinescu [Co65] the author mentions that his results allow applications to solutions $u \geq 0$ of $\Delta u = u^\alpha$ with exponent $\alpha \geq 1$. Details about this claim seem to be unpublished.

References

[An87] Ancona, A., Negatively curved manifolds, elliptic operators and the Martin boundary. Ann. of Math. 125 (1987), 495–536.

[An83] Anderson, M. T., The Dirichlet problem at infinity for manifolds of negative curvature. J. Differential Geom. 18 (1983), 701–721.

[AS85] Anderson, M. T., Schoen, R., Positive harmonic functions on complete manifolds of negative curvature. Ann. of Math. 121 (1985), 429–461.

[BV80] Brezis, H., Véron, L., Removable singularities of some nonlinear equations. Arch. Rational Mech. Anal. 5 (1980), 1–6.

[Co65] Constantinescu, C., An axiomatic theory for the non-linear Dirichlet problem. Rev. Roum. Math. Pures Appl. 10 (1965), 755–764.

[Da93] Dawson, D. A., Measure-valued Markov processes. Ecole d'été de Probabilités de Saint-Flour XXI, 1991. Lecture Notes in Math. 1541, Springer-Verlag, Berlin 1993.

[Dy91] Dynkin, E. B., A probabilistic approach to one class of nonlinear differential equations. Probab. Theory Related Fields 89 (1991), 89–115.

[Dy93] Dynkin, E. B., Superprocesses and partial differential equations. Ann. Probab. 21 (1993), 1185–1262.

[Ki76] Kifer, Yu. I., Brownian motion and harmonic functions on manifolds of negative curvature. Theory Probab. Appl. 21 (1976), 81–95.

[Le93] Leeb, B., Harmonic function along Brownian balls and the Liouville property for solvable Lie groups. Math. Ann. 84 (1993), 577–584.

[Le91] Le Gall, J. F., Brownian excursions, trees and measure-valued branching processes. Ann. Probab. 19 (1991), 1399–1439.

[LN74] Loewner, C., Nirenberg, L., Partial differential equations invariant under conformal or projective transformations. In: Contributions to Analysis (ed. by L. Ahlfors and L. Bers), 245–272, Academic Press, New York 1974.

[Pe90] Perkins, E. A., Polar sets and multiple points for super-Brownian motion. Ann. Probab. 18 (1990), 453–491.

[Su83] Sullivan, D., The Dirichlet problem at infinity for a negatively curved manifold. J. Differential Geom. 18 (1983), 723–732.

Constructions of Dirichlet structures

Nicolas Bouleau

Abstract. We show how the constructions of Dirichlet structures allow to equip the main probabilistic objects with Dirichlet forms. We emphasize the case of local **white** Dirichlet structures on the Poisson space and on the Wiener space. This yields tools for studying the existence of density of functionals of processes with independent increments and of stationary processes.

1991 Mathematics Subject Classification: 31C25.

Introduction

We shall use the expression 'Dirichlet structure' to denote a term $(\Omega, \mathcal{F}, m, \mathbb{D}, \mathcal{E})$ where $(\Omega, \mathcal{F}, m)$ is a measure space, most often a probability space, and $\mathcal{E}$ a Dirichlet form defined on a dense subspace $\mathbb{D}$ of $L^2(m)$, i.e., a closed symmetric positive bilinear form on which contractions operate.

Our aim is simplicity: to construct Dirichlet structures by algebraical methods which follow the classical methods of constructions of usual probability spaces. Our purpose is both pedagogical and methodological to get tools for obtaining results (existence of densities, variational calculus, etc.) on the most fundamental probabilistic objects. This is an extension of Chapter V. *The algebra of Dirichlet structures* of the book written with Francis Hirsch [BH].

We are studying especially **white** structures. For probabilistic structures the concept of whiteness is related to two ideas:

(i) spatial independence: random variables with disjoint sets of indices are independent,

(ii) stationarity: invariance in law under translation of the index set, which is therefore a group.

For Dirichlet structures we reserve the word white for the notion which is the conjunction of the following properties:

(i) spatial independence in the Dirichlet sense, that is in the sense of product of Dirichlet structures,

(ii) translation invariance,

(iii) locality, i.e., we restrict ourselves to local Dirichlet structures, so that the functional calculus and the density criterion hold.

This does not prevent us from using sometimes whiteness in a weaker sense with only (i) and (ii) or even with (i) weakened in spatial orthogonality.

With respect to the existing literature about Dirichlet structures on the Poisson space (or about symmetric semigroups acting on the Poisson space giving rise to a Malliavin calculus), which is numerous ([DKW], [BGJ], [W], [CP], [NV], etc.), the approach is in the spirit of [BGJ] and [W] but with a presentation which allows to study more easily the criterion of existence of density. The structure on the elementary Poisson space (on $\mathbb{R}_+$) defined by Carlen and Pardoux [CP] which is local and possesses the property that the divergence operator coincides with the stochastic integral on previsible processes and satisfies the density criterion (EID) as proved in [BH], Chapter V., § 5.3 (cf. also [P]) but *it is not* white and does not extend easily to define structures for general processes with independent increments (PII's).

About the case of Wiener space, the approach can be seen as an elementary introduction to the white noise theory and to the second quantization.

1. Products, Dirichlet independence

1.1. Notation and definitions

We consider first Dirichlet structures $(\Omega, \mathcal{F}, m, \mathbb{D}, \mathcal{E})$ where $(\Omega, \mathcal{F}, m)$ is a probability space and $(\mathbb{D}, \mathcal{E})$ a Dirichlet form, i.e., a closed symmetric positive bilinear form with dense domain $\mathbb{D}$ in $L^2(m)$ such that

$$f \in \mathbb{D} \quad \Rightarrow \quad f \wedge 1 \in \mathbb{D} \quad \text{and} \quad \mathcal{E}(f \wedge 1) \leq \mathcal{E}(f),$$

and we suppose $1 \in \mathbb{D}$ and $\mathcal{E}(1) = 0$. To give such a structure is equivalent to consider a strongly continuous symmetric semigroup $(P_t)_{t \geq 0}$ on $L^2(m)$ which is Markovian ($0 \leq f \leq 1 \Rightarrow 0 \leq P_t f \leq 1,\ P_t 1 = 1$), cf. [BH].

Three properties concerning Dirichlet structures will be interesting for us:

- (OCC) (Existence of a *carré du champ* operator):

$$\forall f \in \mathbb{D} \cap L^\infty,\ \exists \tilde{f} \in L^1,\ \forall h \in \mathbb{D} \cap L^\infty, \quad 2\mathcal{E}(fh, f) - \mathcal{E}(h, f^2) = \int \tilde{f} h \, dm.$$

Then one sets $\Gamma(f,f) = \tilde{f}$, $\Gamma(f,g)$ is defined by polarization. The operator Γ is a continuous operator from $\mathbb{D} \times \mathbb{D}$ into $L^1(m)$, it is uniquely defined. One has

$$\mathcal{E}(f,g) = \frac{1}{2} \int \Gamma(f,g) \, dm, \quad \forall f,g \in \mathbb{D}.$$

If F is a contraction then

$$\Gamma(F \circ f, F \circ g) \leq \Gamma(f,f), \quad \forall f \in \mathbb{D}.$$

- (L) (Locality):

$$\forall f \in \mathbb{D}, \quad \mathcal{E}(|f+1| - 1) = \mathcal{E}(f),$$

or, equivalently,

$$\forall f,g \in \mathbb{D}, \quad fg = 0 \Rightarrow \mathcal{E}(f,g) = 0.$$

Under (OCC) and (L) a functional calculus holds: If $f \in \mathbb{D}^m$, $g \in \mathbb{D}^n$, F, G Lipschitz and C^1,

$$\Gamma(F(f), G(g)) = \sum_{i,j} F_i'(f) G_j'(g) \Gamma(f_i, g_j);$$

these hypotheses can be weakened if $m = n = 1$:

If F, G are Lipschitz from $\mathbb{R}$ to $\mathbb{R}$ then

$$\Gamma(F(f), G(g)) = F'(f) G'(g) \Gamma(f, g),$$

where F', G' are the derivatives in the sense of Lebesgue; this formula makes sense by the fact that

$$\forall f \in \mathbb{D}, \quad f_*(\Gamma(f,f).m) << \lambda_1 \text{ (Lebesgue measure on } \mathbb{R}).$$

- (EID) (Energy image density property):
  It is the preceding property extended to the case of dimension k:

$$\forall f \in \mathbb{D}^k, \quad f_*(\det(\Gamma(f, f^*)).m) << \lambda_k,$$

where $\Gamma(f, f^*)$ is the matrix $\Gamma(f_i, f_j)$.

This property is satisfied on the Wiener space when equipped with the form associated with the Ornstein–Uhlenbeck semigroup.

1.2. Products

Definition. Let $S_i = (\Omega_i, \mathcal{F}_i, m_i, \mathbb{D}_i, \mathcal{E}_i)$, $i = 1, 2$ be two Dirichlet structures as defined in 1.1. The product structure $S_1 \otimes S_2$ is defined as $(\Omega_1 \times \Omega_2$, $\mathcal{F}_1 \otimes \mathcal{F}_2$, $m_1 \times m_2$, $\mathbb{D}$, $\mathcal{E})$ with

$$\mathbb{D} = \Big\{ f \in L^2(m_1 \times m_2) : f(.,y) \in \mathbb{D}_1 \text{ for } m_2\text{-a.e. } y, f(x,.) \in \mathbb{D}_2 \text{ for } m_1\text{-a.e. } x,$$

$$\int \mathcal{E}_1(f(.,y))\, dm_2(y) + \int \mathcal{E}_2(f(x,.))\, dm_1(x) < +\infty \Big\}$$

and

$$\mathcal{E}(f) = \int \mathcal{E}_1(f(.,y))\, dm_2(y) + \int \mathcal{E}_2(f(x,.))\, dm_1(x).$$

It is easy to show that $S_1 \otimes S_2$ is a Dirichlet structure. If S_1 and S_2 satisfy (OCC) then $S_1 \otimes S_2$ satisfies (OCC), too and

$$\Gamma(f)(x,y) = \Gamma_1(f(.,y))(x) + \Gamma_2(f(x,.))(y).$$

If S_1 and S_2 are local, $S_1 \otimes S_2$ is local, too.

Infinite products

Definition. Let $S_i = (\Omega_i, \mathcal{F}_i, m_i, \mathbb{D}_i, \mathcal{E}_i)$, $i \in \mathbb{N}$ be Dirichlet structures. The product $S = \bigotimes_{i\in\mathbb{N}} S_i$ is defined by $S = (\Omega, \mathcal{F}, m, \mathbb{D}, \mathcal{E})$ with

$$(\Omega, \mathcal{F}, m) = \bigotimes_{i\in\mathbb{N}} (\Omega_i, \mathcal{F}_i, m_i) \quad \text{(product probability space)},$$

$$\mathbb{D} = \Big\{ f \in L^2(m) \ : \ \forall i, \text{ for almost all } \omega_0, \omega_1, \ldots, \omega_{i-1}, \omega_{i+1}, \ldots \quad \text{the map}$$

$$x \mapsto f(\omega_0, \omega_1, \ldots, \omega_{i-1}, x, \omega_{i+1}, \ldots) \text{ belongs to } \mathbb{D}_i \text{ and } \int \sum_{i\in\mathbb{N}} \mathcal{E}_i(f)\, dm \ < +\infty \Big\}$$

and

$$\mathcal{E}(f) = \int \sum_{i\in\mathbb{N}} \mathcal{E}_i(f)\, dm.$$

The fact that this defines a closed form comes from the following argument: Let (f_n) be a Cauchy sequence in $\mathbb{D}$ equipped with the norm $\|\,.\,\|_1$ defined by $\|\,.\,\|_1 = \mathcal{E} + \|\,.\,\|^2_{L^2}$, such that

$$\|f_n - f\|_{L^2} \to 0.$$

There exists a subsequence f_{n_k} such that

$$\sum_k \|f_{n_k} - f\|^2 < +\infty \text{ and } \sum_k (\mathcal{E}(f_{n_{k+1}} - f_{n_k}))^{\frac{1}{2}} < +\infty.$$

Then for every i and for almost every $\omega_0, \omega_1, \ldots, \omega_{i-1}, \omega_{i+1}, \ldots$ (product measure) one has

(a) $\displaystyle\sum_k \int (f_{n_k}(\ldots, \omega_{i-1}, x, \omega_{i+1}, \ldots) - f(\ldots, \omega_{i-1}, x, \omega_{i+1}, \ldots))^2 dm_i(x) < +\infty,$

(b) $\displaystyle\sum_k (\mathcal{E}_i(f_{n_{k+1}} - f_{n_k}))^{\frac{1}{2}} \ < +\infty;$

the last fact follows from the inequalities

$$\int \mathcal{E}_i(g)^{\frac{1}{2}}\, dm \ \leq \ (\int \mathcal{E}_i(g)\, dm)^{\frac{1}{2}} \ \leq \ \mathcal{E}(g)^{\frac{1}{2}}.$$

It follows from the fact that the form $(\mathbb{D}_i, \mathcal{E}_i)$ is closed that for almost every $\omega_0, \omega_1, \ldots, \omega_{i-1}, \omega_{i+1}, \ldots$ the map

$$x \mapsto f(\omega_0, \omega_1, \ldots, \omega_{i-1}, x, \omega_{i+1}, \ldots)$$

is in $\mathbb{D}_i$ and $\mathcal{E}_i(f_{n_k} - f) \to 0$; hence by the Fatou lemma

$$\int \sum_{i\in\mathbb{N}} \mathcal{E}_i(f)\, dm \ = \ \int \sum_{i\in\mathbb{N}} \lim_k \mathcal{E}_i(f_{n_k})\, dm \ \leq \ \liminf_k \mathcal{E}(f_{n_k})$$

and since f_n is Cauchy in $\mathbb{D}$, this last term is finite and $f \in \mathbb{D}$. Similarly,

$$\int \sum_{i\in\mathbf{N}} \mathcal{E}_i(f - f_n)dm = \int \sum_{i\in\mathbf{N}} \lim_k \mathcal{E}_i(f_{n_k} - f_n)dm \le \liminf_k \mathcal{E}(f_{n_k} - f_n),$$

which can be made $\le \varepsilon$ for k and n large enough since f_n is Cauchy in $\mathbb{D}$, so we have $f_n \to f$ in $\mathbb{D}$. After proving the closedness, the fact that the form is Dirichlet follows easily. □

• Like for the finite products, if the S_i's are local then S is local, too.
• If every S_i satisfies (OCC) with carré du champ operator Γ_i, then S satisfies (OCC) and its carré du champ operator Γ is given by

$$\Gamma(f) = \sum_{i\in\mathbf{N}} \Gamma_i(f), \quad f \in \mathbb{D}.$$

• For the property (EID), the passage from finite to infinite dimension (which is the heart of the proof of property (EID) on the Wiener space, cf. [BH]) comes from the following fact:

Proposition. *Let S_i, $i \in \mathbb{N}$ be Dirichlet structures satisfying* (OCC) *and* (L). *If each finite product*

$$S_u = \bigotimes_{i\in u} S_i, \qquad u \textit{ finite part of } \mathbb{N}$$

satisfies (EID), *then the infinite product*

$$S = \bigotimes_{i\in\mathbf{N}} S_i$$

satisfies (EID).

Proof. Let $U \in \mathbb{D}^n$ and the notation is the same as in the preceding proof. Then

$$\Gamma(U, U^*) = \lim_{N\uparrow\infty} \sum_{i=0}^{N} \Gamma_i(U, U^*)$$

is an increasing limit in the sense of semi-definite positive matrices. Hence if B is a Lebesgue negligible Borel set in $\mathbb{R}^n$,

$$\int 1_B \circ U . \det \Gamma(U, U^*)\, dm = \lim_N \int 1_B \circ U . \det (\sum_{i=0}^{N} \Gamma_i(U, U^*))\, dm = 0$$

and the proposition is proved. □

1.3. Images

We are now almost able to define the concept of D-independence. But, as in the case of probability structures, first we need to define the law (here the D-law);

indeed, independence means that the law of a pair is the product of the laws. Thus we need the notion of image structure.

1.3.1. Finite dimensional images

Definition. Let $(\Omega, \mathcal{F}, m, \mathbb{D}, \mathcal{E})$ be a Dirichlet structure.

(a) For $d \in \mathbb{N}^*$ and $U \in \mathbb{D}^d$, let us define

$$\widetilde{\mathbb{D}_U} = \{f \in L^2(U_* m) \ : \ f \circ U \in \mathbb{D}\}, \quad \widetilde{\mathcal{E}_U}(f) = \mathcal{E}(f \circ U).$$

Then $(\mathbb{R}^d, \mathcal{B}(\mathbb{R}^d), U_* m, \widetilde{\mathbb{D}_U}, \widetilde{\mathcal{E}_U})$ is a Dirichlet structure and the set Λ_d of Lipschitz functions from $\mathbb{R}^d$ into $\mathbb{R}$ is in $\widetilde{\mathbb{D}_U}$.

(b) Let $\mathbb{D}_U$ be the closure of Λ_d in $\widetilde{\mathbb{D}_U}$ and $\mathcal{E}_U = \widetilde{\mathcal{E}_U}|_{\mathbb{D}_U}$. Then the Dirichlet structure $(\mathbb{R}^d, \mathcal{B}(\mathbb{R}^d), U_* m, \mathbb{D}_U, \mathcal{E}_U)$ is regular in the sense of Fukushima (i.e., there is a dense subspace of $C_K(\mathbb{R}^d)$ equipped with uniform topology which is also dense in $\mathbb{D}_U$).

The structure $(\mathbb{R}^d, \mathcal{B}(\mathbb{R}^d), U_* m, \mathbb{D}_U, \mathcal{E}_U)$ is called the image of S by U and denoted by $U_* S$.

Remark. There are cases in which $\widetilde{\mathbb{D}_U} \neq \mathbb{D}_U$.

• If S satisfies (OCC) with carré du champ operator Γ, then $U_* S$ satisfies (OCC) with carré du champ operator Γ_U given by

$$\Gamma_U(f)(x) = \mathbb{E}\,[\Gamma(f \circ U)|U = x] \quad U_* m\text{-a.e.},$$

where $\mathbb{E}$ refers to the measure m. In other words, for all bounded g

$$\mathbb{E}\,[g \circ U . \Gamma_U(f)] = \mathbb{E}\,[g \circ U . \Gamma(f \circ U)].$$

• If S is local then $U_* S$ is local, too.

Under (OCC) and (L) for S we have thus $\forall F \in \Lambda_d \cap C^1(\mathbb{R}^d)$

$$\Gamma_U(F)(x) = \sum_{i,j=1}^{d} \mathbb{E}\,[\Gamma(U_i, U_j)|U = x]\, \frac{\partial F}{\partial x_i}(x)\, \frac{\partial F}{\partial x_j}(x).$$

Remark. The fact that Dirichlet structures can be easily transported by images is remarkable, because the associated semigroups and the associated Markov process *do not* correspond in the same way; cf. [BH], Chapter V., § 1.2.

1.3.2. General images

For the existence of the image by U the map U need not be in $\mathbb{D}$. Let $S = (\Omega, \mathcal{F}, m, \mathbb{D}, \mathcal{E})$ be a Dirichlet structure. Let $(W, \mathcal{G})$ be a measurable space and U a measurable map from Ω into W.

Let us consider a set $\mathcal{A}$ of measurable maps from W into $\mathbb{R}$ such that

(a) $\mathcal{A}$ is a dense vector subspace of $L^2(U_*m)$,
(b) for all $f \in \mathcal{A}$ and all $F \in \mathcal{D}(\mathbb{R})$ we have $F \circ f \in \mathcal{A}$,
(c) for all $f \in \mathcal{A}$ we have $f \circ U \in \mathbb{D}$.

Then the form $(\mathcal{A}, \mathcal{E}_{\mathcal{A},U})$ defined by $\mathcal{E}_{\mathcal{A},U}(f) = \mathcal{E}(f \circ U)$ is closable in $L^2(U_*m)$. Let $(\mathbb{D}_{\mathcal{A},U}, \mathcal{E}_{\mathcal{A},U})$ be its closure. The Dirichlet structure

$$U_*^{(\mathcal{A})}S = (W, \mathcal{G}, U_*m, \mathbb{D}_{\mathcal{A},U}, \mathcal{E}_{\mathcal{A},U})$$

is called the image structure of S by U with respect to $\mathcal{A}$; generally it depends on $\mathcal{A}$.

1.4. Dirichlet independence

Let $S = (\Omega, \mathcal{F}, m, \mathbb{D}, \mathcal{E})$ be a Dirichlet structure.

Definition 1.4.1. $U \in \mathbb{D}^p$ and $V \subset \mathbb{D}^q$ are said to be D-independent if

$$(U, V)_*S = (U_*S) \otimes (V_*S),$$

i.e., the image structure by the pair (U, V) is the product of the images by U and V.

Proposition 1.4.2. *U and V are D-independent if and only if for all $\varphi_1, \varphi_2 \in C^1(\mathbb{R}^p)$ and for all $\psi_1, \psi_2 \in C^1(\mathbb{R}^q)$*

$$\mathcal{E}(\varphi_1 \circ U \psi_1 \circ V,\ \varphi_2 \circ U \psi_2 \circ V) =$$
$$= \mathcal{E}(\varphi_1 \circ U, \varphi_2 \circ U)(\psi_1 \circ V, \psi_2 \circ V)_{L^2(m)} + \mathcal{E}(\psi_1 \circ V, \psi_2 \circ V)(\varphi_1 \circ U, \varphi_2 \circ U)_{L^2(m)}.$$

Under (OCC) we have the following criterion:

Proposition 1.4.3. *If S satisfies* (OCC) *with operator Γ, for $U \in \mathbb{D}^p$, $V \in \mathbb{D}^q$ to be independent it is necessary and sufficient that*
(a) *U and V are independent on the probability space $(\Omega, \mathcal{F}, m)$,*
(b) $\mathbb{E}\,[\Gamma(U_i, V_k)|U, V] = 0$ *for all i, k m-a.e.,*
(c) $\mathbb{E}\,[\Gamma(U_i, U_j)|U, V] = \mathbb{E}\,[\Gamma(U_i, U_j)|U]$ *for all i, j m-a.e.,*
(d) $\mathbb{E}\,[\Gamma(V_k, V_l)|U, V] = \mathbb{E}\,[\Gamma(V_k, V_l|V]$ *for all k, l m-a.e..*

Remark 1.4.4. A sufficient condition for U and V to be independent is that
(a) U and V are independent,
(b′) $\Gamma(U_i, V_k) = 0$ for all i, k,
(c′) $(U, \Gamma(U_i, U_j))$ is independent of V,
(d′) $(V, \Gamma(V_i, V_j))$ is independent of U.

The notion of D-independence can be easily extended to infinite families.

2. White structures related to the Poisson space

We shall first construct local Dirichlet structures on a Poisson space over a measurable space $(X, \mathcal{X})$ without other properties for the moment. It will yield white structures in the sense of (i) and (iii) only. After that we shall look at what happens when $X = Y \times \mathbb{R}_+$, getting structures associated with general processes with independent increments.

2.1. Finite Poisson space

Let us first recall the construction of the finite Poisson space, that is, in other words, of the Poisson point process with finite intensity measure.

Let $(X, \mathcal{X})$ be a measurable space and μ a finite positive measure on $(X, \mathcal{X})$. Let us set $\theta = \mu(X)$, $\mu_0 = \frac{1}{\theta}\mu$. Let us consider the product probability space

$$(X, \mathcal{X}, \mu_0)^{\mathbb{N}^*} \times (\mathbb{N}, \mathcal{P}(\mathbb{N}), \mathbb{P}_\theta)$$

where $\mathbb{P}_\theta$ is the Poisson law on $\mathbb{N}$ with parameter θ:

$$\mathbb{P}_\theta\{n\} = e^{-\theta}\frac{\theta^n}{n!}, \quad n \in \mathbb{N}.$$

Let $(X_n)_{n \in \mathbb{N}^*}$, Y be the coordinate mappings from this product space into its factors. So X_n are independent variables with values in X and law μ_0 and Y is independent of the X_n's with law $\mathbb{P}_\theta$. The formula

$$N = \sum_{n=1}^{Y} \delta_{X_n} \qquad (\text{with } \sum_{1}^{0} = 0)$$

defines a random variable N with values in the set M_p of point measures on $(X, \mathcal{X})$, that is sums of Dirac measures, equipped with the σ-field $\mathcal{M}_p$ generated by the maps

$$m \longrightarrow m(A), \quad A \in \mathcal{X},$$

and N has the following properties; cf. Neveu [N]:

(a) For every $A \in \mathcal{X}$, the random variable $N(A)$ has a Poisson law with parameter $\mu(A)$

$$\mathbb{P}(N(A) = k) = e^{\mu(A)}\frac{\mu(A)^k}{k!}.$$

In particular, $\mathbb{E}\, N(A) = \mu(A)$ and hence μ is the intensity measure of N.

(b) For pairwise disjoint sets $B_1, \dots, B_n \in \mathcal{X}$, the corresponding random variables $N(B_1), \dots, N(B_n)$ are independent.

We follow this construction starting from a Dirichlet structure $(X, \mathcal{X}, \mu_0, \mathbf{d}, \mathbf{e})$, which, as before, is supposed to satisfy $1 \in \mathbf{d}$, $\mathbf{e}(1) = 0$, and in addition satisfies (L) and (OCC) with operator γ.

The product Dirichlet structure

$$(\Omega, \mathcal{F}, m, \mathbb{D}, \mathcal{E}) = (X, \mathcal{X}, \mu_0, \mathbf{d}, \mathbf{e})^{\mathbb{N}^*} \times (\mathbb{N}, \mathcal{P}(\mathbb{N}), \mathbb{P}_\theta, L^2(\mathbb{P}_\theta), 0), \tag{2.1.1}$$

as we have seen, satisfies (L) and (OCC). Let still $(X_n)_{n \in \mathbb{N}^*}$ and Y be the coordinate mappings and

$$N = \sum_{n=1}^{Y} \delta_{X_n}$$

the associated Poisson point process.

Lemma 2.1.2. *Let $U = F(Y, X_1, X_2, \ldots, X_n, \ldots)$ be in $L^2(\Omega, \mathcal{F}, m)$;*
(a) *$U \in \mathbb{D}$ if and only if for all $m \in \mathbb{N}$*

$$F(m, X_1, X_2, \ldots) \in \mathbb{D} \text{ and } \mathcal{E}(U) = \int \mathcal{E}(F(m, \ldots))\, d\mathbb{P}_\theta(m) < +\infty;$$

(b) *$U \in \mathbb{D}$ if and only if for all m and k and for $\mu_0^{\otimes \mathbb{N}^*}$-almost every $x_1, \ldots, x_{k-1}, x_{k+1}, \ldots$*

$$F(m, x_1, \ldots, x_{k-1}, \cdot, x_{k+1}, \ldots) \in \mathbf{d} \quad \text{and}$$

$$\mathcal{E}(U) = \frac{1}{2} \int \sum_{k=1}^{\infty} \gamma_k(F(m, x_1, \ldots, x_{k-1}, \cdot, x_{k+1}, \ldots))\, d\mathbb{P}_\theta(m)\, d\mu_0^{\otimes \mathbb{N}^*} < +\infty;$$

(c) *For $U \in \mathbb{D}$, the carré du champ operator Γ of $(\Omega, \mathcal{F}, m, \mathbb{D}, \mathcal{E})$ is given by*

$$\Gamma(F) = \sum_{k=1}^{\infty} \gamma_k(F(Y, X_1, \ldots, X_{k-1}, \cdot, X_{k+1} \ldots))(X_k).$$

This is a direct consequence of properties of product structures.

Lemma 2.1.3. *For $f \in \mathbf{d}$, $N(f)$ is defined and $N(f)$ is in $\mathbb{D}$.*

Proof. Suppose that f is an equivalence class for μ_0 or μ-a.e. equality; now, if $f = g$ μ-a.e.,

$$\mathbb{E}\,|N(f) - N(g)| \leq \mathbb{E}\, N|f - g| = \mu|f - g| = 0,$$

so $N(f) = N(g)$ μ-a.e.

Using the Laplace functional of N (cf. [N]),

$$\mathbb{E}\, e^{-\lambda N(f)} = \exp\left(-\int (1 - e^{-\lambda f})\, d\mu\right), \qquad \lambda \geq 0,$$

we get

$$\mathbb{E}\, N(f)^2 = \int f^2\, d\mu + \left(\int f\, d\mu\right)^2$$

proving that $N(f) \in L^2(m)$ for $f \in L^2(\mu)$.

Then for any $f \in \mathbf{d}$,

$$\Gamma(N(f)) = \sum_{k=1}^{\infty} \gamma_k(\sum_{n=1}^{Y} f(X_n)) = \sum_{k=1}^{\infty} 1_{\{k \leq Y\}}\gamma(f)(X_k) =$$
$$= \sum_{k=1}^{Y} \gamma(f)(X_k) = N(\gamma(f))$$

and $\mathcal{E}(N(f)) = \frac{1}{2}\mu(\gamma(f)) = e(f)$. □

So we have obtained

Lemma 2.1.4. *For all $f, g \in \mathbf{d}$ we have $\Gamma(N(f)) = N(\gamma(f))$ and $\Gamma(N(f), N(g)) = N(\gamma(f, g))$.*

Corollary 2.1.5. *If $fg = 0$ and if $f, g \in \mathbf{d}$, then $N(f)$ and $N(g)$ are D-independent.*

Proof. We know that $N(f)$ and $N(g)$ are independent. By Remark 1.4.4 it suffices to show that

(a) $\gamma(N(f), N(g)) = 0$, which comes from (2.1.4), because $\gamma(f, g) = 0$,

(b) $(N(f), \Gamma(N(f)))$ is independent of $N(g)$, which comes from (2.1.4), too, since $(N(f), \Gamma(N(f))) = (N(f), N(\gamma(f)))$ and $\gamma(f) = 0$ on $\{g \neq 0\}$ by the fact that the structure $(X, \mathcal{X}, \mu_0, \mathbf{d}, \mathbf{e})$ is local. □

Remark 2.1.6. If $(X, \mathcal{X}, \mu_0, \mathbf{d}, \mathbf{e})^m$ satisfies (EID) for every m then also $(\Omega, \mathcal{F}, m, \mathbb{D}, \mathcal{E})$ satisfies (EID).

Remark 2.1.7. The map

$$\widetilde{N} : f \longrightarrow (N(f) - \int f \, d\mu)$$

from $L^2(X, \mathcal{X}, \mu_0)$ into $L^2(\Omega, \mathcal{F}, m)$ is homomorphism which preserves L^2-norms and Dirichlet forms, and is therefore an isometry from $\mathbf{d}$ into $\mathbb{D}$ equipped by any α-norm: $(\alpha\| \cdot \|^2_{L^2(\mu_0)} + \mathbf{e})^{\frac{1}{2}}$ and $(\alpha\| \cdot \|^2_{L^2(m)} + \mathcal{E})^{\frac{1}{2}}$. Indeed, we have for any $f \in \mathbf{d}$

$$\mathbb{E}\left(N(f) - \int f d\mu\right)^2 = \int f^2 d\mu, \quad \mathcal{E}\left(N(f) - \int f d\mu\right) = \mathbf{e}(f).$$

For f and g with disjoint support we obtain images $\widetilde{N}(f)$ and $\widetilde{N}(g)$ which are D-independent. Hence $\widetilde{N}$ is a Wiener–Karhunen measure in the strong Dirichlet sense.

Remark 2.1.8. Let us recall that M_p is the set of point measures on $(X, \mathcal{X})$ with the σ-field $\mathcal{M}_p$ generated by the maps $m \to m(A)$, $A \in \mathcal{X}$. We can consider the image of the structure $(\Omega, \mathcal{F}, m, \mathbb{D}, \mathcal{E})$ by N in the sense of general images of § 1.2.3 with respect to the set $\mathcal{A}$ of measurable functions from $(M_p, \mathcal{M}_p)$ into $(\mathbb{R}, \mathcal{B}(\mathbb{R}))$:

$$\mathcal{A} = \{m \to G(m(f_1), \dots, m(f_d)) \; : \; G \in \mathcal{D}(\mathbb{R}^d), \; f_1, \dots, f_d \in \mathbf{d}, d \in \mathbb{N}^*\}.$$

The required properties on $\mathcal{A}$ are fulfilled, as easily seen. The image structure is

$$(\mathcal{M}_p, \, \mathcal{M}_p, \, \mathbb{P}, \, \mathbb{D}_1, \, \mathcal{E}_1)$$

where $\mathbb{P}$ is the law of a Poisson point process with intensity μ, $(\mathbb{D}_1, \, \mathcal{E}_1)$ is a local Dirichlet form satisfying (OCC) and characterized by the fact that its carré du champ operator Γ_1 satisfies for all $f_i \in \mathbf{d}$, $i = 1, \dots, d$ and all $F \in \mathcal{D}(\mathbb{R}^d)$

$$\Gamma_1(F(N(f_1), \dots, N(f_d))) =$$
$$= \sum_{i,j} F_i'(N(f_1), \dots, N(f_d)) F_j'(N(f_1), \dots, N(f_d)) N(\gamma(f_i, f_j)).$$

2.2. σ-finite Poisson space

The probabilistic definition of a Poisson point process with σ-finite intensity measure μ is easily made by product using property (b) of the Poisson point process.

In the case of Dirichlet structures, the construction cannot be so simple.

Remark 2.2.1. We have defined for the moment Dirichlet structures only on probability spaces. It is possible (cf. [BH]) with minor changes to define structures $(E, \mathcal{F}, m, \mathbb{D}, \mathcal{E})$ with m σ-finite. Hypotheses (L) and (OCC) are expressed in the same way.

Remark 2.2.2. We shall assume that Dirichlet structures $(X, \mathcal{X}, \mu_k, \mathbf{d}_k, \mathbf{e}_k)$ are given where μ_k are σ-finite measures such that $\mu = \sum_k \mu_k$ is σ-finite and that $\cap_k \mathbf{d}_k$ is dense in $L^2(\mu)$. If we define the form $e = \sum_k \mathbf{e}_k$ on $\cap_k \mathbf{d}_k$, it is easy to see that it is closable and therefore defines a Dirichlet structure

$$(X, \mathcal{X}, \mu, \mathbf{d}, \mathbf{e}).$$

Moreover, we suppose that the structures $(X, \mathcal{X}, \mu_k, \mathbf{d}_k, \mathbf{e}_k)$ are local and satisfy (OCC) with the same operator in the sense that their carré du champ operator coincides with that of $(X, \mathcal{X}, \mu, \mathbf{d}, \mathbf{e})$ on $\mathbf{d}$. It is denoted by γ.

By the construction of Remark 2.1.8, with each of these structures, it is possible to define a Poisson point process N_k with intensity measure μ_k defined on a Dirichlet structure S_k given by formula (2.1.1). For the corresponding carré du champ operator Γ_k we have: for $U = F(N_k(f_1), \dots, N_k(f_d))$, $F \in \mathcal{D}(\mathbb{R}^d)$ and $f_i \in \mathbf{d}$

$$\Gamma_k(U) = \sum_{i,j} F_i'(N_k(f_1), \dots, N_k(f_d)) F_j'(N_k(f_1), \dots, N_k(f_d)) N_k(\gamma(f_i, f_j)).$$

Let us consider the product $S = \bigotimes_k S_k$ and set

$$S = (\widetilde{\Omega}, \widetilde{\mathcal{F}}, \widetilde{\mathbb{P}}, \widetilde{\mathbb{D}}, \widetilde{\mathcal{E}}).$$

As it is quite often done in probability theory, to avoid the use of the coordinate mappings in the notation, we shall consider that the N_k's are random variables defined on $\widetilde{\Omega}$ depending only on the coordinate map of order k. So the N_k's are independent and D-independent random variables defined on $S = (\widetilde{\Omega}, \widetilde{\mathcal{F}}, \widetilde{\mathbb{P}}, \widetilde{\mathbb{D}}, \widetilde{\mathcal{E}})$.

Then we define $\widetilde{N}$ by

$$\widetilde{N} = \sum_k N_k;$$

we know already by well known probabilistic arguments that $\widetilde{N}$ is a Poisson point process with intensity measure $\mu = \sum_k \mu_k$.

The structure S possesses a carré du champ operator $\widetilde{\Gamma}$ which is the sum of the Γ_k's acting on the k-th coordinate mapping.

If $U = F(\widetilde{N}(f_1), \dots, \widetilde{N}(f_d))$, $f_i \in \mathbf{d}$, $i = 1, \dots, d$, $F \in \mathcal{D}(\mathbb{R}^d)$, one has (2.2.2.1):

$$\begin{aligned}\widetilde{\Gamma}(U) &= \sum_k \sum_{i,j} F_i'(\widetilde{N}(f_1), \dots, \widetilde{N}(f_d)) F_j'(\widetilde{N}(f_1), \dots, \widetilde{N}(f_d)) N_k(\gamma(f_i, f_j)) = \\ &= \sum_{i,j} F_i'(\widetilde{N}(f_1), \dots, \widetilde{N}(f_d)) F_j'(\widetilde{N}(f_1), \dots, \widetilde{N}(f_d)) \widetilde{N}(\gamma(f_i, f_j)).\end{aligned}$$

Finally, the image of the structure S by $\widetilde{N}$ with respect to the set $\mathcal{A}$ of measurable mappings from $(M_p, \mathcal{M}_p)$ into $(\mathbb{R}, \mathcal{B}(\mathbb{R}))$

$$\mathcal{A} = \{m \to G(m(f_1), \dots, m(f_d)),\ G \in \mathcal{D}(\mathbb{R}^d),\ f_1, \dots, , f_d \in \mathbf{d},\ d \in \mathbb{N}^*\}$$

introduced in 2.1.8, is a Dirichlet structure

$$(\mathcal{M}_p,\ \mathcal{M}_p,\ \mathbb{P},\ \mathbb{D}_1,\ \mathcal{E}_1).$$

It is local, satisfies (OCC) with operator Γ such that, if we denote by N the identity mapping from M_p into M_p,

- N is a Poisson point process with intensity μ,
- for $U = F(N(f_1), \dots, N(f_d))$ $f_i \in \mathbf{d}$, $i = 1, \dots, d$, $F \in \mathcal{D}(\mathbb{R}^d)$,

$$\Gamma(U) = \sum_{i,j} F_i'(N(f_1), \dots, N(f_d)) F_j'(N(f_1), \dots, N(f_d)) N(\gamma(f_i, f_j)).$$

In particular, if $f \in \mathbf{d}$, then $\Gamma(N(f)) = N(\gamma(f))$ and

$$\mathcal{E}(N(f)) = \frac{1}{2}\mathbb{E}\,\Gamma(N(f)) = \frac{1}{2}\int \gamma(f)\,d\mu = \mathbf{e}(f).$$

If $f \in \mathbf{d} \cap L^1(\mu)$ one has

$$\mathbb{E}\left(N(f) - \int f d\mu\right)^2 = \int f^2 d\mu \qquad \text{and} \qquad \mathcal{E}\left(N(f) - \int f d\mu\right) = \mathbf{e}(f).$$

By these relations, the compensated Poisson process $N - \mu$ is defined on $L^2(\mu)$ and is an isometry in the Dirichlet sense as before.

2.3. Dirichlet structures related to a process with independent increments (PII)

We apply the preceding study to the case where

$$X = \mathbb{R}_+ \times Y, \qquad \mathcal{X} = \mathcal{B}(\mathbb{R}_+) \times \mathcal{Y} \qquad (\mathcal{Y} \text{ is a } \sigma\text{-field on } Y)$$

and where

$$\mu = dt \times n$$

with n a σ-finite measure on $(Y, \mathcal{Y})$.

We are given a Dirichlet structure on $(X, \mathcal{X})$, say $(X, \mathcal{X}, \mu, \mathbf{d}, \mathbf{e})$, which is local (L) and possesses a carré du champ operator (OCC) denoted γ. By the preceding construction one obtains a structure $(\mathcal{M}_p, \mathcal{M}_p, \mathbb{P}_\mu, \mathbb{D}, \mathcal{E})$ and the identity mapping N is a Poisson point process with intensity μ.

This contains the case of processes with independent increments (Y_t), if we put

$$N = \sum_{s \in \mathbb{R}_+} \delta_{(s, Y_s - Y_{s-})}.$$

This is a Poisson point process on $\mathbb{R}_+ \times \mathbb{R}^*$ with intensity $dt \times \nu$ where ν is the Lévy measure of the PII (Y_t).

It is now possible to develop the stochastic calculus related to this PII (Y_t) (stochastic integrals, s.d.e., etc.) to study the belonging of the obtained random variables to $\mathbb{D}$, and applying remark 2.1.6 and (EID) to obtain density results.

3. White structures related to the Wiener space

3.1. Wiener integral and Brownian motion

Let us recall the construction of the Wiener integral and that of Brownian motion it gives. Let (χ_n) be an orthonormal basis of $L^2(T, \mathcal{T}, \mu)$, where $(T, \mathcal{T}, \mu)$ is a σ-finite measured space, (g_n) be a sequence of independent reduced Gaussian random variables defined on a probability space $(\Omega, \mathcal{A}, \mathbb{P})$. With $f \in L^2(T, \mathcal{T}, \mu)$ we associate $I(f) \in L^2(\Omega, \mathcal{A}, \mathbb{P})$ by

$$I(f) = \sum_n < f, \chi_n > g_n.$$

Mapping I is a homomorphism from $L^2(T, \mathcal{T}, \mu)$ into $L^2(\Omega, \mathcal{A}, \mathbb{P})$. If f and g are in L^2 and if $\int fg d\mu = 0$, then $I(f)$ and $I(g)$ are independent.

Let us take $T = [0,1]$, $\mathcal{T} = \mathcal{B}([0,1])$, $\mu = dt$, and let us put

$$B(t) = \sum_n \psi_n(t) g_n \quad \text{with} \quad \psi_n(t) = \int_0^1 \chi_n(s)\, ds.$$

This series converges in $\mathcal{C}([0,1])$ a.s. and in $L^p((\Omega, \mathcal{A}, \mathbb{P}), \mathcal{C}([0,1]))$ for $p \in [1, \infty[$ (cf. [BH], Chapter V., Example 1.3.2) and defines a Brownian motion. We put

$$I(f) = \int_0^1 f(s)\, dB_s.$$

3.2. Dirichlet structure

The preceding construction involves the product probability space

$$(\Omega, \mathcal{A}, \mathbb{P}) = (\mathbb{R}, \mathcal{B}(\mathbb{R}), N(0,1))^{\mathbb{N}},$$

the g_n's being the coordinate mappings. If we equip each factor of this product with a Dirichlet form $(\mathbb{R}, \mathcal{B}(\mathbb{R}), N(0,1), \mathbf{d}_n, \mathbf{e}_n)$ which we suppose, as usual, local and with (OCC) satisfied, γ_n, by the theorem on products we obtain on $(\Omega, \mathcal{A}, \mathbb{P})$ a Dirichlet structure

$$(\Omega, \mathcal{A}, \mathbb{P}, \mathbb{D}, \mathcal{E}) = \prod_n (\mathbb{R}, \mathcal{B}(\mathbb{R}), N(0,1), \mathbf{d}_n, \mathbf{e}_n).$$

The structure is local with Γ satisfying (OCC) such that for every $U \in \mathbb{D}$, $U = F(g_0, g_1, \cdots, g_n, \cdots)$

$$\Gamma(U) = \sum_n \gamma_n(F),$$

where the γ_n acts on the n-th argument.

Example 3.2.1. Let us take

$$\gamma_n(u) = u'^2, \quad \mathbf{e}_n(u) = \frac{1}{2}\int u'^2\, dN(0,1) \quad \text{and} \quad \mathbf{d}_n = H^1(\mathbb{R}, N(0,1)).$$

We obtain

$$(\Omega, \mathcal{A}, \mathbb{P}, \mathbb{D}, \mathcal{E}) = \left(\mathbb{R}, \mathcal{B}(\mathbb{R}), N(0,1), H^1(\mathbb{R}, N(0,1)), u \to \frac{1}{2}\int u'^2\, dN(0,1)\right)^{\mathbb{N}}$$

and this structure is an abstract Wiener space equipped with the Ornstein–Uhlenbeck structure since we have

$$\Gamma(\int_0^1 f(s) dB_s) = \Gamma(\sum_n < f, \chi_n > g_n) = \sum < f, \chi_n > \gamma(g_n)$$

$$= \sum < f, \chi_n >= \|f\|^2_{L^2[0,1]} = \int_0^1 f^2(s)\, ds\,.$$

Example 3.2.2. Let us take

$$\chi_n(t) = e^{2i\pi nt}, \quad n \in \mathbb{Z}, \quad \gamma_n(u) = a(n)u'^2,$$
$$\mathbf{e}_n(u) = \frac{1}{2}a(n)\int u'^2\, dN(0,1), \quad \mathbf{d}_n = H^1(\mathbb{R}, N(0,1)).$$

We get

$$(\Omega, \mathcal{A}, \mathbb{P}, \mathbb{D}, \mathcal{E}) = \prod_n (\mathbb{R}, \mathcal{B}(\mathbb{R}), N(0,1), H^1(\mathbb{R}, N(0,1)), u \to \frac{1}{2}a(n)\int u'^2\, dN(0,1)).$$

(a) For $a(n) = 4\pi n^2$ we have for any $f \in L^2[0,1]$

$$f(t) = \sum_{n\in\mathbb{Z}} \hat{f}_n e^{2i\pi nt}, \qquad \int_0^1 f(s)dB_s = \sum \hat{f}_n g_n,$$
$$\Gamma(\int_0^1 f(s)dB_s) = \sum_n \hat{f}_n{}^2 4\pi n^2 = \|f'\|^2_{L^2[0,1]};$$

by the theorem on products we see that $\int_0^1 f(s)dB_s$ belongs to $\mathbb{D}$, if and only if

$$f \in L^2[0,1] \quad \text{and} \quad \sum_n \hat{f}_n{}^2 4\pi n^2 < +\infty,$$

that is f is continuous on the torus T^1 and belongs to the Sobolev space $H^1(T^1)$.

(b) For $a(n) = 2\pi n^{2q}$, $q \in \mathbb{N}$, we obtain, similarly a local Dirichlet structure with carré du champ operator Γ satisfying

$$\Gamma\left(\int_0^1 f(s)dB_s\right) = \int_0^1 f^{(q)^2}(s)ds$$

and $\int_0^1 f(s)dB_s \in \mathbb{D}$ iff $\sum_{n\in\mathbb{Z}} n^{2q}\hat{f}_n{}^2 < +\infty$.

Whiteness. The structures obtained in 3.1.2 are **white** because if f, g are such that $\sum_{n\in\mathbb{Z}} n^{2q}(\hat{f}_n{}^2 + \hat{g}_n{}^2) < +\infty$ and if

$$fg = 0,$$

then $U = \int f(s)dB_s$ and $V = \int g(s)dB_s$ are D-independent. Indeed, by remark 1.4.4, it suffices to show

$$\begin{cases} \Gamma = 0, \\ (U, \Gamma(U)) \text{ is independent of } V, \\ (V, \Gamma(V)) \text{ is independent of } U. \end{cases}$$

But $\Gamma(U, V) = \int f^{(q)}(s)g^{(q)}(s)\, ds = 0$ by the fact that if $u, v \in H^1(T^1)$ $uv = 0 \Rightarrow u'v' = 0$ a.e. and the two other properties are fulfilled because $\Gamma(U)$ and $\Gamma(V)$ are constants.

(EID). Since the energy image density property is fulfilled on finite products, it follows that all structures obtained in 3.1.2 satisfy (EID).

Gradient and semigroup. Let $(\Omega, \mathcal{A}, \mathbb{P}, \widetilde{\mathbb{D}}, \widetilde{\mathcal{E}})$ be the Dirichlet structure obtained in 3.2.1.a). By looking at the multiple Wiener integrals, it is easy to show that this structure possesses a gradient $\widetilde{D}$ (cf. [BH], Chapter V., § 5.2) with the Hilbert space $H = L^2[0,1]$.

Denoting respectively by D, δ, A the gradient, the divergence and the generator of the classical Ornstein–Uhlenbeck structure (cf. Yan [Y]) we have:

$$\widetilde{D} = \frac{d}{dt} D, \qquad \widetilde{\delta} = -\delta \frac{d}{dt}, \qquad \widetilde{A} = \frac{1}{2} \delta \frac{d^2}{dt^2} D.$$

Let p_t be the heat semigroup on T^1; then the semigroup $\widetilde{P}_t$ associated with the structures

$$(\Omega, \mathcal{A}, \mathbb{P}, \widetilde{\mathbb{D}}, \widetilde{\mathcal{E}})$$

is characterized by its action on multiple Wiener integrals

$$\widetilde{P}_t(I_k(f)) = I_k(p_t^{\otimes k} f),$$

where $f \in H^1_{\mathrm{sym}}(T^k)$.

Looking then at the exponential vectors

$$\mathrm{Exp}(h) = \exp\left(\int h(s) dB_s - \frac{1}{2} < h, h >\right)$$

we obtain

$$\widetilde{P}_t \,\mathrm{Exp}(h) = \mathrm{Exp}(p_t h),$$

that is $(\widetilde{P}_t)$ is obtained by the second quantization and we have the Mehler formula (cf. [FLP])

$$\widetilde{P}_t F(\omega) = \mathbb{E}_w[F(p_t \omega + (I - p_{2t})^{\frac{1}{2}} w)].$$

Similar results hold for the other structures of Example 3.2.2.

References

[BGJ] Bichteler K., Gravereaux J. B., Jacod J., Malliavin calculus for processes with jumps. Gordon and Breach, London, 1987.

[BH] Bouleau, N., Hirsch, F., Dirichlet forms and Analysis on Wiener space. De Gruyter Stud. Math. 14, Walter de Gruyter, Berlin 1991.

[CP] Carlen E. A., Pardoux E., Differential calculus and integration by parts on Poisson space. In: Stochastics, algebra and analysis in classical and quantum mechanics, 63–73. Kluwer, Dordrecht 1990.

[DKW] Dermoune A., Kree P., Wu L., Calcul stochastique non adapté par rapport à la mesure aléatoire de Poisson. Séminaire de Probabilité XXII (ed. by J. Azéma et al.), 477–484. Lecture Notes in Math. 1321, Springer-Verlag, Berlin 1988.

[FLP] Feyel, D., La Pradelle, A. de, Opérateurs linéaires gaussiens. Potential Analysis 3 (1994), 89–105.

[N] Neveu, J., Processus ponctuels. Ecole d'été de probabilité VI de Saint-Flour (ed. by J. Hoffmann-Jørgensen), 250–447. Lecture Notes in Math. 598, Springer-Verlag, Berlin 1977.

[NV] Nualard, D., Vives, J., Anticipative calculus for the Poisson process based on the Fock space. Séminaire de Probabilité XXIV (ed. by J. Azéma et al.), 154–165. Lecture Notes in Math. 1426, Springer-Verlag, Berlin 1990.

[P] Privault N., Calcul chaotique et variationnel pour le processus de Poisson. Thesis, Univ. Paris 6, Paris 1994.

[W] Wu L., Construction de l'operateur de Malliavin sur l'espace de Poisson. Séminaire de Probabilité XXI (ed. by J. Azéma et al.), 100–113. Lecture Notes in Math. 1247, Springer-Verlag, Berlin 1987.

[Y] Yan Y. A., Dévelopment des distributions suivant les chaos de Wiener et application à l'analyse stochastique. Séminaire de Probabilité XXI (ed. by J. Azéma et al.), 27–32. Lecture Notes in Math. 1247, Springer-Verlag, Berlin 1987.

Les processus mentaux de la création

Gustave Choquet

1991 Mathematics Subject Classification: 00A00

Science et création

A 30 ans, on cherche et on trouve; un peu plus tard on forme des chercheurs, et plus tard encore on parle des processus mentaux de la création: C'est ce que je me propose de faire ici.

Je ne donnerai pas de recette pour devenir un découvreur, cela s'apprend en découvrant. Mais plusieurs fois j'ai constaté qu'une réflexion sur les raisons d'un succès, améliorait mes recherches ultérieures; mon expérience peut donc être utile.

Il semble d'ailleurs que de plus en plus de chercheurs comprennent qu'*il n'est pas possible de dissocier la science de l'activité humaine qui la crée*; et qu'ainsi les millions de théorèmes qui reposent sur les rayons des bibliothèques ne sont que la cendre refroidie, d'ailleurs précieuse, du feu créateur qui les a fait naître; d'où chez les chercheurs un intérêt grandissant pour l'histoire des sciences et techniques.

Les témoignages de chercheurs, sans être encore nombreux, constituent néanmoins une bonne base d'étude; j'en citerai quelques uns: Celui de René Descartes qui, le 10 Novembre 1619, à 23 ans, dans une nuit d'enthousiasme fait trois rêves exaltants et découvre les fondements "d'une science admirable" (voir Bell, "Men of mathematics"). Le récit d'Henri Poincaré concernant sa découverte des groupes fuchsiens et de leur lien avec certaines formes quadratiques, est célèbre à juste titre.

Jacques Hadamard a réuni en 1959 ses analyses commencées en 1937 sur la psychologie de l'invention (134 pages).

Paul Levy, en 1970, s'est longuement exprimé dans "Quelques aspects de la pensée d'un mathématicien" (Albert Blanchard, épuisé).

En 1976 j'ai, dans le Séminaire Loi, parlé de "La naissance de la théorie des capacités" (publié dans "La vie des Sciences" de l'Académie des Sciences, 1986).

Laurent Schwartz, après avoir en 1982 raconté à l'Université de Patras (Grèce) l'histoire de sa théorie des distributions, a fait en 1987 une conférence sur la découverte en Mathématiques.

André Weil et Alain Connes, plus récemment, se sont aussi exprimés brièvement à ce sujet dans la revue "Pour la Science".

Offprint from: Potential Theory – ICPT 94, Eds.: Král/Lukeš/Netuka/Veselý

Dans les sciences expérimentales, les témoignages abondent: Ceux de Newton, Pasteur, Darwin, Einstein, Fleming, Watson ("La double hélice", 1968). En chimie le demi-rêve de Kekule qui, d'après lui, l'aurait conduit à la structure hexagonal du benzène (un serpent se mordant la queue) est notable car il est exceptionnel que les rêves conduisent à des découvertes importantes.

Le cerveau, organe de notre pensée

Bien que l'introspection ait été longtemps un outil essentiel d'étude des processus mentaux, et qu'elle joue encore un rôle important (le livre d'Hebb de 1949 reste un ouvrage de référence), des outils performants tels que l'imagerie médicale permettront certainement de progresser dans l'étude de ces processus. Le terrain aujourd'hui est encore peu sûr; on sait toutefois que toutes les parties du cerveau sont interdépendantes et que son fonctionnement est extrêmement complexe; ce que j'en dirai aura surtout pour but de souligner cette complexité, et de justifier en partie certaines affirmations.

Une vue fort schématique du cerveau lui attribue une structure ternaire:

• Le cerveau postérieur, dit parfois reptilien, regroupe la moelle épinière, le bulbe rachidien, le mésencéphale. Il est responsable de la vie inconsciente des organes et de pulsions primitives.

• Le système limbique regroupe des structures très diverses, telles que l'hippocampe, l'hypothalamus. Il joue un rôle important dans la mémoire et la genèse des états de motivation. Ses lésions modifient les comportements alimentaires, sexuels, sociaux.

• Le néo-cortex (ou cortex, ou matière grise).

Ces trois parties sont interconnectées et jouent donc un rôle dans le processus de la recherche, ce qui après coup justifie la phrase célèbre de Pascal "L'homme n'est ni ange ni bête". Il est probable que leur plus ou moins grand rôle détermine le style de la recherche.

Les connections cérébrales sont redondantes, à des degrés variés, ce qui explique peut-être que la folie ou la dépression n'empêche pas toujours le travail mathématique, artistique ou littéraire: Rappelons le cas d'André Bloch qui, après avoir tué à coups de hache plusieurs membres de sa famille entra en relation, de son asile-d'aliénés, avec Georges Valiron et découvrit plusieurs beaux théorèmes; et celui plus célèbre encore de Cantor, victime dès l'age de 46 ans de graves crises de dépression, mais qui après ces crises se sentait l'esprit très clair et faisait un excellent travail.

Neurones

Notre cerveau comporte de 50 à 100 milliards de neurones. Chacun d'eux est constitué d'un corps cellulaire contenant le noyau, d'où émergent d'une part de nombreuses dendrites munies de protubérances appelées épines, d'autre part un long axone muni à son extrémité de nombreuses terminaisons axonales.

Chaque neurone est en relation avec d'autres neurones par des *synapses*, les unes électriques, d'autres chimiques, qui unissent épines et terminaisons axonales. Certains neurones peuvent avoir jusqu'à 30 000 synapses avec d'autres neurones; il peut exister aussi certaines relations de neurone à neurone à travers leurs axones.

Bien qu'élevé, le nombre des neurones et des synapses n'expliquerait sans doute pas la multitude de possibilités de notre cerveau: Pensons par exemple à la vision d'un paysage en mouvement et à son enregistrement ne serait-il que partiel. La solution de cette énigme semble résider dans une observation mathématique élémentaire: Un ensemble fini de N éléments contient 2^N sous-ensembles distincts, nombre gigantesque dès que N dépasse 1000; si donc neurones et synapses s'organisent en groupements fonctionnels, le nombre de ces groupements peut être d'un ordre de grandeur incomparablement supérieur au nombre de ces neurones et synapses.

Les plus petits de ces groupements ont une structure en réseaux vaguement rectangulaires; ce sont des sous-unités de traitement (recevant par exemple des informations visuelles, auditives, tactiles) qui peuvent elles-mêmes s'organiser en assemblées plus importantes que j'appellerai ici faisceaux.

Associations d'idées et facilitation

Les connections de neurones dans les réseaux et faisceaux expliquent les précieuses associations d'idées qui sont une base essentielle du fonctionnement, conscient ou non, du cerveau.

Ce sont elles qui se manifestent dans les calembours et coq-à-l'âne; mais les personnes agées en prennent conscience de façon encore plus tangible lorsqu'elles perdent la mémoire des noms propres ou de certains noms communs; elles peuvent souvent les retrouver par un long cheminement conscient, de proche en proche, utilisant analogie ou chronologie. Le fait qu'elles conservent plus longtemps leur maniement de notions complexes, leur possibilité de discourir longuement, s'explique par le fait, justement de cette complexité: Ces activités peuvent s'effectuer de nombreuses façons différentes, grâce à des chaines variées d'associations d'idées, et leur but n'est pas un point précis, mais une zone étendue et assez floue.

Les associations d'idées sont aussi une des explications des rêves, de leur motivation – liée tantôt à notre vie active, tantôt à un passé lointain – , mais aussi de leur mélange de fantaisie et de cohérence: Car si les associations d'idées obéissent au même déterminisme physico-chimique que les neurones, elles dépendent aussi comme eux d'un chaos déterministe qui nous apparaît comme le fruit du hasard, de la fantaisie.

Mais surtout c'est leur vie souterraine, cachée, qui nous donne l'explication du *subconscient* (ou préconscient de Freud) et sans doute de l'inconscient, dans la mesure du moins où celui-ci a une réalité définissable et n'est pas simplement un subconscient profond et refoulé depuis longtemps.

Je vois dans le subconscient la source de ce que j'appellerai *illumination* (pour les mystiques on dit révélation, et pour les poètes et musiciens, inspiration); et aussi un des facteurs de l'*intuition*.

Subconscient, illumination, hasard, intuition

Le subconscient travaille sans relâche; les associations d'idées qui en sont la base ne se font pas arbitrairement; leur dynamique est régie en partie par le hasard, mais surtout par le phénomène de *facilitation* (les interactions entre neurones au niveau des synapses sont plus ou moins grandes, en fonction de leur activité passée). Ce phénomène rend compte de nos habitudes, du petit robot qui semble prendre en charge certaines de nos actions; d'où l'importance considérable, bénéfique ou néfaste, de l'éducation, qu'elle vienne du monde extérieur ou de notre activité cérébrale consciente.

L'activité subconsciente peut impliquer simultanément plusieurs aires corticales: on peut être au même instant amoureux et préoccupé d'un problème mathématique.

Chez un mathématicien qu'un problème a intéressé, récemment ou dans le passé, s'est constitué à son insu un groupement de faisceaux liés à ce problème, et dont l'activité peut s'être progressivement renforcée par des retours conscients au problème. Dans des cas favorables, par exemple grâce à des contacts mathématiques en apparence sans rapport avec le problème, ce groupe de faisceaux acquiert une vie cohérente et assez intense pour qu'un petit événement (par exemple détente après une grande fatigue, changement agréable d'activité, ou une simple tasse de café) déclenche brusquement un passage du subconscient au conscient: C'est l'*illumination*, expérience merveilleuse dont on garde précieusement le souvenir. Elle ne diffère que par la nature de son objet de l'extase mystique, ou de l'inspiration des grands artistes ou poètes qui leur permet de réaliser en un temps très bref des oeuvres qui nous étonnent (Mozart, etc.).

Stendhal donnait à ce phénomène le nom de *cristallisation*, qui évoque le changement brusque d'état d'une solution saline sursaturée ou le gel brutal d'une eau en surfusion.

J'ai raconté déjà dans le Séminaire Loi les illuminations, grandes ou petites qui m'ont conduit, vers 1950, à créer la théorie des capacités; en voici une autre: Vers 1960 j'avais pris nettement conscience qu'il y avait en Analyse des cônes convexes importants sans base compacte (e.g. le cône des fonctions réelles sur $\mathbb{R}$ dont toutes les dérivées sont positives); je ne savais donc pas, en particulier, s'ils possédaient tous des génératrices extrémales. Je cherchais donc à construire de tels cônes sans génératrices extrémales, par un procédé de limite projective à partir de cônes ayant une base compacte. Et voici qu'un matin du printemps 1962, alors qu'avec ma femme je passais une semaine de vacances dans un petit hotel de Barbizon, nous décidons d'aller nous promener en forêt; un seul pas séparait notre chambre du sable de la forêt; je franchis le seuil et "Joie, pleurs de joie": Oui, comme on coupe en biseau une branche pointue avec une lame aiguisée, il faut, de ces cônes détacher un petit copeau compact, convexe ainsi que le reste du cône. En une minute je vois la structure des opérations à effectuer sur ces copeaux, appelés plus tard "chapeaux", et comment les utiliser.

C'est une illumination du même type qui donna à Schwartz sa théorie des distributions: En 1935 nous avions tous deux organisé à l'Ecole Normale Supérieure un baby-Séminaire où je parlais des travaux de Baire et de Cantor, et où lui parlait de son anxiété devant les comportements différents des équations

$$\frac{\partial^2 u}{\partial x^2} + \frac{\partial^2 u}{\partial y^2} = 0 \quad \text{et} \quad \frac{\partial^2 u}{\partial x \partial y} = 0;$$

cette anxiété le poursuivit jusqu'en 1944. Il n'avait pas cessé pendant ces 10 années de s'intéresser aux équations linéaires aux dérivées partielles, sans pouvoir pénétrer le secret de leur vie cachée; mais son subconscient travaillait lui aussi de son côté. C'est alors qu'arrive 1944; je venais, avec Jacques Deny, de terminer un travail sur une caractérisation des fonctions polyharmoniques dans le plan; nous y utilisions les séries de Fourier, ce qui ne s'étendait pas à l'espace à 3 dimensions; je demandai alors à Schwartz s'il connaissait une façon de s'en passer; il la trouva assez rapidement en utilisant des fonctions indéfiniment dérivables à support compact, et peu de temps plus tard, une illumination lui donna au cours d'une seule nuit, en utilisant une idée analogue, la plupart des théorèmes de sa théorie.

Citons encore deux mathématiciens fort différents l'un de l'autre:
Jean Dieudonné, cité dans le livre de Pierre Dugac (Gabay 1995): "Qu'est-ce qui se passe dans la vie d'un chercheur? On est sur un problème; on réussit quelquefois et il y a même ce qu'on appelle des illuminations .... Moi je peux dire que cela m'est arrivé trois ou quatre fois dans ma vie en tout".
Alain Connes: "En revenant de conduire ma femme au lycée, alors que je pensais à tout autre chose, j'eus la certitude absolue, devant un feu rouge, que les calculs longs et pénibles que je faisais depuis 6 mois, s'éclairaient à la lueur d'une astuce mathématique. Tout s'était passé comme si mon inconscient s'était brutalement exprimé".

Certes de grandes illuminations de ce type restent des instants mémorables de notre vie, mais toute activité mathématique qui ne soit pas de pure routine est parsemée de mini-illuminations. C'est par exemple celle de l'élève lent mais obstiné qui, après bien du travail et quelques aides extérieures dit brusquement "J'ai compris" et toute sa journée en est éclairée, et parfois sa vie changée: Il vient de subir une mutation mentale.

J'ai souvent constaté que des mini-illuminations pouvaient être déclenchées par des modifications physiologiques: Se lever; faire quelques pas après une longue station assise; se livrer à une activité très différente, lecture, musique, marche; sortir d'un engourdissement qu'on finirait par confondre avec du travail.

Le hasard. Il est temps de dire clairement ici qu'un chercheur débutant qui attendrait d'une illumination ou d'un hasard heureux la solution d'un problème difficile irait vers des désillusions. On souligne parfois le rôle du hasard dans une découverte scientifique, et on cite le cas de Fleming qui, arrivant un matin dans son laboratoire découvre sur la gélatine d'une boîte de Petri abandonnée un rond, vide de bactéries; une analyse lui montre que le responsable de ce rond – qui fait penser aux "ronds de sorcière" dans certains prés – est une moisissure du genre pénicillum: Il vient de découvrir la pénicilline, le premier d'une longue série de nos antibiotiques.

Mais Fleming était déjà un bactériologiste compétent; depuis plusieurs années il avait mis en évidence dans la salive et dans les larmes des substances antibiotiques; et c'était un esprit curieux, exigeant, qui mis en présence d'un petit fait inattendu, en cherchait l'explication. Un autre chercheur, moins motivé par une longue recherche préalable, et aussi moins exigeant, aurait rageusement jeté à la poubelle cette boîte de Petri gachée, ce que ne fit pas Fleming.

L'anthropologue qui, tête baissée marche lentement sous le soleil dans le vallée de l'Omo, à la recherche du "chaînon manquant" et qui heurte du pied une machoire d'homme fossile, a-t-il mérité sa chance?

Il y a des hasards qui peuvent devenir heureux, mais seulement pour les chercheurs qui l'ont mérité par un long travail préalable, de l'intelligence et une curiosité qui ont dévéloppé en eux sens critique et intuition. "Tu ne me chercherais pas si tu ne m'avais déjà trouvé"; cette phrase de Pascal pourrait conclure ma réflexion sur le hasard, mais je préfère la conclure par une anecdote: Voici quelques années j'accompagnais mon ami Jacques Deny dans un petit bois, à la recherche d'un petit champignon curieux qui ne pousse que dans la chrysalide d'un papillon précis; il en trouva de nombreux spécimens; je n'en trouvai pas un seul, mais il était mycologue et je ne l'étais pas.

L'intuition. Comme les illuminations, elle dépend des associations d'idées; mais contrairement aux illuminations, c'est un état acquis, stable et qui peut s'enrichir par un travail soutenu, bien que son acquisition dépende beaucoup de dons innés.

Elle apparaît comme une connaisance globale et presque immédiate d'un vaste domaine, mais c'est une connaisance non déductive et parfois assez floue; l'intuitif est semblable au jeune bébé qui, d'un seul coup d'oeil, reconnaît sa mère; il est plus un stratège qu'un tacticien qui, lui, a besoin de connaître les détails et leurs relations mutuelles.

• Pour un géomètre intuitif, la division du plan en deux régions par toute courbe fermée est une évidence, même s'il n'en a pas de preuve.

• Le Saint Curé d'Ars avait une telle intuition des problèmes moraux de ses pénitents que son confessionnal ne désemplissait pas.

Au début de mes recherches j'avais l'ambition de bien connaître tous les ensembles fermés plans, ambition irréaliste certes, mais j'avais néanmoins réussi à me sentir proche d'eux et à deviner, souvent avec succès, leurs propriétés topologiques: J'en avais l'intuition; c'est qu'en effet l'intuition se développe *par le travail*, beaucoup de travail conscient; c'est un ensemble déjà bien organisé d'associations d'idées liées à la mémoire parfois ancienne et dont on n'a conservé que celles qu'un jugement subconscient ou partiellement conscient juge bonnes.

Glaeser qui s'est beaucoup occupé de la formation des géomètres différentiels, leur conseillait de se constituer un florilège personnel d'êtres différentiels remarquables, éventuellement pathologiques, dont l'ensemble constituerait une bonne approximation du domaine étudié. On pourrait presque dire que l'intuition, qui doit être une des qualités d'un stratège se construit par un patient travail préparatoire de tacticien. Un tel travail préalable éviterait à certains débutants des échecs cuisants lorsque par exemple ils se lancent dans l'étude d'une structure axiomatique sans avoir vérifié l'existence d'êtres intéressants vérifiant les axiomes.

L'exemple des géomètres algébristes italiens montre clairement tout ce qu'une bonne intuition peut apporter, et aussi ses limites. Ces géomètres avaient étudié tant de courbes et surfaces qu'ils avaient énoncé sans démonstrations des théorèmes généraux d'une grande beauté; mais il fallut attendre pour leur preuve la naissance d'outils nouveaux (A. Weil, etc.).

L'exemple d'Euclide est également éclairant; parmi ses axiomes figure celui-ci "Une ligne droite est une ligne (i.e. "qui a une longueur sans largeur") qui repose également avec les points sur elle même". Euclide avait certainement une grande intuition géomètrique et cet axiome devait avoir pour lui un sens précis, mais aucun commentateur, pas même Proclus n'a bien réussi à le comprendre.

Tous les bons chercheurs ont de l'intuition, qu'elle soit géométrique, algébrique ou combinatoire; sinon leur démarche s'apparenterait à une reptation plutôt qu'au survol indispensable pour attaquer les problèmes aux points sensibles, et planifier leur recherche. Mais il faut, impérativement, vérifier les théorèmes suggérés par l'intuition, par de véritables démonstrations; celui qui se contente de lancer des idées sans prouver des théorèmes n'est pas un mathématicien; il est ce que les Américains appellent un "theoretical mathematician", ce que je traduirai plutôt par "mathématicien flou".

Je possède une intuition géométrique et celà, je crois, depuis la classe de 1ère; j'aimais la géométrie et j'avais pris l'habitude de résoudre de tête, dans l'obscurité, des problèmes assez complexes de géométrie plane ou spatiale. Je pense que mes contributions en Analyse et en Topologie sont toutes nées d'une vision géométrique, souvent très simple, des problèmes: Par exemple en Théorie du potentiel, c'est la géométrie du triangle qui m'a donné une des clefs de l'étude des grands principes de cette théorie.

Je me sens très proche d'Henri Lebesgue; lorsqu'il était jeune normalien et qu'on lui enseignait que les surfaces applicables sur le plan sont réglées, il sortait de sa poche un mouchoir ou un papier bien chiffonné et demandait à ses camarades "Cette surface est-elle réglée?" Ce mouchoir fut le point de départ de sa thèse et de sa théorie de l'intégration.

Plus tard, c'est la vue d'un mur de briques en construction, où il observa que ce mur comportait toujours des points de contact de 4 briques, qui le conduisit à la meilleure définition existante de la dimension topologique d'un continu.

Citons Lebesgue "Je veux dire ici que toutes mes recherches ont ce caractère commun de procéder d'une vue directe, et en quelque sorte géométrique, des problèmes étudiés". Et ailleurs "J'ai toujours été guidé dans mes recherches par des considérations géométriques ... ; il me semble que j'ai fait constamment des applications de la géométrie à l'Analyse".

Mais Lebesgue a fait aussi une autre remarque intéressante: "Non, ce n'est pas l'intuition qui trompe, c'est le fait de ne pas avoir assez d'intuition". Certes, mais on pourrait dire aussi: C'est le fait de ne pas vérifier pas à pas les suggestions de l'intuition; ce qui me conduit à rappeler une erreur célèbre de Lebesgue: Dans un de ses mémoires célèbres, il énonce que la projection de tout ensemble borélien est aussi un ensemble borélien, en utilisant la relation visiblement fausse

$$p(\cap_n X_n) = \cap_n p(X_n);$$

son théorème était faux et l'erreur, relevée par Lusin, conduisit Lusin et Souslin à la belle théorie des ensembles dits analytiques. Erreur fructueuse donc, mais erreur quand même, née d'une insuffisance de vérification.

Les intuitions des bons chercheurs ne sont pas toutes géométriques, même lorsque comme André Lichnerovicz ils sont géomètres différentiels. Michel Talagrand a une intuition combinatoire qui s'est confirmée au cours de sa carrière; ses succès récents dans l'étude des processus gaussiens et des produits infinis de mesures sont dûs à cette intuition.

Laurent Schwartz a un type d'intuition difficile à classer; elle n'est certainement pas géométrique; il ne "voit" pas géométriquement dans l'espace et prétend qu'il ne voit que par le calcul, même s'il ne s'agit, par exemple, que de troncatures de polyèdres convexes réguliers. On pourrait qualifier son intuition de culturelle; il a en effet une vaste culture mathématique, qui figure dans sa conscience comme un ensemble de fils tendus entre de nombreuses théories (concrétisées par ses dossiers). Quand il cherche, il se réjouit de "sécher" en se disant que si ça résiste, c'est que c'est intéressant puisque ce qu'il connaît ne lui apporte pas de réponse.

On oppose parfois abstraction et intuition, en sous-entendant sans doute que l'intuition concerne le concret. Ce n'était certainement pas l'avis de Jean Dieudonné qui déjà, à propos de sa thèse, déclarait: "Dans les démonstrations de beaucoup de propositions de ce travail, je n'ai pas craint de faire appel à l'intuition géométrique". Mais surtout, dans un intéressant article de Dialectica, Vol.29, 1975, il étudie les liens entre abstraction et intuition:

"La qualité essentielle d'un mathématicien est l'imagination; la logique ne sert qu'à mettre les démonstrations sous une forme irréfutable, elle est incapable de les suggérer. L'imagination se fonde sur une sorte d' "intuition" des objets mathématiques étudiés, mais cela n'a que très peu de contact avec ce qu'on appelle d'ordinaire l'intuition sensible, les objets mathématiques considérés étant le plus souvent l'aboutissement d'un long processus d'abstraction qui leur ôte toute possibilité de représentation concrète. Cette "intuition" mathématique est avant tout le résultat d'une longue familiarité avec le sujet étudié; mais en outre il peut s'opérer des "transferts" d'intuition d'une théorie dans une autre; .... La première conclusion que j'en tirerai, c'est qu'il n'y a certainement pas *une* intuition en mathématiques; il y en a toute une série de fort diverses, avec des liens inattendus. La deuxième, c'est que les intuitions mathématiques ne sont pas du tout stables, elles se modifient sans cesse par de nouveaux apports, de nouveaux résultats, de nouvelles idées .... Je crois que les progrès de l'intuition mathématique ... , sont toujours allés de pair avec les progrès de l'abstraction mathématique".

Cette analyse de Jean Dieudonné est remarquable, et surprend aujourd'hui sous la plume de celui qui, onze ans plus tôt, dans l'introduction de son livre "Algèbre linéaire et géométrie élémentaire" (Hermann 1964) soutenait que l'enseignement de la géométrie élémentaire doit être basé sur l'algèbre linéaire. Il n'avait pas encore compris que l'intuition des objets de cette algèbre ne s'obtient que par un "transfert" progressif à partir de l'intuition géométrique; essentiellement visuelle, acquise par le maniement préalable de concepts issus du monde sensible, droites et parallisme, distance, cercles , orthogonalité.

Les transferts d'intuition et de savoir-faire sont importants et même essentiels pour le progrès dans tout genre d'activité humaine; le psychologue Henri Pieron l'a bien souligné dans le "Traité de psychologie appliquée"; mais ils supposent toujours, bien sûr, une bonne maîtrise d'une activité initiale.

Le physicien Anatole Abragam en souligne l'importance, sans utiliser d'ailleurs le mot de "transfert": On sait bien que le renouvellement de la science se fait souvent aux points de contact de deux disciplines jusque-là étrangères; comment alors un chercheur va-t-il utiliser dans une discipline l'intuition acquise dans l'autre; voici ce qu'en dit Abragam:

"Un chercheur qui s'est confirmé dans une discipline et qui en aborde une autre, peut retrouver l'enthousiasme de ses débuts .... Il n'y arrive pas les mains vides. Il apporte avec lui des concepts, des habitudes de pensée, des méthodes et des techniques qui lui permettent d'aborder des problèmes d'une façon nouvelle pour le laboratoire qui l'accueille. Il y a là une cross-fertilisation".

J'ai pu vérifier moi-même cette observation lorsque, en 1980, je commençai à me passionner pour un problème à la frontière de l'Analyse et de la Théorie des nombres, à savoir l'étude de la distribution des points $(\frac{3}{2})^n$ modulo 1 sur le cercle $\mathbb{R}/\mathbb{Z}$, et plus généralement les suites $(k\theta^n)$ modulo 1, où k, θ sont réels et $\theta > 1$.

Stratèges, tacticiens et les deux hémisphères du cortex

Notre cortex est divisé en deux parties appelés hémisphères et réunies par le corps calleux de couleur blanche. Ce corps calleux a une importance considérable; si on le sectionne, les deux hémisphères deviennent presque indépendants; son importance est telle que si à la naissance il est peu développé, d'autres voies se créent entre les deux hémisphères. C'est son existence qui explique les correspondances notées par des écrivains et des poètes:
Baudelaire dans son sonnet "Correspondances"

> "Les parfums, les couleurs et les sons se répondent
> ....
> Il est des parfums frais comme des chairs d'enfant,
> Doux comme les hautbois, verts comme les prairies"

Rimbaud, jeune poète inspiré, dans le sonnet "Voyelles" écrit à 16 ans

> "A noir, E blanc, I rouge, U vert, O blanc, voyelles,
> Je dirai quelque jour vos naissances latentes".

Est-il utile aussi de rappeler les pages célèbres de Marcel Proust sur le flot de souvenirs que réveille le goût d'une madeleine?

Bien qu'il n'y ait pas de certitude absolue sur les rôles respectifs des deux hémisphères, de nombreux neurologues sont assez d'accord sur les faits suivants (chez les droitiers) révélés par l'observation des blessés et par la tomographie:

Relèvent de *l'hémisphère gauche* les processus déductifs linéaires; la pensée logique et analytique; les capacités linguistiques; apprentissage, production et compréhension du langage: Ceux, en gros où excellent les *tacticiens*. Cet hémisphère est bien adapté à la vie quotidienne, mais il est trop critique et sans grande intuition ni créativité. Ce que mesurent les tests de Q.I. sont essentiellement les performances de cet hémisphère.

Relèvent de *l'hémisphère droit* l'imagination, la reconnaissance des formes, la créativité, les associations d'idées multiples, les fonctions intellectuelles non verbales et un traitement global des informations. C'est l'hémisphère des *stratèges*. Il jouerait peut-être un rôle dans l'élaboration des rèves. Cet hémisphère voit le monde sous des couleurs parfois déplaisantes et hostiles.

Le droit invente, le gauche enseigne ces découvertes, en parle rationnellement, les met en forme. Mais cette dichotomie ne doit pas être exagérée; en particulier, il peut y avoir substitution de leurs activités en cas de besoin, par exemple lors de la lésion d'un hémisphère. Et surtout le progrès scientifique exige la collaboration de deux tels hémisphères, qu'ils appartiennent tous deux à un même individu, ou à deux individus qu'on pourrait appeler complémentaires: C'est peut-être là l'explication du succès de la collaboration de certains couples de chercheurs, dont le plus fameux fût sans doute le couple Einstein, Mileva Mariĉ, en 1903-1905, avant même leur mariage; Einstein ayant les idées, Mileva les discutant et apportant sa virtuosité mathématique.

Certes les caractéristiques respectives des deux hémisphères que je viens d'évoquer font un peu trop penser aux affirmations gratuites des voyantes extra-lucides. Toutefois pour chacun d'eux le faisceau de ces caractéristiques semble assez cohérent et suggère l'intérêt d'une étude expérimentale des corrélations entre les compétences supposées d'un même hémisphère et des corrélations croisées entre les deux hémisphères.

Il semblerait en tout cas que chez les bébés et les jeunes enfants ce soit l'hémisphère droit qui domine.

Il est probable aussi que le style des recherches d'un mathématicien révèle la prédominance d'un de ses hémisphères; c'est ainsi que je déciderais volontiers que mon hémisphère dominant est le droit: Goût de la poèsie, de la rêverie, difficultés linguistiques (je conçois facilement, mais souffre pour mettre en forme mes pensées). Il semble que, l'âge venant, mon intuition géométrique s'exerce encore sur de nombreux ilôts assez représentatifs d'un champ assez vaste, mais que la mise en ordre, puis en forme, de ces observations intérieures demanderait des dons de tacticien qui me font défaut. C'est pourquoi la plupart de mes travaux ont concerné des structures si simples à mes yeux que j'ai de chacun d'eux une vision immédiate et facilement communicable: "C'est trivial" ai-je envie de dire quand j'en parle; par contre les deux seuls articles assez techniques que j'aie publiés m'ont coûté de pénibles efforts de rédaction, bien que leur conception ne m'ait pris que quelques heures:

Je souffre assez peu pour préparer une conférence, mais incomparablement plus pour la rédiger!

Facteurs propices à la création

Jeunesse

Neurones et faisceaux de neurones sont encore presque intacts, énergie nerveuse et sexuelle à son summun (pensons à Einstein et Mileva), aptitude à une concentration prolongée, curiosité encore fraîche, un cerveau peu encombré par les préjugés du passé:

Newton à 23 ans, en 1666, fait les trois découvertes qui lui assureront l'immortalité.

Einstein à 26 ans, en 1905, fait lui aussi trois de ses plus belles découvertes.

Motivation

La motivation, née d'une pulsion intérieure, soutenue d'abord par l'approbation d'un maître et l'estime des pairs mais soutenue ensuite par les défis, intérieurs ou extérieurs, par le plaisir de faire fonctionner son cerveau et, pourquoi pas, par l'espoir de laisser un nom célèbre, ainsi que nous le confie André Weil "Dès mon jeune âge, j'ai espéré que mon travail aurait une certaine place dans l'histoire des mathématiques. N'est-ce pas une motivation aussi noble que de prétendre au Prix Nobel?"

En ce qui me concerne je voudrais dire que ma motivation a toujours été le plaisir de créer de belles mathématiques, utiles ou simplement élégantes, et d'avoir l'approbation de mes pairs. Puis-je ajouter que j'ai souvent été étonné que la société me donne un salaire convenable pour une activité qui me procure tant de plaisir, alors que nombre de mes contemporains reçoivent un salaire très inférieur pour un travail répétitif et ennuyeux.

L'enthousiasme, la joie de la création, les instants brefs mais inoubliables des illuminations, sont le véritable aiguillon du chercheur. Que Dieu a dû bien s'amuser pendant les six jours de la création, mais quel ennui après le jour de repos!

En Mai 68, j'ai entendu des étudiants, mécontents de leur passivité dans les salles de cours, dire à leur professeur: "Nous avons tous le droit de faire de la recherche; dites nous ce qu'il faut faire et nous le ferons". Je vois dans cette déclaration à la fois une critique des cours magistraux, et aussi une méconnaissance des facteurs qui soutiennent l'activité du chercheur, en particulier succès et enthousiasme qui s'engendrent mutuellement. Comme le dit le physicien George Charpak "Si la vie d'un chercheur se résume à exécuter des tâches qu'il ne trouve pas intéressantes, autant entrer dans l'industrie". (C'est dûr pour l'industrie!)

Facteurs sociaux, émulation

Je voudrais revenir un instant sur le rôle des facteurs sociaux, de l'émulation et des challenges. Que la recherche en sciences expérimentales, avec ses laboratoires et ses équipes ait besoin d'une société bien organisée est une évidence. Mais c'est vrai aussi de la recherche mathématique, et pas seulement à cause de l'utilisation accrue des ordinateurs et du développement du travail en équipe; c'était vrai déjà du temps d'Euclide et d'Archimède où les Académies, le soutien de princes éclairés, l'échange de lettres, constituaient une stimulation indispensable; c'était vrai au 17e siècle où le Père Mersenne, conscient de ce besoin, se fit la boîte-aux-lettres mathématique de l'Europe. La recherche mathématique a commencé avec la naissance des grandes villes! Un Robinson Crusoe, mathématicien avant son naufrage, et encore productif sur son île, est-il crédible?

Jusqu'à quand notre République étouffera-t-elle l'éclosion des talents dans nos écoles de toutes niveaux en luttant, par la loi et par des mesures de fausse démocratisation contre émulation et compétition; ce n'est pas seulement sur les stades que la compétition est indispensable!

Les excitants

J'ai mentionné déjà l'importance pour Einstein au début de ses travaux, de ses liens amoureux avec Mileva. Il y eut alors une conjonction heureuse de cet amour mutuel et de l'excitation cérébrale apportée par leur collaboration.

Une relation amoureuse peut donc être un facteur favorable à la recherche, mais elle est rarement neutre et peut au contraire avoir un effet déstabilisant.

Puis-je rappeler enfin l'importance du thé et du café? Lorsque mon ami Marcel Brelot, après une période de dure recherche sans succès sentait que cependant il avait en mains tous les atouts, il s'enfermait dans son vaste bureau de l'Institut Fourier de Grenoble et se faisait une tasse de thé fort pour provoquer le déclic, l'illumination cherchée car, disait-il, les découvertes se font lors de pointes de l'activité cérébrale, après un long travail de défrichement. (Mais aucune tasse de thé ne transformera un champion du 100 m en marathonien.)

Est-il nécessaire de souligner ici l'illusion de l'efficacité de "drogues" pour réussir en recherche? Je ne connais aucun chercheur scientifique dont les découvertes aient été faites sous l'emprise de la drogue.

Facteurs défavorables

Age

Les soucis, la fatigue physique ou nerveuse, et l'âge: à 30 ans le travail vous tire en avant, à 70 il faut le tirer, et c'est difficile! Mais si un âge avancé n'est jamais favorable à la découverte, une meilleure hygiène alimentaire, de meilleures relations scientifiques allongent la période de productivité des mathématiciens. Sans donner de chiffres précis, disons seulement qu'à 60 ans ou même un peu plus, ils peuvent encore faire des travaux de qualité honnête et rendre des services à la communauté mathématique s'ils acceptent de ne pas utiliser leur notoriété pour soutenir abusivement des recherches qui prolongent les leurs.

Et puis parfois, une bonne surprise, telle celle d'Apery qui à 61 ans démontra l'irrationalité de $\zeta(3)$.

Les pertes de performance ont des causes exactement opposées à celles qui favorisent la jeunesse:

Détérioration des neurones, perte de mémoire et baisse de l'intuition; moins bonne irrigation sanguine, pulsions amoureuses diminuées. Impression de "déjà vu", d'où motivation moins grande; et après tout, on a déjà une certaine célébrité qu'on ne peut espèrer accroître!

Mais surtout, tous les fils à la patte: Comités de rédaction, assemblées générales, lettres de recommendation, ... qui grignotent le temps et l'énergie. Voyons sur un exemple célèbre comment vieillissait un mathématicien au début de ce siècle; dans une des lettres de Lebesgue à Borel, retrouvées dans les sous-sols de l'Institut H. Poincaré, Lebesgue écrit ceci, alors qu'il n'avait encore que 35 ans:

"Mon cher ami, ce n'est pas par coquetterie que je dis que je baisse. Je sais fort bien que depuis 5 ans je n'ai rien fait de nouveau .... C'est à une place de Recteur que j'aurais dû être candidat."

Il avait fait sa théorie de l'intégration à 27 ans, et son travail sur les fonctions représentables analytiquement à 30 ans.

Mais il faut savoir aussi quelle était sa vie: Mauvaise santé familiale, préoccupations de carrière, gêne financière, début de sa polémique avec Borel.

Blocages, collectifs ou individuels

Il y a des blocages *collectifs*: Celui lié au postulatum d'Euclide n'a commencé à céder que le jour où on a commencé à en douter; il avait duré plus de 2000 ans. Ceux liés à la notion d'infini, actuel et potentiel; les vaines disputes de caractère parfois métaphysique concernant l'infini ne furent balayées que par les premières définitions de Cantor, en particulier celle de bijection.

Tout déblocage est comme l'ouverture d'une vanne, qui laisse couler l'eau à grands flots, l'eau qui apporte une vie nouvelle: Géométrie non-euclidienne et les travaux de Poincaré; la théorie des ensembles et tout un pan des mathématiques modernes.

Les blocages collectifs, et surtout ceux qui durent des siècles sont dûs au conservatisme de l'esprit humain: L'homme a tendance à prendre pour des vérités intangibles ce qu'il a appris dans son enfance et qu'il transmet à ses enfants. Pour sortir du sillon de la tradition, il a besoin de l'aiguillon de la nécessité, les impasses d'une science, la rencontre d'un petit fait qu'aucune théorie admise ne peut expliquer.

Il faut se méfier des modes, des grands gourous, rechercher les fleurs vénéneuses, ... et changer nos méthodes d'enseignement. On peut rêver à ce que serait l'évolution de la science si n'existaient pas ces longs blocages collectifs; mais peut-être Dieu après le 7e jour fût-il un peu machiavélique et a-t-il, depuis l'épisode de la pomme, prévu des blocages comme garde-fous contre une évolution trop rapide de l'humanité!

Mais il y a aussi des blocages *individuels*, même chez les créateurs authentiques; nous retrouvons ici Lebesgue:

L'observation d'un petit fait, son mouchoir chiffonné l'avait conduit, après avoir précisé la théorie de la mesure de Borel, à définir la notion d'intégrale de certaines fonctions mesurables, mais l'histoire serait trop simple s'il avait, du premier coup, défini les fonctions intégrables quelconques. Lui aussi est resté pendant des années esclave de la tradition: L'intégrale de Riemann supposait les fonctions bornées; Lebesgue conserve cette restriction dans sa thèse de 1902; en 1905, à propos des fonctions non bornées il écrit encore à Borel "Il est si difficile d'attaquer ces maudites fonctions"; et alors qu'en 1908 l'obstacle semble tout à fait surmonté, il en reste encore des traces dans ses "Leçons sur l'intégration" de 1928, comme si dans son esprit les fonctions intégrables non bornées étaient un luxe, une curiosité; et pourtant le théorème de Riesz-Fischer qui s'exprime en termes de fonctions non-bornées avait été publié en 1907.

Comment éviter les blocages? L'humanité va-t-elle répéter l'erreur catastrophique qu'elle a faite pendant 15 siècles en acceptant comme canonique l'enseignement d'Aristote, erreur qui n'a cessé qu'après le choc, psychologique, technique ou scientifique qu'a constitué la rencontre avec d'autres civilisations, par les croisades, Marco Polo, Christophe Colomb?

Aujourd'hui deux faits nous protègent peut-être des blocages: communication et information rapides, et l'accroissement considérable du nombre des chercheurs.

La probabilité de blocage existe encore, mais elle restera minime si l'on sait se méfier des modes et des gourous. Le grand gourou mathématicien de notre temps est Bourbaki; son porte-parole le plus connu, Jean Dieudonné, a laissé croire, en insistant sur "Le choix bourbakiste", que c'était l'unique choix, et que les disciplines naissantes ne méritaient pas l'attention tant qu'elles n'étaient pas structurées en un bel échafaudage formalisable.

Or les clefs de l'avenir, celles qui vont peut-être nous donner une nouvelle façon de regarder le monde existent peut-être dès aujourd'hui à l'état de germes; laissons les germer. Ce sont ces concepts nouveaux qui constituent les mutations humaines de notre temps.

Citations

Vérités établies et gourous. Il faut parfois déboulonner les statues trop imposantes, y compris notre statue intérieure. On peut s'aider, un temps, des épaules des plus grands, mais il faut être soi-même dès qu'on se sent assez fort. Plusieurs penseurs, venus d'horizons différents l'ont excellemment dit:

KRISHNAMURTI: "Toute autorité aveugle et tue." Il rejette les gourous et refuse d'en être un; il dit d'un de ses visiteurs "Il avait tant lu qu'il lui était difficile de savoir où commençait sa propre pensée".

MACHADO: "Marcheur, le chemin, ce sont tes traces et rien de plus. Marcheur, il n'y a pas de chemin; c'est toi qui le traces en marchant."

GIDE: "Nathanael, jette mon livre, ne t'y satisfais point. Ne crois pas que ta vérité puisse être trouvée par quelqu'autre".

LAVELLE, lui aussi s'élève contre les dogmes: "Il n'y a rien que nous ayons pensé une fois pour toutes et qui soit tel qu'il suffirait d'en garder la mémoire et de le convertir en règles".

C'est dire qu'il n'y a pas de vérité; c'est sa recherche qui importe. L'architecte et le maçon sont plus précieux que le monument qu'ils ont édifié!

Les dons

Unicité des hommes, dons et génie

Chaque homme est unique, irremplaçable. Nous classons trop souvent nos contemporains suivant des critères utilisant un très petit nombre de paramètres, les maigres et les gros, les honnêtes et les malhonnêtes, les bons en gymnastique, les forts en thème, etc.: classement certes facile mais qui occulte, de chaque individu sa richesse, sa diversité, son unicité en un mot; car on sait maintenant, après la mise en évidence et le comptage des gènes, des groupes sanguins (dont le système HL-A) et des mutations des acides aminés, qu'au sens biologique il n'y a pas deux hommes identiques, même si ce sont de vrais jumeaux, et que ce serait vrai encore si l'humanité était mille fois plus nombreuse. Mais c'est autre chose d'en prendre une pleine conscience.

Combien apparaît misérable, dans cette lumière, le racisme, linguistique, social, religieux, ethnique ou tribal. Il y a racisme parce que l'Autre dérange nos habitudes, et nous obligerait pour le comprendre à sortir de notre cocon, aussi inconfortable soit-il. Si nous en prenons une pleine conscience, nous devons en assumer toutes les conséquences: Une ouverture totale envers tous les êtres, mais aussi puisque chacun de nous constitue une synthèse unique ne serait-ce que par notre sensibilité et notre vision du monde, comprendre que notre devoir envers nous mêmes doit être de faire fructifier au mieux notre unicité.

Puisque chaque homme est unique, il mériterait que tous les chercheurs du monde se penchent sur lui pour l'étudier; mais de cette synthèse unique, ne nous frappent souvent que les facettes les plus brillantes et les plus rares, qu'on appelle des *dons*. Il s'agit de caractéristiques humaines, physiques, intellectuelles ou morales considérées comme souhaitables: Beauté, force, adresse, grâce, amabilité, dévouement, résistance physique, persévérance, mémoire, etc. ... Ce sont ces dons que les bonnes fées des contes apportent dans le berceau des jeunes princesses; mais on sait aussi que de vieilles sorcières jalouses apportent de leur coté des anti-dons, que l'on appelle des mauvais sorts: Paresse, jalousie, esprit brouillon, laideur, etc. ...

Lorsqu'un don se manifeste avec une intensité exceptionnelle, et surtout chez un jeune sujet, il frappe d'étonnement ceux qui en sont les témoins et on use, et abuse, du mot *génie*. Quant à moi, je serais tenté de n'appliquer ce qualificatif qu'à ceux dont l'action a déterminé une mutation de l'humanité: des savants, philosophes, artistes, musiciens, poètes, mais aussi des fondateurs des grandes religions.

Mais où s'arrête l'échelle des génies? William James disait: "Le génie n'est guère plus que la faculté de percevoir sur un mode inhabituel"; à ce compte, tout mathématicien ou scientifique, créateur d'une nouvelle notion féconde serait génial, ce qui certainement ne lui déplairait pas! Je préfère donc parler de dons exceptionnels; ces dons se manifestent presque toujours dans la jeunesse.

Certains enfants manifestent une précocité exceptionnelle pour parler, lire, écrire, mais ce don ne préjuge d'ailleurs nullement de leurs aptitudes d'adultes. D'autres, très jeunes encore telle Anne Frank, tiennent un journal intime qui nous émeut encore. J'ai un collègue qui, dans sa jeunesse, avait un tel besoin de s'exprimer qu'il payait sa soeur cadette pour qu'elle l'écoute parler. La plupart des mongoliens ont reçu, par leurs gènes, le don d'une affection débordante. Certains enfants autistes gardent en mémoire, avec leurs dates, le détail des jours passés. Mais les calculateurs prodiges n'ont presque tous (si l'on excepte Gauss et quelques autres) en dehors du calcul que des capacités intellectuelles moyennes.

Mais revenons aux très grands, dont il est agréable et réconfortant d'égrener quelques noms; ce furent toujours des êtres d'une grande intelligence, en dehors même de leur orientation maîtresse.

Un grand conquérant, Alexandre, élève d'Aristote, meurt à 33 ans après avoir en 12 ans conquis un immense empire, de la Macédoine à l'Indus.

Des musiciens dont les dons se manifestent très tôt: Mozart à 4 ans, Beethoven, Chopin à 7 ans, Prokofiev avant 9 ans.

Des peintres, sculpteurs, architectes: Bruegel l'Ancien, Michel-Ange, Léonard de Vinci, Van Gogh, Picasso, etc. ...

Des philosophes comme Pic de la Mirandole, proclamé à 10 ans "Prince des orateurs et des poètes", mort à 31 ans, laissant une oeuvre immense.

Pascal, mathématicien, physicien et philosophe, célèbre dès l'age de 16 ans, créateur du calcul des probabilités avec Fermat, et de la machine arithmétique.

Descartes, créateur du lien entre géométrie et algèbre, et auteur du "Discours de la Méthode".

Newton, mathématicien et physicien crée entre 22 et 23 ans la théorie moderne des couleurs et le calcul des fluxions pour développer la théorie de l'attraction universelle et poser les bases de la mécanique rationnelle.

Leibniz, créateur universel, qui partage avec Newton la gloire de la création du calcul différentiel et qui, à 15 ans formule ses premières idées sur la logique.

Einstein que l'on cite souvent comme l'exemple d'un génie de première grandeur qui n'a pas manifesté précocement ses dons, publia pourtant, à 26 ans, trois mémoires fondamentaux qui changèrent la physique: Relativité restreinte, effet photo-électrique, mouvement brownien.

En mathématiques, la liste est longue de ceux que nous admirons:

Les Bernoulli, exemple remarquable d'hérédité (8 en 3 générations) dont certains allaient vers les mathématiques comme les alcooliques vers leur drogue.

Euler, élève de Jean Bernoulli qui sera à 16 ans professeur à l'Ecole d'Artillerie de Turin et décidera à 17 ans, malgré son père, de se consacrer aux mathématiques.

Laplace qui, très jeune avait déjà une mémoire prodigieuse.

Monge, géomètre né, professeur de physique à 16 ans.

Gauss, mathématicien, physicien, astronome et calculateur prodige, dont le génie exceptionnel se manifeste très tôt.

Galois mort à 21 ans, et Abel mort à 27 ans, les deux cadets des mathématiques, à la trajectoire courte mais fulgurante.

Poincaré, esprit universel, issu d'une famille distinguée, doué d'une mémoire puissante, et dont la passion mathématique débuta à 15 ans.

Cantor, créateur de la théorie des ensembles, dont les dons se manifestèrent vers 15 ans.

Ramanujan, fils de l'Inde, autodidacte, énonça sans preuve de nombreuses identités remarquables entre séries, si cachées que toutes n'ont pas encore été démontrées.

Cette courte liste ne contient aucun nom de femme parce qu'elle concerne une période où il était encore malséant pour une femme de consacrer son énergie à une oeuvre scientifique ou même artistique, et où de plus l'éducation était le privilège de classes aisées; on ne peut cependant pas oublier la lutte que dût mener la jeune Sophie Germain pour étudier les travaux de Gauss et lui adresser ses propres résultats en théorie des nombres et sur les surfaces élastiques; ni les travaux plus connus encore de Sophie Kovalevskaia, lauréate à 28 ans de l'Académie des Sciences.

Hérédité et génétique des dons

J'entends par hérédité la transmission de certaines caractéristiques d'un couple, animal ou humain, à sa descendance; et par génétique la détermination des caractéristiques d'un individu par le génôme de l'oeuf fécondé d'où il provient. L'hérédité ne détermine qu'en partie ce génôme, puisque la loterie génétique intervient au moins à deux étapes de sa formation.

Cependant n'existerait pas de sélection des animaux domestiques si l'hérédité ne déterminait pas fortement chez ces animaux leurs traits physiques et certains traits du comportement; n'existeraient pas la race des percherons et celle des pur-sang, les teckels ni les Saint-Bernard.

Il n'y a aucune raison pour qu'il en soit autrement dans l'espèce humaine; de fait il est rare que les fils de deux parents longilignes soient petits et trapus, ou inversement; ou, de façon plus frappante encore, qu'une fille de paysans chinois soit une blonde aux yeux bleus. J'ai eu au lycée un camarade qui était petit, rond, de caractère aimable et s'appelait Triboulet; son père, son frère et sa soeur étaient petits, ronds et souriants!

La transmission par hérédité de caractères physiques dans l'espèce humaine ne fait donc pas de doute; c'est là l'origine de la notion de race, basée sur un tout petit nombre de traits physiques.

Il en va tout autrement de la transmission des traits de caractère et surtout de ce que l'on peut appeler les dons intellectuels.

S'il est vrai que pendant les deux derniers siècles s'est produite en Chine une stagnation des innovations scientifiques et techniques, la cause en est culturelle et sociale, mais non raciale; pendant des millénaires sa civilisation avait ébloui les explorateurs, tel Marco Polo, et ses apports à la science et à la technique étaient fondamentaux. Et depuis son récent réveil, elle commence à produire des chercheurs de très haut niveau international, annonciateurs peut-être d'un avenir culturel et scientifique de premier plan.

Il est donc possible, bien qu'une preuve scientifique n'existe pas encore, que tout don intellectuel soit la résultante d'une coexistence harmonieuse d'un petit nombre des gènes, ou encore une synergie non contrariée par l'existence d'anti-dons; on chercherait donc en vain à localiser sur un gène, un don pour les mathématiques ou même un penchant pour les sciences.

Il existe cependant des lignées célèbres de peintres, de musiciens (les Bach, Strauss). En mathématiques, les Bernoulli sont un exemple éclatant et assez mystérieux d'une telle lignée; je ne vois pour l'expliquer sans faire intervenir la vertu de l'exemple, de l'émulation et de l'environnement, que l'existence dans les génôme du premier grand Bernoulli, d'une assemblée de gènes dominants, responsable de ses dons mathématiques, et si fortement soudés que la probabilité de sa transmission sur trois générations n'était pas négligeable.

Certes la probabilité pour qu'un couple ait un enfant doué est plus grande si ce couple est lui même doué que s'il ne l'est pas; mais cette condition n'est ni nécessaire, ni suffisante; il existe des exemples fameux de très grands hommes issus de parents ne brillant par aucun don:

Le père de Newton était "sauvage, extravagant et faible"; sa mère était simplement économe, travailleuse et bien organisée.

Laplace était le fils de pauvres paysans normands.

Le père de Gauss était un homme tout ordinaire; toutefois sa mère était une femme remarquable.

La première épouse et collaboratrice d'Einstein, Mileva Mariè était une femme exceptionnelle par son énergie, son enthousiasme et ses dons mathématiques; et cependant, de leurs deux fils, le premier Hans Albert (né en 1904) devint siplement ingénieur, puis professeur d'hydraulique; et le second, Edward (né en 1910), très doué dans son enfance, devint schizophrène à 19 ans; sa ressemblance *physique* avec son père était frappante; il mourut à 55 ans.

Le développement du cerveau est différent chez l'homme et chez les animaux; chez ceux-ci les premières semaines et même parfois les premières minutes de la vie ont une importance considérable. Les travaux sur les oiseaux de Heinroth (1910) puis de K. Lorenz (1935) ont montré que le premier objet mobile aperçu pendant une période critique de quelques heures ou même de quelques minutes chez certaines espèces peut devenir définitivement un objet-parent ou même un objet sexuel.

Chez les chatons dont on empêche les paupières de s'ouvrir pendant quelques semaines, les connections des neurones optiques dégénèrent ou ne s'établissent pas, et le chaton devient définitivement aveugle.

Ces conclusions ne s'étendent pas sans précautions aux jeunes humains, sans doute parce que dans leur cerveau le nombre de synapses potentielles ou déjà établies est si grand que, par des voies indirectes peuvent s'établir, même tardivement, des faisceaux supplétifs de neurones. Toutefois l'observation des "enfants sauvages" au cours des derniers siècles montre que contrairement au Mawgli de Kipling, ces enfants ont été incapables d'acquérir un développement physique et mental normal.

Il est donc clair que l'éducation familiale d'abord, puis l'école et la société jouent un rôle essentiel dans la formation du cerveau et la maturation d'un don d'origine génétique: "Lorsqu'une semence tombe sur un terrain pierreux, elle lève mais dès que le soleil paraît, elle est brulée et sèche faute de racines".

Mais comment reconnaître les enfants doués puis, les ayant reconnus, quelle éducation leur donner? C'est la grande responsabilité à la fois de leur famille et de l'école.

Les Américains, souvent excessifs, ont créé à côté de leurs écoles publiques assez médiocres, des écoles pour surdoués, dont la réussite est fort controversée.

En France des gouvernements successifs, mus par un faux esprit démocratique ou par démagogie, ont prétendu vouloir donner à tous, les mêmes chances de succès, et ont pour cela rendu obligatoires les classes indifférenciées, néfastes pour les enfants au rythme lent, et désastreuses pour les enfants doués. Il faut se réjouir que la pression conjuguée de certains enseignants et des familles a conduit à certains palliatifs, mais la loi, même absurde, reste la loi: L'erreur a la vie dure. Alors que le but de l'école devrait être de développer au mieux les possibilités, parfois très différentes, des enfants, la loi slogan actuelle pourrait s'énoncer "Tous les enfants doivent recevoir la *même* éducation".

Je ne saurais mieux conclure mon éloge de la "différence" que par cette citation de Lavelle:

"Si l'on veut que tous les hommes soient semblables, ils ne cesseront de se heurter et de se haïr; ... sinon chacun d'eux sera pour tous les autres une révélation et un soutien".

Epanouissement d'un don

Il ne suffit pas d'avoir un don à la naissance; encore faut-il qu'il puisse se développer, fructifier et ne soit pas "laissé sous le boisseau". Que de dons perdus chez les bébés chinois pendant les deux derniers siècles de décadence de l'empire chinois!

Un minimum de mémoire est indispensable; Einstein se plaignait de la sienne, mais apparemment elle était suffisante; et inversement une mémoire non sélective peut encombrer l'esprit, comme le soulignait Krishnamurti parlant d'un de ses visiteurs.

Il faut certes, de la curiosité et de l'imagination, mais aussi de la suite dans les idées, de la persévérance et un travail soutenu – avant et après l'illumination espérée; celle-ci lorsqu'elle se produit doit être une "ivresse maîtrisée".

Une certaine forme de culture est indispensable, acquise non de façon passive, mais avec appétit, avec l'objectif plus ou moins proche de l'utiliser dans la recherche en cours. En un siècle où les mathématiques accèdent à l'unité, où les sciences expérimentales ont à se préserver de certains effets néfastes d'un réductionnisme pourtant nécessaire, une telle façon active d'acquérir de la culture semble être la plus efficace. Topologie et géométrie algébriques, variétés différentiables et groupes de Lie, ont tissé des liens féconds avec la physique et chacune de leurs facettes est vitalisée par les autres; ces liens doivent être présents dans le subconscient et dans la mémoire consciente de tout chercheur, ne fut-ce qu'à l'état d'ébauches.

Certes il n'est ni possible ni souhaitable de définir des règles précises pour développer un don, car l'originalité du chercheur doit aussi consister à découvrir son propre cheminement: "Marcheur; il n'y a pas de chemin; c'est toi qui le traces en marchant".

Poisons de la création

A côté des dons il y a les anti-dons, dont les enseignants et les chercheurs doivent se méfier.

• Soit internes tels que manque de vitalité, manque de confiance en son étoile, fragilité psychologique pour trop de sensibilité aux coups inévitables de l'environnement: Un créateur doit avoir une âme robuste et savoir, quand il le désire, protéger du monde extérieur son activité de recherche.

• Soit externes comme les pressions familiales pour imposer à un jeune doué une profession qui lui déplait. Que de savants illustres n'ont échappé que par miracle ou par leur force de caractère à de telles pressions; mais nous ne connaîtrons jamais ceux qui n'ont pas eu cette chance.

• Poison aussi, bien que plus subtil, l'effet que j'ai déjà souligné, néfaste après avoir été fécond, d'une grande synthèse de concepts présentés comme des vérités immuables. Les gourous, quelle que soit leur valeur personnelle, ne doivent être écoutés que d'une oreille; mais bien sûr, les plus dangereux sont les "petits gourous" et ceux qui se complaisent dans ce rôle.

Il faut savoir *donner congé aux dogmes.*

En présence d'un jeune doué, le devoir d'un maître à quelque niveau que ce soit, doit être seulement de l'aider à se mieux connaître, et à trouver seul sa voie.

La lumière sous le boisseau

Nous avons pris l'habitude de publier dès que possible tout résultat scientifique, avant même parfois d'en avoir une preuve écrite, détaillée et convaincante; la compétition internationale, accentuée par la rapidité des communications électroniques, ainsi que par les règlements des jurys d'évaluation, en sont en grande partie responsables.

Il n'en a pas toujours été ainsi: Sans avoir le goût du secret aussi poussé que l'école Pythagoricienne, les mathématiciens du 17e siècle n'annonçaient souvent leurs découvertes que sous forme de cryptogrammes ou de problèmes signés d'un pseudonyme. Souvent aussi, leur amour-propre d'auteur se contentait de l'estime de quelques amis ou correspondants au courant de leurs recherches. Il faut dire que les revues scientifiques n'existaient pas encore, d'où le rôle important d'un altruiste éclairé comme le Père Mersenne.

C'est peut-être là qu'il faut voir la raison principale du retard de la publication par Newton de sa "Méthode des fluxions" et surtout de ses "Principes mathématiques de la philosophie naturelle" qu'il ne se décida à rédiger que sous l'amicale insistance de Halley, pour paraître en 1686, donc 20 ans après leur création.

Plus de 130 ans plus tard, Gauss allait récidiver. Plusieurs de ses découvertes, telles que les propriétés des fonctions elliptiques ou le critère d'analyticité des fonctions continues, restèrent enfouies, souvent à l'état d'esquisse ou sous forme cryptique dans son journal personnel publié seulement 62 ans après sa mort. Pourquoi tant de richesses cachées? Certes, avant l'âge de 20 ans il avait tant d'idées que le temps lui manquait pour les développer; mais il semble établi qu'il considérait comme secondaire la publication de résultats que seule une motivation intérieure puissante l'avait conduit à formuler et à démontrer. Pareil désintéressement ne semble pas menacer aujourd'hui le monde mathématique et on peut s'en réjouir si l'on désire que continue l'actuel développement explosif des mathématiques. Encore faudra-t-il veiller à ce que la qualité ne soit pas étouffée par la quantité, et à préserver l'unité féconde de l'édifice mathématique en développant un réseau de diffusion et de synthèse accessible à tout bon chercheur.

Amour et idée intérieure

Me voici bientôt parvenu au terme de ma trop longue réflexion sur les processus mentaux de la création: Trop longue mais cependant très incomplète puisque l' "homo creator" est un être complexe et qu'ainsi la seule création mathématique, par exemple, ne peut être bien comprise, à la fois dans un seul cerveau ou dans le mégacerveau de l'humanité tout entière, que par ses interactions avec les autres modes de création, musicale, artistique, poétique, etc. ...

Si l'on définit l'amour d'un être humain comme un élan mental de cet être vers un objet, inerte, animé ou purement conceptuel, on pourrait dire que toute création est mue par un amour. Mais je préfère, lorsque cet objet est purement conceptuel, reprendre pour cet amour le nom d' "idée intérieure": Cette notion a été formulée par Nietzsche, après le peintre allemand Caspard David Friedrich, et reprise récemment par J. Harthong pour l'appliquer à G. Reeb dans son article "Comment j'ai connu et compris Georges Reeb". Friedrich affirmait: "Il existe un sentiment esthétique spontané qui permet à l'artiste de sentir en lui-même ce qu'est *le* beau et ce qu'il *doit* représenter". C'est donc un sentiment personnel, une pulsion subjective qui dicte la voie à suivre; c'est donc lui qui aide le marcheur de Machado déjà cité, à trouver son chemin en marchant.

Le beau-conte de Paulo Coelho "L'alchimiste" est une belle illustration du développement d'une idée intérieure: Les suggestions voilées que son jeune berger reçoit de mystérieux vieillards, y apparaissent comme les pulsions d'une voix intérieure:
"Tu dois essayer de vivre ta légende personnelle."
"Quand tu veux quelque chose, tout l'univers conspire à te permettre de réaliser ton désir."
"On a toujours la possibilité de faire ce que l'on rêve."

Une idée intérieure puissante est en général liée à un don – par exemple une mémoire exceptionnelle dans le cas de calculateurs prodiges. Elle peut exister et se manifester très précocement, comme chez Mozart; mais souvent elle est d'abord imprécise et même multiforme, puis elle se précise et se nourrit par l'expérience, pour se manifester enfin dans une trajectoire éblouissante.

Une étude approfondie de la formation de l'idée intérieure chez les grands hommes du passé permettrait de comprendre mieux les rôles respectifs des dons d'origine génétique et de l'environnement; les matériaux pour une telle étude ne manquent pas; ils illustrent de façon convaincante la puissance et l'efficacité des idées intérieures, par exemple celles des fondateurs de grandes religions, Bouddha, Jésus, Mahomet, ou d'hommes dont le charisme était si puissant qu'il pouvait entraîner tout un peuple.

Aussi les exégètes et les Pères de l'Eglise, surtout préoccupés par l'aspect messianique de Jésus, me semblent-ils avoir trop négligé les facteurs de formation de son idée intérieure: Equilibre, dons de synthèse, vision de plus en plus claire de sa mission, contacts encore mal connus avec les Esséniens, mais surtout l'importance, non plus seulement de l'Ancien Testament, mais d'un amour pour tous les hommes qui ira lorsque l'affrontement avec les prêtres du Temple lui apparaîtra comme inévitable, jusqu'à lui faire préférer sa mort à la violence.

Une étude comparative de la formation des idées intérieures des fondateurs des grandes religions permettrait de mieux comprendre l'évolution de ces religions et leurs facettes actuelles, comme si le présent était la fleur épanouie du bouton qu'elle fut voici plus d'un millénaire.

Les exemples fameux d'idées intérieures puissantes ne manquent pas:

Celles des grands mystiques, des martyrs et des grands saints, mais aussi sous une forme pervertie, celles des fanatiques politiques ou religieux.

Celle de Mahâtma Gândhi qui, à mains nues, mais avec une volonté inflexible, a fait céder le puissant Empire anglais.

L'idée intérieure d'Alexandre le Grand, dont il prit progressivement conscience, au fur et à mesure de sa progression vers l'Est, était le rêve d'une synthèse des cultures grecques et orientales.

Celle de Newton, qu'il a résumée dans cette courte phrase: "Je ne forme pas d'hypothèses", exprimant ainsi sa réaction contre la tendance de nombreux savants de son temps, à vouloir expliquer le monde par des explications a priori et des théories non basées sur l'expérience, telle la théorie des tourbillons de Descartes.

Celle d'Einstein qui, vers 15 ans, cesse d'être simplement un bon élève un peu lent: Trouver des explications *simples* et *belles* aux problèmes du temps, de la lumière et de la gravitation. Il disait souvent que ses grandes idées lui étaient dictées par une voix intérieure.

Plus proche de nous, Lebesgue, dont l'idée intérieure était une vision géométrique des objets mathématiques. Il affirmait "Pour ma part, j'ai toujours été guidé dans mes recherches par des considérations géométriques et, si je ne puis donner aucun de mes mémoires comme une application certaine de l'Analyse à la Géométrie, il me semble que j'ai fait constamment des applications de la Géométrie à l'Analyse".

L'auteur d'une oeuvre scientifique ne prend parfois conscience de ce qu'est l'idée intérieure qui oriente son activité créatrice qu'après plusieurs années de recherche. Ce fut mon cas; je pourrais reprendre en partie à mon compte ce que dit Lebesgue: Dès le début de mes recherches, j'ai toujours cherché à voir simplement les choses simples, à en avoir une vision directe, géométrique et quasiment visuelle. Mais en oûtre, il s'est greffé sur cette approche stratégique des mathématiques, après plusieurs années de recherche où mon seul plaisir était de résoudre les problèmes variés qui se présentaient à moi, une préoccupation qui est vite devenue comme une seconde nature, à savoir de ne m'intéresser qu'à des problèmes difficiles mais d'énoncé simple, et d'adopter pour les résoudre un cheminement propre à faire de la solution trouvée un *outil* commode et utile dans un cadre général.

La morale de cette histoire!

Il n'existe aucune recette que l'on puisse donner à un jeune chercheur pour l'aider à trouver sa voie. On peut lui dire seulement – et c'est aussi un message pour tous les éducateurs –

"Sois toi-même. Développe et fortifie ton idée intérieure. Crois en toi et tu pourras soulever le monde."

Harmonic approximation

Stephen J. Gardiner

Abstract. This survey article concerns uniform, and better-than-uniform, approximation on relatively closed sets by functions harmonic on a given open set in $\mathbb{R}^n$. The main emphasis is on recent developments concerning approximation on non-compact sets and their applications.

1991 Mathematics Subject Classification: 31B05.

1. Introduction

We begin with two problems which have a long history.

Problem 1. The Dirichlet problem with non-compact boundary. Let Ω be an open set in $\mathbb{R}^n (n \geq 2)$ with Euclidean boundary $\partial\Omega$, let $\partial\Omega_{\mathrm{reg}}$ be the set of points in $\partial\Omega$ at which $\mathbb{R}^n \backslash \Omega$ is non-thin, and let $\partial\Omega_{\mathrm{irr}} = \partial\Omega \backslash \partial\Omega_{\mathrm{reg}}$. If Ω is bounded, then the Perron–Wiener–Brelot method can be used to solve the Dirichlet problem on Ω, in the sense that for each continuous function $f : \partial\Omega \to \mathbb{R}$ there is a (unique) harmonic function h_f on Ω such that

$$h_f(X) \to f(Y) \qquad (X \to Y; Y \in \partial\Omega_{\mathrm{reg}}) \tag{1}$$

and

$$\limsup_{X \to Y} |h_f(X)| < +\infty \qquad (Y \in \partial\Omega_{\mathrm{irr}}). \tag{2}$$

However, if Ω is unbounded, it is not obvious whether, for every continuous function $f : \partial\Omega \to \mathbb{R}$, there exists a harmonic function h_f on Ω which satisfies (1) and (2). In 1925 R. Nevalinna [25] showed that, for every continuous function $f : \mathbb{R} \times \{0\} \to \mathbb{R}$ there is a harmonic function h_f in the upper half-plane satisfying (1) (there are no irregular boundary points in this case). Do all open sets Ω have is property? If not, which are the open sets which do?

Problem 2. Non-uniqueness for the Radon transform. Let f be a real- or complex-valued function on $\mathbb{R}^n$ such that f is integrable on each $(n-1)$-dimensional hyperplane P in $\mathbb{R}^n$. The Radon transform $\hat{f}$ of f is defined on the collection $\mathcal{P}^{(n)}$

of all such hyperplanes by

$$\hat{f}(P) = \int_P f\,d\Lambda \qquad (P \in \mathcal{P}^{(n)}),$$

where Λ denotes $(n-1)$-dimensional Lebesgue measure on P. (The original paper of Radon, in which this transform was studied, appeared in 1917 [27].) Does there exist a non-constant continuous function f such that $\hat{f} \equiv 0$ on $\mathcal{P}^{(n)}$?

These two apparently unrelated problems have this in common: they were solved recently using methods of harmonic approximation. In fact, they form only a sample from a growing collection of applications of harmonic approximation to a wide variety of problems. Since this subject has undergone rapid development in the last few years, it seems appropriate to give an introduction to it here. A more complete account may be found in the lecture notes [15]. Sections 2 and 3 below briefly review some of the classical theory (1920–1950). The remainder of the article deals with developments since 1980 with particular emphasis on the last few years.

2. Pole pushing

Some of the earliest work on harmonic approximation is due to Walsh [29] (1929), who adapted ideas from Runge's fundamental paper on rational approximation [28]. The first theorem below is a reformulation of part of his work. If $A \subseteq \mathbb{R}^n$, then we use $\mathcal{H}(A)$ to denote the collection of all functions which are harmonic on some open set which contains A. In particular, if A is open, then $\mathcal{H}(A)$ consists of the functions which are harmonic on A. If Ω is an open set in $\mathbb{R}^n$, then Ω^* will denote the Alexandroff (or one-point) compactification of Ω. We note that, if K is a compact subset of Ω, then $\Omega^* \backslash K$ is connected if and only if each component of $\mathbb{R}^n \backslash K$ contains a point of $\mathbb{R}^n \backslash \Omega$.

Theorem 1. *Let Ω be an open set in $\mathbb{R}^n$, let K be a compact subset of Ω, and suppose that $\Omega^* \backslash K$ is connected. Then, for each u in $\mathcal{H}(K)$ and each positive number ϵ, there exists v in $\mathcal{H}(\Omega)$ such that $|v - u| < \epsilon$ on K.*

The proof of Theorem 1 involves two main ingredients which are outlined in the two lemmas below since they are relevant to subsequent developments. We define $\phi_n : [0, +\infty) \to \mathbb{R} \cup \{+\infty\}$ by $\phi_2(t) = \log(1/t)$ and $\phi_n(t) = t^{2-n} (n \geq 3)$, where $\phi_n(0)$ is interpreted as $+\infty$ in either case. We denote by $B(X, r)$ the open ball of centre X and radius r. Also, we use σ to denote surface area measure and σ_n to denote the surface area of the unit sphere in $\mathbb{R}^n$.

Lemma 1. *Let K be a compact subset of $\mathbb{R}^n$, let $u \in \mathcal{H}(K)$ and $\epsilon > 0$. Then there are points $Y_1, Y_2, \ldots, Y_m$ in $\mathbb{R}^n \backslash K$ and constants $a_1, a_2, \ldots, a_m$ such that*

$$\left| u(X) - \sum_{k=1}^{m} a_k \phi_n\big(|X - Y_k|\big) \right| < \epsilon \qquad (X \in K).$$

To see this, we use Green's identity to observe that

$$u(X) = \frac{1}{\sigma_n \max\{n-2, 1\}} \int_{\partial U} \left\{ \phi_n\big(|X - Y|\big) \frac{\partial u}{\partial n_Y}(Y) \right.$$
$$\left. - u(Y) \frac{\partial}{\partial n_Y} \phi_n\big(|X - Y|\big) \right\} d\sigma(Y) \qquad (X \in K),$$

where U is a smoothly bounded open set containing K such that $u \in \mathcal{H}(\overline{U})$, and n_Y denotes the outer unit normal to ∂U at Y. The above integral can be approximated uniformly on K by a Riemann sum. Further, for a fixed choice of Y in ∂U, the derivative $X \mapsto (\partial / \partial n_Y) \phi_n\big(|X - Y|\big)$ can be uniformly approximated on K by a difference quotient. Hence Lemma 1 holds.

Lemma 2. *If h is harmonic on $\mathbb{R}^n \backslash \overline{B(Y, a)}$ and $b > a$ then, for each $\epsilon > 0$, there exists H in $\mathcal{H}(\mathbb{R}^n \backslash \{Y\})$ such that $|H - h| < \epsilon$ on $\mathbb{R}^n \backslash B(Y, b)$.*

This follows by suitably truncating the singular part of the Laurent expansion of h.

To prove Theorem 1, it is enough to check (in view of Lemma 1) that, if $Y_0 \in \mathbb{R}^n \backslash K$, then the function $h(X) = \phi_n\big(|X - Y_0|\big)$ can be uniformly approximated on K by functions in $\mathcal{H}(\Omega)$. If Y_0 belongs to a bounded component V of $\mathbb{R}^n \backslash K$ then, since $\Omega^* \backslash K$ is connected, there exists Z in $V \backslash \Omega$ and thus balls $B(Y_1, r_1), \ldots, B(Y_m, r_m)$ in V, where $Y_m = Z$, such that $Y_{k-1} \in B(Y_k, r_k)$ for each k in $\{1, 2, \ldots, m\}$. By repeated application of Lemma 2 we obtain v in $\mathcal{H}(\mathbb{R}^n \backslash \{Z\})$ such that $|v - h| < \epsilon$ on K. A similar argument applies when Y_0 belongs to the unbounded component of $\mathbb{R}^n \backslash K$, so Theorem 1 follows.

3. Thin sets and approximation

The next main development in harmonic approximation came in work of Keldyš [22] (1941) and Deny [10] (1949), who investigated when functions in $C(K) \cap \mathcal{H}(K^\circ)$ can be uniformly approximated by functions in $\mathcal{H}(K)$. The sets K for which such approximation is always possible have a characterization in terms of thin sets.

Theorem 2. *Let K be a compact subset of $\mathbb{R}^n$. The following are equivalent:*
(a) *for each u in $C(K) \cap \mathcal{H}(K^\circ)$ and each positive number ϵ, there exists v in $\mathcal{H}(K)$ such that $|v - u| < \epsilon$ on K;*

(b) *$\mathbb{R}^n\backslash K$ and $\mathbb{R}^n\backslash K^\circ$ are thin at the same points (of K).*

To illustrate the above thinness condition we give below two examples.

Examples 1. (i) Let $\{Y_k : k \in \mathbb{N}\}$ be a dense subset of $[0,1]^{n-1} \times (0,1]$, let $u(X) = \sum 2^{-k}\phi_n(|X - Y_k|)$ on $\mathbb{R}^n$ and

$$K = ([0,1]^{n-1} \times [-1,0]) \cup \{(X', x_n) \in [0,1]^{n-1} \times (0,1] : u(X', x_n) \leq \phi_n(x_n)\}.$$

Then u is superharmonic on $\mathbb{R}^n$, and the lower semicontinuity of u ensures that K is closed and so compact. Clearly $K^\circ = (0,1)^{n-1} \times (-1,0)$ since $u(Y_k) = +\infty$ for each k. If $Z \in (0,1)^{n-1} \times \{0\}$, then $\mathbb{R}^n\backslash K^\circ$ is non-thin at Z. However, if $x_n > 0$, then $u(X) > \phi_n(x_n) \geq \phi_n(|X - Z|)$ on $(0,1)^n\backslash K$ and, since the Riesz measure associated with u does not charge $\{Z\}$, it follows that $\mathbb{R}^n\backslash K$ is thin at Z. Thus condition (b) of Theorem 2 fails to hold for this set K.
(ii) In $\mathbb{R}^3$ let E_c be the Lebesgue spine defined by

$$E_c = \{(x,y,z) : x > 0 \text{ and } y^2 + z^2 < \exp(-c/x)\} \qquad (c > 0),$$

and let $K = \overline{B(0,1)}\backslash E_2$. Then $\mathbb{R}^3\backslash K$ is thin at O, and it is easy to see that O is the only point of ∂K at which $\mathbb{R}^3\backslash K$ is thin. Further, $B(O,1)\backslash K^\circ \subset E_1 \cup \{O\}$ so $\mathbb{R}^3\backslash K^\circ$ is also thin at O. Thus $\mathbb{R}^3\backslash K$ and $\mathbb{R}^3\backslash K^\circ$ are thin at precisely the same points (namely, points of $K^\circ \cup \{O\}$), and so condition (b) of Theorem 2 holds for this choice of K, even though K° has an irregular boundary point.

The proof of Theorem 2 is rather simple in the special case where $K^\circ = \emptyset$. For, if (b) holds, let $f \in C(K)$, let B be an open ball containing K, and let $U_m = \{X : \operatorname{dist}(X,K) < 1/m\}$ when $m \geq 1$. Then we can approximate f uniformly on K by the difference, $u_1 - u_2$, of two positive continuous superharmonic functions on B and observe from (b) (which now says that $\mathbb{R}^n\backslash K$ is nowhere thin) that

$$R_{u_k}^{B\backslash U_m}(X) \uparrow R_{u_k}^{B\backslash K}(X) = u_k(X) \qquad (X \in K; k = 1,2;\ m \to \infty) \tag{3}$$

(reductions relative to B). The convergence in (3) is uniform on K by Dini's theorem, so (a) follows. The converse in this special case can be proved by assuming that (a) holds and taking f to be a bounded continuous potential w on B with the property that a set A is thin at a point Y in B if and only if $\hat{R}_w^A(Y) < w(Y)$.

Now let us drop the assumption that $K^\circ = \emptyset$. Suppose that (b) holds and let $u \in C(K) \cap \mathcal{H}(K^\circ)$. It follows that $\mathbb{R}^n\backslash\partial K$ is nowhere thin, and so the above special case allows us to approximate u uniformly on ∂K by a function in $\mathcal{H}(\partial K)$. Further, similar reasoning shows that u may be locally uniformly approximated on K° by functions in $\mathcal{H}(K)$. These two facts are readily combined to yield (a). The converse can again be proved using the above function w, suitably modified to belong to $C(K) \cap \mathcal{H}(K^\circ)$. (See [15, Chap. 1].)

We mention here that Theorem 2 has been extended in several respects. For example, there are several equivalent formulations of condition (b): see [21] and the references given there. Debiard and Gaveau [9] have characterized, in terms of fine harmonicity, the functions on a compact set K which can be uniformly approximated by members of $\mathcal{H}(K)$. Also, generalizations to more abstract settings can be found in, for example, [26] and [7]. However, many applications require us to be able to approximate on non-compact sets, so it is in that direction that we shall now move.

4. Fusion of harmonic functions

The next landmark in this line of development can be found in work of Gauthier, Goldstein and Ow [18], [19] and Gauthier and Hengartner [20] in the early 1980s. Inspired by work of Alice Roth on "fusion" of rational functions, they developed corresponding fusion results for harmonic functions with isolated singularities. The idea is that, if two such functions are close in value on a given compact set, one can join them together (in an approximate sense) to form a single such function. If $A \subseteq \mathbb{R}^n$, we use $\mathcal{I}(A)$ to denote the collection of functions which are harmonic, apart from isolated singularities, on an open set containing A. By a *Greenian domain* Ω we mean a connected open set Ω with Green function $G(\cdot\,,\cdot)$. We fix a point Q in Ω and define $g(\cdot) = \min\{1, G(Q,\cdot)\}$. Thus g has limit 0 at all regular boundary points of Ω. The following fusion result, due to Armitage and Goldstein [6], is a refinement of the work cited above.

Theorem 3. *Let K and E_1 be compact subsets of a Greenian domain Ω, and let E_2 be a relatively closed subset of Ω which is disjoint from E_1. Then there is a positive constant C with the following property: if $u_1, u_2 \in \mathcal{I}(\Omega) \cap \mathcal{H}(K)$ and $|u_1 - u_2| < \epsilon$ on K, then there exists u in $\mathcal{I}(\Omega)$ such that*

$$\left|(u - u_k)(X)\right| < C\epsilon g(X) \qquad (X \in K \cup E_k; \quad k = 1, 2). \tag{4}$$

We first note that, in the proof of the above result, it is enough to consider the case where $u_2 \equiv 0$. For, if this case is established, then in the general case we can find v in $\mathcal{I}(\Omega)$ such that

$$\left|(v - (u_1 - u_2))(X)\right| < C\epsilon g(X) \qquad (X \in K \cup E_1)$$

and

$$\left|v(X)\right| < C\epsilon g(X) \qquad (X \in K \cup E_2).$$

Thus (4) is satisfied by defining $u = v + u_2$.

Next, let ω_1, ω_2 be smoothly bounded open sets such that $E_1 \subset \omega_1$, $\overline{\omega_1} \subset \omega_2$ and $\overline{\omega_2} \subset \Omega \backslash E_2$. We choose ϕ in $C^\infty(\mathbb{R}^n)$ such that $0 \le \phi \le 1$ on $\mathbb{R}^n$, $\phi = 1$ on ω_1 and $\phi = 0$ on $\mathbb{R}^n \backslash \omega_2$, and define

$$\psi = \begin{cases} \phi u_1 & \text{on } \omega_2 \\ 0 & \text{on } \mathbb{R}^n \backslash \omega_2. \end{cases}$$

Thus $\psi = u_1$ on E_1, $\psi = 0$ on E_2 and $|\psi| \le |u_1| < \epsilon$ on K. The idea is then to use Green's identity to approximate ψ on $K \cup E_1 \cup E_2$ by a function u in $\mathcal{I}(\Omega)$.

(The details are quite lengthy.) The role of the function g is partly explained by the following elementary lemma based on Harnack's inequalities.

Lemma 3. *Let $K \subset \omega$ and $\overline{\omega} \subset \Omega$, where K is compact and ω is open. Then*

$$G(X,Y) + |\nabla_2 G(X,Y)| \leq C(K,\omega,\Omega,Q)g(X) \qquad (X \in \Omega\backslash\omega;\ Y \in K),$$

where $C(a,b,\dots)$ denotes a positive constant, depending at most on $a,b,\dots$, and ∇_2 denotes the gradient operator with respect to the second argument.

5. Local connectedness and approximation

The inspiration for the material in this section comes from the following celebrated result of Arakelyan (1968) (see [1] or [2]) concerning uniform holomorphic approximation. Let Hol(A) denote the collection of functions which are holomorphic on an open set containing A, where $A \subseteq \mathbb{C}$.

Theorem A. *Let Ω be an open set in $\mathbb{C}$ and E be a relatively closed subset of Ω. The following are equivalent:*
(a) *for each f in $C(E) \cap \mathrm{Hol}(E^\circ)$ and each positive number ϵ, there exists g in $\mathrm{Hol}(\Omega)$ such that $|g - f| < \epsilon$ on E;*
(b) *$\Omega^*\backslash E$ is connected and locally connected.*

Recall that a topological space is called *locally connected* if, for each point X in the space and each neighbourhood ω of X, there is a connected neighbourhood ω' of X such that $\omega' \subseteq \omega$. The set $\Omega^*\backslash E$ in Theorem A can only fail to satisfy this condition in the case where X is the ideal point $\mathcal{A}$. A set A will be called *Ω-bounded* if $\overline{A}$ is a compact subset of Ω. Thus $\Omega^*\backslash E$ is locally connected if and only if for each compact subset K of Ω there is a compact subset L of Ω which contains every Ω-bounded component of $\Omega\backslash(E \cup K)$. We give below an example of a pair (Ω, E) such that $\Omega^*\backslash E$ is connected but not locally connected.

Example 2. If

$$S = \bigcup_{k=1}^{\infty} \left\{ (x_1, x_2) : \frac{1}{2k+1} < x_1 < \frac{1}{2k} \text{ and } x_2 < k \right\},$$

then $(\mathbb{R}^2)^*\backslash\partial S$ is connected. However, $(\mathbb{R}^2)^*\backslash\partial S$ is not locally connected: the set $(\mathbb{R}^2)^*\backslash(\partial S \cup K)$, where $K = [0,1] \times \{0\}$, is a neighbourhood of $\mathcal{A}$ in $(\mathbb{R}^2)^*\backslash\partial S$ which does not contain any connected neighbourhood of $\mathcal{A}$. (There is no compact set L which contains all the bounded components of $(\mathbb{R}^2)^*\backslash(\partial S \cup K)$.)

The relevance of this local connectedness condition to harmonic approximation is due to the following result. We assume for the remainder of this section that

E is a relatively closed subset of a Greenian domain Ω, and that g is the function defined in §4.

Theorem 4. *If $\Omega^*\backslash E$ is connected and locally connected, then for each u in $\mathcal{I}(\Omega)\cap\mathcal{H}(E)$ and each positive number ϵ, there exists v in $\mathcal{H}(\Omega)$ such that*

$$|(v-u)(X)| < \epsilon g(X) \qquad (X \in E).$$

To see why this result is true, let $u \in \mathcal{I}(\Omega)\cap\mathcal{H}(E)$ and let Y be a singularity of u. It follows from the connectedness hypotheses that there is an open subset T of $\Omega\backslash E$ such that $u \in \mathcal{H}(T\backslash\{Y\})$ and T contains a path connecting Y to $\mathcal{A}$. (That is, there is a continuous function $f : [0,+\infty) \to T$ such that $f(0) = Y$ and $f(t) \to \mathcal{A}$ as $t \to +\infty$.) A pole pushing argument, along the lines indicated for the proof of Theorem 1 above, can now be used to remove the singularities of u one by one. (It is not hard to show that the error bound ϵ in Lemma 2 can be replaced by $\epsilon g(X)$ in the present context.) The local connectedness of $\Omega^*\backslash E$ ensures that only finitely many of these paths meet any given compact subset of Ω, and so the resulting sequence of functions in $\mathcal{I}(\Omega)$ converges locally uniformly on Ω to a function v in $\mathcal{H}(\Omega)$.

Theorem 5. *If $\Omega^*\backslash E$ is connected and locally connected, then for each u in $\mathcal{H}(E)$ and each positive number ϵ, there exists v in $\mathcal{H}(\Omega)$ such that*

$$|(v-u)(X)| < \epsilon g(X) \qquad (X \in E). \tag{5}$$

Theorem 5 follows immediately from Theorem 4 and the following result.

Lemma 4. *For each u in $\mathcal{H}(E)$ and each positive number ϵ, there exists v in $\mathcal{I}(\Omega)\cap\mathcal{H}(E)$ such that* (5) *holds.*

It is in the proof of Lemma 4 that the fusion result of §4 comes into play. The idea is as follows. Let $u \in \mathcal{H}(E)$, let $\epsilon > 0$, and let (K_m) be an exhaustion of Ω by compact sets such that $K_m \subseteq K^\circ_{m+1}$ for each m. Also, let $F_m = K_m \cap E$, and let C_m be the constant of Theorem 3 corresponding to the assignments $K = F_{m+1}$, $E_1 = K_m$ and $E_2 = E\backslash K^\circ_{m+1}$. We may assume that $C_m \geq 1$, and we define (δ_m) to be a decreasing sequence of positive numbers such that $\delta_m \leq 2^{-m-2}\epsilon/C_m$. It follows from Lemma 1 that, for each m, there exists q_m in $\mathcal{I}(\mathbb{R}^n)$ such that

$$|(q_m-u)(X)| < \delta_m g(X) \qquad (X \in F_{m+1}), \tag{6}$$

whence $q_m, q_{m+1} \in \mathcal{H}(F_{m+1})$ and

$$|(q_{m+1}-q_m)(X)| < 2\delta_m \qquad (X \in F_{m+1}).$$

Thus we can apply Theorem 3 to obtain r_m in $\mathcal{I}(\Omega)$ such that

$$|(r_m-q_m)(X)| \leq 2\delta_m C_m g(X) \leq 2^{-m-1}\epsilon g(X) \qquad (X \in K_m \cup F_{m+1}) \tag{7}$$

and

$$\left|(r_m - q_{m+1})(X)\right| \le 2^{-m-1}\epsilon g(X) \qquad (X \in E). \tag{8}$$

It follows from (7) that the series

$$\sum_{k=m}^{\infty} (r_k - q_k)(X)$$

defines a harmonic function on K_m°. Thus, if we define

$$v = q_1 + \sum_{k=1}^{\infty} (r_k - q_k),$$

we see that $v \in \mathcal{I}(K_m^\circ)$ for any m, and hence $v \in \mathcal{I}(\Omega)$. The proof is then completed by using (6)–(8) to observe that $|v-u| < \epsilon g$ on E, and thus $v \in \mathcal{H}(E)$.

Theorems 4 and 5 were originally proved by Gauthier, Goldstein and Ow [18], [19] without the decay function g. (In this case the assumption that Ω is Greenian is unnecessary.) The introduction of g is due to Armitage and Goldstein [6].

6. Relating the error to the set E

The error bound $\epsilon g(X)$ in the results of §5 takes no account of the set E on which we approximate. In fact, information concerning E can be used to improve the error bound as the following theorem and example show. Let $\mathcal{S}^+(A)$ denote the class of functions which are positive and superharmonic on an open set containing A.

Theorem 6. *Let E be a relatively closed proper subset of a connected open set Ω. If $\Omega^*\backslash E$ is connected and locally connected, then for each u in $\mathcal{H}(E)$ and each s in $\mathcal{S}^+(E)$, there exists v in $\mathcal{H}(\Omega)$ such that $0 < v - u < s$ on E.*

Example 3. Let $\Omega = \mathbb{R}^n$, let $\omega = (-1,1)^{n-1} \times \mathbb{R}$ and $\alpha = (\pi/2)(n-1)^{1/2}$. Then given any closed set E such that $E \subset \omega$ and the hypotheses of Theorem 6 are satisfied, any u in $\mathcal{H}(E)$ and any positive number ϵ, there exists v in $\mathcal{H}(\mathbb{R}^n)$ such that

$$0 \le v(X) - u(X) < \epsilon \exp(-\alpha|x_n|) \qquad (X = (x_1, \dots, x_n) \in E).$$

This follows from the fact that the function

$$s(X) = \cos(\pi x_1/2) \dots \cos(\pi x_{n-1}/2) \exp(-\alpha|x_n|)$$

belongs to $\mathcal{S}^+(E)$. It can also be shown that the assertion of the previous paragraph becomes false if α is replaced by any larger number.

Theorem 6 is a special case of a result of the author [13]. It is proved by repeated application of Theorem 5 with different choices of Ω.

7. Hole filling

The connectedness conditions in Arakelyan's theorem (Theorem A in §5) are necessary and sufficient for uniform holomorphic approximation on E. Theorems 4–6 show that these conditions are also sufficient for uniform harmonic approximation. However, they are not necessary in this case. The complete solution of the uniform harmonic approximation problem turns out to be somewhat more delicate, and involves both a topological and a potential theoretic condition.

The topological condition is as follows. Let Ω be a connected open set and E be a relatively closed proper subset of Ω. We will say that the pair (Ω, E) *satisfies the (K, L)-condition* if, for each compact subset K of Ω, there is a compact subset L of Ω which contains every Ω-bounded component of $\Omega\backslash(E \cup K)$ whose closure intersects K.

Let $\hat{E}$ denote the union of E with the Ω-bounded components of $\Omega\backslash E$. It is not hard to see that, if (Ω, E) satisfies the (K, L)-condition, then $\Omega^*\backslash\hat{E}$ is locally connected. (Thus, for example, if S is the set defined in Example 2, then $(\mathbb{R}^2, \partial S)$ does not satisfy the (K, L)-condition.) However, $\Omega^*\backslash E$ need not be locally connected: for, if

$$E = \bigcup_{k=1}^{\infty} \partial B\big((3k, 0, \dots, 0), 1\big),$$

then $(\mathbb{R}^n, E)$ satisfies the (K, L)-condition, but $(\mathbb{R}^n)^*\backslash E$ is not locally connected.

Theorem 7. *The following are equivalent:*
(a) *for each u in $\mathcal{H}(E)$ and each positive number ϵ there exists v in $\mathcal{H}(\Omega)$ such that $|v - u| < \epsilon$ on E;*
(b) *for each u in $\mathcal{H}(E)$ and each s in $\mathcal{S}^+(E)$ there exists v in $\mathcal{H}(\Omega)$ such that $0 < v - u < s$ on E;*
(c) *$\Omega\backslash\hat{E}$ and $\Omega\backslash E$ are thin at the same points of E, and (Ω, E) satisfies the (K, L)-condition.*

We have already discussed the topological condition which occurs in (c) above. For the thinness condition in (c) to hold, it is necessary that each component W of $\hat{E}\backslash E$ is regular and satisfies $\partial W \subseteq \partial\hat{E}$. However, when $n \geq 3$, these conditions on W are not sufficient for the thinness condition to hold, as the following example shows.

Example 4. Let $\Omega = \mathbb{R}^n$, where $n \geq 3$, let $\{Y'_k : k \in \mathbb{N}\}$ be a dense subset of $[0, 1]^{n-1}$, and define

$$u(X') = \sum_{k=1}^{\infty} 2^{-k} \phi_{n-1}\big(|X' - Y'_k|\big) \qquad (X' \in \mathbb{R}^{n-1}),$$

$$E = \partial\big([0, 1]^{n-1} \times [-1, 0]\big) \cup \big\{(X', x_n) \in [0, 1]^{n-1} \times (0, 1] : u(X') \leq \phi_n(x_n)\big\},$$

and

$$F_y = \{X' \in \mathbb{R}^{n-1} : (X', y) \notin E\} \qquad (y > 0).$$

Then u is a superharmonic function on $\mathbb{R}^{n-1}$, and the lower semicontinuity of u ensures that E is closed and hence compact. Also, since

$$\{Y'_k : k \in \mathbb{N}\} \subseteq F_y \subseteq F_z \qquad (0 < y < z \leq 1),$$

it is clear that $\mathbb{R}^n \backslash E$ has exactly one bounded component, namely the cube $V = (0,1)^{n-1} \times (-1, 0)$. Let $v(X', x_n) = u(X')$, so that v is superharmonic on $\mathbb{R}^n$. If $Z \in (0,1)^{n-1} \times \{0\}$, then

$$v(X', x_n) = u(X') > \phi_n(x_n) \geq \phi_n(|(X', x_n) - Z|) \quad ((X', x_n) \in (0,1)^n \backslash E).$$

Since the Riesz measure associated with u does not charge $\{Z\}$, it follows that $(0,1)^n \backslash E$, and hence $\mathbb{R}^n \backslash \hat{E}$, is thin at Z. On the other hand, $\mathbb{R}^n \backslash E$ contains V and so is certainly not thin at Z. Thus condition (b) of Theorem 7 fails to hold, yet the only bounded component V of $\mathbb{R}^n \backslash E$ is regular for the Dirichlet problem and satisfies $\partial V \subseteq \partial \hat{E}$.

Labrèche [23] has used fusion of harmonic functions to extend the Keldyš-Deny theorem (Theorem 2) to the case where K is a relatively closed subset of some open set Ω. If we combine this extended result with Theorem 7 we can obtain the following harmonic analogue of Arakelyan's theorem (Theorem A).

Theorem 8. *The following are equivalent:*

(a) *for each u in $C(E) \cap \mathcal{H}(E^\circ)$ and each positive number ϵ there exists v in $\mathcal{H}(\Omega)$ such that $|v - u| < \epsilon$ on E;*

(b) *for each u and s in $C(E) \cap \mathcal{H}(E^\circ)$, where $s > 0$, there exists v in $\mathcal{H}(\Omega)$ such that $0 < v - u < s$ on E;*

(c) *$\Omega \backslash \hat{E}$ and $\Omega \backslash E^\circ$ are thin at the same points of E, and (Ω, E) satisfies the (K, L)-condition.*

Example 5. Let $\Omega = \mathbb{R}^n$ and let E be a closed subset of $\mathbb{R}^{n-1} \times [0, +\infty)$ such that condition (c) of Theorem 8 holds. Then, for each u in $C(E) \cap \mathcal{H}(E^\circ)$, each positive number ϵ, and each continuous function $f : \mathbb{R}^{n-1} \to (0, 1]$, there exists v in $\mathcal{H}(\mathbb{R}^n)$ such that

$$0 < v(X) - u(X) < \epsilon \left(\frac{1 + x_n}{1 + |X|^n} \right) \qquad (X \in E;\ x_n > 0)$$

and

$$0 < v(X) - u(X) < f(x_1, \ldots, x_{n-1}) \qquad (X \in E;\ x_n = 0).$$

Thus the error in the approximation can be arranged to decay arbitrarily quickly on the hyperplane boundary.

The above results simplify considerably in the case where $n = 2$, as the following theorem shows.

Theorem 9. *Let $n = 2$. The following are equivalent:*
(a) *for each u in $\mathcal{H}(E)$ and each positive number ϵ there exists v in $\mathcal{H}(\Omega)$ such that $|v - u| < \epsilon$ on E;*
(b) *for each u and s in $C(E) \cap \mathcal{H}(E^\circ)$, where $s > 0$, there exists v in $\mathcal{H}(\Omega)$ such that $0 < v - u < s$ on E;*
(c) *$\partial\hat{E} = \partial E$ and (Ω, E) satisfies the (K, L)-condition.*

Theorems 7–9 are due to the author [12], [13], [14].

8. Carleman approximation

In 1927 Carleman [8] proved the following remarkable generalization of the classical Weierstrass approximation theorem: given any continuous function $f : \mathbb{R} \to \mathbb{C}$ and any continuous "error" function $\epsilon : \mathbb{R} \to (0, 1]$, there is an entire function g such that $|g - f| < \epsilon$ on $\mathbb{R}$. Thus the function $\epsilon(x)$ can be chosen to decay arbitrarily quickly as $|x| \to \infty$. In the same spirit one can ask which pairs (Ω, E) have the property that, for each u in $C(E) \cap \mathcal{H}(E^\circ)$ and each continuous function $\epsilon : E \to (0, 1]$ there exists v in $\mathcal{H}(\Omega)$ such that $|v - u| < \epsilon$ on E. The answer is outlined below. (The corresponding result for holomorphic functions is due to Nersesyan [24].)

We will say that (Ω, E) satisfies the *long islands condition* if, for each compact subset K of Ω, there is a compact subset L of Ω which contains every component of E° that intersects K. This condition was first considered by Gauthier [17]. It is trivially satisfied if $E^\circ = \emptyset$. Also, it clearly implies that every component of E° is Ω-bounded. The following example shows that it implies rather more than this. It also motivates the terminology.

Example 6. Let

$$E = (\{0\} \times [0, +\infty)) \cup \left(\bigcup_{k=1}^{\infty} \left[\frac{1}{2k+1}, \frac{1}{2k} \right] \times [0, k] \right).$$

Then $(\mathbb{R}^2, E)$ does not satisfy the long islands condition, as can be seen by choosing K to be $[0, 1]^2$.

Theorem 10. *Let Ω be a connected open set and E be a relatively closed proper subset of Ω. The following are equivalent:*
(a) *for each u in $C(E) \cap \mathcal{H}(E^\circ)$ and each continuous function $\epsilon : E \to (0, 1]$, there exists v in $\mathcal{H}(\Omega)$ such that $|v - u| < \epsilon$ on E;*
(b) *$\Omega \backslash \hat{E}$ and $\Omega \backslash E^\circ$ are thin at the same points of E, and (Ω, E) satisfies both the (K, L)-condition and the long islands condition.*

Theorem 10 is due to Goldstein and the author [16]. Another approach can be found in [13], where the following lemma is used to show that (b) implies (a).

Lemma 5. *If the pair* (Ω, E) *satisfies Theorem* 10 *condition* (b) *and* $\epsilon : E \to (0,1]$ *is continuous, then there exists* s *in* $\mathcal{S}^+(E)$ *such that* $s < \epsilon$ *on* E.

In the opposite direction, (a) can be shown to imply the long islands condition using the following result of Armitage, Bagby and Gauthier [4].

Lemma 6. *Let* ω *be an unbounded connected open set in* $\mathbb{R}^n$. *Then there exists a continuous function* $\epsilon_\omega : [0, +\infty) \to (0,1]$ *with the following property: if* $u \in \mathcal{H}(\omega)$ *and* $|u(X)| \leq \epsilon_\omega(|X|)$ *on* ω, *then* $u \equiv 0$.

9. The Dirichlet problem with non-compact boundary

We now return to the first problem mentioned in §1. Which open sets Ω have the property that, for each continuous function $f : \partial\Omega \to \mathbb{R}$ there is a harmonic function h_f on Ω such that (1) and (2) hold? The answer is given in the following theorem of the author [11].

Theorem 11. *Let* Ω *be an open set in* $\mathbb{R}^n$. *The following are equivalent:*
(a) *for every* f *in* $C(\partial\Omega)$ *there is a harmonic function* h_f *on* Ω *which satisfies* (1) *and* (2);
(b) *for each compact set* K *in* $\mathbb{R}^n$ *there is a compact set* L *which contains the bounded components of* $\Omega \backslash K$ *whose closure intersects* K.

Condition (b) of Theorem 11, which is equivalent to saying that $(\mathbb{R}^n, \mathbb{R}^n \backslash \Omega)$ satisfies the (K, L)-condition, is illustrated below.

Example 7. Let $n = 2$. Then condition (b) of Theorem 11 fails to hold if Ω is replaced by either

$$\Omega_1 = \bigcup_{m=1}^{\infty} \left[\left(\frac{1}{2m+1}, \frac{1}{2m} \right) \times (0, m) \right] \text{ or } \Omega_2 = \Omega_1 \cup \left[\mathbb{R} \times (-\infty, 1) \right].$$

We note that, if $f \in C(\partial\Omega_1)$ and we solve the Dirichlet problem separately in each component V_m of Ω_1, then the resultant function h_f satisfies (1) at each point Y of $\cup_m \partial V_m$, but not necessarily at points Y of $\{0\} \times \mathbb{R}$

The idea of the proof that (b) implies (a) in Theorem 11 is as follows. Given f in $C(\partial\Omega)$, we cannot use the standard approach of integrating f against harmonic measure, since f need not be integrable. However, it is possible to modify Ω by the introduction of certain "slits" in such a way that these slits do not meet $\partial\Omega$ and that f is integrable with respect to harmonic measure for the modified open set Ω'. This leads to a harmonic function u on Ω' with the right boundary behaviour near $\partial\Omega$. We then apply Theorem 5 to each component of Ω separately (with appropriate choices of ϵ) to obtain h_f in $\mathcal{H}(\Omega)$ such that (1) and (2) hold. The topological condition in (b) of the theorem corresponds to when this general strategy can be carried through in detail.

10. Non-uniqueness for the Radon transform

Problem 2 of §1 concerns the Radon transform $\hat{f}$ of f, defined on the collection $\mathcal{P}^{(n)}$ of all hyperplanes by

$$\hat{f}(P) = \int_P f\, d\Lambda \qquad (P \in \mathcal{P}^{(n)}).$$

The uniqueness question is resolved by the following result of Armitage and Goldstein [5]. (When $n = 2$ the problem was solved by Zalcman [30] using holomorphic approximation.)

Theorem 12. *There exists a non-constant harmonic function h on $\mathbb{R}^n$ such that $\hat{h} \equiv 0$ on $\mathcal{P}^{(n)}$.*

The idea of the proof here is to define

$$W = \bigcup_{t \geq 0} B((t, t^2, \dots, t^n), 1),$$

and $E = (\mathbb{R}^n \backslash W) \cup \{O\}$. One then uses harmonic approximation arguments to construct h in $\mathcal{H}(\mathbb{R}^n)$ such that $h(O) > 0$ and

$$|h(X)| < (1 + |X|)^{-n-1} \qquad (X \in E). \tag{9}$$

(This can be done quite easily by a modified form of the pole pushing argument described in §2: see [3].) If $Y \in \partial B(O, 1)$, then we define the $(n-1)$-dimensional hyperplane $P(Y, a) = \{X : \langle X, Y \rangle = a\}$ for each a in $\mathbb{R}$. It is straightforward to check from (9) that h is integrable on each hyperplane, and that the function

$$t \mapsto \int_{P(Y,t)} |h|\, d\Lambda \qquad (t \in \mathbb{R})$$

is locally bounded. Standard arguments then show that $\hat{h}(P(Y, t))$ is a polynomial in t of degree at most 1. However, it can also be shown from (9) that $\hat{h}(P(Y, t)) \to 0$ either as $t \to +\infty$ or as $t \to -\infty$. Hence $\hat{h}(P(Y, t)) \equiv 0$ as a function of t. This holds for all Y in $\partial B(O, 1)$, so Theorem 12 holds.

References

[1] Arakeljan, N .U., Uniform and tangential approximations by analytic functions. Izv. Akad. Nauk Armjan. SSR Ser. Mat. 3 (1968), 273–286. English Translation: Amer. Math. Soc. Transl. (2) 122 (1984), 85–97.

[2] — Approximation complexe et propriétés des fonctions analytiques. Actes, Congrès intern. Math., 1970, Tome 2, 595–600.

[3] Armitage, D .H., A non-constant continuous function on the plane whose integral on every line is zero. Amer. Math. Monthly 101 (1994), 892–894.

[4] Armitage, D .H., Bagby, T., Gauthier, P .M., Note on the decay of solutions of elliptic equations. Bull. London Math. Soc. 17 (1985), 554–556.

[5] Armitage, D .H., Goldstein, M., Nonuniqueness for the Radon transform. Proc. Amer. Math. Soc., 117 (1993), 175–178.

[6] — Tangential harmonic approximation on relatively closed sets. Proc. London Math. Soc. (3) 68 (1994), 112–126.

[7] Bliedtner, J., Hansen, W., Simplicial cones in potential theory II (Approximation theorems). Invent. Math. 46 (1978), 255–275.

[8] Carleman, T., Sur un théorème de Weierstrass. Ark. Mat. Astronom. Fys. 20B (1927), 1–5.

[9] Debiard, A., Gaveau, B., Potentiel fin et algèbres de fonctions analytiques I. J. Funct. Anal. 16 (1974), 289–304.

[10] Deny, J., Systèmes totaux de fonctions harmoniques. Ann. Inst. Fourier (Grenoble) 1 (1949), 103–113.

[11] Gardiner, S. J., The Dirichlet problem with non-compact boundary. Math. Z. 213 (1993), 163–170.

[12] — Superharmonic extension and harmonic approximation. Ann. Inst. Fourier (Grenoble) 44 (1994), 65–91.

[13] — Uniform and tangential harmonic approximation. In: Classical and Modern Potential Theory, (ed. by K. GowriSankaran et al.), NATO ASI Series Vol. 430, 185–198. Kluwer, Dordrecht, 1994.

[14] — Tangential harmonic approximation on relatively closed sets. Illinois J. Math. 39 (1995), 143–157.

[15] — Harmonic approximation. London Math. Soc. Lecture Notes 221, Cambridge Univ. Press, 1995.

[16] Gardiner, S. J., Goldstein, M., Carleman approximation by harmonic functions. Amer. J. Math. 117 (1995), 245–255.

[17] Gauthier, P. M., Tangential approximation by entire functions and functions holomorphic in a disc. Izv. Akad. Nauk Armjan. SSR Ser. Mat. 4 (1969), 319–326.

[18] Gauthier, P. M., Goldstein, M., Ow, W. H., Uniform approximation on unbounded sets by harmonic functions with logarithmic singularities. Trans. Amer. Math. Soc. 261 (1980), 169–183.

[19] — Uniform approximation on closed sets by harmonic functions with Newtonian singularities. J. London Math. Soc. (2) 28 (1983), 71–82.

[20] Gauthier, P. M., Hengartner, W., Approximation uniforme qualitative sur des ensembles non bornées. Séminaire de Mathématiques Supérieures, Press. Univ. Montreal, Montreal 1982.

[21] Hedberg, L. I., Approximation by harmonic functions, and stability of the Dirichlet problem. Exposition. Math. 11 (1993), 193–259.

[22] Keldyš, M. V., On the solvability and stability of the Dirichlet problem. Uspehi Mat. Nauk 8 (1941), 171–231. English Translation: Amer. Math. Soc. Transl. 51 (1966), 1–73.

[23] Labrèche, M., De l'approximation harmonique uniforme. Doctoral thesis, Université de Montréal, Montréal 1982.

[24] Nersesyan, A. A., Carleman sets. Izv. Akad. Nauk Armjan. SSR Ser. Mat. 6 (1971), 465–471. English Translation: Amer. Math. Soc. Transl. (2) 122 (1984), 99–104.

[25] Nevanlinna, R., Über eine Erweiterung des Poissonschen Integrals. Ann. Acad. Sci. Fenn. Ser. A. 24, No. 4 (1925), 1–15.

[26] de la Pradelle, A., Approximation et caractère de quasi-analyticité dans la théorie axiomatique des fonctions harmoniques. Ann. Inst. Fourier (Grenoble) 17, 1 (1967), 383–399.

[27] Radon, J., Über die Bestimmung von Funktionen durch ihre Integralwerte längs gewisser Mannigfaltigkeiten. Ber. Verh. Sächs. Akad. Wiss. Leipzig, Math.-Nat. Kl. 69 (1917), 262–277.

[28] Runge, C., Zur Theorie der eindeutigen analytischen Funktionen. Acta Math. 6 (1885), 228–244.

[29] Walsh, J. L., The approximation of harmonic functions by harmonic polynomials and by harmonic rational functions. Bull. Amer. Math. Soc. (2) 35 (1929), 499–544.

[30] Zalcman, L., Uniqueness and non-uniqueness for the Radon transform. Bull. London Math. Soc. 14 (1982), 241–245.

Restricted mean value property and harmonic functions

Wolfhard Hansen

Abstract. Let f be a real function on a domain U in $\mathbb{R}^d$, $d \geq 1$, and assume that for every $x \in U$ there exists a ball B_x in U with center x such that $f(x)$ is the volume average of f on B_x (the spherical average of f on the boundary of B_x resp.). We shall discuss which additional properties imply that f is harmonic.

1991 Mathematics Subject Classification: 31A05, 31B05, 31C35, 35J10, 60J15, 60J65.

0. Introduction

In the following U will always denote a non-empty domain in $\mathbb{R}^d$, $d \geq 1$. Let λ be Lebesgue measure and let σ be the $(d-1)$-dimensional Hausdorff measure on $\mathbb{R}^d$. For every $x \in \mathbb{R}^d$ and $r > 0$ let

$$B(x,r) = \{y \in \mathbb{R}^d : \|y - x\| < r\}, \quad S(x,r) = \{y \in \mathbb{R}^d : \|y - x\| = r\},$$

and let

$$\lambda_{x,r} = (\lambda(B(x,r)))^{-1} 1_{B(x,r)} \lambda, \quad \sigma_{x,r} = (\sigma(S(x,r)))^{-1} 1_{S(x,r)} \sigma,$$

i.e., for functions f on $B(x,r)$ (on $S(x,r)$ resp.) $\lambda_{x,r}(f)$ will be the volume mean of f on $B(x,r)$ (the spherical mean of f on $S(x,r)$ resp.). It is well known that harmonic functions on U, i.e., functions $h \in \mathcal{C}^2(U)$ satisfying $\Delta h = 0$, can be characterized by mean value properties, for example: A locally bounded measurable function f on U is harmonic if and only if $\lambda_{x,r}(f) = f(x)$ ($\sigma_{x,r}(f) = f(x)$ resp.) for every $x \in U$ and every $r > 0$ such that $\overline{B(x,r)} \subset U$. For the history of these equivalences (as well as other material related to mean value properties) see the recent survey [NV].

In this paper we shall discuss to what extent harmonicity of f is already a consequence of knowing that for every $x \in U$ there exists *one* radius $r > 0$ such that $\lambda_{x,r}(f) = f(x)$ ($\sigma_{x,r}(f) = f(x)$ resp.). To have a suitable terminology we shall say that a real function f on U is (λ, r)-*median* ((σ, r)-*median* resp.) if r is a strictly positive real function on U such that $B(x, r(x)) \subset U$ ($\overline{B(x, r(x))} \subset U$ resp.) and

$$\lambda_{x,r(x)}(f) = f(x) \qquad (\sigma_{x,r(x)}(f) = f(x) \text{ resp.})$$

for every $x \in U$ (where we implicitly assume that f has the necessary measurability and integrability properties). We note that obviously every continuous (λ, r)-median function is (σ, r')-median for some function $0 < r' < r$.

Knowing that f is a (λ, r)-median or (σ, r)-median function on U we have to discuss which additional properties of U, r, and f imply (or do not imply) that f is harmonic.

Volterra [Vol] and Kellogg [Kel] settled the problem for bounded U and continuous functions f on the closure of U. At least if U is regular there is a very elementary proof (cf. [Bur]): Assume that f is (σ, r)-median and let g be the difference between f and the solution of the Dirichlet problem with boundary value f. If $g \neq 0$, say $a = \sup g(U) > 0$, choose $x \in \{g = a\}$ having minimal distance to the boundary. Then clearly $\sigma_{x,r(x)}(g) < g(x)$ contradicting the fact that g is (σ, r)-median. Thus $g = 0$, f is harmonic on U. In fact, a general minimum principle leads to the corresponding result for arbitrary harmonic spaces and for arbitrary representing measures $\mu_x \neq \varepsilon_x$ for harmonic functions on U.

Concerning the one-dimensional situation there is a negative result in the book of Courant–Hilbert [CH]: Chapter V contains an illustration of a zig-zag function f on the open unit interval such that $0 \leq f \leq 1$ and f is (σ, r)-median for a suitable choice of the function r. Note that the case $U = \mathbb{R}$ is as hopeless: The function $x \longmapsto \sin x$ is $(\sigma, 2\pi n)$-median for any natural n.

A first non-trivial positive result is due to Huckemann [Huc]:

Theorem 0.1. *Let f be a continuous (λ, r)-median function on the open interval $]-1, +1[$ and assume that the limits $\lim_{x \uparrow 1} \int_0^x f d\lambda$ and $\lim_{x \downarrow -1} \int_0^x f d\lambda$ exist in $\overline{\mathbb{R}}$. Then f is affine (i.e., harmonic).*

Note that the additional assumption on f is satisfied if f is bounded from below or bounded from above!

Remark 0.2. Without continuity of f or a restriction on r there is a trivial counterexample for any dimension d (cf. [Vee2], [Hea]): Let H be a hyperplane such that $U \cap H \neq \emptyset$ and take $f = 0$ on one side of H, $f = 1$ on the other side of H, and $f = 1/2$ on $U \cap H$.

Moreover, some boundedness of f is needed: Take $U =]0, 1[$, let (a_n) be a sequence in $]\frac{1}{2}, 1[$ which is strictly increasing to 1 and denote $A = \{a_n : n \in \mathbb{N}\}$. Define $r(a_n) = a_{n+2} - a_n$ and for $x \in U \setminus A$ choose any $r(x) > 0$ such that $B(x, r(x)) \cap A = \emptyset$. Proceeding by recurrence it is easily shown that there is a (unique) continuous function f on U such that f is (locally) affinely linear on $U \setminus A$, $f = 0$ on $]0, a_1[$, $f(a_2) = 1$, and $\lambda_{a_n, r(a_n)}(f) = f(a_n)$ for every $n \in \mathbb{N}$. Then f is r-median, but of course not harmonic. (A similar construction in $\mathbb{R}^d$, $d \geq 2$, would lead to a rotation invariant continuous function f on the unit ball U which oscillates at the boundary ∂U and is (λ, r)-median but not harmonic.)

In a booklet [Lit] on some problems in real and complex analysis which was published in 1968 J. E. Littlewood stated the following two questions (expressed in our terminology):

(1) Is every continuous bounded (σ, r)-median function on the open unit disk harmonic?

(2) Is every continuous bounded (λ, r)-median function on the open unit disk harmonic?

In this survey article we first consider (σ, r)-median functions where fairly little is known. We shall see, however, that the answer to (1) is negative [HN3]. Then we shall discuss (λ, r)-median functions which are bounded or at least harmonically bounded [HN1], [HN2], [HN4], [HN5]. In particular, an affirmative answer to (2) will be given. In a final section we want to study positive (λ, r)-median functions [HN6], [HN7], [HN8].

1. (σ, r)-median functions

Let us first recall some positive results on spherical means. In 1973 D. Heath proved the following [Hea]:

Theorem 1.1. *Let $U \neq \mathbb{R}^d$, $d \geq 2$, and let f be a continuous bounded (σ, r)-median function on U. Suppose that r is a Lipschitz function with Lipschitz constant $\alpha < 1$. Then f is harmonic.*

The next results for the plane and the unit disk are due to P. C. Fenton [Fen1], [Fen2], [Fen3]:

Theorem 1.2. *Let f be a continuous (σ, r)-median function on $\mathbb{R}^2$ such that f is majorized by a harmonic function h on $\mathbb{R}^2$, suppose that r is continuous and that there is a point $x_0 \in \mathbb{R}^2$ such that $\{x \in \mathbb{R}^2 : \|x - x_0\| < r(x)\}$ is bounded. Then f is harmonic (in fact, $f - h$ is constant).*

Theorem 1.3. *Let f be a continuous bounded (σ, r)-median function on $\mathbb{R}^2$ and suppose that r is bounded. Then f is harmonic, i.e., f is constant.*

Theorem 1.4. *Let f be a continuous (σ, r)-median function on $U = B(0, 1)$ having a harmonic minorant g_1 and a harmonic majorant g_2 on U. Suppose that $\lim_{x \to z} f(x)$ exists for σ-a.e. $z \in \partial U$ and that $\limsup_{r \uparrow 1} (g_2 - g_1)(rz) < \infty$ for every $z \in \partial U$. Then f is harmonic.*

Further contributions involving the boundary behavior of f can be found in [Gon], [Fol], [OS].

In contrast to these positive statements joint work with N. Nadirashvili showed that the answer to Littlewood's first question is NO. In order to obtain a continuous bounded (σ, r)-median function on the open unit disk which is not harmonic we proved a stronger "annulus theorem". Given $x \in \mathbb{R}^2$ and real numbers s, t with $0 < s < t$ let

$$A(x, s, t) = \{y \in \mathbb{R}^2 : s < \|y - x\| < t\}$$

and

$$\lambda_{x,s,t} = (\lambda(A(x,s,t)))^{-1} 1_{A(x,s,t)} \lambda.$$

Theorem 1.5. *Let U be the unit disk and $0 < \alpha < 1$. Then there exist continuous functions s, t and f on U such that $0 \le f \le 1$, $0 < s < t \le \alpha \operatorname{dist}(\cdot, \partial U)$, and*

$$\lambda_{x,s(x),t(x)}(f) = f(x)$$

for every $x \in U$, but f is not harmonic.

Corollary 1.6. *Let U be the unit disk and $0 < \alpha < 1$. Then there exists a continuous function f on U which is not harmonic, but such that for every $x \in U$ the equality $\sigma_{x,r(x)}(f) = f(x)$ holds for some $0 < r(x) \le \alpha \operatorname{dist}(x, U^c)$.*

For the proof of Theorem 1.5 fix $0 < \alpha \le \frac{1}{2}$ and, for every open subset V of the unit disk U, let $\mathcal{A}(V)$ denote the set of all (Markov) kernels P on U such that

$$P(x, \cdot) = \begin{cases} \lambda_{x,s(x),t(x)} & \text{for } x \in V, \\ \varepsilon_x & \text{for } x \in U \setminus V \end{cases}$$

with continuous real functions $0 < s < t \le \alpha\rho$ where

$$\rho = \operatorname{dist}(\cdot, U^c) = 1 - \|\cdot\|.$$

Then the proof of Theorem 1.5 consists essentially in the construction of a kernel $P \in \mathcal{A}(U)$ and a subset B of U (having very small area) such that the corresponding random walk has the following properties: (i) Starting at 0 it stays in B and converges to ∂U with probability almost 1, (ii) starting at some point $x_0 \in U \setminus B$ close to 0 it hits B with small probability. Then

$$f(x) := \inf_n P^x \left[T_{B \cap \{\rho \le \alpha^n\}} < \infty\right] \qquad (x \in U)$$

is the desired function. It cannot be harmonic since $f(0) - f(x_0)$ is almost 1.

To be more precise let us first introduce the concepts which are already used in the formula above: Let $\Omega = U^{\mathbb{N} \cup \{0\}}$, $X_i(\omega) = \omega_i$ and let $\mathcal{M}$ denote the σ-algebra generated by X_i, $i \in \mathbb{N} \cup \{0\}$. Given a Markov kernel P on U and $x \in U$, let P^x denote the probability measure on $(\Omega, \mathcal{M})$ such that $(\Omega, \mathcal{M}, P^x)$ is the random walk starting at x having transition kernel P, i.e., for all $n \in \mathbb{N}$ and Borel subsets $B_0, B_1, \ldots, B_n$ of U

$$\begin{aligned} P^x[X_0 \in B_0, X_1 \in B_1, X_2 \in B_2, &\ldots, X_n \in B_n] \\ &= \varepsilon_x(B_0) \int_{B_1} P(x, dx_1) \int_{B_2} P(x_1, dx_2) \cdots \int_{B_n} P(x_{n-1}, dx_n). \end{aligned}$$

For every Borel subset B of U let

$$T_B := \inf\{i \in \mathbb{N} \cup \{0\} : X_i \in B\}.$$

Fix $\varepsilon \in]0, \frac{1}{3}]$ and choose $\delta \in]0, \frac{1}{2}[$ such that $|h(0) - h(x)| < \varepsilon$ for every harmonic function $h : U \longrightarrow [0,1]$ and all $x \in B(0,\delta)$. Define $P_0 \in \mathcal{A}(U)$ by

$$P_0(x, \cdot) = \lambda_{x,\alpha\rho(x)/2,\alpha\rho(x)}$$

for $x \in U$. Then there exists a sequence (δ_k) of strictly positive real numbers such that $P_0^x[0 < T_B < \infty] < \varepsilon$ for every $x \in U$ and every Borel subset B of U satisfying $\lambda(B \cap \{\rho \leq \alpha^{k-1}\}) \leq \delta_k$ for every $k \in \mathbb{N}$. This step of the proof is fairly easy and we may of course assume without loss of generality that $\delta_1 \leq \delta^2$ and $\delta_{k+1} \leq \alpha\delta_k$ for every $k \in \mathbb{N}$. It is much more delicate to find an open set V in U and a kernel $P_\infty \in \mathcal{A}(V)$ such that $\lambda(V \cap \{\rho \leq \alpha^{k-1}\}) < \delta_k$ for every $k \in \mathbb{N}$ and $P_\infty^0[X_i \in V \text{ for all } i \in \mathbb{N}, \rho(X_i) \to 0] = 1$. Then there exists a subset B of V which is closed relative to U such that $0 \in V$ and $P_\infty^0[X_i \in B \text{ for all } i \in \mathbb{N}] > 1 - \varepsilon$. For $x \in V$ we have $P_\infty(x, \cdot) = \lambda_{x,s_\infty(x),t_\infty(x)}$ with continuous functions s_∞, t_∞ on V satisfying $0 < s_\infty < t_\infty \leq \alpha\rho$. So we may choose continuous functions s, t on U such that $0 < s < t \leq \alpha\rho$ on U, $s = s_\infty, t = t_\infty$ on B and $s = \alpha\rho/2$, $t = \alpha\rho$ on $U \setminus V$. Defining

$$P(x, \cdot) = \lambda_{x,s(x),t(x)}$$

for $x \in U$ we obviously have $P \in \mathcal{A}(U)$ and it is easily seen that

$$P^x[T_V < \infty] = P_0^x[T_V < \infty] < \varepsilon$$

for every $x \in U \setminus V$ and

$$P^0[T_{U\setminus B} = \infty, \lim_{i\to\infty} \rho(X_i) = 0] = P_\infty^0[T_{U\setminus B} = \infty, \lim_{i\to\infty} \rho(X_i) = 0] > 1 - \varepsilon.$$

For every $n \in \mathbb{N}$ define a function $f_n : U \to [0,1]$ by

$$f_n(x) = P^x[T_{B\cap\{\rho\leq\alpha^n\}} < \infty].$$

Obviously, $[T_{B\cap\{\rho\leq\alpha^n\}} < \infty] \subset [T_V < \infty]$. On the other hand, if $\omega \in \Omega$ such that $X_i(\omega) \in B$ for all $i \geq 0$ and $\lim_{i\to\infty} \rho(X_i(\omega)) = 0$, then clearly $T_{B\cap\{\rho\leq\alpha^n\}}(\omega) < \infty$. Therefore

$$f_n < \varepsilon \text{ on } U \setminus V, \qquad f_n(0) > 1 - \varepsilon.$$

If $x \in \{\rho \geq \alpha^{n-1}\}$ then $P(x, \{\rho \leq \alpha^n\}) = 0$ since $t(x) \leq \alpha\rho(x)$, hence $Pf_n(x) = f_n(x)$. Evidently the sequence (f_n) is decreasing to a function f such that $0 \leq f \leq 1$, $f < \varepsilon$ on $U \setminus V$, and $f(0) \geq 1 - \varepsilon$. Knowing that $Pf_n = f_n$ on $\{\rho \geq \alpha^{n-1}\}$ for every $n \in \mathbb{N}$ we obtain that $Pf = f$. In particular, f is continuous since P is a strong Feller kernel. Moreover, since $\lambda(V) < \delta_1 \leq \delta^2$, there exists a point $x_0 \in B(0,\delta) \setminus V$ and we have $f(x_0) < \varepsilon$, hence

$$f(0) - f(x_0) > 1 - 2\varepsilon > \varepsilon.$$

This shows that f is not harmonic.

Let us finally indicate how the open set V in U and the kernel $P \in \mathcal{A}(V)$ are constructed: For every open subset W of U let $\mathcal{F}(W)$ denote the class of all locally finite countable families of disjoint open disks in W. For every $\mathcal{D} \in \mathcal{F}(W)$ let $U(\mathcal{D})$ denote the union of all $D \subset \mathcal{D}$ and let $r(\mathcal{D})$ be the sum of the radii $r(D)$ of all disks $D \in \mathcal{D}$. The crucial fact for our construction is the following property of Brownian motion in the plane: For every open subset W of U and for every $\delta > 0$ there exists $\mathcal{D} \in \mathcal{F}(W)$ such that $r(\mathcal{D}) < \delta$ and Brownian motion on W hits $U(\mathcal{D})$ with probability 1 before leaving W. It is no problem to get a kernel $P \in \mathcal{A}(W \setminus \overline{U(\mathcal{D})})$ such that the corresponding random walk has the same property, i.e., such that $P^x[T_{U(\mathcal{D})} < \infty] = 1$ for every $x \in W$.

We start with a disk $V_1 = B(0, \delta_1)$, a family $\mathcal{D}_1 \in \mathcal{F}(V_1)$ and a kernel $P_1 \in \mathcal{A}(V_1 \setminus \overline{U(\mathcal{D}_1)})$ such that $r(\mathcal{D}_1) < \delta_2/(12\pi)$ and $P^x[T_{U(\mathcal{D})} < \infty] = 1$ for every $x \in U_1$. Having defined $V_n, \mathcal{D}_n, P_n$ we select a disk $D_n \in \mathcal{D}_n$. Let x_n and r_n denote the center and the radius of D_n, respectively. We define an open annulus A_n by

$$A_n = A(x_n, \alpha\rho(x_n) - 3r_n, \alpha\rho(x_n))$$

and take

$$V_n' := A_n \setminus \overline{V}_n, \qquad V_{n+1} := (V_n \setminus \partial D_n) \cup V_n'.$$

We note that $\lambda(V_n') \le \lambda(A_n) \le 6\pi\, r_n$. Finally, we choose $\mathcal{D}_n' \in \mathcal{F}(V_n')$ and $P_n' \in \mathcal{A}(V_n' \setminus \overline{U(\mathcal{D}_n')})$ such that $r(\mathcal{D}_n') < \delta_{n+2}/(12\pi)$ and $(P_n')^x\,[T_{U(\mathcal{D}_n')} < \infty] = 1$ for every $x \in V_n'$, and define

$$\mathcal{D}_{n+1} := (\mathcal{D}_n \setminus \{D_n\}) \cup \mathcal{D}_n',$$

$$P_{n+1}(x,\,\cdot\,) = \begin{cases} P_n(x,\,\cdot\,), & x \in U \setminus (D_n \cup V_n'), \\ \lambda_{x,\alpha\rho(x_n)-2r_n,\alpha\rho(x_n)-r_n}, & x \in D_n, \\ P_n'(x,\,\cdot\,), & x \in V_n'. \end{cases}$$

Of course $\mathcal{D}_{n+1} \in \mathcal{D}(V_{n+1})$ and it is easily verified that $P_{n+1} \in \mathcal{A}(V_{n+1} \setminus \overline{U(\mathcal{D}_{n+1})})$ and $P_{n+1}^x[T_{U(\mathcal{D}_{n+1})} < \infty] = 1$ for every $x \in V_{n+1}$. Moreover, we have

$$\lambda(V_n \cap \{\rho \le \alpha^{k-1}\}) < \delta_k$$

for all $k \in \mathbb{N}$.

It is possible to make these choices in such a way that we obtain a kernel $P_\infty \in \mathcal{A}(V)$ for $V := \bigcup_{n=1}^\infty V_n$ satisfying $P_\infty(x,\,\cdot\,) = P_n(x,\,\cdot\,)$ for every $x \in V_n \setminus \overline{U(\mathcal{D}_n)}$ and $\lim_{i\to\infty} \rho(X_i) = 0$ P_∞^0-a.s. In particular $P_\infty^x[T_{U\setminus V} < \infty] = 0$ for every $x \in V$ and $\lambda(V \cap \{\rho \le \alpha^{k-1}\}) \le \delta_k$ for all $k \in \mathbb{N}$.

2. Bounded (λ, r)-median functions

If $U \ne \mathbb{R}^d$ let ρ denote the distance to U^c, i.e.,

$$\rho(x) := \inf\{\|x - y\| : y \in U^c\}$$

for $x \in U$. If $U = \mathbb{R}^d$, fix $M_0 \in \mathbb{R}^+$ and define

$$\rho(x) := \|x\| + M_0.$$

A real function r on U will be called *admissible* if $0 < r \leq \rho$. Let us note that an admissible function r on a proper subdomain U of $\mathbb{R}^d$ obviously satisfies a growth condition $r \leq M_0 + \|\cdot\|$ (take $x_0 \in U^c$ and $M_0 = \|x_0\|$).

Using methods from ergodic theory and results from W. Feller [Fel], M. A. Akcoglu and R. W. Sharpe (1968) [AS] showed the following:

Theorem 2.1. *If $U = B(0,1) \subset \mathbb{R}^2$ and $r = \rho$ then every bounded (λ, r)-median function is harmonic.*

Proceeding similarly J. R. Baxter (1972) [Bax1] went a little further:

Theorem 2.2. *Let $\overline{U}$ be a compact C^1-manifold in $\mathbb{R}^2$ and r an admissible function on U such that $r \geq \alpha\rho$ for some real $\alpha > 0$. Then every bounded (λ, r)-median function on U is harmonic.*

Then D. Heath (1973) [Hea] proved:

Theorem 2.3. *If $U \neq \mathbb{R}^d$ and $r : U \longrightarrow \mathbb{R}$ such that $\alpha\rho \leq r \leq (1-\alpha)\rho$ for some $\alpha > 0$ then every bounded (λ, r)-median function is harmonic.*

Let us say that a function f on U is *harmonically bounded* if there exists a harmonic function h on U such that $|f| \leq h$. Using different probabilistic methods W. A. Veech (1973) [Vee2] obtained the following (it was observed later by A. Ancona that the use of the Lipschitz character of U in the proof of Veech can be avoided):

Theorem 2.4. *Suppose that U is a bounded (Lipschitz) domain and that the admissible function r is locally bounded away from zero, i.e., $\inf r(K) > 0$ for every compact K in U. Then every harmonically bounded (λ, r)-median function f is harmonic.*

Note, however, that these results did not answer Littlewood's second question on (λ, r)-median continuous bounded functions on the unit disk! That the answer to this question is YES is a very special case of general results obtained by joint work with N. Nadirashvili. In [HN1] we solved the problem for continuous harmonically bounded r-median functions on bounded domains and obtained the result of W. A. Veech (without Lipschitz assumption) as a corollary. In [HN2] we considered more general means and open sets $U \neq \mathbb{R}^d$ (U^c non-polar if $d = 2$), and in [HN4], [HN7] we discussed the question for $U = \mathbb{R}^d$. Simple observations in [HN5] allow us to treat simultaneously the two cases: a) f is continuous, no condition on r, b) r is locally bounded away from zero, no continuity condition on f. Combining [HN5] with supplements in [HN7] we have the following:

Theorem 2.5. *Let r be an admissible function on U and let f be a harmonically bounded (λ, r)-median function on U. Assume that f is continuous or that r is locally bounded away from zero. Then f is harmonic.*

If U is not a Green domain, we know even more:

Theorem 2.6. *Let $U = \mathbb{R}$ or $U \subset \mathbb{R}^2$ such that U^c is polar, let r be an admissible function on U and let f be a lower bounded numerical function on U. If f is l.s.c. and $\lambda_{x,r(x)}(f) \le f(x)$ for every $x \in U$, then f is constant. If r is locally bounded away from zero and $\lambda^*_{x,r(x)}(f) \le f(x)$ for every $x \in U$, then $f = \inf f(U)$ λ-a.e.*

The method for our proofs is purely analytic, but some of the steps have a probabilistic meaning. To understand the essential ideas let us first consider the case $U = \mathbb{R}^d$, $d \ge 3$. So let r be an admissible function on $\mathbb{R}^d$, i.e.,

$$0 < r \le \rho(x) = M_0 + \|\cdot\|.$$

As already noted by W. A. Veech (see [Vee2] or [HN1]) we may assume that r is Borel measurable. Define

$$Q_0 = B(0, M_0), \quad Q_1 = (2Q_0) \setminus Q_0, \quad Q_n = 2^{n-1} Q_1 \quad \text{for } n \in \mathbb{N}.$$

It will be convenient to write $|A|$ instead of $\lambda(A)$ for the Lebesgue measure of a Borel set $A \subset \mathbb{R}^d$. Given a lower bounded Lebesgue measurable function f on $\mathbb{R}^d$ we define a function $\overline{f}$ on $\mathbb{R}^d$ by

$$\overline{f}(x) = \sup\left\{b \in \mathbb{R} : |\{f \ge b\} \cap Q_n| \ge \frac{1}{2}|Q_n|\right\}$$

for $n = 0, 1, 2, \ldots$ and $x \in Q_n$. So for each n the function $\overline{f}$ is constant on Q_n and $|\{f \ge \overline{f}\} \cap Q_n| \ge \frac{1}{2}|Q_n|$. Let $Y = \mathbb{R}^d \cup \{z_\infty\}$ be the one point compactification of $\mathbb{R}^d$ and define

$$\overline{f}(z_\infty) = \limsup_{x \to \infty} \overline{f}(x).$$

We shall say that f is *(λ, r)-supermedian* if $\lambda_{x,r(x)}(f) \le f(x)$ for every x. The basic result is then the following:

Theorem 2.7. *Let f be a lower bounded (λ, r)-supermedian function on $\mathbb{R}^d$ such that f is l.s.c. or r is locally bounded away from zero. Then $f \ge \overline{f}(z_\infty)$.*

Defining

$$\underline{f}(x) := \inf\left\{b \in \mathbb{R} : |\{f \le b\} \cap Q_n| \ge \frac{1}{2}|Q_n|\right\}$$

and

$$\underline{f}(z_\infty) = \liminf_{x \to z_\infty} \underline{f}(x)$$

for $n = 0, 1, 2, \ldots$ and $x \in Q_n$ we obviously have $\underline{f} \le \overline{f}$ on Y and $\underline{f} = -\overline{(-f)}$ (if f is continuous then $\underline{f} = \overline{f}$ on $\mathbb{R}^d$). So we immediately get the desired result:

Corollary 2.8. *Let f be a bounded (λ, r)-median function on $\mathbb{R}^d$ and suppose that f is continuous or that r is locally bounded away from zero. Then f is constant.*

Indeed, applying (2.7) to f and $-f$ we obtain that $f \geq \overline{f}(z_\infty)$ and $-f \geq \overline{(-f)}(z_\infty) = -\underline{f}(z_\infty)$, i.e., $f \leq \underline{f}(z_\infty)$. Since $\underline{f}(z_\infty) \leq \overline{f}(z_\infty)$ we conclude that $f = \underline{f}(z_\infty) = \overline{f}(z_\infty)$.

Our proof of Theorem 2.7 involves the Schrödinger equation $\left(\Delta - \delta 1_A \rho^{-2}\right) u = 0$ for subsets A of $\mathbb{R}^d$ and very small constants $\delta > 0$. The reason is the following: If $b < \overline{f}(z_\infty)$ and

$$A = \{f \geq b\}$$

then

$$|A \cap Q_n| \geq \frac{1}{2}|Q_n| \tag{*}$$

for infinitely many n by definition of $\overline{f}(z_\infty)$. If we can show that the (transfinite) sweeping induced by the Markov kernel P defined by

$$P(y,\,\cdot\,) = \begin{cases} \lambda_{y,r(y)} & \text{for } y \in \mathbb{R}^d \setminus A, \\ \varepsilon_y & \text{for } y \in A \cup \{z_\infty\} \end{cases}$$

sweeps for every $x \in \mathbb{R}^d$ the measure ε_x on A, the fact that our function f is (λ, r)-supermedian, hence P-supermedian (taking $f(z_\infty) = 0$), will imply that $f(x) \geq b$ for every $x \in \mathbb{R}^d$. This means that $f \geq \overline{f}(z_\infty)$ finishing the proof.

In probabilistic terms it is clear that knowing $(*)$ for infinitely many n Brownian motion will hit A almost surely (Brownian motion on $\mathbb{R}^d$, $d \geq 3$, goes to infinity and passing a shell Q_n with $|A \cap Q_n| \geq \frac{1}{2}|Q_n|$ it will hit A with a probability which is at least some $\varepsilon > 0$ independent of n because of scaling invariance). Unfortunately, the random walk generated by P has obviously less chances to hit A than Brownian motion. However, the Schrödinger operator $\Delta - \delta 1_A \rho^{-2}$ with small $\delta > 0$ offers a mild killing on A. We show that on the one hand the corresponding continuous process, i.e., Brownian motion killed at a rate $\delta\rho^{-2}$ while it passes the set A, will be killed almost surely whatever $\delta > 0$, whereas the random walk associated with P will now have more chances to hit A if δ is sufficiently small.

Formally (and in a purely analytic way) we proceed as follows: Define

$$U_m = 2^m Q_0 = \bigcup_{n=0}^{m} Q_n, \qquad m \in \mathbb{N},$$

and let G denote the Green function on $\mathbb{R}^d$ ($G(x,y) = \kappa_d \|x - y\|^{2-d}$, $\Delta G(\,\cdot\,, y) = -\varepsilon_y$). Then we prove (for the proofs we do not give here see [HN1]):

Lemma 2.9. *Let $V \geq 0$ be a locally bounded Borel measurable function on $\mathbb{R}^d$, let $L_m = \Delta - 1_{U_m} V$, $m \in \mathbb{N}$, and define $K_m f = \int_{U_m} f(y) G(\,\cdot\,, y) V(y)\,\lambda(dy)$ ($m \in \mathbb{N}$, $f \geq 0$ Borel measurable). Then, for every $m \in \mathbb{N}$, there exists a unique Borel function $g_m \geq 0$ on $\mathbb{R}^d$ such that $g_m + K_m g_m = 1$. The functions g_m are continuous and satisfy $L_m g_m = 0$. The sequence (g_m) is decreasing. Moreover, $\lim_{m\to\infty} g_m = 0$ if and only if there is no bounded continuous solution $u > 0$ of the Schrödinger equation $\Delta u - Vu = 0$ on $\mathbb{R}^d$.*

Proposition 2.10. *Let A be a Borel subset of $\mathbb{R}^d$, let $\alpha > 0$ and assume that $|A \cap Q_n| \geq \alpha |Q_n|$ for infinitely many $n \in \mathbb{N}$. Then, for every $\delta > 0$, every solution $u > 0$ of the Schrödinger equation $\Delta u - \delta 1_A \rho^{-2} u = 0$ is unbounded.*

Proposition 2.11. *Let $\beta, \eta \in]0,1[$. Then there exists a constant $\delta > 0$ such that for every ball $B = B(x,r)$, $x \in \mathbb{R}^d$, $r > 0$, for every harmonic function $g \geq 0$ on B, and for all Borel sets $A' \subset A \subset \mathbb{R}^d$ satisfying $A' \cap B(x, \eta r) = \emptyset$ or $|A \cap B| \geq \beta |B|$ the following holds: If*

$$0 \leq V \leq \delta d(\cdot\,, B^c)^{-2} 1_{A'}$$

is a Borel measurable function on B and if $0 \leq u \leq g$ is a continuous solution of $\Delta u - Vu = 0$ on B then

$$u(x) \geq \frac{1}{|B|} \int_{B \setminus A} u \, d\lambda.$$

Note that the function u in (2.11) trivially satisfies $u(x) \leq |B|^{-1} \int_B u \, d\lambda$ since u is subharmonic. Of course it would be nicer to have (2.11) with $A' = A$ being an arbitrary Borel set. Unfortunately, however, there is no $\delta > 0$ leading to $u(x) \geq \int_{A^c} u \, d\lambda_{x,r}$ if we allow that $A = B(x, \varepsilon r)$ with arbitrarily small $\varepsilon > 0$.

So we have to omit a subset E from the set $A = \{f \geq b\}$ we are interested in such that $A' := A \setminus E$ and A satisfy the assumption of (2.11) for every $x \in \mathbb{R}^d$. But in addition we want to apply (2.10) with A' instead of A. Therefore we have to be sure that $|E \cap Q_n| < \gamma |Q_n|$ for every $n \in \mathbb{N}$ with a constant $\gamma < \frac{1}{2}$. This is possible by a covering argument. In [HN1], [HN2] and [HN4] we used the rather sophisticated theorem of Besicovitch. We are indebted to A. Grigor'yan for the hint that the most elementary covering theorem for balls will do. To see this let us write down the proof of the following:

Proposition 2.12. *Let A be a Borel subset of $\mathbb{R}^d$, $\beta, \eta \in]0,1[$ and let E denote the set of all $y \in A$ such that $|A \cap B(x, r(x))| \leq \beta |B(x, r(x))|$ and $\|x - y\| < \eta r(x)$ for some $x \in \mathbb{R}^d$. Then for every $n \in \mathbb{N}$*

$$|E \cap Q_n| \leq \beta \left(\frac{9}{1-\eta} \right)^d |Q_n|.$$

Proof. Fix $n \in \mathbb{N}$ and let

$$Q'_n = \{x \in \mathbb{R}^d : \|x\| \leq (2^{n+1} + 1) M_0\}.$$

Then $|Q'_n| \leq 3^d |Q_n|$ and $B(y, \rho(y)) \subset Q'_n$ for every $y \in Q_n$. For every $y \in E \cap Q_n$ choose $y' \in \mathbb{R}^d$ such that $|A \cap B(y', r(y'))| \leq \beta |B(y', r(y'))|$ and $\|y' - y\| < \eta r(y')$. Then

$$r(y') \leq \rho(y') = \|y'\| + M_0 \leq \|y' - y\| + \rho(y),$$

hence

$$s(y) := (1-\eta) r(y') < \rho(y) \quad \text{and} \quad B(y, s(y)) \subset B(y', r(y')) \cap Q'_n.$$

Let K be a compact subset of $E \cap Q_n$. It is easy to see that there exist finitely many points $y_1, \ldots, y_m \in K$ such that the balls $B(y_j, s(y_j)/3)$, $1 \le j \le m$, are disjoint and $K \subset \bigcup_{j=1}^m B(y_j, s(y_j))$ (cf. [Rud, p. 164]). Then

$$\begin{aligned}
|K| &\le \sum_{j=1}^m |E \cap B(y_j, s(y_j))| \le \sum_{j=1}^m |A \cap B(y_j', r(y_j'))| \\
&\le \beta \sum_{j=1}^m |B(y_j', r(y_j'))| = \beta \left(\frac{3}{1-\eta}\right)^d \sum_{j=1}^m |B(y_j, s(y_j)/3)| \\
&= \beta \left(\frac{3}{1-\eta}\right)^d |\bigcup_{j=1}^m B(y_j, s(y_j)/3)| \le \beta \left(\frac{3}{1-\eta}\right)^d |Q_n'| \le \beta \left(\frac{9}{1-\eta}\right)^d |Q_n|.
\end{aligned}$$

Given a Borel subset A of $\mathbb{R}^d$ we define a Markov kernel P on Y as before by $P(x, \cdot) = \lambda_{x,r(x)}$ for $x \in \mathbb{R}^d \setminus A$ and $P(x, \cdot) = \varepsilon_x$ for $x \in A \cup \{z_\infty\}$. For every $x \in \mathbb{R}^d$ let $\mathcal{M}_P(x)$ be the smallest w^*-closed and P-stable set of probability measures on Y containing ε_x. By [HN1] there exists a unique $\sigma_x \in \mathcal{M}_P(x)$ such that $\sigma_x P = \sigma_x$. If r is locally bounded away from zero then simply $\mathcal{M}_P(x) = \{P^m(x, \cdot) : m = 0, 1, 2, \ldots, \infty\}$ where $P^\infty = \lim_{m\to\infty} P^m$, hence $\sigma_x = P^\infty(x, \cdot)$ and, for every Borel subset B of A, the sequence $(P^m(x, B))$ is increasing to $\sigma_x(B)$ (see [HN5]). This observation will allow us to treat the case of a l.s.c. function f and the case of a locally bounded function r^{-1} simultaneously.

Proposition 2.13. *Assume that there exists $\alpha > 0$ such that $|A \cap Q_n| \ge \alpha |Q_n|$ for infinitely many n. Then $\sigma_x(A) = 1$ for every $x \in \mathbb{R}^d$.*

Proof. Fix $x \in \mathbb{R}^d$ and let $\sigma = \sigma_x$. By [HN1] σ is supported by the base $A \cup \{z_\infty\}$ of P. So we only have to show that $\sigma(\{z_\infty\}) = 0$. To that end we consider the sub-Markov kernel P' defined by

$$P'(y, \cdot) = \begin{cases} 1_{A^c} P(y, \cdot) & \text{for } y \in \mathbb{R}^d \setminus A, \\ \varepsilon_y & \text{for } y \in A \cup \{z_\infty\}. \end{cases}$$

By [HN1] the P'-invariant measure $\sigma' \in \mathcal{M}_{P'}(x)$ satisfies $\sigma'(\{z_\infty\}) = \sigma(\{z_\infty\})$. So the proof is finished once we know that $\sigma'(\{z_\infty\}) = 0$.

To that end fix $\eta \in]0, 1[$, say $\eta = \frac{1}{2}$, choose

$$\beta = \frac{1}{4} 18^{-d},$$

define E as in (2.12), and take

$$A' = A \setminus E.$$

Then $|E \cap Q_n| \le \frac{1}{4}|Q_n|$ for every $n = 0, 1, 2, \ldots$, hence

$$|A' \cap Q_n| \ge \frac{1}{4}|Q_n|$$

for every n such that $|A \cap Q_n| \geq \frac{1}{2}|Q_n|$ and

$$A' \cap B(y, \eta r(y)) = \emptyset \quad \text{or} \quad \lambda_{y,r(y)}(A) \geq \beta$$

for every $y \in \mathbb{R}^d$.

We choose $\delta > 0$ according to (2.11) and take $V = \delta\rho^{-2}1_{A'}$. Then every continuous solution $u > 0$ of $\Delta u - Vu = 0$ on $\mathbb{R}^d$ is unbounded by (2.10). For every $m \in \mathbb{N}$ let $g_m \in C^+(\mathbb{R}^d)$ such that $g_m + K_m g_m = 1$ where $K_m g_m = \int_{U_m} g_m(y)G(\cdot, y)V(y)\,\lambda(dy)$ (see (2.9)). Then $\lim_{m\to\infty} g_m = 0$ by (2.9) whereas by (2.11), for every $m \in \mathbb{N}$ and every $y \in U$,

$$g_m(y) \geq \int_{A^c} g_m \, d\lambda_{y,r(y)},$$

i.e., $P'g_m(y) \leq g_m(y)$ for every $y \in \mathbb{R}^d$. For every $m \in \mathbb{N}$, the potential $K_m g_m$ is harmonic on $\overline{U}_m^c$, hence $\lim_{\|y\|\to\infty} K_m g_m(y) = 0$, $\lim_{\|y\|\to\infty} g_m(y) = 1$. So we may extend g_m to a continuous function on Y defining $g_m(z_\infty) = 1$. Then $P'g_m \leq g_m$ on Y, hence $\sigma'(g_m) \leq g_m(x_0)$ by Proposition 3.3 in [HN1]. Thus

$$\sigma'(\{z_\infty\}) \leq \lim_{m\to\infty} \sigma'(g_m) = 0$$

finishing the proof. □

As already indicated we are now able to prove Theorem 2.7: Clearly, it suffices to consider positive functions. So let $f \geq 0$ be a (λ, r)-supermedian function on $\mathbb{R}^d$ and assume that f is l.s.c. or that r is locally bounded away from zero. Fix $b < \overline{f}(z_\infty)$. Then there are infinitely many n such that the set $A := \{f \geq b\}$ satisfies $|A \cap Q_n| \geq \frac{1}{2}|Q_n|$. Fix $x \in \mathbb{R}^d$, and define P and $\sigma_x \in \mathcal{M}_P(x)$ as before. Then $\sigma_x(A) = 1$ by (2.13). Extend f to a function on Y taking $f(z_\infty) = 0$. Then $Pf \leq f$, hence $P^m f \leq f$ for every $m \in \mathbb{N}$. If r is locally bounded away from zero this is sufficient to conclude that $\sigma_x(f) \leq f(x)$. If f is l.s.c., we consider the set $\mathcal{N}$ of all $\nu \in \mathcal{M}_P(x)$ such that $\nu(f) \leq f(x)$. Then $\varepsilon_x \in \mathcal{N}$, $\mathcal{N}$ is w^*-stable and if $\nu \in \mathcal{N}$ then $(\nu P)(f) = \nu(Pf) \leq \nu(f) \leq f(x)$, hence $\nu P \in \mathcal{N}$. Thus $\mathcal{N} = \mathcal{M}_x(P)$, in particular $\sigma_x(f) \leq f(x)$. Knowing that $f \geq b$ on A we finally conclude that $f(x) \geq \sigma_x(f) \geq \sigma_x(b1_A) = b$.

In Theorem 2.6 we first assume that f is l.s.c. Then the case $U = \mathbb{R}$ is simple [HN7]: Adding functions $x \mapsto \varepsilon|x - M_0|$ to f a general minimum principle shows that $f \geq \inf f[-M_0, 0]$. Similarly $f \geq \inf f[0, M_0]$. There exists a point $x_0 \in [-M_0, M_0]$ such that $f(x_0) \leq f$ on $[-M_0, M_0]$, hence $f(x_0) \leq f$ on $\mathbb{R}$. The set $\{f = f(x_0)\}$ is closed since f is l.s.c. and it is open since $\lambda_{x,r(x)}(f) \leq f(x)$ for every $x \in \mathbb{R}$. Thus $f = f(x_0)$. – The case $U \subset \mathbb{R}^2$ and U^c polar is much more delicate. We reduce it to the case of domains having non-polar complement by removing a suitable small disk V where the value of f is almost maximal (assuming without loss of generality that f is bounded). This requires to replace measures $\lambda_{x,r(x)}$ for $x \in U \setminus \overline{V}$ with $B(x, r(x)) \cap V \neq \emptyset$ by more general means μ_x and to apply the general results of [HN2]. The basic idea is fairly simple, but the details are very technical (see [HN4]). If we have no continuity hypothesis for f, but know instead that r is locally bounded away from zero, we may consider the greatest l.s.c. minorant of f, show that f is r'-supermedian for an admissible function r' and apply the preceding result on g to get the desired statement for f.

Let us next discuss the essential modifications which are necessary if $U \neq \mathbb{R}^d$ and U^c is non-polar for $d = 2$ (for the details see [HN1] and [HN2]).

As a substitute for the sequence (Q_n) in $\mathbb{R}^d$ we take a Whitney decomposition $\mathcal{Q}$ for U. $\mathcal{Q}$ is a partition of U into countably many disjoint (dyadic) cubes of the form

$$Q = \prod_{i=1}^{d} [m_i 2^{-m}, (m_i+1)2^{-m}[\qquad (m \in \mathbb{Z}, m_i \in \mathbb{Z})$$

such that, for every $Q \in \mathcal{Q}$, the diameter $\delta(Q)$ of Q satisfies

$$1 \leq \frac{\inf \rho(Q))}{\delta(Q)} < 3.$$

Then for every $Q \in \mathcal{Q}$

$$\delta(Q) \leq \rho < 4\delta(Q) \quad \text{on } Q.$$

The lower bound for $d(Q, U^c)/\delta(Q)$ implies that there exists a constant $C > 0$ (depending only on the dimension d) such that

$$\sup g(Q) \leq C \inf g(Q)$$

for every $Q \in \mathcal{Q}$ and every harmonic function $g \geq 0$ on U. For every $m \in \mathbb{N}$ let U_m denote the interior of the union of all $Q \in \mathcal{Q}$ such that $\delta(Q) \geq 2^{-m}$.

Next we fix a harmonic function $h \geq 0$ on U and $0 < \alpha \leq 1/2$. A numerical function f on U is called *lower h-bounded, upper h-bounded, h-bounded* if there exists $c \in \mathbb{R}_+$ such that $f \geq -ch$, $f \leq ch$, $|f| \leq ch$, respectively. Given a Lebesgue measurable lower h-bounded numerical function f on U, we define a function $\overline{f}$ on U by

$$\overline{f}(x) = \sup \{a \in \mathbb{R} : |\{f \geq ah\} \cap Q| \geq \alpha |Q|\} \qquad (x \in Q \in \mathcal{Q}).$$

For our purpose it would be sufficient to take $\alpha = 1/2$ as before. Allowing, however, that α is small we shall obtain in addition a new proof for the existence of minimal fine limits of harmonic functions and the connection with the Perron–Wiener–Brelot solution for the Martin compactification.

Note that $\overline{f}$ is lower bounded on U, constant on each $Q \in \mathcal{Q}$, and that

$$\inf \frac{f}{h}(Q) \leq \overline{f} \leq \sup \frac{f}{h}(Q)\ , \quad |\{f \geq \overline{f}h\} \cap Q| \geq \alpha |Q|.$$

If f is a Lebesgue measurable upper h-bounded function on U, we define similarly

$$\underline{f}(x) = \inf\{a \in \mathbb{R} : |\{f \leq ah\} \cap Q| \geq \alpha |Q|\} \qquad (x \in Q \in \mathcal{Q}).$$

Obviously, $\underline{f} = -\overline{(-f)}$ and $\underline{f} \leq \overline{f}$ as before on $\mathbb{R}^d$.

It should be no surprise that the suitable compactification is now the Martin compactification of U (for $\mathbb{R}^d$ it is the one point compactification): Let $\partial^M U$ denote the Martin boundary of U, let $\partial_1^M U$ be the set of minimal points in $\partial^M U$, and let $K : U \times (U \cup \partial^M U) \longrightarrow [0, \infty]$ be a Martin function for U (see [Hel], [Doo]). Let

χ denote the (unique) measure on $\partial^M U$ such that $\chi(\partial^M U \setminus \partial_1^M U) = 0$ and

$$h(x) = \int K(x,z)\,\chi(dz) \qquad (x \in U).$$

Given any measurable function φ on $\partial^M U$ such that φ^+ or φ^- is χ-integrable, we define a numerical function $H\varphi$ on U by

$$H\varphi(x) = \int K(x,z)\varphi(z)\,\chi(dz) \qquad (x \in U).$$

If φ is χ-integrable then $H\varphi$ is harmonic on U. We recall that the Martin compactification is h-resolutive ([Doo, p. 110]), i.e., that, for every bounded measurable function φ on $\partial^M U$,

$$\begin{aligned} H\varphi &= \inf\{s\colon s \text{ superharmonic on } U, \liminf_{x\to z} s(x)/h(x) \ge \varphi(z) \text{ for every } z \in \partial^M U\} \\ &= \sup\{t\colon t \text{ subharmonic on } U, \limsup_{x\to z} t(x)/h(x) \le \varphi(z) \text{ for every } z \in \partial^M U\}. \end{aligned}$$

For every $z \in \partial_1^M U$, let $\mathcal{N}_z$ denote the filter of the intersections of minimal fine neighborhoods of z with U, i.e.,

$$\mathcal{N}_z = \{V \subset U : U \setminus V \text{ is minimal thin at } z\}.$$

Recall that a subset A of U is not minimal thin at z if

$${}^U R^A_{K(\cdot\,,z)} = K(\cdot\,,z)$$

(where ${}^U R^A_v = \inf\{s : s \text{ superharmonic } \ge 0 \text{ on } U,\ s \ge v \text{ on } A\}$), i.e., in probabilistic terms, if for every $y \in U$ the probability ${}^{K(\cdot\,,z)}R_1^A(y)$ that Brownian motion on U starting at y and conditioned by $K(\cdot\,,z)$ will hit A before exiting at z is equal to 1 (see [Doo]).

Now suppose that f is a lower h-bounded numerical function on U which is Lebesgue measurable. We extend $\overline{f}$ to a function on $U \cup \partial^M U$ defining

$$\overline{f}(z) = \begin{cases} \limsup_{\mathcal{N}_z} \overline{f}, & z \in \partial_1^M U, \\ 0, & z \in \partial^M U \setminus \partial_1^M U. \end{cases}$$

By [Doo, p. 216], $\overline{f}$ is Borel measurable on $\partial_1^M U$ and we note that

$$\overline{f}(z) = \sup\{a \in \mathbb{R} : \{\overline{f} \ge a\} \text{ not minimal thin at } z\} \qquad (z \in \partial_1^M U).$$

If f is an upper h-bounded Lebesgue measurable function on U, we define similarly

$$\underline{f}(z) = \begin{cases} \liminf_{\mathcal{N}_z} \underline{f}, & z \in \partial_1^M U, \\ 0, & z \in \partial^M U \setminus \partial_1^M U. \end{cases}$$

Obviously, $\underline{f} = -\overline{(-f)}$ on $U \cup \partial^M U$ and if $f : U \longrightarrow \mathbb{R}$ is h-bounded and Lebesgue measurable, then $\underline{f} \le \overline{f}$. Theorem 2.7 now reads as follows:

Theorem 2.14. *Let r be an admissible function on U, let f be a lower h-bounded (λ, r)-supermedian function and suppose that f is l.s.c. or that r is locally bounded away from zero. Then $f \geq H\overline{f}$.*

Applying (2.14) on f and $-f$ in exactly the same way as we did with (2.7) we obtain

Corollary 2.15. *Let r be an admissible function on U, let f be an h-bounded (λ, r)-median function and suppose that f is continuous or that r is locally bounded away from zero. Then f is harmonic.*

We already mentioned that an application of (2.14) with $\alpha \downarrow 0$ yields the following well known result (see e.g. [Doo]):

Corollary 2.16. *If f is an h-bounded harmonic function on U, then $\varphi(z) := \lim_{\mathcal{N}_z} f/h$ exists for χ-a.e. $z \in \partial_1^M U$ and $f = H\varphi$.*

In order to see where the proof of (2.14) really differs from the proof of (2.7) and where it is more or less the same, let us give a brief outline: By Lusin's theorem there exists a continuous function φ on $U \cup \partial^M U$ which is nearly $\overline{f}$ on $\partial^M U$. Fix $x \in U$ and $\varepsilon > 0$. Let t be a subharmonic function on U such that $\limsup_{y \to z} t(y)/h(y) \leq \varphi(z)$ for every $z \in \partial^M U$ and $t(x) > H\varphi(x) - \varepsilon$. There exists $m \in \mathbb{N}$ such that $t/h - \varepsilon < \varphi$ on $U \setminus U_m$. Choosing a suitable superharmonic function $s \geq 0$ on U such that $s(x)/h(x) < \varepsilon$ we obtain a subset

$$A = (U \setminus U_m) \cap \{\frac{f+s}{h} > \varphi - \varepsilon\}$$

of U such that the union of all $Q \in \mathcal{Q}$ satisfying $|A \cap Q| \geq \alpha |Q|$ is not minimal thin at the points $z \in \partial_1^M U$. As a consequence, for every $\delta > 0$, every continuous solution $u > 0$ of the Schrödinger equation $\Delta u - \delta \rho^{-2} 1_A u = 0$ on U is unbounded. This fact allows us to show that starting with unit mass at x the iterated sweeping induced by the means $h/h(y)\lambda_{y,r(y)}$, $y \in U$, and stopped at A does not reach the boundary of U. The (transfinite) limit measure σ is supported by A and satisfies $\sigma(f/h) \leq f(x)/h(x)$, $\sigma(s/h) \leq s(x)/h(x)$, and $\sigma(t/h) \geq t(x)/h(x)$. Since $(f+s)/h \geq t/h - 2\varepsilon$ on A, we conclude that $f(x)/h(x) + \varepsilon \geq f(x)/h(x) + s(x)/h(x) \geq t(x)/h(x) - 2\varepsilon \geq H\varphi(x)/h(x) - 3\varepsilon$. Thus $f \geq H\overline{f}$.

3. Positive (λ, r)-median functions

In this section let $\widetilde{\rho}$ be a strictly positive function on the domain U in $\mathbb{R}^d, d \geq 1$, such that $\widetilde{\rho} \leq \rho$ and

$$|\widetilde{\rho}(x) - \widetilde{\rho}(y)| \leq \|x - y\|$$

for all $x, y \in U$ (for example, $\widetilde{\rho} = \rho$). Work of W. A. Veech [Vee3], J. R. Baxter [Bax2], A. Cornea and J. Veselý [CV] led to the following result:

Theorem 3.1. *Let U be a Green domain and let r be a real function on U such that $\alpha\widetilde{\rho} \le r \le (1-\alpha)\widetilde{\rho}$ for some real $\alpha > 0$. Then every (λ, r)-median function $f \ge 0$ is harmonic.*

Whereas Veech and Baxter used (quite different) probabilistic tools, Cornea and Veselý recently gave a purely analytic proof. In [Vee3] and [CV] an essential step is the proof of a Harnack inequality for positive (λ, r)-median functions. It implies that Choquet theory for compact convex sets is applicable and hence an investigation of extremal positive (λ, r)-median functions is sufficient. A careful study of the discrete Martin kernel corresponding to the transition kernel $P : (x, A) \to \lambda_{x,r(x)}(A)$ then allowed the authors to show that every extremal positive (λ, r)-median function is harmonic.

In [Vee3] W. A. Veech formulated the

Conjecture. Let U be a bounded domain in $\mathbb{R}^d$ and let r be an admissible function on U which is locally bounded away from zero. Then every (λ, r)-median function $f \ge 0$ on U is harmonic.

Previous work by F. Huckemann [Huc] shows that this is true if $d = 1$. In [HN6], however, it is proven that the conjecture fails already for open balls in any $\mathbb{R}^d$, $d \ge 2$. In fact, even a weakened version of the conjecture where r and f are assumed to be smooth is wrong:

Theorem 3.2. *Let U be the unit ball in $\mathbb{R}^d$, $d \ge 2$. Then for every $0 < \alpha < 1$ there exist strictly positive C^∞-functions r and f on U such that $r \le \alpha\rho$ and f is (λ, r)-median, but not harmonic on U.*

Let us roughly describe the construction (which works for any Green domain U in $\mathbb{R}^d$, $d \ge 2$, such that $\emptyset \ne U \ne \mathbb{R}^d$). We fix $0 < \alpha < 1/4$, define

$$\alpha_n := (1-\alpha)^n \alpha, \qquad \text{for } n = -1, 0, 1, 2, \dots$$

and for every $n = 0, 1, 2, \dots$

$$\xi_n := \sum_{j=0}^{n-1} \alpha_j = 1 - (1-\alpha)^n, \qquad x_n := (\xi_n, 0, \dots, 0) \in \mathbb{R}^d$$

and

$$r_n = \frac{\alpha_{n-1} + \alpha_n}{2}.$$

Choosing real numbers $c_n > 0$ such that $c_n \le \inf(\alpha/5, 2^{-n}) r_n$ and defining

$$C_n := B(x_n, c_n), \qquad n = 0, 1, 2, \dots,$$

we obtain that, for every $x \in C_0$,

$$B(x, r_0/2) \cap C_1 = B(x, r_0) \cap C_2 = \emptyset, \qquad C_1 \subset B(x, r_0)$$

and, for every $x \in C_n$, $n \geq 1$,

$$B(x, r_n/2) \cap C_{n+1} = B(x, r_n) \cap (C_{n-1} \cup C_{n+2}) = \emptyset, \qquad C_{n+1} \subset B(x, r_n).$$

Moreover, $r_n \leq 2\alpha(1 - |x|)$ for every $n \in \mathbb{N}_0$ and every $x \in C_n$. We take $0 < b_0 < c_0/2$, $a_0 = b_0/2$, and define

$$A_0 = \overline{B(0, a_0)}, \qquad B_0 = B(0, b_0).$$

In order to understand the idea of our counterexample let us assume for a moment that we have chosen real numbers $0 < a_n < c_n$, $n \in \mathbb{N}$. Then we take $A_n = \overline{B(x_n, a_n)}$, $\tilde{\gamma}_n = \prod_{j=1}^n \frac{r_{j-1}^d - a_{j-1}^d}{a_j^d}$, and we may define a measurable function $\tilde{f} \geq 0$ on U by

$$\tilde{f} = \begin{cases} \tilde{\gamma}_n & \text{on } A_n, n \geq 0, \\ 0 & \text{on } U \setminus \bigcup_{n=0}^\infty A_n. \end{cases}$$

Obviously, $\tilde{f}$ is not harmonic, but it is $(\lambda, \tilde{r})$-median if we define

$$\tilde{r}(x) = \begin{cases} r_n, & x \in A_n, n \geq 0, \\ \inf(\alpha(1 - |x|), \operatorname{dist}(x, \bigcup_{n=0}^\infty A_n)), & x \in U \setminus \bigcup_{n=0}^\infty A_n . \end{cases}$$

Of course, $\tilde{r}$ is not locally bounded away from zero. We may, however, modify this construction and obtain a (λ, r)-median function for which r is locally bounded away from zero.

In order to see how this can be achieved let us note that $\widetilde{f}$ and the random walk given by the kernel $\widetilde{P} : (x, A) \longmapsto \lambda_{x, \widetilde{r}(x)}$ are connected in the following way: For every $n \in \mathbb{N}$ and every $x \in U \setminus \bigcup_{k=n+1}^\infty A_k$,

$$\tilde{f}(x) = \tilde{\gamma}_n \tilde{P}^x[T_{A_n} < \infty]$$

and

$$\widetilde{P}^x[T_{A_n} < \infty, X_{T_{A_n}+1} \in A_{n+1}] = \left(1 - \left(\frac{a_n}{r_n}\right)^d\right) \widetilde{P}^x[T_{A_{n+1}} < \infty]$$

(where of course $[T_{A_n} < \infty, X_{T_{A_n}+1} \in A_{n+1}] \subset [T_{A_{n+1}} < \infty]$).

We shall arrange our modification in such a way that the corresponding random walk has almost no chance to get to A_{n+1} except to go to A_n first and then to hit A_{n+1} at the next step: We choose a C^∞ function $0 < s \leq r_0$ on $U \setminus \{x_n : n \in \mathbb{N}\}$ such that $s = r_0$ on A_0 and

$$s(x) \leq \inf(\alpha\rho(x), \frac{1}{3} \inf_{n \in \mathbb{N}} \|x - x_n\|)$$

for every $x \in U \setminus (B_0 \cup \{x_n : n \in \mathbb{N}\})$, take

$$C_n' = B(x_n, c_n/2), \qquad C' = \bigcup_{n=1}^\infty C_n',$$

and consider the Markov kernel Q given by

$$Q(x,\cdot)=\begin{cases}\lambda_{x,s(x)} & \text{for } x\in U\setminus C',\\ \varepsilon_x & \text{for } x\in C'.\end{cases}$$

Then for every $n\in\mathbb{N}$

$$\beta_n := \inf\{Q^x[T_{A_0}<\infty] : x\in C_n\setminus C_n'\}>0.$$

Now let $n\in\mathbb{N}$ and suppose that real numbers $a_j\in]0,c_j/4]$, $j=0,\dots,n-1$, have already been chosen. Define

$$\gamma_{n-1}=\prod_{j=1}^{n-1}\left(\frac{r_{j-1}}{a_j}\right)^d$$

(in particular, $\gamma_0=1$). Then there exists $b_n\in]0,c_n/2[$ such that $B_n:=B(x_n,b_n)$ satisfies

$$\begin{aligned}R_1^{B_n} &:= \inf\{s : s\geq 0 \text{ superharmonic on } U,\ s\geq 1 \text{ on } B_n\}\\ &\leq 2^{-(n+2d)}\beta_n/\gamma_{n-1}.\end{aligned}$$

Take

$$a_{n+1}:=\frac{b_{n+1}}{2},\quad A_{n+1}:=\overline{B(x_{n+1},a_{n+1})}.$$

By Harnack's inequality there exists $c\in\mathbb{R}^+$ such that

$$h(x_2)\leq c\,h(x_1)$$

for every harmonic function $h\geq 0$ on U and we clearly may assume that b_2 and hence a_2 are chosen so small that

$$\gamma_2>(e^4\gamma_1+1)c.$$

Having obtained the sequences of balls (A_n) and (B_n) we define $r:U\longrightarrow]0,1[$ by

$$r(x):=\begin{cases}s(x) & \text{for } x\in U\setminus\bigcup_{n=1}^{\infty}B_n,\\ r_n & \text{for } x\in A_n, n\in\mathbb{N},\\ r_n/2 & \text{for } x\in B_n\setminus A_n, n\in\mathbb{N}.\end{cases}$$

Then $r\leq 2\alpha\rho$ and $\inf r(K)>0$ for every compact K in U. Using the random walk associated with the Markov kernel $P:(x,A)\longrightarrow\lambda_{x,r(x)}(A)$ we define measurable functions f_n on U, $n=0,1,2,\dots$, by

$$f_n(x)=\gamma_n P^x[T_{A_n}<\infty]$$

for every $x\in U$. The strong Markov property yields that

$$\begin{aligned}f_n(x)=\gamma_{n+1}\left(\frac{a_{n+1}}{r_n}\right)^d P^x[T_{A_n}<\infty] &= \gamma_{n+1}\int_{[T_{A_n}<\infty]}P^{X_{T_{A_n}}}[X_1\in A_{n+1}]\,dP^x\\ &= \gamma_{n+1}P^x[T_{A_n}<\infty, X_{T_{A_n}+1}\in A_{n+1}].\end{aligned}$$

Since $[T_{A_n} < \infty, X_{T_{A_n}+1} \in A_{n+1}] \subset [T_{A_{n+1}} < \infty]$, we conclude that $f_n \leq f_{n+1}$. On the other hand it can be shown that our choice of b_{n+1} and r implies that

$$P^{\cdot}[T_{A_{n+1}} < \infty] \leq \left(1 + 2^{2-n}\right) P^{\cdot}[T_{A_n} < \infty, T_{A_n+1} \in A_{n+1}] \quad \text{on } U \setminus C_{n+1}$$

and hence

$$f_{n+1} \leq \left(1 + 2^{2-n}\right) f_n \quad \text{on } U \setminus C_{n+1}.$$

Defining

$$f := \lim_{n \to \infty} f_n$$

we therefore know that $f \leq e^4 f_n$ on $U \setminus \bigcup_{j=n+1}^{\infty} C_j$. In particular, f is locally bounded and $f(x_2) > cf(x_1)$, so f is not harmonic. Moreover, $Pf_n = f_n$ on $U \setminus A_n$ for every $n \in \mathbb{N}$ and hence $Pf = f$ on U, i.e., f is (λ, r)-median. So we have a counterexample to Veech's conjecture.

In order to prove Theorem 3.2 (involving C^∞-functions) we take $\varepsilon_n \in]0, b_n/8[$ choose a function $r' \in C^\infty(U)$ such that $0 < r' \leq \alpha\rho$ and $r' = r$ on the complement of the set

$$\bigcup_{n=1}^{\infty} \{y \in B_n \setminus A_n : \|y - x_n\| < a_n + \varepsilon_n \text{ or } \|y - x_n\| > b_n - \varepsilon_n\},$$

and define

$$f'_n(x) = \gamma_n (P')^x [T_{A_n} < \infty], \quad f' = \lim_{n \to \infty} f'_n.$$

If the real numbers ε_n are sufficiently small, we obtain that

$$f'_n \leq f' \leq e^4 f'_n + 1 \quad \text{on } U \setminus \bigcup_{j=n+1}^{\infty} C_j$$

and conclude as before that f' is (λ, r')-median, but not harmonic. Since r' is a C^∞-function, we finally get that f' as well is a C^∞-function. For the details see [HN6].

Considering (3.1) and (3.2) the natural question arises if perhaps a lower bound $\alpha\rho$ for r is sufficient to guarantee harmonicity of positive (λ, r)-median functions. Recent work with N. Nadirashvili [HN8] shows that this is true indeed.

Theorem 3.3. *Let U be a domain in $\mathbb{R}^d$, $d \geq 1$, and let r be a real function on U such that $\alpha\widetilde{\rho} \leq r \leq \widetilde{\rho}$ for some real $\alpha > 0$. Then every (λ, r)-median function $f > 0$ is harmonic.*

Our (purely analytical) proof is quite different from the proofs for the (weaker) results in [Vee3], [Bax2], [CV]. Expressed in probabilistic terms, we show that hitting of cubes Q in U by the random walk associated with kernels $P : (x, A) \longmapsto \lambda_{x,r(x)}(A)$ is uniformly comparable with the hitting of Q by Brownian motion as long as $r \leq \widetilde{\rho}$ and the diameter $\delta(Q)$ of Q satisfies $\delta(Q) \geq \gamma \inf \widetilde{\rho}(Q)$ (where $\gamma > 0$ denotes an arbitrary constant fixed in advance). Using an elementary Harnack inequality for discrete Green functions this allows us to show in a direct way that every positive (λ, r)-median function is comparable to a harmonic function and hence harmonic itself.

To be more precise we first restrict our attention to Green domains U. This is possible because of (2.6). Moreover, we may assume that r is Borel measurable (see [Vee2], [HN1]).

Let Q be a cube in U such that $\overline{Q} \subset U$ and let e_Q denote the (classical) equilibrium potential for Q with respect to U. For every $x \in U$ let $e_Q^r(x)$ denote the mass obtained on Q by (transfinite) sweeping of the unit mass ε_x using the kernel $P : (x, A) \longmapsto \lambda_{x,r(x)}(A)$. If r is locally bounded away from zero (which we shall have assuming later on that $r \geq \alpha\widetilde{\rho}$) then simply

$$e_Q^r(x) = \lim_{n \to \infty} (P_Q)^n (x, Q)$$

where P_Q is the kernel defined by $P_Q(x, \cdot) = P(x, \cdot)$ for $x \in U \backslash Q$ and $P_Q(x, \cdot) = \varepsilon_x$ for $x \in Q$. The function e_Q^r is P-supermedian, i.e., it is Borel measurable, positive, and $Pe_Q^r \leq e_Q^r$. Moreover, $Pe_Q^r = e_Q^r$ on $U \setminus Q$, $e_Q^r = 1$ on Q, and e_Q^r is majorized by any l.s.c. P-supermedian function which is equal to 1 on Q. In particular,

$$e_Q^r \leq e_Q.$$

(If r is locally bounded away from zero, then e_Q^r is the smallest P-supermedian function which is equal to 1 on Q. In general, however, this is not true!)

Using once more the relation between P and solutions of Schrödinger equations $\Delta u - Vu = 0$, $V \leq \delta 1_{A'} \text{dist}\,(\cdot, B^c)^{-2}$ (in Proposition 2.11 we may take $A' = A = Q$!) we obtain the following result:

Theorem 3.4. *Let γ be a strictly positive real number. Then there exists a constant $c_0 > 0$ (depending only on γ and the dimension d) such that*

$$e_Q^r \leq e_Q \leq c_0 e_Q^r$$

for every real function r on U with $0 < r \leq \widetilde{\rho}$ and for every cube Q in U satisfying $\delta(Q) \geq \gamma \inf \widetilde{\rho}(Q)$.

Moreover, it is easily verified that there exists a constant $c_1 > 0$ such that under the assumption made in (3.4)

$$Pe_Q^r \leq e_Q^r \leq cPe_Q^r$$

(we already noted the first inequality) and hence

$$Pe_Q^r \leq e_Q \leq c_0 c_1 Pe_Q^r.$$

Let g denote the discrete Green function corresponding to P, i.e.,

$$g = \sum_{n=1}^{\infty} p_n$$

where the sequence (p_n) is defined recursively by

$$p_1(x,y) = \frac{1}{\lambda(B(x,r(x)))} 1_{B(x,r(x))}(y), \quad p_{n+1}(x,y) = \int p_n(x,z)p_1(z,y)\,\lambda(dz)$$

for $x, y \in U$. Obviously, $g\lambda = \sum_{n=1}^{\infty} P^n$. A P-supermedian function $s < \infty$ is a P-potential if $\lim_{n\to\infty} P^n s = 0$. For every measure ν on U the function g^ν defined by

$$g^\nu(x) = \int g(x,y)\,\nu(dy)$$

is a P-potential provided that $g^\nu < \infty$. If s is a P-potential then it is immediately verified that Ps is a P-potential and

$$Ps = g^{(s-Ps)\lambda}.$$

If r is locally bounded away from zero, then every P-supermedian function on U which is bounded by a (classical) continuous real potential on U is a P-potential, hence in particular

$$Pe_Q^r = g^{\sigma_Q^r} \quad \text{with } \sigma_Q^r = \left(e_Q^r - Pe_Q^r\right)\lambda$$

for every cube Q in U such that $\overline{Q} \subset U$.

Considering the integral of the functions $p_n(x,\,\cdot\,)$ with respect to $\lambda_{y,\alpha\tilde{\rho}(y)/3}$ it is an elementary exercise to show the following Harnack inequality for g:

Proposition 3.5. *Suppose that $\alpha \in]0,1]$ and $\alpha\tilde{\rho} \le r \le \tilde{\rho}$. Then there exists a constant c_2 (depending only on α and the dimension d) such that*

$$g(x,y) \le c_2 g(x,y')$$

for all $x, y, y' \in U$ such that $\|y - y'\| \le \alpha\tilde{\rho}(y)/3$.

From now on we shall fix $0 < \alpha \le 1$ and assume that $\alpha\tilde{\rho} \le r \le \tilde{\rho}$. We choose a decomposition $\mathcal{Q}_\alpha$ of U into half-open cubes Q such that

$$\frac{\alpha}{7} < \frac{\delta(Q)}{\inf \tilde{\rho}(Q)} \le \frac{\alpha}{3}$$

for every $Q \in \mathcal{Q}_\alpha$. Fixing a cube $Q_0 \in \mathcal{Q}_\alpha$ and taking $\lambda_0 = 1_{Q_0}\lambda$ we define a discrete Martin kernel k by

$$k(x,y) = \frac{g(x,y)}{\lambda_0(g(\,\cdot\,,y))}$$

for $x, y \in U$. If $Q \in \mathcal{Q}_\alpha$ and μ, ν are measures supported by $\overline{Q}$ with $\|\mu\| = \|\nu\|$ then we know by (3.5) that

$$k^\mu \le c_2^2 k^\nu$$

(where of course $k^\tau(x) := \int k(x,y)\,\tau(dy)$). Recall that $Pe_Q^r = g^{\sigma_Q^r}$ and hence

$$Pe_Q^r = k^{\tau_Q^r} \quad \text{with } \tau_Q^r = \left(\int g(x,\,\cdot\,)\,\lambda_0(dx)\right)\sigma_Q^r$$

for $Q \in \mathcal{Q}_\alpha$. Moreover,

$$e_Q = K^{\tau_Q}$$

for every $Q \in \mathcal{Q}_\alpha$ if we define

$$K(x,y) = \frac{G(x,y)}{\lambda_0(G(\cdot\,,y))}, \qquad \tau_Q = \left(\int G(x,\,\cdot\,)\,\lambda_0(dx)\right)\sigma_Q$$

where G is the Green function for U and σ_Q the equilibrium distribution for Q with respect to U, i.e., $G^{\sigma_Q} = e_Q$.

Take

$$C = c_0 c_1 c_2^4$$

where c_0, c_1 are chosen for $\gamma = \frac{\alpha}{7}$. Then the inequality $Pe_Q^r \le e_Q \le c_0 c_1 P e_Q^r$ implies that

$$C^{-1} K^{\tau_Q} \le k^\mu \le C K^{\tau_Q}$$

for every $Q \in \mathcal{Q}_\alpha$ and for every measure μ on $\overline{Q}$ with $\|\mu\| = \|\tau_Q\|$.

It is now easy to accomplish the proof of Theorem 3.3: Given a (λ, r)-median function $f \ge 0$ on U we fix a continuous real potential $q > 0$ on U and define

$$f_n = \inf(f, nq).$$

Then (Pf_n) is an increasing sequence of P-potentials and $\lim_{n\to\infty} Pf_n = Pf = f$. For every $n \in \mathbb{N}$,

$$Pf_n = k^{\mu_n} \quad \text{with } \mu_n = \left(\int g(x,\,\cdot\,)\,\lambda_0(dx)\right)(f_n - Pf_n)\lambda,$$

and defining

$$\nu_n = \sum_{Q\in\mathcal{Q}_\alpha} \frac{\mu_n(Q)}{\|\tau_Q\|}\tau_Q$$

we obtain that

$$C^{-1}K^{\nu_n} \le k^{\mu_n} \le C K^{\nu_n}.$$

In particular,

$$\|\nu_n\| = \lambda_0(K^{\nu_n}) \le C\lambda_0(k^{\mu_n}) \le C\lambda_0(f) < \infty.$$

Thus there exists a subsequence (ν_n') of (ν_n) converging weakly to a finite measure ν on the (classical) Martin compactification $U \cup \partial^M U$ of U and $\lim_{n\to\infty} K^{\nu_n'} = K^\nu$, hence

$$C^{-1}K^\nu \le f \le C K^\nu.$$

Finally, it is easily seen that $\lim_{n\to\infty} \mu_n(Q) = 0$ for every $Q \in \mathcal{Q}_\alpha$ and hence $\nu(U) = 0$. So K^ν is harmonic on U. We may now apply (2.5) to conclude that f is harmonic on U.

Let us note, however, that we can completely avoid to use the previous result on harmonically bounded functions: Knowing that for every positive (λ, r)-median function there exists a positive harmonic function h on U such that $C^{-1}h \le f \le Ch$ it is obvious that positive (λ, r)-median functions satisfy Harnack inequalities.

So we may apply Choquet theory and it is sufficient to consider extremal (λ, r)-median functions. But for any extremal (λ, r)-median function $f \geq 0$ the inequality $C^{-1}h \leq f$, h harmonic on U, implies that f is a multiple of h and hence harmonic.

Note added in proof. It is easily seen that the functions r satisfying the hypothesis of (3.3) are exactly the admissible functions r on U such that $r(y) \geq \alpha(r(x) - \|x - y\|)$ for all $x \in U$ and $y \in B(x, r(x))$.

References

[AS] Akcoglu, M. A., Sharpe, R. W., Ergodic theory and boundaries. Trans. Amer. Math. Soc. 132 (1968), 447–460.

[Bax1] Baxter, J. R., Restricted mean values and harmonic functions. Trans. Amer. Math. Soc. 167 (1972), 451–463.

[Bax2] — Harmonic functions and mass cancellation. Trans. Amer. Math. Soc. 245 (1978), 375–384.

[BH] Bliedtner, J., Hansen, W., Potential Theory — An Analytic and Probabilistic Approach to Balayage (Springer Universitext), Springer-Verlag, Berlin–Heidelberg–New York–Tokyo 1986.

[Bur] Burckel, R. B., A strong converse to Gauss's mean value theorem. Amer. Math. Monthly 87 (1980), 819–820.

[CV] Cornea, A., Veselý, J., Martin compactification for discrete potential theory. Potential Analysis 4 (1995) (to appear).

[CH] Courant, R., Hilbert, D., Methoden der mathematischen Physik II (Grundlehren Math. Wiss. 48). Springer-Verlag, Berlin–Heidelberg 1937.

[Doo] Doob, J. L., Classical Potential Theory and its Probabilistic Counterpart (Grundlehren Math. Wiss. 262), Springer-Verlag, Berlin–Heidelberg–New York 1984.

[Fel] Feller, W., Boundaries induced by nonnegative matrices. Trans. Amer. Math. Soc. 83 (1956), 19–54.

[Fen1] Fenton, P. C., Functions having the restricted mean value property. J. London Math. Soc. 14 (1976), 451–458.

[Fen2] — On sufficient conditions for harmonicity. Trans. Amer. Math. Soc. 253 (1979), 139–147.

[Fen3] — On the restricted mean value property. Proc. Amer. Math. Soc. 100 (1987), 477–481.

[Fol] Föllmer, H., Ein Littlewood-Kriterium für gleichmäßig integrable Martingale und insbesondere für Dirichlet-Lösungen. In: Elliptische Differentialgleichungen, Band II, 113–117. Akademie-Verlag, Berlin 1971.

[Gon] Gong, X., Functions with the restricted mean value property. J. Xiamen Univ. Nat. Sci. 27 (1988), 611–615.

[HN1] Hansen, W., Nadirashvili, N., A converse to the mean value theorem for harmonic functions. Acta Math. 171 (1993), 139–163.

[HN2] — Mean values and harmonic functions. Math. Ann. 297 (1993), 147–156.

[HN3] — Littlewood's one circle problem. J. London Math. Soc. 50 (1994), 349–360.

[HN4] — Liouville's theorem and the restricted mean value property. J. Math. Pures Appl. 74 (1995), 185–198.

[HN5] — On the restricted mean value property for measurable functions. In: Classical and Modern Potential Theory and Applications (NATO ASI Series, eds. K. GowriSankaran et al.), 267–271. Kluwer, Dordrecht 1994.

[HN6] — On Veech's conjecture for harmonic functions. Ann. Scuola Norm. Sup. Pisa Cl. Sci. 22 (1995), 137–153.

[HN7] — Restricted mean value property on $\mathbb{R}^d$, $d \leq 2$. Exposition. Math. 13 (1995), 93–95.

[HN8] — Restricted mean value property for positive functions. Preprint 1994.

[Hea] Heath, D., Functions possessing restricted mean value properties. Proc. Amer. Math. Soc. 41 (1973), 588–595.

[Hel] Helms, L. L., Introduction to Potential Theory. Wiley, New York 1969.

[Huc] Huckemann, F., On the 'one circle' problem for harmonic functions. J. London Math. Soc. 29 (1954), 491–497.

[Kel] Kellogg, O. D., Converses of Gauss's theorem on the arithmetic mean. Trans. Amer. Math. Soc. 36 (1934), 227–242.

[Lit] Littlewood, J. E., Some Problems in Real and Complex Analysis (Heath. Math. Monographs), Lexington, Massachusetts 1968.

[NV] Netuka, I., Veselý, J., Mean value property of harmonic functions. In: Classical and Modern Potential Theory and Applications (NATO ASI Series, eds. K. GowriSankaran et al.), 359–398. Kluwer, Dordrecht 1994.

[OS] Øksendal, B., Stroock, D. W., A characterization of harmonic measure and Markov processes. Ann. Inst. Fourier (Grenoble) 32 (4) (1982), 221–232.

[Rud] Rudin, W., Real and Complex Analysis. McGraw Hill, New York 1974.

[Vee1] Veech, W. A., A converse to Gauss' theorem. Bull. Amer. Math. Soc. 78 (1971), 444–446.

[Vee2] — A zero–one law for a class of random walks and a converse to Gauss' mean value theorem. Ann. of Math. 97 (1973), 189–216.

[Vee3] — A converse to the mean value theorem for harmonic functions. Amer. J. Math. 97 (1975), 1007–1027.

[Vol] Volterra, V., Alcune osservazioni sopra proprietà atte individuare una funzione. Atti della Reale Acad. Lincei 18 (1909), 263–266.

Approximation properties of harmonic vector fields and differential forms

*Victor P. Havin**

Abstract. This talk is a survey of recent results about uniform approximation by harmonic vector fields (i.e., curl-free and divergence-free fields) and, more generally, by harmonic differential forms. These results are parallel to some well known theorems on rational approximation in $\mathbb{C}$. However, being inspired by classical examples, they require new approaches, tools, and points of view.

1991 Mathematics Subject Classification: 58A10, 58A14, 41A30, 31C12.

1. Introductory remarks

A smooth differential form ω defined in an open subset of $\mathbb{R}^n$ is called *harmonic* if it is closed and coclosed:

$$d\omega = 0, \qquad \delta\omega = 0. \tag{1}$$

We denote by d the exterior differential, by δ the codifferential $*d*$; the star operator $*$ is defined as follows:

$$*a\, dx^{\alpha_1} \wedge \cdots \wedge dx^{\alpha_p} = \delta^{1\ldots n}_{\alpha_1\ldots\alpha_p,\beta_1\ldots\beta_{n-p}} a\, dx^{\beta_1} \wedge \cdots \wedge dx^{\beta_{n-p}},$$

(α_j), (β_j) are disjoint families of positive integers not exceeding n (see the details in [1], [2]). Operators d and δ (and harmonic differential forms) can be defined in "coordinate-free" terms and in a more general setting (on any oriented Riemannian manifold), but we will stay in $\mathbb{R}^n$.

The notion of harmonic differential form does not belong to any specific area of mathematics. It belongs to mathematics as a whole. It appears naturally and works successfully in Mathematical Physics, Geometry, Real and Complex Analysis (see, e. g., [1], [3]–[7]). The starting point of the present talk is the following observation: Complex Analysis in $\mathbb{C}$ can be defined as the theory of harmonic differential forms of degree one in $\mathbb{R}^2$.

---

* This work was supported by the Russian Fund of Fundamental Research, grant No. 94-01-01628.

2. Two examples

Consider a smooth differential form (of degree one) $\omega = P\,dx^1 + Q\,dx^2 + R\,dx^3$ in a domain $O \subset \mathbb{R}^3$. We identify it with the vector field $v := (P, Q, R)$ in O. In this situation the system (1) takes the following form:

$$\operatorname{curl} v = 0, \qquad \operatorname{div} v = 0 \tag{2}$$

or, what is the same, v is locally the gradient of a harmonic function. In other words, for any $p \in O$ there exist a neighbourhood $V_p \subset O$ and a function $H_p \in C^\infty(V_p)$ such that

$$v|V_p = \nabla H_p, \qquad \Delta H_p = 0.$$

If O is simply connected, then this representation holds globally: $v = \nabla H$ where H is a function harmonic in O.

System (2) describes many ideal stationary fields of mathematical physics (in gravitation theory, electrostatics, hydrodynamics). Its higher dimensional generalizations correspond to harmonic differential forms of degree one in $\mathbb{R}^n$ and can be described as mappings from an open $O \subset \mathbb{R}^n$ into $\mathbb{R}^n$ whose Jacobi matrix is symmetric and of zero trace. Such mappings (harmonic vector fields) play an essential role in the modern real approach to the Hardy classes (see [6]).

The following example is especially important for us. Let $\omega = P\,dx + Q\,dy$ be a smooth differential form in a domain $O \subset \mathbb{R}^2$. Then (1) becomes

$$\frac{\partial P}{\partial y} - \frac{\partial Q}{\partial x} = 0, \qquad \frac{\partial P}{\partial x} + \frac{\partial Q}{\partial y} = 0,$$

or, what is the same,

$$P - iQ \quad \text{is complex analytic.}$$

This observation is one of the most fundamental reasons for the very existence of Complex Analysis, the discipline which can be viewed as a tool of investigation of curl-free and divergence-free vector fields in $\mathbb{R}^3$ satisfying a severe restriction: they must be parallel to a coordinate plane M and constant in the direction orthogonal to M. Such fields can be interpreted as complex analytic functions of one complex variable, so that the joint power of complex multiplication and conformal mapping can be used to create what is known as Complex Analysis (in $\mathbb{C}$). But if we turn to general harmonic vector fields in $\mathbb{R}^3$ (not to mention $\mathbb{R}^n$) and drop that strange (and even ugly) restriction, then this beautiful theory does not work anymore. It rather becomes a source of many appealing questions suggested by two-dimensional experience and addressed to harmonic vector fields (and harmonic differential forms) in higher dimensions. It is very tempting to generalize uniqueness theorems (which abound in Complex Analysis) to this larger context. The same can be said about numerous "maximum principles" (in the spirit of Phragmen-Lindelöf, or normal families).

The trinity "uniqueness – normal families – approximation" plays a prominent organizing role in Complex Analysis and its applications to Harmonic Analysis. It seems that the last term of the above triad is the easiest of the three to be generalized to harmonic differential forms. This is the theme of the present talk. The first two look much harder (see Chapter 5 of Part II of [8], and [9] for further information (including striking counterexamples due to T. Wolff)).

3. Relation to elliptic systems. Approximation by harmonic gradients

a) Creators of approximation theory were motivated by the natural wish *to improve* bad function (say, an arbitrary continuous function on a segment) replacing it by a uniformly close "nice" function (say, a polynomial). Nowadays the mood of approximation theorists is (very often) quite different. They rather try to spoil a "nice" object (a solution of an elliptic system) making it imitate the behavior of something bad. Their aim is to prove that these nice objects are supple enough to reproduce the twists of an arbitrary continuous vector function on a given set. Many facts concerning approximation properties of analytic and harmonic functions have been generalized to solutions of elliptic systems (and not only in a uniform norm, but in many other classical norms); see the survey article of N. Tarkhanov [10]. But these generalizations refer to systems with "surjective symbols" only.

This means, roughly speaking, the possibility to represent *any* solution as a convolution with the fundamental solution *which has a single one-point singularity.* E.g., a generic analytic function in $\mathbb{C}$ looks like $T * \frac{1}{z}$; a generic harmonic function in $\mathbb{R}^n$, $n \geq 3$, like $T * |x|^{2-n}$, etc. But *system* (1) *does not enjoy this property.* There is a lot of its solutions which cannot be split (even approximately) into a sum of "elementary" solutions with one-point singularity (see section 4 below).

b) In [11], [12] P. Paramonov investigated the C^m-approximation by harmonic functions in $\mathbb{R}^n$. These interesting works contain important information pertaining to our theme. Namely, in the case $m = 1$ Paramonov's results are closely related to the uniform approximation by gradients of harmonic functions ("harmonic gradients"). However, harmonic gradients form a proper subclass of harmonic vector fields. This is why Paramonov's theme does not coincide with ours. A complete solution of basic approximation problems for harmonic vector fields (and harmonic differential forms) cannot be obtained in "local" terms successfully used by Paramonov (see Sect. 8). Note that in $\mathbb{R}^2$ (and for forms of degree one) our theme is *identical* with the uniform rational approximation whereas the uniform approximation by harmonic gradients in $\mathbb{R}^2$ is something different.

4. Substitutes for rational functions in $\mathbb{R}^n$, $n \geq 3$

a) Rational approximation in $\mathbb{C}$ starts with the Runge theorem. It can be stated as follows: given a compact set $K \subset \mathbb{C}$, an open set O intersecting all components of $K' := \mathbb{C} \setminus K$, a function f analytic in a neighbourhood of K, and a positive number ε, there exists a linear combination r of the Cauchy kernels with poles in O satisfying

$$|r - f| < \varepsilon \quad \text{on } K \qquad \left(r = \sum_{k=1}^{N} \lambda_k (z - a_k)^{-1}, \quad a_1, a_2, \ldots, a_N \in \mathbb{C} \right).$$

(Actually $a_k \in O$ could be *fixed* in advance, but then r includes not only the Cauchy kernels, but their powers $(z - a_k)^{-n_k}, n_k \in \mathbb{Z}_+$.) The Runge theorem may be interpreted in terms more appropriate for our theme: an arbitrary harmonic vector field defined near K almost coincides (uniformly on K) with a linear combination of "elementary" harmonic vector fields corresponding to $(z - a)^{-1}$, $a \in O$; these "elementary" fields have one-point singularities ("sources").

It is therefore our first task to understand what is a good multidimensional substitute for "elementary fields" $(z - a)^{-1}$.

b) Identifying a harmonic vector field (P, Q) in $\mathbb{R}^2$ with the analytic function $P - iQ$ we observe that the Cauchy kernel $\frac{1}{z}$ corresponds to $-\nabla \log \frac{1}{|z|}$, the gradient of the fundamental solution of the Laplace equation in $\mathbb{R}^2$. This remark suggests that the harmonic vector field $\operatorname{Coul} a := \nabla(|x - a|^{2-n})$ is a good candidate for an "elementary" harmonic vector field in $\mathbb{R}^n$, $n \geq 3$; the only singularity of $\operatorname{Coul} a$ is $\{a\}$, a singleton. Our notation alludes to the Coulomb law describing the electrostatic field created in $\mathbb{R}^3 \setminus \{a\}$ by a unit charge located at $\{a\}$. (For $n = 2$ we put $\operatorname{Coul} a := -\nabla(\log |x - a|)$.)

But these fields do not suffice for a good Runge theorem (even in $\mathbb{R}^2$). Note that in $\mathbb{R}^2$ ($\simeq \mathbb{C}$) we actually need linear combinations (with real coefficients) not only of kernels $(z - a)^{-1}$, but also of kernels $i(z - a)^{-1}$ which behave quite differently.

To be more precise, consider harmonic vector fields $\vec{v}_a = (P_a, Q_a)$, $\vec{w}_a = (Q_a, -P_a)$ corresponding to $(z - a)^{-1}$ and $i(z - a)^{-1}$ (respectively) according to Sect. 2. It is easy to see that $\vec{v}_a$ is *exact* (i.e., it is a gradient) in $\mathbb{C} \setminus \{a\}$ whereas $\vec{w}_a$ is not; at the same time $\vec{w}_a = *\vec{v}_a$, so that $\vec{w}_a$ is *coexact*.

Let $\vec{v}$ be a smooth vector field in a domain $O \subset \mathbb{R}^2$, $\vec{\gamma}$: $[0, l] \to O$, ($\vec{\gamma}(0) = \vec{\gamma}(l)$, $|\vec{\gamma}'| \equiv 1$ on $[0, l]$) be a positively oriented smooth closed path in O (a loop). Put

$$\operatorname{Circ}_{\vec{\gamma}} \vec{v} := \int_{\vec{\gamma}} P\,dx + Q\,dy = \int_0^l < \vec{v}, \vec{\tau} > ds,$$

the circulation of $\vec{v}$ along $\vec{\gamma}$ ($<\, ,\, >$ denotes the scalar product in $\mathbb{R}^2$, $\vec{\tau} := \vec{\gamma}'$, the unit tangent vector of $\vec{\gamma}$; ds is the element of the arclength on $\vec{\gamma}$).

We also need *the flux of $\vec{v}$ through $\vec{\gamma}$*

$$\mathrm{Flux}_{\vec{\gamma}}\,\vec{v} := \mathrm{Circ}_{\vec{\gamma}}(*\vec{v}) = \int_{\vec{\gamma}} -Q\,dx + P\,dy = \int_0^l <\vec{v},\vec{n}>ds,$$

$\vec{n}$ being the unit normal vector on $\vec{\gamma}$ ($\vec{n} = i\vec{\tau}$ in the "complex" notation).

Using these terms we can express the difference between $\vec{v}_a$ and $\vec{w}_a$ saying that $\mathrm{Circ}_{\vec{\gamma}}\,\vec{v}_a = 0$, $\mathrm{Flux}_{\vec{\gamma}}\,\vec{w}_a = 0$ for any loop $\vec{\gamma}$ in $\mathbb{R}^2 \setminus \{a\}$ whereas $\mathrm{Circ}_{\vec{\gamma}}\,\vec{w}_a \neq 0$, $\mathrm{Flux}_{\vec{\gamma}}\,\vec{v}_a \neq 0$ if $\vec{\gamma}$ is not homotopic to zero in $\mathbb{R}^2 \setminus \{a\}$.

Let us look at $\vec{v}_a$, $\vec{w}_a$ (i.e., $(z-a)^{-1}$, $i(z-a)^{-1}$) from $\mathbb{R}^3$.

c) Imagine an infinite string L in $\mathbb{R}^3$: $L = \{(x,y,t);\, x = y = 0,\, t \in \mathbb{R}\}$. A charge homogenously distributed on L (with the unit density) creates the Coulomb field $\vec{E}$ (in $\mathbb{R}^3 \setminus L$) with the (formal) Newtonian potential $\int_{-\infty}^{+\infty} |p - (0,0,t)|^{-1}dt$, $p = (p_1,p_2,p_3) \in \mathbb{R}^3 \setminus L$:

$$\vec{E}(p) = \int_{-\infty}^{+\infty} \nabla_p(|p-(0,0,t)|^{-1})dt = \int_{-\infty}^{+\infty} \mathrm{Coul}_{(0,0,t)}(p)dt. \tag{3}$$

This field does not depend on p_3 and is parallel to the (x,y)-plane $\mathbb{R}^2$. An easy computation shows that it can be represented in $\mathbb{R}^2$ ($\simeq \mathbb{C}$) as $\frac{1}{\bar{z}}$ (which we associate with the analytic function $\frac{1}{z}$, see Sect. 2). According to b), $\frac{1}{\bar{z}}$ is created by one-point sources $(0,0,t)$ situated on L:

$$\frac{1}{\bar{z}} \approx \sum_k \mathrm{Coul}_{(0,0,t_k)}\,\Delta t_k,$$

every element $\mathrm{Coul}_{(0,0,t)}\,dt$ of the last integral in (3) being itself a harmonic vector field (in $\mathbb{R}^3$).

But the nature and origin of $\frac{i}{\bar{z}}$ are totally different. This field stems not from electricity, but from magnetism. It is related to the field

$$\vec{B}(p) = \mathrm{curl}_p \int_{-\infty}^{\infty} \frac{\vec{k}\,dt}{|p-(0,0,t)|} := \int_{-\infty}^{\infty} \mathrm{curl}_p \frac{\vec{k}}{|p-(0,0,t)|}\,dt, \qquad (p \in \mathbb{R}^3 \setminus L)$$

where $k = (0,0,1)$ is the unit vector of the t-axis (so that $\vec{B}$ is the curl of a (formal) vectorial Newton potential). The field $\vec{B}$ is also parallel to $\mathbb{R}^2$ and does not depend on p_3. It can be interpreted as $\frac{i}{\bar{z}}$. But unlike $\vec{E}$ *elements* $\mathrm{curl}_p \frac{\vec{k}}{|p-(0,0,t)|}\,dt$ *are no longer harmonic vector fields* (integrals $\int_A^B \mathrm{curl}_p \frac{\vec{k}}{|p-(0,0,t)|}\,dt$ for finite A, B are not curl-free). We must consider *the whole of L* as "the source" of $\vec{B}$ (see also Sect. d) below). This field can be viewed as the magnetic field created by a constant circuit along L, a closed conductor (passing through infinity).

d) Now we define a Biot-Savart field of a closed loop. Consider a Lipschitz vector field $\vec{\gamma} : s \to \vec{\gamma}(s)$ $(s \in [0,l])$ in $\mathbb{R}^3$. Suppose that

$$\begin{aligned} &\text{(a)} \quad \vec{\gamma}(0) = \vec{\gamma}(l); \\ &\text{(b)} \quad |\vec{\gamma}'| = 1 \quad \text{a.e. on } [0,l]. \end{aligned} \tag{4}$$

Put

$$BS_{\vec{\gamma}}(p) := \operatorname{curl}_p \int_0^l \frac{\vec{\gamma}'(s)}{|\vec{\gamma}(s) - p|}\, ds \qquad (p \in \mathbb{R}^3 \setminus \Gamma). \tag{5}$$

We call $BS_{\vec{\gamma}}$ *the Biot-Savart field of the loop* $\vec{\gamma}$. It is obvious that $BS_{\vec{\gamma}}$ is divergence-free in $\mathbb{R}^3 \setminus \Gamma$ ($\Gamma := \vec{\gamma}([0,l])$). An easy computation shows that $BS_{\vec{\gamma}}$ is also curl-free in $\mathbb{R}^3 \setminus \Gamma$, see [13]. Thus $BS_{\vec{\gamma}}$ *is a harmonic vector field.* Note that closedness of $\vec{\gamma}$ (i.e., condition (a) in (4)) is very important. Replacing $[0,l]$ by $[a,b] \subsetneqq [0,l]$ in the right side of (5) we lose harmonicity. Even more can be said.

Put $\Gamma_1 := \vec{\gamma}([0,c])$, $\Gamma_2 := \vec{\gamma}([c,l])$ where $c \in [0,l]$ and suppose $\mathbb{R}^3 \setminus \Gamma_j$ $(j = 1,2)$ are simply connected (as is the case for a simple loop $\vec{\gamma}$). Then *it is impossible to find harmonic vector fields v_j in $\mathbb{R}^3 \setminus \Gamma_j$ $(j = 1,2)$ satisfying*

$$BS_{\vec{\gamma}} = v_1 + v_2 \quad \text{in} \quad \mathbb{R}^3 \setminus \Gamma.$$

Indeed, being curl-free, v_j must coincide with the gradient of a function in $\mathbb{R}^3 \setminus \Gamma_j$ (its circulation along a cycle linked with $\vec{\gamma}$ being non-zero, see formula (7) below).

We see that (like in c)) the loop $\vec{\gamma}$ is *as a whole* "a source" of the field $BS_{\vec{\gamma}}$ which cannot be split into a sum of harmonic fields with "lesser" singularities.

Hopefully we have convinced the reader that fields $\vec{v}$ of the form

$$\vec{v} = \sum_{k=1}^{N} \lambda_k \operatorname{Coul}_{a_k} + \sum_{k=1}^{N} \mu_k BS_{\vec{\gamma}_k} \quad (\lambda_k, \mu_k \in \mathbb{R}) \tag{6}$$

harmonic in $\mathbb{R}^3 \setminus (\{a_1, \ldots, a_N\} \cup \Gamma_1 \cup \cdots \cup \Gamma_N)$ are right substitutes for rational functions $r = \sum \frac{\lambda_k}{z - a_k}$ of the complex variable z. But the most convincing reason for this belief is the validity of the Runge theorem of Sect. 6.

Let us emphasize that neither Coulomb fields nor Biot-Savart fields suffice to approximate any field $\vec{V}$ harmonic near an arbitrary compact set $K \subset \mathbb{R}^3$. If K is, say, a circle and the circulation of $\vec{V}$ along K does not vanish, then we need BS-fields (since Coulomb fields being gradients have zero circulation). If K is a sphere, we cannot approximate $\vec{V}$ by BS-fields only, if the flux of $\vec{V}$ through K does not vanish (since BS-fields are curls, and the flux of any curl through K is zero).

e) Now we drop the "vectorial" three-dimensional language and turn to the general case of harmonic differential forms in $\mathbb{R}^n$, $n = 3, 4, \ldots$

Take $r \in \mathbb{Z}_+$, $0 < r < n$, and consider a cycle Γ in $\mathbb{R}^n$ of dimension $n - r$ (i.e., a finite singular chain with real coefficients and with zero boundary, $\partial\Gamma = 0$; see, e.g., [1]).

We need a special differential form $\omega_{\Gamma,r}$ whose coefficients are "Newton potentials of Γ":

$$\omega_{\Gamma,r}(x) := \frac{1}{c_n} \sum_{\alpha_1<\cdots<\alpha_r} \delta^{1,\ldots,n}_{\alpha,\beta} \int_\Gamma \frac{d\xi^{\beta_1} \wedge \cdots \wedge d\xi^{\beta_{n-r}}}{|\xi - x|^{n-2}} dx^{\alpha_1} \wedge \cdots \wedge dx^{\alpha_r}$$

(c_n denotes the $(n-1)$-dimensional measure of the unit sphere $\mathbb{S}^{n-1}$). Thus $\omega_{\Gamma,r}$ is defined outside of Γ, and $\deg \omega_{\Gamma,r} = r = n - \dim \Gamma$; the second lower index in the left side alludes to the degree of ω).

If $r = n$, we define $\omega_{\Gamma,n}$ for a one-point cycle $\Gamma = \{\xi\}$ ($\dim \Gamma = 0$) as follows:

$$\omega_{\Gamma,n}(x) := \frac{1}{c_n} \frac{dx^1 \wedge \cdots \wedge dx^n}{|x - \xi|^{n-2}},$$

(the definition for a general zero-dimensional chain Γ is clear).

Now our main definitions come. For a cycle $\Gamma = \Gamma_{n-r-1}$ of dimension $n-r-1$ we put

$$BS_{\Gamma_{n-r-1}} := \delta\omega_{\Gamma,r+1}$$

and call BS_Γ *the Biot-Savart form corresponding to* Γ (BS-*form* for short). If $\dim \Gamma = r - 1$, then we put

$$\mathrm{Coul}_{\Gamma_{r-1}} := *\delta\omega_{\Gamma,n-r+1}$$

and call this differential form *the Coulomb form corresponding to* Γ.

Note that

(1) $\deg BS_{\Gamma_{n-r-1}} = \deg \mathrm{Coul}_{\Gamma_{r-1}} = r \quad (r = 1, \ldots, n-1)$ (a harmonic differential form of degree 0 or n is locally constant, and we won't consider such forms).

(2) BS- and Coulomb forms are harmonic (off the generating cycle Γ).

(3) Coulomb forms are exact (i.e., are differentials).

(4) BS-forms are coexact (i.e., are codifferentials), and

$$\int_\gamma BS_\Gamma = (-1)^{r+1} v(\gamma, \Gamma) \tag{7}$$

where $\gamma = \gamma_r$, $\Gamma = \Gamma_{n-r-1}$ are cycles with no point in common (of dimension r and $n-r-1$ respectively), and $v(\gamma,\Gamma)$ denotes their linking number. The proof of (7) is implicit in [1] (a particular case $n = 3$, $\dim\gamma = \dim\Gamma = 1$ is in [14]). Thus BS_Γ is (in general) not exact (its periods along cycles γ linked with Γ do not vanish).

(5) $*\,\mathrm{Coul}_{\Gamma_{r-1}} = (-1)^{r(n-1)} BS_{\Gamma_{r-1}}$. Thus Coulomb forms are not coexact (periods of $*\,\mathrm{Coul}_\Gamma$ coincide with periods of $\pm BS_\Gamma$).

5. On the Alexander–Pontryagin duality

Denote by $H_l(\mathcal{M})$ the vector space of l-dimensional singular homologies (with real coefficients) of a manifold $\mathcal{M}$; A^* will stand for the dual of a vector space A.

We need the following particular case of the Alexander–Pontryagin duality theorem (see [14], [15]). *Suppose K is a compact smooth n-dimensional manifold of $\mathbb{R}^n$ (with border). Then $H_r(K)$ is finite dimensional, and*

$$(H_r(K))^* \simeq H_{n-r-1}(\mathbb{R}^n \setminus K) \quad (r = 1, \dots, n-1).$$

Here $\simeq$ stands for the natural isomorphism induced by the pairing $(\sigma, \Delta) \to v(\sigma, \Delta)$ of r-dimensional cycles in K and $(n-r-1)$-dimensional cycles in $\mathbb{R}^n \setminus K$. The following fact is an immediate corollary.

Given homologically independent cycles $\Delta_1, \dots, \Delta_N$ in $\mathbb{R}^n \setminus K$ forming a base in $H_{n-r-1}(\mathbb{R}^n \setminus K)$ there exist homologically independent cycles $\sigma_1, \dots, \sigma_N$ in K forming a base of $H_r(K)$ such that $v(\sigma_j, \Delta_k) = \delta_k^j$ $(j, k = 1, \dots, N)$.

These theorems can be easily deduced from basic facts on De Rham's theory of currents combined with the Cauchy formula for harmonic differential forms (see [16]).

6. The Runge theorem

a) A harmonic differential form $\omega = \mathrm{Coul}_{\Gamma_{r-1}} + BS_{\Gamma_{n-r-1}}$ will be called *elementary* (corresponding to cycles Γ_r and Γ_{n-r-1}). The harmonic vector field (6) may be identified with the elementary 1-form ω corresponding to the zero-cycle $\Gamma_0 := \sum_{k=1}^N \lambda_k a_k$ and 1-cycle $\Gamma_1 := \sum_{k=1}^N \mu_k \vec{\gamma}_k$ (we interpret loops $\vec{\gamma}_k$ as 1-cycles).

Let E be a compact set in $\mathbb{R}^n$, $E' := \mathbb{R}^n \setminus E$, Ω be a harmonic differential form defined in a neighbourhood of E. The Runge theorem asserts that Ω can be approximated uniformly on E by elementary forms. Its precise statement (to be formulated in b) below) contains also some information concerning the structure and location of the cycles Γ_{r-1} and Γ_{n-r-1} (i.e., of singularities of the elementary form ω uniformly close to Ω on E).

The data necessary for the statement are as follows: a harmonic differential form Ω of degree r in a neighbourhood $\mathcal{V}$ of E and a compact smooth n-dimensional manifold K (with border) contained in $\mathcal{V}$ and containing E in its interior. We also need two systems of cycles in $K' := \mathbb{R}^n \setminus K$:

$\left(\bar{\Delta}_{r-1}^s\right)_{s=1}^{M_1}$, a homologically independent system of $(r-1)$-cycles forming a base of $(r-1)$-homologies of K'; (8)

$\left(\bar{\bar{\Delta}}_{n-r-1}^s\right)_{s=1}^{M_2}$, a homologically independent system of $(n-r-1)$-cycles forming a base of $(n-r-1)$-homologies of K'. (9)

For a differential r-form $\beta = \sum_I \beta_I \, dx^I$, $I = (i_1, \dots, i_r)$, $1 \le i_1 < \dots < i_r \le n$, defined (at least) on a set $A \subset \mathbb{R}^n$ we put $|\beta|_A := \sup_{I, x \in A} |\beta_I(x)|$.

Now we are ready to state the main result of this section.

b) *Given any open set $O \subset K'$ intersecting every component of K', any two systems of cycles* (8) *and* (9), *and an $\varepsilon > 0$ there exist two families of real numbers $(\bar{\lambda}_s)_{s=1}^{M_1}$, $(\bar{\bar{\lambda}}_s)_{s=1}^{M_2}$, and cycles Γ_{r-1}, Γ_{n-r-1} in O such that*

$$|\Omega - \mathrm{Coul}_{\bar{\gamma}} - BS_{\bar{\bar{\gamma}}} - \mathrm{Coul}_{\Gamma_{r-1}} - BS_{\Gamma_{n-r-1}}|_E < \varepsilon,$$

where

$$\bar{\gamma} := \sum_{s=1}^{M_1} \bar{\lambda}_s \bar{\Delta}_{r-1}, \quad \bar{\bar{\gamma}} := \sum_{s=1}^{M_2} \bar{\bar{\lambda}}_s \bar{\bar{\Delta}}_{n-r-1}.$$

Moreover, each of Γ_{r-1}, Γ_{n-r-1} can be chosen to be zero.

c) Thus we are free to prescribe the basic cycles $\bar{\Delta}^s$ and $\bar{\bar{\Delta}}^s$ (see (8), (9)); after they have been chosen we are free to choose any open $O \subset K'$ (provided it meets every component of K') to contain Γ_{r-1}, Γ_{n-r-1}. If Ω is exact (or coexact), then $\bar{\bar{\gamma}}$, Γ_{n-r-1} (resp. $\bar{\gamma}$, Γ_{r-1}) are not needed (i.e., we may take $\bar{\bar{\lambda}}_s \equiv 0$, $\Gamma_{n-r-1} = 0$, resp. $\bar{\lambda}_s \equiv 0$, $\Gamma_{r-1} = 0$). It is interesting to note that we do not need Γ_{r-1} even for exact Ω's and can manage with $BS_{\Gamma_{n-r-1}}$ only (which is not exact in general); for coexact Ω's we can manage with $\mathrm{Coul}_{\Gamma_{r-1}}$ only (which is in general not coexact). This phenomenon is well known in electrodynamics. E.g., the Biot-Savart magnetic field of a circuit running along a loop $\vec{\gamma}$ almost coincides with the Coulomb field of a dipole at points sufficiently far from $\vec{\gamma}$ (see, e.g., [17]); this observation gives rise to the notion of a "magnetic dipole".

d) There are two proofs of the Runge theorem. The first one uses the Hahn-Banach theorem and thus is not constructive (see [18], [19]). Another one is constructive and gives us an effective procedure to obtain the approximating elementary forms (see [16]). Both proofs use the Alexander–Pontryagin duality. First we find bases $(\bar{\sigma}_i)_{i=1}^{M_1}$, $(\bar{\bar{\sigma}}_i)_{i=1}^{M_2}$ of $H_r(K)$ and $H_{n-r}(K)$ (respectively) dual to the bases (8) and (9); this is possible according to Sect. 5. Then we define number families $(\bar{\lambda}_s)$, $(\bar{\bar{\lambda}}_s)$ such that all periods and coperiods of $\Omega' := \Omega - \mathrm{Coul}_{\bar{\gamma}} - BS_{\bar{\bar{\gamma}}}$ along $\bar{\sigma}_i$ and, resp. $\bar{\bar{\sigma}}_i$ vanish, so that Ω' is exact and coexact in K. The construction (of the second proof) can be roughly described as follows. We express Ω' in the interior of K (and thus on E) by a Cauchy integral (over ∂K) and then transform this integral so as to get the representation

$$\Omega' = \int_{\mathcal{S}_{r-1}(\partial K)} \mathrm{Coul}_\gamma \, d\bar{\mu}(\gamma) + \int_{\mathcal{S}_{n-r-1}(\partial K)} BS_\gamma \, d\bar{\bar{\mu}}(\gamma) \tag{10}$$

where $\bar{\mu}$ and $\bar{\bar{\mu}}$ are scalar measures on $\mathcal{S}_{r-1}$ (resp. $\mathcal{S}_{n-r-1}$), and $\mathcal{S}_j(A)$ stands for the set of all j-cycles in A. From this moment on we can essentially repeat the usual Runge procedure, replacing integrals in (10) by their integral sums and moving "poles" (i.e., cycles generating Coulomb and BS-forms) to O.

7. Good and invisible sets

a) We call a compact set $K \subset \mathbb{R}^n$ *r-good* if for any differential form Ω of degree r continuous on K and any $\varepsilon > 0$ there is an r-form ω harmonic in a neighbourhood of K such that $|\Omega - \omega|_K < \varepsilon$. (By the Runge theorem we may assume ω to be elementary.) A compact set K is 1-good iff any continuous vector field on K is the uniform limit (on K) of a sequence of vector fields harmonic in a neighbourhood of K. A class of r-good sets is described in [18], [19]; for $n = 2, r = 1$ this description coincides with the Hartogs-Rosenthal theorem ($C(K) = R(K)$ if $\text{Area}(K) = 0$; $R(K)$ denotes the closure in $C(K)$ of all rational function with poles off K). Here we only discuss 1-good sets (which we call just *good*).

b) The harmonicity of a form does not depend on the choice of an orthonormal base and we make use of this fact. We consider a Euclidean space $\mathcal{E}$ of dimension $n \geq 2$ and harmonic differential forms therein. We call a set $A \subset \mathcal{E}$ *invisible* if there is an orthonormal base of $\mathcal{E}$ such that the area of the orthogonal projection of A into any two-dimensional coordinate plane is zero.

Note that there are invisible sets which become "visible" after a change of coordinates (whence the appearance of a base in the definition). Any set of zero two-dimensional Hausdorff measure (area) is invisible, but not vice versa. There exist invisible sets of infinite area.

c) *Any invisible compact set is good.* This theorem is a particular case of a result from [18], [19]. In general (i.e., for r-good sets) the notion of invisibility has to be changed and adjusted to r.

As in 6d), there are two proofs of this theorem. The first is again based on the Hahn–Banach theorem ([18], [19]), the second is constructive [16] and uses a transformation of the Cauchy-Green formula for smooth differential forms.

The articles [18], [19] contain some other variants of these results (e.g., on approximation by exact or coexact harmonic differential forms).

8. On Bishop's locality principle

We call a compact set $K \subset \mathbb{R}^n$ *locally good* if for any $p \in K$ there is a neighbourhood $\mathcal{V}(p)$ of p such that the closure of $\mathcal{V}(p) \cap K$ is good. The Bishop's principle asserts that *if $n = 2$, then any locally good set is good* (see [20], [21]). Vitushkin's characterization of good sets in terms of the analytic capacity makes this principle explicitly explaining which local property of a set is responsible for the goodness. But in $\mathbb{R}^{2+k}$, $k = 1, 2, \ldots$, Bishop's principle fails as is shown by a counterexample in [22]. This counterexample is the graph $K := \{(x, y, z) \in \mathbb{R}^3 : z = F(x, y), x^2 + y^2 \leq 1\}$ of a continuous function F vanishing on $bK := \{(x, y, 0) : x^2 + y^2 = 1\}$ and such that no pair of points p, $q \in K$ can be joined by a rectifiable arc in K unless p, $q \in bK$. Clearly, K is not good (since the circulation of any harmonic vector field along bK is zero), but it is locally good. The proof is based on a theorem concerning solenoids (i.e., compactly supported divergence-free vector measures in $\mathbb{R}^3$; the divergence is understood in the sense of distributions). Namely, it is proved in [22] that the support of any non-zero solenoid contains a non-degenerate rectifiable arc (a much stronger result is in [23] where a Choquet decomposition of solenoids is given).

A description of good sets in $\mathbb{R}^n$, $n \geq 3$, in the spirit of Vitushkin's theorem is an open problem. But it is clear that it needs new terms, not just a capacity measuring point sets, but rather a new quantitative characterization of families of curves.

Acknowledgements. I am grateful to the Organizing Committee for a possibility to participate in ICPT – 94 and especially for a generous help in the technical preparation of this text. I am also indebted to Ye. V. Malinnikova for many helpful discussions and critical remarks.

References

[1] De Rham, G., Variétés Différentiables. Hermann, Paris 1955.

[2] Schwartz, L., Variedades Analiticas Complejas. Equaciones Diferenciales Parciales. Univ. de Colombia, Bogota 1956.

[3] Weyl, H., Method of orthogonal projection in potential theory. Duke Math. J. 7 (1940), 417–444.

[4] Hodge, W. V. D., The theory and applications of harmonic integrals. Cambridge Univ. Press, Cambridge 1952.

[5] Yano, K., Bochner, S., Curvature and Betti Numbers. Princeton Univ. Press, Princeton 1953.

[6] Fefferman, Ch., Stein, E. M., H^p spaces of several variables. Acta Math. 7, 3–4 (1972), 137–193.

[7] Springer, G., Introduction to Riemann Surfaces. Addison-Wesley, Reading 1957.

[8] Havin, V., Jöricke, B., The Uncertainty Principle in Harmonic Analysis. Springer-Verlag, Berlin 1994.

[9] Aleksandrov, A. B., Kargaev, P. P., Hardy classes of functions harmonic in a half-space. Algebra i Analiz 5, 2 (1993), 1–73 (Russian).

[10] Tarkhanov, N. N., Approximation on compacta by solutions of systems with a surjective symbol. Uspekhi Mat. Nauk 48, 5 (1993), 107–146 (Russian).

[11] Paramonov, P. V., On harmonic approximations in the C^1-norm. Matem. Sb. 181, 10 (1990), 1341–1365 (Russian).

[12] Paramonov, P. V., C^m-approximations by harmonic polynomials on compact subsets in $\mathbb{R}^n$. Matem. Sb. 184, 2 (1993), 105–128 (Russian).

[13] Dubrovin, B. A., Novikov, S. P., Fomenko, A. T., Modern Geometry, Methods and Applications. Nauka, Moscow 1986 (Russian).

[14] Pontryagin, L. S., Selected Works, vol. 1. Nauka, Moscow 1988 (Russian).

[15] Husemoller, D., Fibre Bundles. McGraw Hill, New York 1966.

[16] Malinnikova, Ye. V., Havin, V. P., Approximation by harmonic differential forms. A constructive approach. (To appear.)

[17] Purcell, E. M., Electricity and Magnetism, Berkeley Physics Course, vol. 2. McGraw Hill, New York 1965.

[18] Havin, V., Presa Sagué, A., Approximation Properties of Harmonic Vector Fields and Differential Forms, Methods of Approximation Theory in Complex Analysis and Math. Physics. In: Lecture Notes in Math. 1550, 149–156, Springer-Verlag, Berlin 1993.

[19] — Uniform Approximation by Harmonic Differential Forms. Algebra i Analiz (to appear).

[20] Gamelin, T. M., Uniform Algebras. Prentice Hall, Englewood Cliffs 1969.

[21] Gaier, D., Lectures on Complex Approximation. Birkhäuser, Basel 1987.

[22] Havin, V. P., Smirnov, S. K., Approximation and Extension for Some Classes of Vector Fields. Preprint IHES/M/92/67 (1992), Institut des Hautes Etudes Scientifiques.

[23] Smirnov, S. K., Decomposition of Solenoidal Vector Charges into Elementary Solenoids and the Structure of Normal One-Dimensional Currents. Algebra i Analiz 5, 4 (1993), 206–238 (Russian).

Nonlinear potentials and quasilinear PDE's

Jan Malý

Abstract. Quasilinear elliptic equations of type

$$-\operatorname{div} A(x,u,\nabla u) + B(x,u,\nabla u) = \mu$$

are studied. Estimates of subsolutions and supersolutions in terms of integrals

$$\int_0^R \left(\frac{\operatorname{cap}_p(B(z,r)\setminus\Omega,r)}{r^{n-p}}\right)^{1/(p-1)} \frac{dr}{r}$$

(Wiener type conditions) and

$$\int_0^R \left(\frac{\mu(B(x,r))}{r^{n-p}}\right)^{1/(p-1)} \frac{dr}{r}$$

are discussed.

1991 Mathematics Subject Classification: 35J67, 35J70, 35J65.

1. Introduction

The aim of this lecture is to discuss results in which criteria of Wiener type and comparison with nonlinear potentials are used to describe pointwise behavior of solutions and supersolutions of quasilinear elliptic equations of type

$$-\operatorname{div} \mathbf{A}(x,u,\nabla u) + \mathbf{B}(x,u,\nabla u) = 0 \tag{1.1}$$

(precise conditions depending on a growth exponent $p \in (1,\infty)$ will be given later). We consider problems with a measure on right hand side, Dirichlet problem and obstacle problems.

The model equation for (1.1) is

$$-\operatorname{div}\left(|\nabla u|^{p-2}\nabla u\right) + \lambda|u|^{p-2}u = 0 \tag{1.2}$$

* Research supported by the grant No. 201/93/2174 of Czech Grant Agency and by the grant No. 354 of Charles University.

with $\lambda \in \mathbb{R}$. We define the p-Laplacian Δ_p as the differential operator

$$\Delta_p = \operatorname{div}\left(|\nabla u|^{p-2}\nabla u\right).$$

If $\lambda \geq 0$, then (1.2) is the Euler equation of the functional

$$\frac{1}{p}\int \left(|\nabla u|^p + \lambda |u|^p \, dx\right). \tag{1.3}$$

This indicates that this equation is of great importance for the theory of Sobolev spaces $W^{1,p}$.

If $p = 2$, then the equation (1.2) is linear; if μ is a finite nonnegative Radon measure, we obtain a generalized entire solution of

$$-\Delta u + \lambda u = \mu$$

by convolving μ with the fundamental solution u_0 of

$$-\Delta u_0 + \lambda u_0 = \delta_0$$

(δ_0 denotes the Dirac measure at the origin). No such integral representation is available when $p \neq 2$.

Let $\Omega \subset \mathbb{R}^n$ be a bounded open set and μ be a nonnegative Radon measure on $\mathbb{R}^n$. Let us consider the equation

$$-\operatorname{div}\mathbf{A}(x, u, \nabla u) + \mathbf{B}(x, u, \nabla u) = \mu. \tag{1.4}$$

Under some additional hypotheses, the equation has a unique weak solution $u \in W_0^{1,p}(\Omega)$ provided that μ belongs to the dual space $\left(W_0^{1,p}(\Omega)\right)^*$. For example, the weak solution of

$$-\operatorname{div}\left(|\nabla u|^{p-2}\nabla u\right) + \lambda |u|^{p-2}u = \mu$$

with $\lambda \geq 0$ is obtained as the minimizer of

$$\frac{1}{p}\int_\Omega \left(|\nabla u|^p + \lambda |u|^p \, dx\right) - \int_\Omega u \, d\mu. \tag{1.5}$$

However, the Dirac measures and many others do not belong to $\left(W_0^{1,p}(\Omega)\right)^*$. Then, the only way how to obtain "solutions" is to solve approximate problems, e.g., with regularized μ, and then pass to a limit. The existence results for very weak solutions (also for more general equations with right hand part a measure) were established by Boccardo and Gallouët [7], [8], see also Rakotoson [59], [60], Boccardo and Murat [9] and Kilpeläinen and Malý [34]. The variational approach is also available; the concept of (very) weak minimizers was introduced by Iwaniec and Sbordone [31]. The uniqueness problem for measures of infinite p-energy remains so far open. Some partial cases are solved: discrete measures (Kichenassamy [32]) and absolutely continuous measures (Bénilan, Boccardo, Gallouët, Gariepy, Pierre and Vazquez [6], Lions and Murat [44]).

Notice that by means of convolution with the fundamental solution we obtain a unique "canonical" generalized solution of

$$-\Delta u + \lambda u = \mu,$$

no matter whether the 2-energy of μ is finite or not. The absence of this approach is deeply felt in the nonlinear potential theory.

We will work with the integrals

$$\mathbf{w}_p(x, E) = \int_0^1 \Big(\frac{\operatorname{cap}_p(E \cap B(x,r), r)}{r^{n-p}}\Big)^{1/(p-1)} \frac{dr}{r} \tag{1.6}$$

and

$$\mathbf{W}_p^\mu(x) = \int_0^1 \Big(\frac{\mu(B(x,r))}{r^{n-p}}\Big)^{1/(p-1)} \frac{dr}{r}. \tag{1.7}$$

The Wiener-type condition says that a point $z \in \partial\Omega$ is regular for the Dirichlet problem if and only if $\mathbf{w}_p(z, \mathbb{R}^n \setminus \Omega) = \infty$. In terms of Wiener integrals we also find conditions for continuity of solutions of obstacle problems. The function $\mathbf{W}_p^\mu$ is a kind of nonlinear potential of the measure μ. We present pointwise estimates for solutions of (1.4) in terms of $\mathbf{W}_p^\mu$.

Almost all the material of the lecture will be contained with detailed proofs in the forthcoming monograph [50] by Malý and Ziemer.

In what follows, Ω is an open subset of $\mathbb{R}^n$ and p is an exponent in $(1, n]$. We write C, C' etc. for various constants (they may differ from line to line). We denote by $B(z,r)$ the open ball in $\mathbb{R}^n$ with center at z and radius r. If $B = B(z,r)$, then $2B$ means the ball $B(z, 2r)$. We denote by $C_c^\infty(\Omega)$ the set of all infinitely differentiable functions with a compact support in Ω. The norm in the Lebesgue space $L^p(\Omega)$, resp. in the Sobolev space $W^{1,p}(\Omega)$ is denoted by $\|...\|_p$, resp. $\|...\|_{1,p}$. We use $|E|$ for the Lebesgue measure of the set E.

A value k is said to be an approximate limit of a function f at $z \in \mathbb{R}^n$, which we denote by

$$k = \operatorname*{app-lim}_{x \to z} f(x),$$

if there is a set $M \subset \mathbb{R}^n$ such that

$$\lim_{x \to z,\, x \in M} f(x) = k$$

and

$$\lim_{r \to 0} \frac{|B(z,r) \cap M|}{|B(z,r)|} = 1.$$

Similarly we introduce app-lim sup and app-lim inf.

Elements of Sobolev spaces are "functions" determined up to a set of measure zero. If we want to emphasize that u is "pointwise defined" we speak about *representatives*.

2. Capacity and thinness

We define the *p-capacity* of a set $E \subset \mathbb{R}^n$ by $\operatorname{cap}_p E = \operatorname{cap}_p(E, 1)$, where

$$\operatorname{cap}_p(E,r) = \inf\Big\{ \int_{\mathbb{R}^n} (|\nabla u|^p + r^{-p}|u|^p)\,dx : \\ u \in W^{1,p}(\mathbb{R}^n),\ \liminf_{x\to z} u(x) \ge 1 \text{ for all } z \in E \Big\}.$$

A set $U \subset \mathbb{R}^n$ is said to be *p-quasiopen* if for each $\varepsilon > 0$ there is an open set $G \subset \mathbb{R}^n$ such that $\operatorname{cap}_p G < \varepsilon$ and $U \cup G$ is open. Similarly, a function u is said to be *p-quasicontinuous* on Ω if for each $\varepsilon > 0$ there is an open set $G \subset \mathbb{R}^n$ such that $\operatorname{cap}_p G < \varepsilon$ and $u|\Omega \setminus G$ is continuous.

We use the abbreviation p-q.e. (*p-quasi everywhere*) for the phrase "except a set of p-capacity zero". We say that a set $E \subset \mathbb{R}^n$ is *p-thin* at a point $z \in \mathbb{R}^n$ if the *Wiener integral*

$$\mathbf{w}_p(x,E) = \int_0^1 \Big(\frac{\operatorname{cap}_p(E \cap B(x,r), r)}{r^{n-p}}\Big)^{1/(p-1)} \frac{dr}{r}$$

converges. The *p-fine closure* adds to each set E the set of all points where E is not p-thin. This introduces the *p-fine topology*.

We recall two useful classical results.

Proposition 2.1 (Federer and Ziemer [14], Maz'ya and Khavin [48], Meyers [52], Frehse [17]). *Every $u \in W^{1,p}_{\mathrm{loc}}(\Omega)$ has a p-quasicontinuous representative.*

Proposition 2.2 (Fuglede [19], Brelot [10], Hedberg and Wolff [25]). *A function u on Ω is p-quasicontinuous if and only if it is p-finely continuous p-q.e.*

We say that a function u is *p-finely represented* if u is defined everywhere in Ω, p-quasicontinuous and

$$p\text{-fine-}\liminf_{x\to z} u(x) \le u(z) \le p\text{-fine-}\limsup_{x\to z} u(x)$$

for each $z \in \Omega$. Each element of $W^{1,p}_{\mathrm{loc}}(\Omega)$ can be p-finely represented: if u is a p quasicontinuous representative, then

$$y \mapsto p\text{-fine-}\limsup_{x\to y} u(x) \tag{2.1}$$

is an example of a p-fine representative. Let us mention a pleasant property of a p-finely represented function u: if k is a constant and $u \le k$ a.e. in an open set $G \subset \Omega$, then $u \le k$ everywhere in G; the same holds for $\ge$ and $=$.

3. Nonlinear potentials

Denote by G_1 the Bessel kernel defined by means of its Fourier transform

$$\hat{G}_1(\xi) = (1 + |\xi|^2)^{-1/2}.$$

The following result shows the equivalence of norms.

Proposition 3.1 (Calderón [11]). *Suppose that* $1 < p < \infty$.

(a) *The operator*

$$f \mapsto G_1 * f$$

is one-to-one and maps $L^p(\mathbb{R}^n)$ *onto* $W^{1,p}(\mathbb{R}^n)$.

(b) *There are constants* C, $C' > 0$ (*depending on* p *and* n) *such that*

$$C'\|f\|_p \leq \|G_1 * f\|_{1,p} \leq C\|f\|_p$$

for each $f \in L^p(\mathbb{R}^n)$.

Now we may introduce a variational problem similar to (1.5) on $\Omega = \mathbb{R}^n$; we rewrite (1.5) as

$$\frac{1}{p}\|u\|_{1,p}^p - \int u\, d\mu,$$

but instead of the $1,p$-norm $\|G_1 * f\|_{1,p}$ we use the equivalent norm $\|f\|_p$. We arrive at minimization of

$$\mathcal{F}(u) = \frac{1}{p}\int_{\mathbb{R}^n} |f|^p\, dx - \int_{\mathbb{R}^n} G_1 * f\, d\mu$$

with $u = G_1 * f$. This problem is solved by

$$f = (G_1 * \mu)^{1/(p-1)}, \qquad u = G_1 * (G_1 * \mu)^{1/(p-1)}.$$

The function

$$\mathbf{V}_p^\mu = G_1 * (G_1 * \mu)^{1/(p-1)}$$

is called the $\mathbf{V}_p$-potential of μ. These potentials provide a natural nonlinear counterpart to Newtonian potentials. Their theory was developed e.g. by Maz'ya and Khavin [48], and Meyers [51]. Notice that for $p = 2$ the potential $\mathbf{V}_p^\mu$ satisfies the equation $-\Delta u + u = (2\pi)^{n/2}\mu$. This fact easily follows using the Fourier transform.

Now, let us define the $\mathbf{W}_p$-potential

$$\mathbf{W}_p^\mu(x; R) = \int_0^R \left(\frac{\mu B(x,r)}{r^{n-p}}\right)^{1/(p-1)} \frac{dr}{r}.$$

These potentials were introduced by Adams and Meyers [3], Hedberg [24] and Hedberg and Wolff [25]. For more information about $\mathbf{V}_p$ and $\mathbf{W}_p$ potentials we refer to the recent monograph by Adams and Hedberg [2].

The *p-energy* of μ is defined by the formula

$$\mathcal{E}_p(\mu) = \int_{\mathbb{R}^n} (G_1 * \mu)^{p/(p-1)}\, dx.$$

It is easily seen that

$$\mathcal{E}_p(\mu) = \int_{\mathbb{R}^n} \mathbf{V}_p^\mu \, d\mu.$$

It is important to observe that there are constant C, C' such that for any nonnegative measure

$$C' \, \|\mu\|^{p/(p-1)}_{(W^{1,p}(\mathbb{R}^n))^*} \le \mathcal{E}_p(\mu) \le C \, \|\mu\|^{p/(p-1)}_{(W^{1,p}(\mathbb{R}^n))^*}$$

(see e.g. [68], Proof of Theorem 4.7.5).

Proposition 3.2 (Hedberg and Wolff [25]). *There are constants C and C' such that for any nonnegative Radon measure μ on $\mathbb{R}^n$ we have*

$$C' \int_{\mathbb{R}^n} \mathbf{W}_p^\mu \, d\mu \le \int_{\mathbb{R}^n} \mathbf{V}_p^\mu \, d\mu \le C \int_{\mathbb{R}^n} \mathbf{W}_p^\mu \, d\mu.$$

The preceding result justifies to understand also the $\mathbf{W}_p$-potentials as natural nonlinear potentials associated with $W^{1,p}$.

Theorem 3.3 (Hedberg and Wolff [25]). *A set $E \subset \mathbb{R}^n$ is p-thin at a point $z \in \overline{E} \setminus E$ if and only if there is a Radon measure μ on $\mathbb{R}^n$ such that*

$$\mathbf{W}_p^\mu(z) < \liminf_{x \to z,\, x \in E} \mathbf{W}_p^\mu(x).$$

Remark 3.4. We see that all $\mathbf{W}_p$ potentials are lower semicontinuous (this is clear) and p-finely continuous (this is an immediate consequence of the preceding theorem). We will see soon that an analogous result holds for supersolutions of (1.1).

4. Quasilinear PDE's

Let $\Omega \subset \mathbb{R}^n$ be an open set. We are interested in problems related to the equation (1.1). Throughout Sections 4–7 we suppose that the functions $\mathbf{A}: \mathbb{R}^n \times \mathbb{R} \times \mathbb{R}^n \to \mathbb{R}^n$ and $\mathbf{B}: \mathbb{R}^n \times \mathbb{R} \times \mathbb{R}^n \to \mathbb{R}$ are Borel measurable and satisfy the following structure conditions:

$$\begin{aligned} |\mathbf{A}(x,\zeta,\xi)| &\le a_1|\xi|^{p-1} + a_2|\zeta|^{p-1} + a_3, \\ |\mathbf{B}(x,\zeta,\xi)| &\le b_1|\xi|^{p-1} + b_2|\zeta|^{p-1} + b_3 + b_0|\xi|^p, \\ \mathbf{A}(x,\zeta,\xi)\cdot\xi &\ge c_1|\xi|^p - c_2|\zeta|^p - c_3, \quad c_1 > 0. \end{aligned} \tag{4.1}$$

For simplicity we assume that a_i, b_i, c_i are nonnegative constants. The methods work also for a_i, b_i and c_i in appropriate Lebesgue classes (see [22]). The model example $\mathbf{A}(x,\zeta,\xi) = |\xi|^{p-2}\xi$, $\mathbf{B}(x,\zeta,\xi) = \lambda|\zeta|^{p-2}\zeta$ leads to (1.2).

We say that u is a *solution* (frequently termed a "weak solution") of (1.1) in Ω if $u \in W^{1,p}_{\text{loc}}(\Omega)$ and

$$\int_\Omega \big(\mathbf{A}(x,u,\nabla u)\cdot\nabla\varphi + \mathbf{B}(x,u,\nabla u)\,\varphi\big)\,dx \geq 0 \tag{4.2}$$

for all "test functions" $\varphi \in C_c^\infty(\Omega)$. A function $u \in W^{1,p}_{\text{loc}}(\Omega)$ is said to be a *supersolution* of (1.1) in Ω if (4.2) holds for all *nonnegative* test functions $\varphi \in C_c^\infty(\Omega)$. Similarly we define *subsolutions* using *nonpositive* test functions.

The solutions of (1.1) have locally Hölder continuous representatives. This regularity result was established for $p = 2$ by De Giorgi [13], Nash [57] and Moser [55]. A Harnack inequality was proved by Moser in [56]. The regularity problems were systematically studied by Ladyzhenskaya and Ural'tseva in the sixties, e.g. [39], mainly in the book [40]. Serrin [63] proved the Harnack inequality for $p \neq 2$. Supersolutions and subsolutions are not continuous in general. Harnack type inequalities for subsolutions and supersolutions were established by Trudinger [65]. Michael and Ziemer [53] showed that supersolutions are lower semicontinuous and p-finely continuous.

In the terminology of Section 7, the following lemma studies regularity of supersolutions with a constant upper obstacle. Although the formulation seems to be new, all the work required for the proof is done by Gariepy and Ziemer in [22]. The history of obstacle problems will be mentioned in Section 7.

Lemma 4.1. *Let $u \in W^{1,p}(\Omega)$ be bounded and ℓ be a constant. Suppose that u is p-finely represented, $u \leq \ell$ and*

$$\int_\Omega \big(\mathbf{A}(x,u,\nabla u)\cdot\nabla\varphi + \mathbf{B}(x,u,\nabla u)\,\varphi\big)\,dx \geq 0 \tag{4.3}$$

for all "test functions" $\varphi \in W_0^{1,p}(\Omega)$ such that $0 \leq \varphi \leq \ell - u$ on Ω. Then u is lower semicontinuous and p-finely continuous.

Proof. Choose $k < \ell$. Let $\varphi_0 \in W_0^{1,p}(\Omega)$ be vanishing on $\{u \geq k\}$, nonnegative and bounded. Then there is $\varepsilon > 0$ such that $\varepsilon\varphi_0 \leq \ell - u$ and thus we may use $\varphi = \varphi_0$ as a test function for (4.3). Consider $B(z,r) \subset \Omega$ with $0 < r < 1$. Let $\eta \in C_c^\infty(\mathbb{R}^n)$ be a nonnegative function supported in $B(z,r)$ such that $|\nabla\eta| < 5/r$. Denote $b = b_0/c_1$, $u_k = (k-u)^+$. We write

$$v = \sup_{B(z,r)} u_k - u_k + h(r),$$

where $h(r)$ is a positive function on $(0,\infty)$ to be specified later. Let $\beta > 0$. Then

$$\varphi_1 = \Big(v^{-\beta} - \big(\sup_{B(z,r)} u_k + h(r)\big)^{-\beta}\Big)\,e^{bu_k}\,\eta^p$$

and

$$\varphi_2 = u_k\, e^{bu_k}\, \eta^p$$

are legitimate test functions. Now we may proceed exactly as in the proof of Theorem 2.1 of [22]: If h is computed from the structure as in [22] (in our case of constant a_i's, b_i's and c_i's and $r < 1$ it is possible to find h in the form

$$h(r) = C_h\, r$$

for a suitable constant C_h), we obtain

$$\begin{aligned} r^{p-n} &\int_{B(z,r)\cap\{u<k\}} |\nabla(\eta v)|^p\, dx \\ &\le C\big(\sup_{B(z,r)} u_k + h(r)\big)\Big(\inf_{B(z,r/2)} u - \inf_{B(z,r)} u + h(r)\Big)^{p-1}. \end{aligned} \tag{4.4}$$

(Notice that this part of the proof represents a lot of work; modifications of proofs of Theorems 4.3 and 4.4 appear as parts of the calculation. One cannot use directly the assertions of Theorems 4.3 and 4.4 because of the upper obstacle complication.)

The lower semicontinuity follows from (4.4) as in Remark 4.2 of [67], see also [22], Theorem 2.2. The fine continuity is also a consequence of (4.4), as in Corollary 4.4 of [53], see also [22], Theorem 2.4. □

Now, we need to study supersolutions of (1.1) on $B(z,r)$ with a lower bound $u \ge m \ge 0$. We simplify the situation assuming that $r < r_0$ and constants may depend on r_0. On the other hand, we shall observe how the estimates depend on m. Let us consider a positive function $h(r)$ and a point z. Set

$$\begin{aligned} \tilde{\mathbf{A}}(y,\zeta,\xi) &= \mathbf{A}(z+ry, r\zeta + m - h(r), \xi), \\ \tilde{\mathbf{B}}(y,\zeta,\xi) &= r\,\mathbf{B}(z+ry, r\zeta + m - h(r), \xi). \end{aligned} \tag{4.5}$$

If $u \ge m$ is a supersolution of (1.1) in $B(z,r)$, then

$$w\colon y \mapsto \frac{u(z+ry) - m + h(r)}{r}$$

is a nonnegative supersolution of

$$-\operatorname{div} \tilde{\mathbf{A}}(y,w,\nabla w) + \tilde{\mathbf{B}}(y,w,\nabla w) = 0$$

on $B(0,1)$. It is easily verified that $\tilde{\mathbf{A}}$, $\tilde{\mathbf{B}}$ satisfy the structure inequalities

$$\begin{aligned} |\tilde{\mathbf{A}}(x,\zeta,\xi)| &\le \tilde{a}_1|\xi|^{p-1} + \tilde{a}_2|\zeta|^{p-1}, \\ |\tilde{\mathbf{B}}(x,\zeta,\xi)| &\le \tilde{b}_1|\xi|^{p-1} + \tilde{b}_2|\zeta|^{p-1} + \tilde{b}_0|\xi|^p, \\ \tilde{\mathbf{A}}(x,\zeta,\xi)\cdot\xi &\ge \tilde{c}_1|\xi|^p - \tilde{c}_2|\zeta|^p, \quad c_1 > 0, \end{aligned}$$

provided

$$h(r) = C_h\,(r + mr) \tag{4.6}$$

with C_h large enough. (This is our case if a_i, b_i and c_i are constants; available generalizations with variable a_i, b_i and c_i would change the form of h). Notice that the constant C_h depends only on n, p, r_0 and the structure. In what follows, we keep the symbol h for the function introduced by (4.6).

The following estimate is a starting point of the iteration technique by Moser [55], [55], Serrin [63] and Trudinger [65], and a universally important tool.

Theorem 4.2. *Let u be a supersolution of* (1.1) *in $B(z,r)$, $0 < r \le r_0$. Suppose that $u \ge m \ge 0$ on $B(z,r)$. Let $\eta \in C^\infty(\mathbb{R}^n)$ be a nonnegative function supported in $B(z,r)$ such that $|\nabla\eta| < 5/r$. Denote $b = b_0/c_1$. If $b_0 \ne 0$ suppose that u is bounded. Let h be given by* (4.6) *and $v = u - m + h(r)$. Let $\beta > 0$. Then*

$$\beta^p \int_{B(z,r)} v^{-1-\beta} |\nabla u|^p \eta^p \, dx \le C\,(1+\beta)^p \int_{B(z,r)} v^{p-1-\beta}(r^{-p}\eta^p + |\nabla\eta|^p)\, dx$$

where C depends only on r_0, n, p, the bound of b_0u and the structure.

Proof. We use the transformation (4.5) which reduces the situation to the case when $a_3 = b_3 = c_3 = 0$, $r = 1$ and $m = h = 0$. Now we refer to the formula (1.20) of [65]. □

Theorem 4.3 (Trudinger [65]). *Let u be a supersolution of* (1.1) *in $B(z,r)$, $0 < r \le r_0$, and $\gamma \in \big(0, n(p-1)/(n-p)\big)$. Suppose that u is p-finely represented and $u \ge m \ge 0$ on $B(z,r)$. Let $v = u - m + h(r)$, where h is given by* (4.6). *If $b_0 \ne 0$ suppose that u is bounded. Then*

$$\Big(r^{-n} \int_{B(z,r/2)} v^\gamma \, dx\Big)^{1/\gamma} \le C \inf_{B(z,r/2)} v,$$

where the constant C depends only on r_0, n, p, ℓ, γ, the bound of b_0u and the structure.

Proof. As in the preceding proof we use the transformation (4.5) which reduces the situation to the case when $a_3 = b_3 = c_3 = 0$, $r = 1$ and $m = h = 0$. Now we use Theorem 1.2 of [65]. □

Lemma 4.4 (Gariepy and Ziemer [22]). *Let u be a supersolution of* (1.1) *in $B(z,r)$, $0 < r \le r_0$. Suppose that u is p-finely represented and $u \ge m \ge 0$ on $B(z,r)$. If $b_0 \ne 0$ suppose that u is bounded. Let $h(r)$ be given by* (4.6) *and $v = u - m + h(r)$. Let $\eta \in C^\infty(\mathbb{R}^n)$ be a nonnegative function supported in $B(z,r/2)$ such that $|\nabla\eta| < 5/r$. Then*

$$\int_{B(z,r/2)} |\nabla u|^{p-1} \eta^{p-1} (\eta + |\nabla\eta|)\, dx \le C\, r^{n-p} \big(\inf_{B(z,r/2)} v\big)^{p-1}, \tag{4.7}$$

where the constant C depends only on r_0, n, p, ℓ, γ, the bound of b_0u and the structure.

Proof. The estimate can be found in the proof of Theorem 2.1 in [22], its simplified version is Lemma 3.4 in [34]. We sketch the proof here to see how the dependence on r and m looks like. Find $\beta \in (0, p-1)$ such that $(n-p)\beta < n$. Using Hölder's inequality, Theorem 4.2, Theorem 4.3 and the estimate $r^{-p}|\eta|^p + |\nabla\eta|^p \le Cr^{-p}$ we obtain

$$\begin{aligned}
&\int_{B(z,r/2)} |\nabla u|^{p-1}\eta^{p-1}(\eta + |\nabla\eta|)\,dx \\
&\le C\Big(\int_{B(z,r/2)} |\nabla u|^{p-1} v^{-\beta-1}\eta^p\,dx\Big)^{(p-1)/p} \\
&\quad \times \Big(\int_{B(z,r/2)} v^{(\beta+1)(p-1)}(\eta^p + |\nabla\eta|^p)\,dx\Big)^{1/p} \\
&\le C\Big(\int_{B(z,r/2)} v^{p-\beta-1}(r^{-p}\eta^p + |\nabla\eta|^p)\,dx\Big)^{(p-1)/p} \\
&\quad \times \Big(\int_{B(z,r/2)} v^{(\beta+1)(p-1)}(\eta^p + |\nabla\eta|^p)\,dx\Big)^{1/p} \\
&\le C\Big(r^{n-p}(\inf_{B(z,r/2)} v)^{p-\beta-1}\Big)^{(p-1)/p} \\
&\quad \times \Big(r^{n-p}(\inf_{B(z,r/2)} v)^{(\beta+1)(p-1)}\Big)^{1/p} \\
&\le C r^{n-p}(\inf_{B(z,r/2)} v)^{p-1}.
\end{aligned}$$

□

We are going to present results which study behavior of solutions of

$$-\operatorname{div} \mathbf{A}(x, u, \nabla u) + \mathbf{B}(x, u, \nabla u) = \mu \tag{4.8}$$

near z using knowledge of the function $r \to \mu B(z, r)$. Related, but different results are due to Rakotoson and Ziemer [61] and Lieberman [41]. We will compare solutions of (4.8) with $\mathbf{W}_p$-potentials. The link between nonlinear potential theories and PDE's also motivates the paper [1] by Adams.

The following result was proved by Kilpeläinen and Malý [34] for the equation (8.2). Now we present its version holding for the general structure (4.1).

Theorem 4.5. *Let $u \in W^{1,p}(\Omega)$ be nonnegative and bounded and $\mu \in (W_0^{1,p}(\Omega))'$ be a nonnegative Radon measure on Ω. Suppose that u is p-finely represented and*

$$\int_\Omega \big(\mathbf{A}(x, u, \nabla u)\cdot\nabla\varphi + \mathbf{B}(x, u, \nabla u)\,\varphi\big)\,dx \ge \int_\Omega \varphi\,d\mu \tag{4.9}$$

for all nonnegative "test functions" $\varphi \in W_0^{1,p}(\Omega)$. Let $B(z, R) \subset \Omega$, $0 < R \le r_0$. Let h be given by (4.6). Then there is a constant C depending only on n, p, r_0, the bound of $b_0\,u$ and the structure such that

(a)

$$\mu(B(z, r/4)) \le Cr^{n-p}\Big(\inf_{B(z,r/2)} u - \inf_{B(z,r)} u + h(r)\Big)^{p-1}$$

for all $r \in (0, R)$,

(b)

$$\mathbf{W}_p^\mu(z, R/4) \le C\big(u(z) - \inf_{B(z,R)} u + R + Ru(z)\big)\,.$$

Proof. (a) We denote $B = B(z,r)$, $b = b_0/c_1$, $\ell = \inf_{B(z,r/2)} u$, $m = \inf_{B(z,r)} u$ and $v = u - m + h(r)$. Let $\eta \in C^\infty(\mathbb{R}^n)$ be a nonnegative function supported in $B(z,r/2)$ such that $|\nabla\eta| < 5/r$ and $\eta \ge 1$ on $B(z,r/4)$. We use the test function

$$\varphi = e^{-bu}\eta^p$$

and obtain

$$\begin{aligned} -\int_B \mathbf{A}(x,u,\nabla u)\cdot b\,e^{-bu}\,\nabla u\,\eta^p\,dx + p\int_B \mathbf{A}(x,u,\nabla u)\cdot e^{-bu}\eta^{p-1}\nabla\eta\,dx \\ + \int_B \mathbf{B}(x,u,\nabla u)e^{-bu}\eta^p\,dx \ge \int_B e^{-bu}\eta^p\,d\mu. \end{aligned}$$

Using the structure (4.1) and obvious estimates it follows

$$\begin{aligned} \int_{B(z,r/4)} e^{-bu}\,d\mu &\le b\int_B (c_2|u|^p + c_3)e^{-bu}\,\eta^p\,dx \\ &+ \int_B (a_1+b_1)|\nabla u|^{p-1}(\eta + p|\nabla\eta|)\,e^{-bu}\eta^{p-1}\,dx \\ &+ \int_B \big((a_2+b_2)|u|^{p-1} + (a_3+b_3)\big)(\eta + p|\nabla\eta|)\,e^{-bu}\eta^{p-1}\,dx. \end{aligned}$$

After some simplification we get

$$\begin{aligned} \mu(B(z,r/4)) &\le C\int_{B(z,r)} (|\nabla u|^{p-1} + |u|^{p-1} + 1)(\eta + p|\nabla\eta|)\eta^{p-1}\,dx \\ &\le C\Big(\int_{B(z,r/2)} |\nabla u|^{p-1}(\eta + p|\nabla\eta|)\eta^{p-1}\,dx \\ &\quad + r^{-1}\int_{B(z,r/2)} (m^{p-1} + v^{p-1} + 1)\Big). \end{aligned} \tag{4.10}$$

By Theorem 4.3 we have

$$r^{-1}\int_{B(z,r/2)} v^{p-1}\,dx \le Cr^{n-1}(\inf_{B(z,r/2)} v)^{p-1}. \tag{4.11}$$

Using (4.10), (4.11) and (4.7) (Lemma 4.4) we obtain

$$\begin{aligned} \mu(B(z,r/4)) &\le Cr^{n-p}\Big(\inf_{B(z,r/2)} v + r + mr\Big)^{p-1} \\ &\le Cr^{n-p}\Big(\inf_{B(z,r/2)} u - m + h(r)\Big)^{p-1} \end{aligned}$$

as required.

(b) Denote

$$m(r) = \inf_{B(z,r)} u.$$

By the part (a) we have

$$\left(r^{p-n}\mu(B(z,r/4))\right)^{1/(p-1)} \le C\bigl(m(r/2) - m(r) + r + rm(r)\bigr).$$

Multiplying both sides by $1/r$ and integrating from R' to R, $0 < R' < R/2$, we get

$$\begin{aligned}
&\int_{R'}^{R} \Bigl(\frac{\mu(B(z,r/4))}{r^{n-p}}\Bigr)^{1/(p-1)} \frac{dr}{r} \\
&\le C \int_{R'}^{R} \Bigl(m(r/2) - m(r) + r + rm(r)\Bigr) \frac{dr}{r} \\
&= C\Bigl(\int_{R'/2}^{R'} \frac{m(r)\,dr}{r} - \int_{R/2}^{R} \frac{m(r)\,dr}{r} + \int_{R'}^{R} (r + rm(r)) \frac{dr}{r}\Bigr) \\
&\le C(u(z) - m(R) + R + Ru(z)).
\end{aligned}$$

□

The following result has been previously established by Kilpeläinen and Malý [36] for the equation (8.1), see Section 8. The estimate is in some sense opposite to the previous lemma.

Theorem 4.6 (Malý [49], Malý and Ziemer [50]). *Let $u \in W^{1,p}(\Omega)$ be a nonnegative function and $\mu \in (W_0^{1,p}(\Omega))^*$ be a nonnegative Radon measure on Ω. Suppose that*

$$\int_\Omega (\mathbf{A}(x,u,\nabla u) \cdot \nabla\varphi + \mathbf{B}(x,u,\nabla u)\,\varphi)\,dx \le \int_\Omega \varphi\,d\mu \tag{4.12}$$

for all nonnegative "test functions" $\varphi \in W_0^{1,p}(\Omega)$. Let $z \in \Omega$ and

$$p - 1 < \gamma < \frac{n(p-1)}{n-p+1}.$$

Then there are $r_0 > 0$ and $C > 0$ such that

$$\operatorname*{app-lim\,sup}_{x \to z} u(x) \le C(R^{-n} \int_{B(z,R)} u^\gamma\,dx)^{1/\gamma} + C\mathbf{W}_p^\mu(z,R)$$

for all $R \in (0, r_0)$ such that $B(z,2R) \subset \Omega$; r_0 and C depend only on n, p, γ and the structure.

5. Boundary behavior of solutions

In this section we assume for simplicity that Ω is bounded. Let $f \in W^{1,p}(\Omega)$ and u be a solution of (1.1). We say (temporarily) that u solves the Dirichlet problem with boundary data f if $u - f \in W_0^{1,p}(\Omega)$. A point $z \in \partial\Omega$ is said to be regular for the equation (1.1) if

$$\lim_{x\to z} u(x) = f(z)$$

whenever $u \in C(\Omega)$ is a solution of the Dirichlet problem with boundary data $f \in W^{1,p}(\Omega)\cap C(\overline{\Omega})$. Wiener [66] showed that z is regular for the Laplace equation if and only if the classical Wiener criterion is satisfied. This more or less says that z is regular for the Laplace equation if and only if the Wiener integral $\mathbf{w}_2(z,\mathbb{R}^n\setminus\Omega)$ diverges. Littman, Stampacchia, Weinberger [45] proved that the same condition applies to linear elliptic divergence form equations with discontinuous bounded measurable coefficients. If $p\neq 2$, we say that the Wiener condition is satisfied at z if $\mathbf{w}_p(z,\mathbb{R}^n\setminus\Omega)$ diverges, i.e., if $\mathbb{R}^n\setminus\Omega$ is not p-thin at Ω. Maz'ya [47] established the sufficiency of the Wiener criterion under a simplified structure. Gariepy and Ziemer [22] proved the sufficiency in the general case of equation (1.1). We present here the whole statement of their result, which applies also to the case of arbitrary continuous data. (Notice that not every continuous function on $\partial\Omega$ is a trace of a function in $W^{1,p}(\Omega)$.)

Let u be a function on Ω, $z\in\partial\Omega$ and $\ell\in\mathbb{R}$. Following [22], we say that $u\leq\ell$ (weakly) at z if for all $k>\ell$ there is a $r>0$ such that for each $\psi\in C_c^\infty(B(z,r))$ we have $(u-k)^+\,\psi\in W_0^{1,p}(\Omega\cap B(z,r))$. We will omit the word "weakly" in our terminology. Analogously we introduce the condition $u\geq\ell$ at z. We say that $u=\ell$ at z if $u\leq\ell$ at z and $u\geq\ell$ at z simultaneously. Now, we are ready to improve the definition of the Dirichlet problem. Let $f\in C(\partial\Omega)$ and u be a solution of (1.1). We say that u solves the Dirichlet problem with boundary data f if $u=f(z)$ holds at each $z\in\partial\Omega$.

Theorem 5.1 (Gariepy and Ziemer [22]). *Let $u\in W^{1,p}_{\mathrm{loc}}(\Omega)$ be a p-finely represented supersolution of* (1.1). *Suppose that $u\geq\ell$ at a point $z\in\partial\Omega$. If $\mathbb{R}^n\setminus\Omega$ is not p-thin at z, then*

$$\liminf_{x\to z,\ x\in\Omega} u(x)\geq\ell.$$

Proof. Choose a $k<\ell$ and find a $r>0$ such that for each $\psi\in C_c^\infty(B(z,2r))$ we have $(k-u)^+\,\psi\in W_0^{1,p}(\Omega\cap B(z,2r))$. Set

$$u_k=\begin{cases}\min(u,k) & \text{on } \Omega,\\ k & \text{on } B(z,r)\setminus\Omega.\end{cases}$$

Let $\varphi\in W_0^{1,p}(B(z,r))$. Suppose that φ satisfies $0\leq\varphi\leq k-u_k$. If we prove that

$$\int_{B(z,r)}\left(\mathbf{A}(x,u_k,\nabla u_k)\cdot\nabla\varphi+\mathbf{B}(x,u_k,\nabla u_k)\,\varphi\right)dx\geq 0, \tag{5.1}$$

Lemma 4.1 shows that u_k is lower semicontinuous and p-finely continuous on $B(z,r)$. Since u_k approximate u_ℓ uniformly, it follows that u_ℓ is lower semicontinuous and p-finely continuous at z. Now we use the Wiener condition: since $\mathbb{R}^n\setminus\Omega$ is not p-thin at z and u_ℓ is p-finely continuous at z, it is easily seen that $u_\ell(z)=\ell$.

Using the lower semicontinuity we obtain

$$\ell = u_\ell(z) \le \liminf_{x\to z} u_\ell(x) \le \liminf_{x\to z,\, x\in\Omega} u(x).$$

It remains to prove (5.1). But we postpone this step to a more complicate situation of fine domains, see the proof of Theorem 6.4. □

The next theorem describes behavior near an irregular point. It is closely related to Theorem 9.1 (Kilpeläinen and Malý [36]).

Theorem 5.2 (Malý [49], Malý and Ziemer [50]). *Suppose that $\mathbb{R}^n \setminus \Omega$ is p-thin at $z \in \partial\Omega$. Let u be a bounded nonnegative subsolution of* (1.1) *on Ω. Let*

$$p - 1 < \gamma < \frac{n(p-1)}{n-p+1}.$$

Then there are $r_0 > 0$ and $C > 0$ such that

$$\begin{aligned}\operatorname{app-lim\,sup}_{x\to z} u(x) &\le C\Big(R^{-n}\int_{\Omega\cap B(z,R)} u^\gamma\,dx\Big)^{1/\gamma}\\ &\quad + C(1+\|u\|_\infty)\int_0^R \Big(\frac{\operatorname{cap}_p(B(z,r)\setminus\Omega, r)}{r^{n-p}}\Big)^{1/(p-1)}\frac{dr}{r}\end{aligned}$$

for all $R \in (0, R_0)$; r_0 and C depend only on n, p, γ and the structure.

6. Fine domains

It seems sometimes to be useful to investigate Dirichlet problem on p-finely open sets. Indeed, if $f \in C_c^\infty(\mathbb{R}^n)$ and U is a bounded open set, then the solution of the Dirichlet problem for the p-Laplace equation and boundary data f is obtained by minimizing $\int_U |\nabla u|^p\,dx$ over all functions $u \in W^{1,p}(\mathbb{R}^n)$ with $u - f \in W_0^{1,p}(U)$. Bagby's theorem [4] says that $u - f \in W_0^{1,p}(U)$ if and only if $u \in W^{1,p}(\mathbb{R}^n)$ and the p-quasicontinuous representatives of u are equal to f p-q.e. on the complement of U. In this setting the minimization problem has a sense for all (not only open) bounded sets U. It can be shown that the solution is the same for the set U as for its p-fine interior. Hence it is enough to investigate the Dirichlet problem only on p-finely open sets.

Our exposition is based on the approach by Kilpeläinen and Malý [35], where the fine Sobolev spaces are introduced and the p-fine Dirichlet problem is studied for the equation (8.2). The idea of Sobolev spaces on finely open sets appears also in earlier works [15] and [16] by Feyel and De La Pradelle and [21] by Fuglede. The fine Dirichlet problem for linear equations was introduced and solved by Fuglede; we refer to his monograph [20].

In this section (and nowhere else) Ω will be a bounded *p-finely* open set. A function $g\colon \Omega \to \mathbb{R}^n$ is called a p-fine weak derivative of a function $u\colon \Omega \to \mathbb{R}$, if there are p-finely open subsets V_j of Ω covering Ω up to a set of zero p-capacity, open sets $U_j \supset V_j$ and $u_j \in W^{1,p}(U_j)$ such that $u_j = u$ a.e. on V_j and $\nabla u_j = g$ a.e. on V_j for each j . The p-fine weak derivative of u is denoted by ∇u; if it exists, it is unique up to a set of zero measure. The $W^{1,p}(\Omega)$-norm of u is defined as usually by

$$\|u\|_{1,p} = \|u\|_p + \|\nabla u\|_p$$

where $\|...\|_p$ denotes the norm in $L^p(\Omega)$. We define $W^{1,p}(\Omega)$ as the set of all functions on Ω whose p-fine weak derivative exists and $W^{1,p}(\Omega)$-norm is finite.

The definition of the space $W_0^{1,p}(\Omega)$ is easier: we say that a function u belongs to $W_0^{1,p}(\Omega)$ if u extended by zero outside Ω belongs to $W_0^{1,p}(U)$ for each open set $U \supset \Omega$.

There are many ways how to characterize these Sobolev spaces, and for open sets the new definitions are equivalent to the ordinary ones. For these properties we refer to [35].

Now, a function u is said to be a p-fine solution of (1.1) in Ω if for each point $y \in \Omega$ there is a p-finely open set V such that $y \in V \subset \Omega$, $u \in W^{1,p}(V)$ and

$$\int_V (\mathbf{A}(x,u,\nabla u)\cdot\nabla\psi + \mathbf{B}(x,u,\nabla u)\psi)\,dx \geq 0 \tag{6.1}$$

holds for each $\psi \in W_0^{1,p}(V)$. Analogously we define p-fine supersolutions (subsolutions, resp.) by means of nonnegative (nonpositive, resp.) test functions.

A routine approximation argument shows that any "ordinary" solution (supersolution) on an open set is a p-fine solution (supersolution).

We define the properties $u \leq \ell$ at z, $u \geq \ell$ at z and $u = \ell$ at z in the same way as in Section 5: e.g., we say that $u \leq \ell$ at z if for all $k > \ell$ there is a $r > 0$ such that for each $\psi \in C_c^\infty(B(z,r))$ we have $(u-k)^+\,\psi \in W_0^{1,p}(\Omega \cap B(z,r))$.

Now we are ready to introduce the fine Dirichlet problem: Let $f \in \mathcal{C}(\partial\Omega)$. A function u on Ω is said to solve the p-fine Dirichlet problem for the equation (1.1) and boundary data f if u is a p-fine solution of (1.1) in Ω and $u = f(z)$ at each point $z \in \overline{\Omega} \setminus \Omega$.

Example. The problem of minimization of $\int_\Omega |\nabla u|^p\,dx$ among all functions u with $u - f \in W_0^{1,p}(\Omega)$ has an unique solution H_f for any $f \in C^1(\mathbb{R}^n)$. Now, given $f_0 \in \mathcal{C}(\partial\Omega)$, there are C^1 functions f_k, $k = 1, 2, \ldots$, such that $f_k \to f$ uniformly on $\partial\Omega$. Due to comparison principles, there is a uniform limit u_0 of the sequence H_{f_k}. Standard arguments show that u_0 is a (unique) solution of the p-fine Dirichlet problem for the p-Laplace equation and boundary data f_0.

Let us recall some useful properties of p-finely open sets and p-fine Sobolev spaces.

Proposition 6.1. *Suppose that u is a function on Ω and the family of all p-finely open sets V with $u \in W^{1,p}(V)$ covers Ω. Then there is a p-quasicontinuous function* *w which equals u a.e. Such a function w is unique up to a set of zero p-capacity and admits a p-fine representative.*

Proof. This is clear from the proof of Theorem 2.2 in [35]. □

Proposition 6.2 ([35], Theorem 2.10). *Let* $u \in W^{1,p}(\Omega)$. *Then* $u \in W_0^{1,p}(\Omega)$ *if and only if there is a p-quasicontinuous function* v *on* $\mathbb{R}^n$ *such that* $v = u$ *a.e. in* Ω *and* $v = 0$ *on* $\mathbb{R}^n \setminus \Omega$.

Proposition 6.3. *Let* φ *be a nonnegative function from* $W_0^{1,p}(\Omega)$. *Let* V_ι *be a family of finely open subsets of* Ω *covering* Ω *up to a set of zero p-capacity. Then there are nonnegative p-quasicontinuous functions* $\psi_{i,j}$, $\varphi_j \in W_0^{1,p}(\Omega)$ $(i = 1, \dots, j$, $j = 1, \dots, \infty)$ *on* Ω *such that* $\varphi_j = \sum_{i \le j} \psi_{i,j}$, *each* $\psi_{i,j}$ *has support in* V_ι *for some* ι *and* $\varphi_j \to \varphi$ *in* $W_0^{1,p}(\Omega)$.

Proof. This follows from Theorem 2.10 and Lemma 2.4 of [35]. □

The following result is a fine counterpart of Theorem 5.1 and implies a regularity result for solutions of p-fine Dirichlet problems.

Theorem 6.4. *Let* $u \in W_{\mathrm{loc}}^{1,p}(\Omega)$ *be a p-finely represented p-fine supersolution of* (1.1). *Suppose that* $u \ge \ell$ *at a point* $z \in \overline{\Omega} \setminus \Omega$. *If* $\mathbb{R}^n \setminus \Omega$ *is not p-thin at* z, *then*

$$\liminf_{x \to z\ x \in \Omega} u(x) \ge \ell.$$

Proof. Choose a $k < \ell$ and find a $r > 0$ such that for each $\psi \in C_c^\infty(B(z, 2r))$ we have $(k - u)^+ \psi \in W_0^{1,p}(\Omega \cap B(z, 2r))$. Set

$$u_k = \begin{cases} \min(u, k) & \text{on } \Omega, \\ k & \text{on } B(z, r) \setminus \Omega. \end{cases}$$

Let $\varphi \in W_0^{1,p}(B(z, r))$. Suppose that φ satisfies $0 \le \varphi \le k - u_k$. Let us assume that φ is p-finely represented (Proposition 6.1) and u_k is represented by the formula (2.1). Then $\Omega' := B(z, r) \cap \{u_k < k\}$ is a p-finely open subset of $B(z, r) \cap \Omega$. Since $0 \le \varphi \le k - u_k$, $\varphi \in W_0^{1,p}(\Omega')$. (To prove the last assertion, we can use Proposition 6.2.) We need to verify (5.1). With each $y \in \Omega \cap B(z, r)$ we associate a p-finely open set $V = V_y$ such that $y \in V \subset \Omega$, $u \in W^{1,p}(V)$ and (6.1) holds for each nonnegative $\psi \in W_0^{1,p}(V)$. We use Proposition 6.3 to find a sequence φ_j of nonnegative functions from $W_0^{1,p}(\Omega')$ such that $\varphi_j \to \varphi$ in $W_0^{1,p}(\Omega')$ and

$$\int_{\Omega'} \big(\mathbf{A}(x, u_k, \nabla u_k) \cdot \nabla \varphi_j + \mathbf{B}(x, u_k, \nabla u_k)\, \varphi_j\big)\, dx$$
$$= \int_{\Omega'} \big(\mathbf{A}(x, u, \nabla u) \cdot \nabla \varphi_j + \mathbf{B}(x, u, \nabla u)\, \varphi_j\big)\, dx \ge 0$$

(we have employed that $u = u_k$ on Ω'). Since $u_k \in W^{1,p}(\Omega')$, using the structure assumptions (4.1) we can pass to the limit to obtain

$$\int_{B(z,r)} \big(\mathbf{A}(x,u_k,\nabla u_k)\cdot\nabla\varphi + \mathbf{B}(x,u_k,\nabla u_k)\,\varphi\big)\,dx$$
$$= \int_{\Omega'} \big(\mathbf{A}(x,u,\nabla u)\cdot\nabla\varphi + \mathbf{B}(x,u,\nabla u)\,\varphi\big)\,dx \geq 0.$$

(Notice that $\varphi = 0$ on $B(z,r)\setminus\Omega'$.) This verifies (5.1). The rest of the proof follows the lines of the proof of Theorem 5.1. □

7. Obstacle problems

We introduce the *double obstacle problem* with a lower obstacle t and an upper obstacle s. As special cases we obtain single obstacle problems: lower obstacle problem (given lower obstacle t is supplemented by $s = +\infty$) and upper obstacle problem ($t = -\infty$). Let s and t be arbitrary functions on Ω. We denote

$$\mathcal{K} = \{u \in W^{1,p}_{\mathrm{loc}}(\Omega)\colon\ u \text{ is } p\text{-finely represented, } t \leq u \leq s \text{ q.e.}\}.$$

By a *test function* we understand a function $\varphi \in W^{1,p}_0(\Omega)$ with a compact support in Ω such that $u + \varphi \in \mathcal{K}$. The collection of all such test functions will be denoted by $\mathcal{T}$. If a function u on Ω satisfies

$$\begin{cases} u \in \mathcal{K}, \\ \displaystyle\int_{\Omega} \big(\mathbf{A}(x,u,\nabla u)\cdot\nabla\varphi + \mathbf{B}(x,u,\nabla u)\,\varphi\big)\,dx \geq 0 \text{ for all } \varphi \in \mathcal{T}, \end{cases} \tag{7.1}$$

then u is called a *solution* of (7.1). Let us underline that it is not necessary to repeat the assumption of p-fine representative at each occurance because the definition of the solution u of (7.1) includes already that $u \in \mathcal{K}$ and thus u is p-finely represented.

One of first regularity results for the obstacle problems stated that if the obstacle is locally Hölder continuous, then the solution of the lower obstacle problem is locally Hölder continuous as well. This was done for linear equations by Beirao da Veiga [5]. There is a large class of problems with discontinuous obstacles whose solutions remain continuous. An appropriate condition uses a Wiener type integral (details will be explained below). For linear equations it was used by Frehse and Mosco [18], a necessity part is due to Mosco [54]. Michael and Ziemer [53] proved sufficiency of the Wiener condition for the equation (1.1). Heinonen and Kilpeläinen [28] studied lower obstacle problems for the equation (8.2). They obtained necessity for $p > n - 1$ and various relations to potential theory of (8.2). The Wiener criterion for double obstacle problems was established by Dal Maso, Mosco and Vivaldi [12] in the linear case. Kilpeläinen and Ziemer [37] studied the double obstacle problem for the equation (1.1). They obtained sufficiency of the Wiener condition under some additional assumptions. Also, they discussed necessity of Wiener's criterion for $p > n-1$ or linear case. Kilpeläinen and Malý [36] considered the obstacle problems for the equation (8.1) and showed necessity of the Wiener criterion for all p's.

Consider the double obstacle problem with an upper obstacle s and a lower obstacle t. We denote

$$\begin{aligned} s_f(z) &= p\text{-fine-}\liminf_{x\to z} s(x),\\ s_q(z) &= \lim_{r\to 0} \inf\{\lambda\in\mathbb{R}: \mathrm{cap}_p(\{x\in B(z,r): s(x)<\lambda\})>0\},\\ t_f(z) &= p\text{-fine-}\limsup_{x\to z} t(x),\\ t_q(z) &= \lim_{r\to 0} \sup\{\lambda\in\mathbb{R}: \mathrm{cap}_p(\{x\in B(z,r): t(x)>\lambda\})>0\}. \end{aligned}$$

Obviously $s_q \le s_f$ and $t_q \ge t_f$. We say that z is a Wiener point for the obstacle (t,s) if $t_q(z) \le t_f(z) \le s_f(z) \le s_q(z)$.

Examples. The sharpness of Wiener's condition can be demonstrated by the following examples.

1. Assume that $p < n$, Ω is bounded and contains the point 0. Let $u = \frac{x_1}{|x|}$, $t = s = u$. Then u solves the double obstacle problem no matter what is the equation. We see that u is discontinuous at $z = 0$. In fact, a similar situation occurs whenever $s_q(z) < t_q(z)$.

2. Let u be the function mentioned in Theorem 8.4. Then u is the solution of the lower obstacle problem with lower obstacle $t = u$. It is discontinuous at z. Notice that $t_f(z) < t_q(z)$.

Theorem 7.1. *Let u be a solution of* (7.1) *and $z \in \Omega$ be a Wiener point for the double obstacle* (t,s). *Then u is continuous at z.*

Proof. We denote $v_k = \min(u,k)$, $w_k = \max(u,k)$. Let $k < s_q(z)$. Then there is $r > 0$ such that $B(z,r) \subset \Omega$ and $s > k$ p-q.e. on $B(z,r)$. Let $\varphi \in W_0^{1,p}(B(z,r))$, $0 \le \varphi \le k - v_k$. Then $t \le u \le u + \varphi$, $u + \varphi = v_k + \varphi \le k \le s$ p-q.e. on $B(z,r) \cap \{u \le k\}$ and $u + \varphi = u \le s$ p-q.e. on $B(z,r) \cap \{u > k\}$. It follows that φ extended by 0 outside $B(z,r)$ is an admissible test function for (7.1). Of course, $\varphi = 0$ wherever u differs from v_k. We have

$$\begin{aligned} &\int_{B(z,r)} \big(\mathbf{A}(x,v_k,\nabla v_k)\cdot\nabla\varphi + \mathbf{B}(x,v_k,\nabla v_k)\,\varphi\big)\,dx\\ &\quad = \int_\Omega \big(\mathbf{A}(x,u,\nabla u)\cdot\nabla\varphi + \mathbf{B}(x,u,\nabla u)\,\varphi\big)\,dx \ge 0. \end{aligned}$$

This verifies the assumption of Lemma 5.1. It follows that v_k are lower semicontinuous and p-finely continuous at z for all $k < s_q(z)$, in fact for each $k \le s_q(z)$. The p-fine continuity guarantees that $v_k(z) = k$ for all $k \le t_f(z) = t_q(z)$, which means that $u(z) \ge t_q(z)$. Analogously we prove that w_k are upper semicontinuous and p-finely continuous at z for all $k \geq t_q(z)$ and $u(z) \leq s_q(z)$. Write $\ell = u(z)$. Since $\ell \leq s_q(z)$, we have

$$u(z) = v_\ell(z) \leq \liminf_{x \to z} v_\ell(x) \leq \liminf_{x \to z} u(x)$$

and similarly $u(z) \geq \limsup_{x \to z} u(x)$. This proves that u is continuous at z. □

Remark. There is a close interplay between the Dirichlet and the obstacle problems. An analysis of the proofs of Theorems 5.1 and 6.4 shows that main step consists in Lemma 4.1, which concerns an obstacle problem. On the other hand, we can give an alternative proof of Theorem 7.1 using the Dirichlet problem knowledge only. However, this is essential that we use *p-fine* Dirichlet problem. Namely, if u is a solution of (7.1) represented by (2.1) and k is a constant such that $s > k$ p-q.e. in a ball B, then u is a p-fine supersolution of (1.1) in $\Omega' := B \cap \{u < k\}$ which is a p-finely open set and $u \geq k$ at each $z \in B \cap \overline{\Omega}' \setminus \Omega'$. Then we may use Theorem 6.4 instead of Lemma 4.1.

We can even combine Dirichlet and obstacle conditions. Using Lemma 4.1 or the above mentioned fine approach it can be verified that the following result is true:

Theorem 7.2. *Suppose that Ω is bounded. Let u be a solution of* (7.1), $z \in \partial\Omega$ *and $\ell \in \mathbb{R}$. Suppose that*

$$\lim_{r \to 0} \inf\{\lambda \in \mathbb{R}: \operatorname{cap}_p(\{x \in \Omega \cap B(z,r): s(x) < \lambda\}) > 0\} \geq \ell.$$

If either

$$p\text{-fine-}\liminf_{x \to z,\, x \in \Omega} t(x) \geq \ell,$$

or $\mathbb{R}^n \setminus \Omega$ is not p-thin at z and $u \geq \ell$ at z, then

$$\liminf_{x \to z,\, x \in \Omega} u(x) \geq \ell.$$

8. A-superharmonic functions

As an important special case of (1.1) we consider sometimes the equation

$$-\operatorname{div} \mathrm{A}(x, \nabla u) = 0 \tag{8.1}$$

where

$$\begin{aligned} &|\mathrm{A}(x,\xi)| \leq a_1 |\xi|^{p-1}, \\ &\mathrm{A}(x,\xi) \cdot \xi \geq c_1 |\xi|^p, \\ &(\mathrm{A}(x,\xi_2) - \mathrm{A}(x,\xi_2))(\xi_2 - \xi_1) > 0 \text{ (monotonicity)}. \end{aligned} \tag{8.2}$$

For such a structure the *theory of monotone operators* can be applied to obtain satisfactory existence results and a *comparison principle* with many consequences.

The potential theory for the equation (8.1), (8.2) has been developed by Granlund, Lindqvist and Martio [23], Lindqvist [42], Lindqvist and Martio [43], Heinonen and Kilpeläinen [26], [27], [28], Heinonen, Kilpeläinen and Martio [29], Kilpeläinen [33], Kilpeläinen and Malý [34], [35], [36] etc. Many results are collected in the book by Heinonen, Kilpeläinen and Martio [30], where the situation is generalized to a weighted case.

A key notion of the potential theory for the equation (8.1) is that of A-superharmonic function. We say that a function $u: \Omega \to (-\infty, +\infty]$ is A-superharmonic in Ω if u is lower semicontinuous, a.e. finite, and the implication

$$h \le u \text{ on } \partial G \Rightarrow h \le u \text{ on } G$$

holds whenever G is a bounded open set with $\overline{G} \subset \Omega$ and $h \in C(\overline{G})$ is a solution of (8.1) in G (in the weak sense like in (4.2)).

The relation between A-superharmonic functions and supersolutions is explained by the following assertions.

Theorem 8.1 (Heinonen and Kilpeläinen [26]).
(a) *If u is a supersolution of* (8.1), *then there is a unique A-superharmonic representative of u.*
(b) *If $u \in W^{1,p}_{\mathrm{loc}}(\Omega)$ is A-superharmonic, then u is a supersolution of* (8.1).
(c) *If u is an A-superharmonic function, then* $\min(u, k)$ *are supersolutions of* (8.1) *for all $k \in \mathbb{R}$.*

Theorem 8.2 (Kilpeläinen and Malý [34]). *Let u be an A-superharmonic function on Ω. Then* $\mathrm{A}(x, \nabla u) \in L^1_{\mathrm{loc}}(\Omega)$ *and there is a unique Radon measure μ such that*

$$\int \mathrm{A}(x, \nabla u) \cdot \nabla \varphi = \int \varphi \, d\mu \tag{8.3}$$

for all $\varphi \in C^\infty_c(\Omega)$.

Theorem 8.3 (Kilpeläinen and Malý [34] (part (8.4)), [36] (part (8.5))). *Let u be a nonnegative A-superharmonic function on $B(z, 3r_0)$ and μ be the measure associated with u by* (8.3). *Then there are constants $C > 0, C' > 0$ depending only on n, p and on the structure such that*

$$u(x) \ge C' \mathbf{W}^{\mu}_{p}(x, \rho), \tag{8.4}$$

$$u(x) \le C\big(\inf_{B(z,r_0)} u + \mathbf{W}^{\mu}_{p}(x, r_0)\big) \tag{8.5}$$

for all $x \in B(z, r_0)$.

Notice that A-superharmonic functions are p-finely continuous; in view of Theorem 8.1 this is a consequence of the fine continuity result by Michael and Ziemer [53]; see also [30]. In [29] it was proved that the p-fine topology is the coarsest topology making all A-superharmonic functions continuous. The following theorem contains a stronger assertion. It is an important consequence of (8.5).

Theorem 8.4 (Kilpeläinen and Malý [36]). *Let $E \subset \mathbb{R}^n$ be p-thin at $z \subset \overline{E} \setminus E$. Then there is an A-superharmonic function u on a neighborhood of z such that*

$$u(z) < \liminf_{x \to z,\, x \in E} u(x).$$

Proof. The function u is constructed as a solution of the lower obstacle problem on $B(z,r)$, where r is small enough, the zero Dirichlet boundary data are imposed and the lower obstacle t is given by the formula

$$t(x) = \begin{cases} 1 & \text{on } E \cap B(z, r/4), \\ 0 & \text{elsewhere.} \end{cases}$$

(We need the monotonicity assumption for the existence.) Using (8.5) it can be shown that u has the required properties. The details are done in [36].

9. Necessity of the Wiener condition

The Wiener criteria established by Wiener [66] and Littman, Stampacchia, Weinberger [45] (cf. Section 5) were presented as both necessary and sufficient. On the other hand, the sufficient condition by Maz'ya [47] waited a longer time for its necessity counterpart. For a special class of equations (not the structure (8.2)) some necessary conditions differring in an exponent from the sufficient conditions were proved by Skrypnik [64]. The necessity of the Wiener condition for the equation (8.1), (8.2) was shown by Lindqvist and Martio [43] with the restriction $p > n - 1$. For all $p \in (1, \infty)$ it was proved in [36]:

Theorem 9.1 (Kilpeläinen and Malý [36]). *Suppose that $\mathbb{R}^n \setminus \Omega$ is p-thin at $z \in \partial\Omega$. Then there is a $f \in C^1(\mathbb{R}^n)$ and a continuous solution h of* (8.1) *such that $h - f \in W_0^{1,p}(\Omega)$ and*

$$p\text{-fine-lim}_{x \to z,\, x \in \Omega} h(x) < f(z).$$

Remark. Suppose that $\mathbb{R}^n \setminus \Omega$ is p-thin at $z \in \partial\Omega$. The singleton $\{z\}$ has zero capacity. Hence, given $\rho > 0$ and $\varepsilon > 0$, we find a C^1-function f on $\mathbb{R}^n$ supported in $B(z,\rho)$ such that $f(z) = 1$ and $\int_{B(z,\rho)} |\nabla f|^p < \varepsilon$. Let h be a continuous solution of (8.1) such that $h - f \in W_0^{1,p}(\Omega)$ (it exists!). We fix a ball B containing $\overline{\Omega \cup B(z,\rho)}$ and extend h by f on $B \setminus \Omega$. Using $f - h$ as a test function we obtain

$$\int_\Omega \mathrm{A}(x, \nabla h) \cdot \nabla h \, dx = \int_\Omega \mathrm{A}(x, \nabla h) \cdot \nabla f \, dx.$$

It easily follows

$$\int_B |\nabla h|^p \, dx \le C \int_B |\nabla f|^p \, dx < C\varepsilon.$$

Since $h = 0$ on a set with nonempty interior in B, a well known form of Poincaré's inequality yields

$$\Big(\int_B |h^\gamma|\,dx\Big)^{p/\gamma} \le C\int_B |\nabla h|^p\,dx \le C\varepsilon$$

where γ is as in Theorem 5.2. Using this theorem it follows that there is $r_0 > 0$ such that

$$\begin{aligned}\operatorname*{app-lim\,sup}_{x\to z} h(x) &\le C_1\Big(R^{-n}\int_{B(z,R)} h^\gamma\,dx\Big)^{1/\gamma}\\ &+ C_2\int_0^R\Big(\frac{\operatorname{cap}_p(B(z,r)\setminus\Omega,r)}{r^{n-p}}\Big)^{1/(p-1)}\frac{dr}{r}\end{aligned}$$

for all $R \in (0, r_0)$. Since $\mathbb{R}^n \setminus \Omega$ is p-thin at z, we can find $R \in (0, r_0)$ such that

$$C_2\int_0^R\Big(\frac{\operatorname{cap}_p(B(z,r)\setminus\Omega,r)}{r^{n-p}}\Big)^{1/(p-1)}\frac{dr}{r} < \frac{1}{3}.$$

Then we can specify the choice of ε so that

$$C_1\Big(R^{-n}\int_\Omega h^\gamma\,dx\Big)^{1/\gamma} \le \frac{1}{3}.$$

We obtain that

$$\operatorname*{app-lim\,sup}_{x\to z} h(x) < 1 = f(z),$$

hence z is not regular. This shows that necessity of the Wiener criterion for (8.1) is a consequence of Theorem 5.2.

References

[1] Adams, D. R., L^p potential theory techniques and nonlinear PDE. In: Potential Theory (Ed. M. Kishi), 1–15, Walter de Gruyter, Berlin 1992.

[2] Adams, D. R., Hedberg, L. I., Function spaces and potential theory (in preparation).

[3] Adams, D. R., Meyers, N. G., Thinness and Wiener criteria for non-linear potentials. Indiana Univ. Math. J. 22 (1972), 169–197.

[4] Bagby, T., Quasi topologies and rational approximation, J. Funct. Anal. 10 (1972), 259–268.

[5] Beirao da Veiga, H., Sulla holderianita delle soluzioni di alcune disequazioni variazionali con condizioni unilatere al bordo. Annali Mat. Pura Appl. 84 (1969), 73–112.

[6] Bénilan, P., Boccardo, L., Gallouët, Th., Gariepy, R., Pierre, M., Vazquez, J. L., An L^1-theory of existence and uniqueness of solutions of nonlinear elliptic equations. Preprint 1993.

[7] Boccardo, L., Gallouët, T., Non-linear elliptic and parabolic equations involving measure data. J. Funct. Anal. 87 (1989), 149–169.

[8] — Non-linear elliptic equations with right hand side measures. Comm. Partial Differential Equations 17 (1992), 641–655.

[9] Boccardo, L., Murat, F., A property of nonlinear elliptic equations with the source term a measure. Potential Analysis 3 (1994), 257–263.

[10] Brelot, M. On Topologies and Boundaries in Potential Theory. Lecture Notes in Math. 175, Springer-Verlag, Berlin 1971.

[11] Calderón A. P., Lebesgue spaces of differentiable functions and distributions. Proc. Symp. Pure Math. 4 (1961), 33–49.

[12] Dal Maso, G., Mosco, U., Vivaldi, M. A., A pointwise regularity theory for the two-obstacle problem. Acta Math. 63 (1989), 57–107.

[13] De Giorgi, E., Sulla differenziabilità e l'analiticità delle estremali degli integrali multipli regolari. Memorie della Reale Accademia delle Science di Torino, Classe di Science Fisiche, Matematiche e Naturali 3 (1957), 25–43.

[14] Federer, H., Ziemer, W. P., The Lebesgue set of a function whose partial derivatives are p-th power summable. Indiana Univ. Math. J. 22 (1972), 139–158.

[15] Feyel, D., De La Pradelle, A., Espaces de Sobolev sur les ouverts fin. C. R. Acad. Sci. Paris, Ser. A 280 (1975), 1125–1127.

[16] — Le role des espaces de Sobolev en topologie fine. In: Seminaire de Theorie du Potentiel Paris, no. 2, 62–99, Lecture Notes in Math. 563, Springer-Verlag, Berlin 1976.

[17] Frehse, J., Capacity methods in the theory of partial differential equations. Jber. Deutsch. Math. Verein. 84 (1982), 1–44.

[18] Frehse, J., Mosco, U., Variational inequalities with one-sided irregular obstacles. Manuscripta Math. 28 (1979), 219–233.

[19] Fuglede, B., The quasi topology associated with a countably subadditive set function. Ann. Inst. Fourier Grenoble 21.1 (1971), 123–169.

[20] — Finely Harmonic Functions. Lecture Notes in Math. 289, Springer-Verlag, Berlin 1972.

[21] — Fonctions BLD et fonctions finement surharmoniques. In: Seminaire de Theorie du Potentiel Paris, no. 6, 126–157, Lecture Notes in Math. 906, Springer-Verlag, Berlin 1982.

[22] Gariepy, R., Ziemer, W. P., A regularity condition at the boundary for solutions of quasilinear elliptic equations. Arch. Rat. Mech. Anal. 67 (1977), 25–39.

[23] Granlund, S. P., Lindqvist, P., Martio, O., Conformally invariant variational integrals. Trans. Amer. Math. Soc. 277 (1983), 43–73.

[24] Hedberg, L. I., Nonlinear potentials and approximation in the mean by analytic functions. Math. Z. 129 (1972), 299–319.

[25] Hedberg, L. I., Wolff, Th. H., Thin sets in nonlinear potential theory. Ann. Inst. Fourier 33.4 (1983), 161–187.

[26] Heinonen, J., Kilpeläinen, T., $\mathcal{A}$-superharmonic functions and supersolutions of degenerate elliptic equations. Ark. Mat. 26 (1988), 87–105.

[27] — Polar sets for supersolutions of degenerate elliptic equations, Math. Scand. 63 (1988), 136–150.

[28] — On the Wiener criterion and quasilinear obstacle problems. Trans. Amer. Math. Soc. 310 (1988), 239–255.

[29] Heinonen, J., Kilpeläinen, T., Martio, O., Fine topology and quasilinear elliptic equations. Ann. Inst. Fourier 39.2 (1989), 293–318.

[30] — Nonlinear potential theory of degenerate elliptic equations. Oxford University Press, Oxford 1993.

[31] Iwaniec, T., Sbordone, C., Weak minima of variational integrals. Preprint 1992.

[32] Kichenassamy, S., Quasilinear problems with singularities. Manuscripta Math. 57 (1987), 281–313.

[33] Kilpeläinen, T., Potential theory for supersolutions of degenerate elliptic equations. Indiana Univ. Math. J. 38 (1989), 253–275.

[34] Kilpeläinen, T., Malý, J., Degenerate elliptic equations with measure data and nonlinear potentials. Ann. Scuola Norm. Sup. Pisa. Cl. Science, Ser. IV 19 (1992), 591–613.

[35] — Supersolutions to degenerate elliptic equations on quasi open sets, Comm. Partial Differential Equations 17 (1992), 371–405.

[36] — The Wiener test and potential estimates for quasilinear elliptic equations. Acta Math. 172 (1994), 137–161.

[37] Kilpeläinen, T., Ziemer, W. P., Pointwise regularity of solutions to nonlinear double obstacle problems. Arkiv Mat. 29,1 (1991), 83–106.

[38] Kinderlehrer, D., Stampacchia, G., An introduction to variational inequalities and their applications. Academic Press, New York 1980.

[39] Ladyzhenskaya, O. A., Ural'tseva, N. N., Quasilinear elliptic equations and variational problems with many independent variables. Uspekhi Mat. Nauk 16 (1961), 19–92 (Russian). English translation: Russian Math. Surveys 16 (1961), 17–91.

[40] — Linear and Quasilinear Elliptic Equations. Academic Press, New York 1968.

[41] Lieberman, G. M., Sharp forms of estimates for subsolutions and supersolutions of quasilinear elliptic equations with right hand side a measure. Comm. Partial Differential Equations 18 (1993), 1991–2112.

[42] Lindqvist, P., On the definition and properties of p-superharmonic functions. J. Reine Angew. Math. 365 (1986), 67–79.

[43] Lindqvist, P., Martio, O., Two theorems of N. Wiener for solutions of quasilinear elliptic equations. Acta Math. 155 (1985), 153–171.

[44] Lions, P. L., Murat, F., Sur les solutions renormalisées d'équations elliptiques non linéaires. C. R. Acad. Sci. Paris (to appear).

[45] Littman, W., G. Stampacchia, G., Weinberger, H. F., Regular points for elliptic equations with discontinuous coefficients. Ann. Scuola Norm. Sup. Pisa. Serie III. 17 (1963), 43–77.

[46] Lukeš, J., Malý, J., Thinness, Lebesgue density and fine topologies (An interplay between real analysis and potential theory). Proceedings of the Summer School on Potential Theory, Joensuu 1990. University of Joensuu, Publication in Sciences 26, Joensuu 1992.

[47] Maz'ya, V. G., On the continuity at a boundary point of solutions of quasi-linear elliptic equations. Vestnik Leningrad. Univ. 25 (1970), 42–55 (Russian). English translation: Vestnik Leningrad Univ. Math. 3 (1976), 225–242.

[48] Maz'ya, V. G., Khavin, V. P., Nonlinear potential theory. Uspekhi Mat. Nauk 27,6 (1972), 67–138. English translation: Russian Math. Surveys 27 (1972), 71–148.

[49] Malý, J., Pointwise estimates of nonnegative subsolutions of quaslinear elliptic equations at irregular boundary points. In preparation.

[50] Malý, J., Ziemer, W. P., Fine Regularity of Solutions of Elliptic Equations. Book in preparation.

[51] Meyers, N. G., A theory of capacities for potentials of functions in Lebesgue classes. Math. Scand. 26 (1970), 255–292.

[52] — Continuity properties of potentials. Duke Math. J. 42 (1975), 157–166.

[53] Michael, J. H., Ziemer, W. P., Interior regularity for solutions to obstacle problems. Nonlin. Anal. TMA 10 (1986), 1427–1448.

[54] Mosco, U., Wiener criterion and potential estimates for the obstacle problem. Indiana Univ. Math. J. 36 (1987), 455–494.

[55] Moser, J., A new proof of De Giorgi's theorem concerning the regularity problem for elliptic differential equations. Comm. Pure Appl. Math. 13 (1960), 457–468.

[56] — On Harnack's theorem for elliptic differential equations. Comm. Pure Appl. Math. 14 (1961), 577–591.

[57] Nash, J., Continuity of solutions of parabolic and elliptic equations. Amer. J. Math. 80 (1958), 931–954.

[58] Rakotoson, J. M., Equivalence between the growth of $\int_{B(x,r)} |\nabla u|^p dy$ and T in the equation $P[u] = T$. J. Differential Equations 86 (1990), 102–122.

[59] — Quasilinear elliptic problems with measures as data. Diff. and Integral Equations 4 (1991), 449–457.

[60] — Generalized solutions in a new type of sets for problems with measures as data. Differential Integral Equations 6 (1993), 27–36.

[61] Rakotoson, J. M., Ziemer, W. P., Local behavior of solutions of quasilinear elliptic equations with general structure. Trans. Amer. Math. Soc. 319 (1990), 747–764.

[62] Serrin, J., Pathological solutions of elliptic differential equations. Ann. Scuola Norm. Sup. Pisa (1964), 385–387.

[63] — Local behavior of solutions of quasi-linear equations. Acta Math. 111 (1964), 247–302.

[64] Skrypnik, I. V., Nonlinear elliptic boundary value problems. Teubner Verlag, Leipzig 1986.

[65] Trudinger, N. S., On Harnack type inequalities and their application to quasilinear elliptic equations. Comm. Pure Appl. Math. 20 (1967), 721–747.

[66] Wiener, N., Certain notions in potential theory. J. Math. Phys. 3 (1924), 24–51. Reprinted in: Norbert Wiener: Collected works. Vol. 1 (1976), MIT Press, 364–391.

[67] Ziemer, W. P., Mean values of subsolutions of elliptic and parabolic equations. Trans. Amer. Math. Soc. 279,2 (1983), 555–566.

[68] — Weakly differentiable functions: Sobolev spaces and functions of bounded variation. Springer-Verlag, New York 1989.

Removability, geometric measure theory, and singular integrals

Pertti Mattila

Abstract. The basic question here is: what sets are removable for bounded analytic functions in the complex plane and for Lipschitz harmonic functions in $\mathbb{R}^n$. We discuss the relations of this problem to the rectifiability properties of sets and to the behaviour of certain singular integral operators.

1991 Mathematics Subject Classification: 28A75, 30E20, 31B05, 42B20.

1. Introduction

Removable sets have been studied extensively for many different function classes. If $\mathcal{F}$ is a class of functions defined on open subsets of $\mathbb{R}^n$, a compact set $E \subset \mathbb{R}^n$ is called removable for $\mathcal{F}$ if for all open sets U containing E every function on $U \setminus E$ which belongs to $\mathcal{F}$ has an extension to U which also belongs to $\mathcal{F}$. Usually there is a natural capacity related to $\mathcal{F}$ whose null-sets are exactly the removable sets for $\mathcal{F}$. However, often understanding this capacity geometrically is difficult and the problem of finding a geometric characterization of the removable sets remains. This is so in the cases of bounded analytic functions and Lipschitz harmonic functions which we shall discuss below. These questions are related to the behaviour of certain singular integral operators on E and the rectifiability properties of E.

For surveys and various developments related to this topic, see e.g. [C1], [D2], [G2], [J], [M] and the other articles in this volume on pp. 485–506, [M4], [Mu2], [O], [V2], [Vi2] and [Z].

2. Removable sets and capacities

Let $E \subset \mathbb{C}$ be compact. According to the above, E is removable for bounded analytic functions if for every open set U containing E every bounded analytic function $f\colon U\setminus E \to \mathbb{C}$ extends to an analytic function $g\colon U \to \mathbb{C}$; the boundedness of g is then automatic since it is easy to see that such a set E cannot have interior points. A simple application of the Cauchy integral formula shows that it is enough to check this for $U = \mathbb{C}$, cf. e.g. [Z]. Thus by Liouville's theorem, E is removable if and only if every bounded analytic function $f\colon \mathbb{C}\setminus E \to \mathbb{C}$ is constant. Moreover, subtracting $f(\infty)$, we may restrict to functions f with $f(\infty) = 0$.

To define the capacity related to this class, we write $f\colon \mathbb{C}\setminus E \to \mathbb{C}$, $f \not\equiv 0$, with $f(\infty) = 0$ as the Laurent series;

$$f(z) = a_k z^{-k} + a_{k+1} z^{-k-1} + \cdots, \quad a_k \neq 0,\ k \geq 1.$$

Multiplying by z^{k-1} we see that if such a function f exists we have also a non-zero analytic function $\mathbb{C}\setminus E \to \mathbb{C}$ vanishing at infinity for which $a_1 \neq 0$. It is natural to maximize this coefficient;

$$\gamma(E) = \sup_f \lim_{z\to\infty} |zf(z)|$$

where the supremum is taken over all analytic functions $f\colon \mathbb{C}\setminus E \to \mathbb{C}$ with $f(\infty) = 0$ and $|f| \leq 1$. This is the *analytic capacity* defined by Ahlfors [A] in 1947. Its null-sets are exactly the removable sets for bounded analytic functions, as is almost immediate from the above discussion. Unfortunately this gives practically no geometric insight. Before starting to pursue such insight, let us also look at harmonic functions.

The removable sets for bounded harmonic functions are characterized as the null-sets of the classical capacity. This is also geometrically a fairly satisfactory solution. But the problem which is close to the above complex analysis problem is that of Lipschitz harmonic functions. Why this is so will hopefully become clearer later. Since Lipschitz functions always extend we can say that a compact set $E \subset \mathbb{R}^n$ is removable for Lipschitz harmonic functions if for all open sets U containing E, every Lipschitz function $u\colon U \to \mathbb{R}$ which is harmonic in $U\setminus E$ is also harmonic in U.

For $n = 2$ every removable set for bounded analytic functions is also removable for Lipschitz harmonic functions: use the fact that if u is Lipschitz harmonic, then $\partial_z u$ is bounded analytic. The converse is not known. So it might be that in $\mathbb{C}$ these two classes are the same.

The capacities κ_n and κ_n' related to C^1 and Lipschitz, respectively, harmonic functions were introduced by Paramonov in [P]. We shall only discuss κ_n'. It is defined for compact sets $E \subset \mathbb{R}^n$ as

$$\kappa_n'(E) = \sup\langle \Delta u, 1\rangle$$

where the supremum is taken over all locally Lipschitzian functions $u\colon \mathbb{R}^n \to \mathbb{R}$ such that

$$\operatorname{spt}\Delta u \subset E, \quad \|\nabla u\|_\infty \leq 1 \quad \text{and} \quad \nabla u(\infty) = 0.$$

Here $\langle \Delta u, \varphi\rangle$ denotes the action of the distribution Δu on the smooth function φ and spt stands for the support. Note that the test functions u are harmonic in $\mathbb{R}^n \setminus E$ and the conditions required mean that

$$u = \Phi_n * (\Delta u) + \text{constant}$$

where Φ_n is the fundamental solution of the Laplace equation in $\mathbb{R}^n$. That is,

$$\Phi_n(x) = \begin{cases} c_2 \log(1/|x|) & \text{for } n = 2, \\ c_n |x|^{2-n} & \text{for } n \geq 3. \end{cases}$$

It is easy to see, cf. [MP], that $E \subset \mathbb{R}^n$ is removable for Lipschitz harmonic functions if and only if $\kappa'_n(E) = 0$. A quantitative form of the question asked above is: are γ and κ'_2 comparable?

3. Hausdorff measures and Riesz capacities

We now define two set functions which are closely connected with the above, and many other, removability questions. The s-dimensional *Hausdorff measure* $\mathcal{H}^s$ is defined for $A \subset \mathbb{R}^n$ as

$$\mathcal{H}^s(A) = \lim_{\delta \to 0} \mathcal{H}^s_\delta(A)$$

where for $0 < \delta \leq \infty$,

$$\mathcal{H}^s_\delta(A) = \inf \left\{ \sum_{i=1}^{\infty} d(B_i)^s : A \subset \bigcup_{i=1}^{\infty} B_i,\ d(B_i) \leq \delta \right\},$$

and $d(B)$ is the diameter of B. Trivially, $\mathcal{H}^s$ and $\mathcal{H}^s_\delta$ have the same null-sets. For $s = n$, $\mathcal{H}^n = \mathcal{H}^n_\delta$ is a constant multiple of the Lebesgue measure in $\mathbb{R}^n$.

The *Riesz s-capacity* C_s is defined for $A \subset \mathbb{R}^n$ by

$$C_s(A) = \sup \left\{ \mu(A) : \operatorname{spt} \mu \subset A, \int |x-y|^{-s}\, d\mu y \leq 1 \quad \text{for } x \in \mathbb{R}^n \right\}.$$

Here $\operatorname{spt} \mu$ is the support of the (positive) Radon measure μ. The well-known relations for Borel sets $A \subset \mathbb{R}^n$ between $\mathcal{H}^s$ and C_s are, see e.g. [Ca] or [M4],

$$\mathcal{H}^s(A) < \infty \Rightarrow C_s(A) = 0 \Rightarrow \mathcal{H}^t(A) = 0 \quad \text{for } t > s. \tag{3.1}$$

Thus both have the same critical index $\dim A$, the *Hausdorff dimension* of A,

$$\dim A = \inf\{s : \mathcal{H}^s(A) = 0\} = \inf\{s : C_s(A) = 0\}.$$

The Hausdorff measures characterize the removable sets of analytic functions which are Hölder continuous with exponent α, $0 < \alpha \leq 1$. The condition is: $\mathcal{H}^{1+\alpha}(E) = 0$, see [Do] and [U2]. In fact, quantitatively it is the Hausdorff content $\mathcal{H}^{1+\alpha}_\infty$ which is comparable to the related capacity. The limiting case $\alpha \to 0$ is not that of bounded analytic functions but BMO where the characterization is given by $\mathcal{H}^1(E) = 0$. This was proved by Král [Kr], see also [K]. There are corresponding results for harmonic functions for $0 < \alpha < 1$, see [Ca]. Both for analytic and harmonic functions these results are fairly easy when $0 < \alpha < 1$. The case $\alpha = 1$ of Lipschitz analytic functions is rather complicated, but there is still a characterization which depends only on the size of E. We shall see that this is not so for bounded analytic and Lipschitz harmonic functions which makes these classes much more difficult to deal with.

We now give a necessary and a sufficient condition, which are well-known, for the removability. Recall from (3.1) that $\dim E > n-1$ implies $C_{n-1}(E) > 0$.

Theorem 3.2. 1. *Let $E \subset \mathbb{C}$ be compact.*
(a) *If $C_1(E) > 0$, then $\gamma(E) > 0$.*
(b) *If $\mathcal{H}^1(E) = 0$, then $\gamma(E) = 0$.*

2. *Let $E \subset \mathbb{R}^n$ be compact.*
(a) *If $C_{n-1}(E) > 0$, then $\kappa_n'(E) > 0$.*
(b) *If $\mathcal{H}^{n-1}(E) = 0$, then $\kappa_n'(E) = 0$.*

Proof. 1 (a): If $C_1(E) > 0$ there is a non-zero Radon measure μ such that $\operatorname{spt}\mu \subset E$ and $\int |x-y|^{-1}\,d\mu y \le 1$ for $x \in \mathbb{C}$. Define

$$f(z) = \int \frac{1}{\zeta - z}\,d\mu\zeta.$$

Then f is a non-constant bounded analytic function in $\mathbb{C}\setminus E$, whence $\gamma(E) > 0$.

For 2 (a) one has μ with $\int |x-y|^{1-n}\,d\mu y \le 1$ and one defines

$$u(x) = \int \Phi_n(x-y)\,d\mu y \quad \text{for } x \in \mathbb{R}^n.$$

Then u is a C^1 function on $\mathbb{R}^n$ with

$$\nabla u(x) = c\int \frac{x-y}{|x-y|^n}\,d\mu y.$$

Thus $|\nabla u(x)| \le c$ and so u is Lipschitz. Moreover, $\Delta u = \mu$ in the distributional sense so that u is harmonic in $\mathbb{R}^n \setminus E$ but not in $\mathbb{R}^n$. Hence $\kappa_n'(E) > 0$.

To prove parts (b) assume $\mathcal{H}^{n-1}(E) = 0$, and for $k = 1, 2, \ldots$ cover E with closed balls $B_{k,1}, \ldots, B_{k,m_k}$ such that

$$\sum_{i=1}^{m_k} d(B_{k,i})^{n-1} < 1/k. \tag{3.3}$$

Denote $G_k = \cup_{i=1}^{m_k} B_{k,i}$.

If $n = 2$ and $f\colon \mathbb{C}\setminus E \to \mathbb{C}$ is analytic with $\|f\|_\infty \le 1$ and $f(\infty) = 0$, we can write by the Cauchy integral formula

$$f(z) = -\frac{1}{2\pi i}\int_{\partial G_k} \frac{f(\zeta)}{\zeta - z}\,d\zeta \quad \text{for } z \notin G_k.$$

Then (3.3) yields

$$|f(z)| \le c(z)/k \to 0 \quad \text{as } k \to \infty,$$

and we conclude $f \equiv 0$. Thus $\gamma(E) = 0$.

For 2 (b) let $u\colon \mathbb{R}^n \to \mathbb{R}$ be a Lipschitz function which is harmonic in $\mathbb{R}^n \setminus E$ such that $\|\nabla u\|_\infty \le 1$ and $\nabla u(\infty) = 0$. Let u_k be a smooth function with $u = u_k$ in a neighborhood of $\mathbb{R}^n \setminus G_k$, e.g. a standard convolution approximation $u_k = \varphi_{\varepsilon_k} * u$ is such for ε_k small enough. Green formula gives for $x \in \mathbb{R}^n \setminus G_k$ (as $\Delta u_k = \Delta u = 0$ on $\mathbb{R}^n \setminus G_k$),

$$\begin{aligned} u(x) = u_k(x) &= \int_{G_k} \Phi_n(x-y)\, \Delta u_k(y)\, dy \\ &= \int_{G_k} \left(\operatorname{div}_y(\Phi_n(x-y)\, \nabla_y u_k(y)) - \nabla_y \Phi_n(x-y) \cdot \nabla_y u_k(y) \right) dy \\ &= \int_{\partial G_k} \Phi_n(x-y) \frac{\partial u}{\partial \nu_y}\, dy + \int_{G_k} \nabla \Phi_n(x-y) \cdot \nabla u(y)\, dy \end{aligned} \tag{3.4}$$

where ν_y is the outer unit normal to ∂G_k. Letting $k \to \infty$ and using (3.3) we conclude $u(x) = 0$. Hence $\kappa_n'(E) = 0$.

Central concepts which are present in the above proof are the integral transforms which are obtained by convolving with the kernels $1/z$, Φ_n and $\nabla \Phi_n$. Their role will be clearer when we now study the case where $\mathcal{H}^{n-1}(E) < \infty$ instead of $\mathcal{H}^{n-1}(E) = 0$. Let us first consider $n = 2$ and analytic functions.

So assume $\mathcal{H}^1(E) < \infty$. For each $k = 1, 2, \dots$, cover now E with a finite union $G_k = \cup_{i=1}^{m_k} B_{k,i}$ of discs with $d(B_{k,i}) < 1/k$ and

$$\sum_{i=1}^{m_k} d(B_{k,i}) \leq 2\mathcal{H}^1(E) + 1/k. \tag{3.5}$$

If $f \colon \mathbb{C} \setminus E \to \mathbb{C}$ is analytic with $|f| \leq 1$ and $f(\infty) = 0$ one derives from the Cauchy integral formula that for $z \notin G_k$

$$f(z) = -\frac{1}{2\pi i} \int_{\partial G_k} \frac{f(\zeta)}{\zeta - z}\, d\zeta = \int \frac{1}{\zeta - z}\, d\sigma_k$$

where σ_k is the complex measure $-(2\pi i)^{-1} f(\zeta)\, d\zeta$ on ∂G_k. Using (3.5) one finds that the total variations of the measures σ_k are bounded, so there is a weakly converging subsequence $\sigma_{k_j} \to \sigma$. Then $\operatorname{spt} \sigma \subset E$ and

$$f(z) = \int \frac{d\sigma\zeta}{\zeta - z} \quad \text{for } z \in \mathbb{C} \setminus E.$$

With a little bit more work, cf. [M4], one can show that σ is absolutely continuous with respect to $\mathcal{H}^1$ with Radon–Nikodym derivative bounded by 1. Thus there is a Borel function $g \colon E \to \mathbb{C}$ such that $|g| \leq 1$ and

$$f(z) = \int_E \frac{g(\zeta)}{\zeta - z}\, d\mathcal{H}^1 \zeta.$$

Hence

$$\lim_{z \to \infty} |z f(z)| = \lim_{z \to \infty} \left| \int_E \frac{g(\zeta)}{1 - \zeta/z}\, d\mathcal{H}^1 \zeta \right| \leq \mathcal{H}^1(E).$$

This proves the first part of the following theorem. The second part is obtained by similar arguments starting from (3.4).

Theorem 3.6.
(1) $\gamma(E) \leq \mathcal{H}^1(E)$ *for compacts* $E \subset \mathbb{C}$.
(2) $\kappa_n'(E) \leq c\mathcal{H}^{n-1}(E)$ *for compacts* $E \subset \mathbb{R}^n$.

(In fact, $\mathcal{H}^{n-1}$ can be replaced by $\mathcal{H}^{n-1}_\infty$.)

Note that from the above argument we have the following information: if $E \subset \mathbb{C}$ is compact with $\mathcal{H}^1(E) < \infty$ and $f\colon \mathbb{C} \setminus E \to \mathbb{C}$ is analytic with $|f| \leq 1$ and $f(\infty) = 0$, then there is a Borel function $g\colon E \to \mathbb{C}$ with $|g| \leq 1$ and

$$f(z) = \int_E \frac{g(\zeta)}{\zeta - z}\, d\mathcal{H}^1\zeta \quad \text{for } z \in \mathbb{C} \setminus E. \tag{3.7}$$

Similarly: if $E \subset \mathbb{R}^n$ is compact with $\mathcal{H}^{n-1}(E) < \infty$ and $u\colon \mathbb{R}^n \to \mathbb{R}$ is a Lipschitz function such that u is harmonic on $\mathbb{R}^n \setminus E$, $\|\nabla u\|_\infty \leq 1$ and $\nabla u(\infty) = 0$, then there is a Borel function $g\colon E \to \mathbb{R}$ with $|g| \leq 1$ and

$$u(x) = \int_E \Phi_n(x-y)\, g(y)\, d\mathcal{H}^{n-1}y \quad \text{for } x \in \mathbb{R}^n \setminus E,$$

whence

$$\nabla u(x) = c\int_E \frac{x-y}{|x-y|^n}\, g(y)\, d\mathcal{H}^{n-1}y \quad \text{for } x \in \mathbb{R}^n \setminus E. \tag{3.8}$$

The integrals in (3.7) and (3.8) do not usually exist for $x \in E$. But they often exist in the principal value sense and define singular operators on E. Their behaviour is closely related to the above removability questions, as we shall see.

4. Singular integrals and removability

If we want to show that $\gamma(E) > 0$, we should be able to produce a non-zero analytic function $f\colon \mathbb{C} \setminus E \to \mathbb{C}$ with $|f| \leq 1$ and $f(\infty) = 0$. We now restrict to the case $0 < \mathcal{H}^1(E) < \infty$. (Then $C_1(E) = 0$ so we cannot use the crude information from Theorem 3.2.) We know that such an f has the representation (3.7). So the problem is equivalent to producing g. In some cases this is easy. For example, if $E \subset \mathbb{R}$ we can first define

$$h(z) = \int_E \frac{1}{x-z}\, dx.$$

This is not bounded but it has bounded imaginary part; in fact, $|\operatorname{Im} h(z)| \leq \pi$. Mapping the strip $|\operatorname{Im} w| \leq \pi$ conformally onto the unit disc and composing with h, we find f and also g with an explicit formula. For E on sufficiently smooth curves, $C^{1+\varepsilon}$ is enough, this can still be done constructively, see [I] and [G2]. However for C^1 curves no constructive method is known but then singular integrals and duality methods come to help.

This problem of showing that a compact subset of a C^1 or a rectifiable (i.e., a Lipschitz image of an interval) curve has $\gamma(E) = 0$ if and only if $\mathcal{H}^1(E) = 0$ has been known as the Denjoy conjecture and it has a long history. By approximation it can easily be reduced to Lipschitz graphs, even with a small Lipschitz constant. Based on earlier results of Havin, Havinson and Davie [H], [HH], [D], the last fact needed to complete the story of the Denjoy conjecture was Calderón's theorem [C] saying that the Cauchy singular integral operator is bounded on L^2 on Lipschitz graphs with a sufficiently small Lipschitz constant. Let us have a look at what this means.

Let Γ be a Lipschitz graph in $\mathbb{C}$, that is,

$$\Gamma = \{x + i\psi(x) : x \in I\}$$

where ψ is a real-valued Lipschitz function on an interval $I \subset \mathbb{R}$. Define

$$C_\Gamma g(z) = \lim_{\varepsilon \to 0} \int_{|z-\zeta|>\varepsilon} \frac{g(\zeta)}{\zeta - z}\, d\mathcal{H}^1\zeta$$

provided the limit exists.

Theorem 4.1. For $g \in L^1(\Gamma)$, $C_\Gamma g(z)$ exists for $\mathcal{H}^1$ almost all $z \in \Gamma$. Moreover,

$$C_\Gamma\colon L^p(\Gamma) \to L^p(\Gamma) \quad \text{is bounded for } 1 < p < \infty.$$

For $p = 1$, there is the weak type inequality

$$\mathcal{H}^1\{z \in \Gamma : |C_\Gamma g(z)| > \lambda\} \le C\lambda^{-1} \int_\Gamma |g|\, d\mathcal{H}^1. \tag{4.2}$$

The essential ingredient is to prove that the operator C_Γ defined in some reasonable sense, not necessarily via the principal values, is bounded on $L^2(\Gamma)$. The rest follows then from the standard Calderón–Zygmund theory, see [S] and [CW].

Theorem 4.1 was first proved by Calderón [C] for ψ with small Lipschitz constant and then extended by Coifman, McIntosh and Meyer in [CMM]. Now there are many proofs, see [Mu2]. It is also valid for more general curves. In fact, David [D1] showed that it is valid exactly for curves Γ which are *Ahlfors regular* in the sense that there is a constant C such that

$$\mathcal{H}^1(\Gamma \cap B(z,r)) \le Cr \quad \text{for } z \in \mathbb{C},\ r > 0.$$

From Theorem 4.1 (which holds for Ahlfors regular curves) we have

Theorem 4.3. *Let Γ be a rectifiable curve in $\mathbb{C}$ and $E \subset \Gamma$ compact.*
(1) *$\gamma(E) = 0$ if and only if $\mathcal{H}^1(E) = 0$.*
(2) *If γ is Ahlfors regular, then γ and $\mathcal{H}^1$ are comparable on Γ;*

$$\gamma(E) \lesssim \mathcal{H}^1(E) \lesssim C(\Gamma)\,\gamma(E).$$

If $E \subset \Gamma$ with $\mathcal{H}^1(E) > 0$, then, by simple approximation, $\mathcal{H}^1(E \cap \Gamma') > 0$ for some Ahlfors regular curve (even Lipschitz graph) Γ'. Hence it suffices to prove (2). There are various duality methods based on the Hahn–Banach theorem for this, see [G2], [M] and [C1]. The basic idea in an elegant method due to Uy [U1–3] is the following: use the weak type inequality (4.2) and the Hahn–Banach theorem to show that C_Γ maps some bounded non-zero function g vanishing on $\Gamma \setminus E$ to a bounded function $C_\Gamma g$ on Γ. Then it is easy to see that $C_\Gamma g$ is also bounded and analytic on $\mathbb{C}\setminus E$ proving that $\gamma(E)>0$. For details, see [C1].

The above argument forgets completely that Γ is a curve and it proves much more about the relations between analytic capacity and Cauchy singular integral operator. For $0<m<n$ call a closed set $F\subset\mathbb{R}^n$ AD$_m$-*regular* (AD = Ahlfors–David) if there is a positive and finite constant C such that

$$r^m/C\le\mathcal{H}^m(F\cap B(x,r))\le Cr^m\quad\text{for } x\in F,\ 0<r<d(F). \tag{4.4}$$

We would like to talk about the L^2-boundedness of the Cauchy integral operator C_F on such AD$_1$-regular subsets of $\mathbb{C}$. But there is no reason why $C_F g(z)$ should exist in the principal value sense even for smooth functions g. Still there are ways to define the operator C_F, cf. [C1], but since we can easily define the L^2-boundedness of C_F without really defining C_F, we only do that. For later purposes, we use more general kernels.

Let $0<m<n$ and $K\colon\mathbb{R}^n\setminus\{0\}\to\mathbb{R}^n$ (or $\mathbb{C}$) be an odd C^∞ function such that for some $C(j)<\infty$

$$|\nabla^j K(x)|\le C(j)\,|x|^{-m-j}\quad\text{for } x\ne0,\ j=0,1,2,\dots,$$

where $\nabla^0K=K$ and, for $j\ge1$, $\nabla^jK(x)$ is a vector formed by the partial derivatives of K of order j. Denote by $\mathcal{K}_m(\mathbb{R}^n)$ the set of such kernels K. The basic examples are the Cauchy kernel $K(z)=1/z$ (then $m=1$) and the Riesz kernel $K(x)=|x|^{-m-1}x$. Let $F\subset\mathbb{R}^n$ be an AD$_m$-regular set. We now define the L^2-boundedness of the formal singular integral operator T_K:

$$T_Kg(x)=\int_F K(x-y)\,g(y)\,d\mathcal{H}^my.$$

Definition 4.5. Let K and F be as above. We say that T_K is bounded on $L^2(F)$ if there is $C<\infty$ such that

$$\int_F\Big|\int_{F\setminus B(x,\varepsilon)}K(x-y)\,g(y)\,d\mathcal{H}^my\Big|^2\,d\mathcal{H}^mx\le C\int_F|g|^2\,d\mathcal{H}^m$$

for all $g\in L^2(F)$ and $\varepsilon>0$.

According to the standard singular integral theory this agrees with other natural definitions for the L^2-boundedness and it implies the boundedness in L^p for $1<p<\infty$ and the weak type L^1 inequality, cf. [CW].

Returning to the Cauchy integral operator we have (cf. [C1] and [C2])

Theorem 4.6. *Let $F\subset\mathbb{C}$ be AD$_1$-regular and $C_F=T_K$ with $K(z)=1/z$.*
(1) *If C_F is bounded on $L^2(F)$, then $\gamma(F)>0$.*
(2) *Conversely, if $\gamma(F)>0$, there exists an AD$_1$-regular set F' such that $\mathcal{H}^1(F\cap F')>0$ and $C_{F'}$ is bounded on $L^2(F')$.*

The proof described above for Theorem 4.3 yields also (1). The part (2) is a kind of a $T(b)$-theorem: since $\gamma(F)>0$ there is a bounded analytic function in $\mathbb{C} \setminus F$. This has a Cauchy integral representation, recall (3.7), which implies that C_F maps some non-zero bounded function to a bounded function. Thus we are in a situation where a suitable $T(b)$-theorem could be hoped to give the L^2-boundedness. In fact, Christ proves in [C2] such a general $T(b)$-theorem.

From the above discussion, recalling also (3.6), we can conclude that the positiveness of the analytic capacity and the L^2-boundedness of the Cauchy integral operator are both linked with the existence of a non-zero L^∞-function g for which

$$\int_F \frac{g(\zeta)}{\zeta - z}\, d\mathcal{H}^1\zeta$$

is bounded. For Lipschitz harmonic functions in $\mathbb{R}^n$ there is essentially the same story but now we are looking at an $\mathrm{AD}_{(n-1)}$-regular set F, the operator T_K with $K(x) = |x|^{-n}x$ and studying whether

$$\int_F \frac{x-y}{|x-y|^n}\, g(y)\, d\mathcal{H}^{n-1}y$$

could be bounded for some non-zero L^∞-function g. The analogue of Theorem 4.1 is then valid on $(n-1)$-dimensional Lipschitz graphs, see [MP] and [U3].

5. Positive measure but zero capacity

In the previous chapter we saw that γ and $\mathcal{H}^1$ vanish simultaneously on rectifiable curves, and the same is true for κ_n' and $\mathcal{H}^{n-1}$ on hypersurfaces. But do they always vanish simultaneously? This was answered in the negative by Vitushkin [Vi1] already in 1959. We shall discuss various examples below.

Perhaps the simplest example of a set with positive length but zero analytic capacity is the product of the $(1/2)$-dimensional standard Cantor set with itself. It can be constructed as follows. Take a closed unit square Q. Let $Q_{1,1}, \dots, Q_{1,4} \subset Q$ be the closed squares in the corners of Q with side-length $1/4$. Next let $Q_{2,i}$, $i = 1, \dots, 16$, be the squares in the corners of the $Q_{1,j}$'s with side-lenght $1/16$, and so on. Then

$$C = \bigcap_{k=1}^{\infty} \bigcup_{i=1}^{4^k} Q_{k,i}. \tag{5.1}$$

It is easy to see that $0 < \mathcal{H}^1(C) < \infty$. Garnett [G1], and independently Ivanov, showed that $\gamma(C) = 0$. Now there are many proofs for that, see [C1], [J], [M1] and [Mu1]. Let us see how one could proceed starting from (3.7). We should show that there is no bounded Borel function $g\colon C \to \mathbb{C}$, which does not vanish $\mathcal{H}^1$ almost everywhere on C, and for which

$$\int_C \frac{g(\zeta)}{\zeta - z}\, d\mathcal{H}^1\zeta$$

is bounded. It is clear that no g with $0 < c \leq g(z) \leq 1$ for $z \in C$ can work, because the integral $\int_C (\zeta - z)^{-1}\, d\mathcal{H}^1\zeta$ is easily seen to diverge at the corner points z of the squares $Q_{k,i}$. By a slight modification, no continuous g can work. Using Lusin's theorem and a density point argument it is possible to proceed along these lines for a general g. This argument is given for more general sets in [M1] and in a different form in [M4]. The latter is based on tangent measures to which we come soon.

Since $\gamma(C) = 0$, also $\kappa_2'(C) = 0$. The corresponding construction for C with $0 < \mathcal{H}^{n-1}(C) < \infty$ can of course be done in any $\mathbb{R}^n$ with squares replaced by cubes. Using similar arguments as Garnett involving Green's formula, Uy [U4] showed that $\kappa_n'(C) = 0$. A different argument applying to a more general class of sets can be found in [MP]. Paramonov and I showed in the same paper that if $E \subset \mathbb{R}^n$, $n \geq 2$, then

$$C^{-1}\kappa_n'(E) \leq \kappa_n'(E \times [0,1]) \leq C\kappa_n'(E). \tag{5.2}$$

In particular, $\kappa_3'(E \times [0,1]) = 0$ for any set $E \subset \mathbb{C}$ with $\gamma(E) = 0$.

So the null-sets for Hausdorff measures do not characterize the null-sets of analytic and harmonic capacities. But could there be some other geometric characterization? For general sets there is none in sight. Vitushkin asked in [Vi2] whether the null-sets of γ could be the same as the null-sets of the Favard integralgeometric measure $\mathcal{I}^1$. For a compact set $E \subset \mathbb{C}$, $\mathcal{I}^1(E) = 0$ if and only if $\mathcal{H}^1(\operatorname{proj}_\theta E) = 0$ for almost all $\theta \in [0, \pi]$, where $\operatorname{proj}_\theta$ is the orthogonal projection onto the line $\{te^{i\theta} : t \in \mathbb{R}\}$. This turned out be false; I showed in [M2] that the class of the null-sets for $\mathcal{I}^1$ is not conformally invariant as opposed to that of γ and κ_2'. Jones and Murai [JM] constructed a set with $\gamma(E) > 0$ and $\mathcal{I}^1(E) = 0$. I do not know if their argument can be modified to work also for κ_2'. It is not known whether $\gamma(E) = 0$ implies $\mathcal{I}^1(E) = 0$.

The compact null-sets of $\mathcal{I}^{n-1}$ in $\mathbb{R}^n$ are those which project into zero $\mathcal{H}^{n-1}$ measure on almost all hyperplanes through the origin. For no n can κ_n' and $\mathcal{I}^{n-1}$ have the same class of compact null-sets. To see this, recall from (5.2) that $\kappa_n'(E) = 0$ if and only if $\kappa_{n+1}'(E \times [0,1]) = 0$. It is easy to verify that also $\mathcal{I}^{n-1}(E) = 0$ if and only if $\mathcal{I}^n(E \times [0,1]) = 0$. Hence we can proceed by induction from the case $n = 2$.

Although a geometric characterization of the null-sets of γ and κ_n' might be a hopeless task, there is some hope for those with $\mathcal{H}^{n-1}(E) < \infty$ ($n = 2$ for γ). This would not be so bad since by Theorem 3.2 we can in any case restrict to sets of Hausdorff dimension $n - 1$; among them those with non-σ-finite $\mathcal{H}^{n-1}$ measure look most problematic. The conjecture is

Conjecture 5.3. Suppose $E \subset \mathbb{R}^n$ is compact with $\mathcal{H}^{n-1}(E) < \infty$. Then $\kappa_n'(E) = 0$ (or $\gamma(E) = 0$ for $n = 2$) if and only if $\mathcal{H}^{n-1}(E \cap M) = 0$ for every $(n-1)$-dimensional Lipschitz graph M.

Such sets E are central in geometric measure theory and they are called purely $(n-1)$-unrectifiable. We return to them later. Lipschitz graphs can be replaced by other C^1 or Lipschitz surfaces in their definition. Thus the conjecture for γ says that, provided $\mathcal{H}^1(E) < \infty$, $\gamma(E)$ should be zero exactly when $\mathcal{H}^1(E \cap \Gamma) = 0$ for all rectifiable curves Γ. Clearly, it is necessary to assume here that $\mathcal{H}^1(E) < \infty$: a modification of the construction (5.1) yields a purely 1-unrectifiable set E with $\dim E > 1$, whence $\gamma(E) > 0$.

It is not difficult to show that the Cantor set in (5.1) is purely 1-unrectifiable. The same is true about the other examples of sets with zero capacity that will be given later.

6. Rectifiable sets and tangent measures

In this section we study some aspects of rectifiability, for more see [F] and [M4]. Let m be an integer, $0 < m < n$, and E an $\mathcal{H}^m$ measurable subset of $\mathbb{R}^n$ with $\mathcal{H}^m(E) < \infty$. We say that E is *m-rectifiable* if there exist m-dimensional C^1 surfaces (e.g. submanifolds) $M_1, M_2, \ldots$ such that $\mathcal{H}^m(E \setminus \cup_{i=1}^{\infty} M_i) = 0$. If $\mathcal{H}^m(E \cap M) = 0$ for every m-dimensional C^1 surface M, E is said to be *purely m-unrectifiable*. Again we could use Lipschitz graphs or Lipschitz images of $\mathbb{R}^m$ without changing the definition. There are various characterizations of rectifiability and unrectifiability. One particularly interesting is due to Besicovitch and Federer: E is purely m-unrectifiable if and only if the orthogonal projection of E on almost all m-planes through the origin has $\mathcal{H}^m$ measure zero. Another one can be stated in terms of the existence of tangent planes, which have to be defined in an approximate, measure-theoretic, way. We give a variant of this in terms of tangent measures. They were introduced by Preiss in [Pr]; some basic facts on them are given also in [M4].

For E with $\mathcal{H}^m(E) < \infty$, a basic density estimate is

$$2^{-m} \leq \limsup_{r \to 0} (2r)^{-m} \mathcal{H}^m(E \cap B(x,r)) \leq 1 \tag{6.1}$$

for $\mathcal{H}^m$ almost all $x \in E$, cf. [F] or [M4]. The lower limit may be zero everywhere, but we restrict to sets for which it is positive:

$$\liminf_{r \to 0} (2r)^{-m} \mathcal{H}^m(E \cap B(x,r)) > 0 \quad \text{for } \mathcal{H}^m \text{ almost all } x \in E. \tag{6.2}$$

This is valid in all of our examples, which in fact will be even AD_m-regular.

We say that ν is a tangent measure for E at $a \in \mathbb{R}^n$, $\nu \in \mathrm{Tan}(E,a)$, if there exists a sequence $r_i \to 0$, $r_i > 0$, such that

$$(2r_i)^{-m} T_{a,r_i}(\mathcal{H}^m|E) \to \nu \quad \text{weakly.}$$

Here $T_{a,r}x = (x-a)/r$ so that

$$T_{a,r}(\mathcal{H}^m|E)(A) = \mathcal{H}^m(E \cap (rA + a)).$$

Hence to build $\mathrm{Tan}(E,a)$ we are blowing up the balls $B(a,r)$ for small r to $B(0,1)$, also blowing up the measure $\mathcal{H}^m|E$ under these maps $T_{a,r}$, normalizing and taking all possible weak limits corresponding to sequences $r_i \to 0$ for which there is convergence. Thus knowing $\mathrm{Tan}(E,a)$ we know how E looks near a. Here we are using the special normalization constants $(2r_i)^{-m}$ motivated by (6.1) and (6.2). Under the assumption (6.2) there is no essential difference to [Pr] and [M4].

It is not hard to see that if E is a nice m-dimensional surface having a tangent plane T_a at $a \in E$, then $\mathrm{Tan}(E,a)$ contains only $\mathcal{H}^m|(T_a - a)$. For the Cantor set C in (5.1), $\mathrm{Tan}(C,a)$ contains infinitely many measures all living on unbounded Cantor sets which look locally as C itself. We also note that if E satisfies (6.2) then for $\mathcal{H}^m$ almost all $a \in E$, there is $C(a)$, $0 < C(a) < \infty$, such that for all $\nu \in \mathrm{Tan}(E,a)$,

$$r^m/C(a) \leq \nu B(x,r) \leq C(a)\, r^m \tag{6.3}$$

for $0 < r < \infty$ and $x \in \operatorname{spt}\nu$. This means that $\nu = \varphi_\nu \mathcal{H}^m|\operatorname{spt}\nu$ where $\varphi_\nu\colon \operatorname{spt}\nu \to \mathbb{R}$ is a Borel function with $0 < C_1(a)^{-1} \leq \varphi_\nu \leq C_1(a) < \infty$. To check (6.3) for $x = 0$, note that

$$(2r_i)^{-m}\, T_{a,r_i}(\mathcal{H}^m|E)(B(0,r)) = (2rr_i)^{-m}\, \mathcal{H}^m(E \cap B(a,rr_i))$$

and use (6.1), (6.2) and the definition of tangent measures. For general x there are a few more technical details involved, see [M4].

Theorem 6.4. *Suppose E is $\mathcal{H}^m$ measurable with $\mathcal{H}^m(E) < \infty$ such that* (6.2) *holds. Then the following are equivalent:*
(1) *E is m-rectifiable.*
(2) *For $\mathcal{H}^m$ almost all $a \in E$ there is an m-plane V_a through* 0 *such that*

$$\mathrm{Tan}(E,a) = \{\mathcal{H}^m|V_a\}.$$

(3) *For $\mathcal{H}^m$ almost all $a \in E$,*

$$\mathrm{Tan}(E,a) \subset \{c\mathcal{H}^m \mid V : 0 < c < \infty, V \text{ an } m\text{-plane with } 0 \in V\}.$$

(4) *For $\mathcal{H}^m$ almost all $a \in E$, $\operatorname{spt}\nu$ is an m-plane for all $\nu \in \mathrm{Tan}(E,a)$.*

So rectifiability means that locally E looks like an m-plane. The plane $V_a + a$ is the approximate tangent plane of E at a. The conditions (3) and (4) are often convenient for verifying the rectifiability of E: we need to show that locally E looks like an m-plane, but we do not have to know that it is the same plane at all scales. This then follows from the equivalence of (2) and (3) or (4). The proof of Theorem 6.4 can be found in [M4]. The condition (4) is not explicitly stated there but it follows by the same methods.

7. Tangent measures and removability

First we see how tangent measures can be used to check that a compact set $E \subset \mathbb{C}$ has zero analytic capacity. For simplicity, assume that E is AD_1-regular, recall (4.4). According to (3.7) we can start assuming that there is a non-zero Borel function $g\colon E \to \mathbb{C}$ with $|g| \le 1$ such that

$$\left|\int_E \frac{g(\zeta)}{\zeta - z}\, d\mathcal{H}^1\zeta\right| \le 1 \quad \text{for } z \in \mathbb{C}\setminus E.$$

By some simple estimates, see [M4], one derives a bit more. Namely, there is $C < \infty$ such that

$$\left|\int_{\varepsilon<|\zeta-z|\le\delta} \frac{g(\zeta)}{\zeta - z}\, d(\mathcal{H}^1\lfloor E)\,\zeta\right| \le C \quad \text{for } z \in \mathbb{C},\ 0<\varepsilon<\delta. \tag{7.1}$$

For tangent measures this turns into the following information: for $\mathcal{H}^1$ almost all $a \in \{a \in E : g(a) \ne 0\}$ there is $C(a)$, $0 < C(a) < \infty$, such that for all $\nu \in \operatorname{Tan}(E,a)$

$$\left|\int_{r<|\zeta-z|\le R} \frac{1}{\zeta - z}\, d\nu\zeta\right| \le C(a) \quad \text{for } z \in \mathbb{C},\ 0<r<R<\infty. \tag{7.2}$$

If $g \equiv 1$ this follows from

$$\begin{aligned}
&\left|\int_{r\le|\zeta-z|\le R} \frac{1}{\zeta - z}\, d((2r_i)^{-1}T_{a,r_i}(\mathcal{H}^1\lfloor E))\,\zeta\right| \\
&= \left|(2r_i)^{-1}\int_{r\le|\frac{\xi-a}{r_i}-z|\le R} \frac{1}{\frac{\xi-a}{r_i} - z}\, d(\mathcal{H}^1\lfloor E)\,\xi\right| \\
&= \frac{1}{2}\left|\int_{rr_i\le|\xi-(a+r_iz)|\le Rr_i} \frac{1}{\xi-(a+r_iz)}\, d(\mathcal{H}^1\lfloor E)\,\xi\right| \\
&\le C
\end{aligned}$$

by (7.1). This argument also works for continuous g. Some more details involving Lusin's theorem and density point theorem are needed for general g, cf. [M4]. Remember that ν also satisfies (6.3) with $m = 1$.

From this it is easy to check that $\gamma(C) = 0$ for the Cantor set C of (5.1): for every $\nu \in \operatorname{Tan}(C,a)$, $\operatorname{spt}\nu$ is again a Cantor set at whose corner points, using (6.3), it is trivial to check that (7.2) cannot hold. Thus $\mathcal{H}^1\{a \in C : g(a) \ne 0\} = 0$, which yields $\gamma(C) = 0$.

Let us look at another example where this method works. Start from a closed segment J of length l. Replace it by two segments J_1 and J_2 of lenght $l/2$, each having one end-point at the different end-points of J and making an angle α_1 with J. Continue this construction letting $\alpha_k \to 0$ so that at the kth stage you have 2^k segments of length $l2^{-k}$. At the limit one gets an AD_1-regular set E. By the Pythagoras theorem, E is contained in a rectifiable curve if $\sum \alpha_k^2 < \infty$; otherwise E is purely 1-unrectifiable. We assume this second case: $\sum \alpha_k^2 = \infty$. As the angle α_k tends to zero, the support of every tangent measure is contained in some line. But by Theorem 6.4 we know that at almost all points $a \in E$ there are tangent measures whose support is not the full line. So we can find $\nu \in \operatorname{Tan}(E,a)$, a line $L \supset \operatorname{spt}\nu$, $b \in \operatorname{spt}\nu$ and an open line segment $I \subset L\setminus\operatorname{spt}\nu$ with b as an end-point. Then using (6.3) it is trivial to check that (7.2) is false at $z = b$.

This last example was also treated by Fang in [F2] by a different method.

For the following general theorem see [M4] and [MP]; the case of κ'_n in [MP] was formulated and proved without tangent measures but it is essentially the same.

Theorem 7.3. *Let $E \subset \mathbb{R}^n$ be compact with $\mathcal{H}^{n-1}(E) < \infty$. If for $\mathcal{H}^{n-1}$ almost all $a \in E$,*

$$\liminf_{r\to 0}(2r)^{1-n}\mathcal{H}^{n-1}(E\cap B(a,r)) > 0$$

and for every $\nu \in \operatorname{Tan}(E,a)$, $\operatorname{spt}\nu$ is not contained in a hyperplane, then $\kappa'_n(E) = 0$, and in the case $n = 2$ also $\gamma(E) = 0$.

We still give a construction for a purely 1-unrectifiable set for which the tangent measure method does not apply, cf. [M4, 14.2 (3)]. Choose $m_1 = 1/r_1$ discs of radius r_1 inside the closed unit disc $B(0,1)$ touching the boundary of $B(0,1)$ at the points $e^{j2\pi i/m_1}$, $j = 1,\ldots,m_1$. Let r_k be the radius of each disc at the kth stage and let each of these k-level discs contain m_{k+1} $(k+1)$-level discs uniformly distributed along the boundary. We assume that $m_{k+1} = r_k/r_{k+1} \to \infty$ as $k \to \infty$. Then the resulting Cantor set is AD_1-regular. But now the support of every tangent measure is either a line or a countable union of circles, and the estimate (7.2) does hold. So we cannot conclude anything from that. However, Fang [F1] has developed the method of Jones from [J] (or [C1]) together with some harmonic measure estimates to prove that this set has zero analytic capacity.

8. Rectifiable sets and singular integrals

We have seen that the problems on removable sets are linked with singular integrals and rectifiability properties of sets. We now discuss some relations between these two latter topics. First, the rectifiability can often be characterized in terms of the principal values.

Theorem 8.1. *Let $E \subset \mathbb{R}^n$ be $\mathcal{H}^m$ measurable with $\mathcal{H}^m(E) < \infty$ and*

$$\liminf_{r\to 0}(2r)^{-m}\mathcal{H}^m(E\cap B(x,r)) > 0$$

for $\mathcal{H}^m$ almost all $x \in E$. Then E is m-rectifiable if and only if the limit

$$\lim_{\varepsilon\to 0}\int_{E\setminus B(x,\varepsilon)}\frac{x-y}{|x-y|^{m+1}}\,d\mathcal{H}^m y \in \mathbb{R}^n$$

exists for $\mathcal{H}^m$ almost $x \in E$.

That the rectifiability follows from the existence of these principal values was proved in [M3] for $m = 1$, $n = 2$ and in [MPr] for general m and n. The proofs make use of tangent measures. The converse for $m = 1$, $n = 2$ was proved in [MM]. Later Verdera derived this with an elegant duality method in [V1]; his proof works for general m and n. See also [M4].

The proof in [M3] and [MPr] goes as follows, see also [M4]. Recall how (7.1) led to (7.2). Now instead of the boundedness in (7.1), with the kernel $|x|^{-m-1}x$, we have that these integrals tend to zero as ε and δ tend to zero. This implies generically for tangent measures ν that

$$\int_{r<|x-y|\leq R} \frac{x-y}{|x-y|^{m+1}}\,d\nu y = 0 \quad \text{for } x \in \operatorname{spt}\nu,\ 0<r<R<\infty,$$

which is easily seen to be equivalent with

$$\frac{1}{\nu(B(x,r))}\int_{B(x,r)} y\,d\nu y = x \quad \text{for } x \in \operatorname{spt}\nu,\ r>0.$$

Such measures were called symmetric in [M3] and [MPr]. One can then show that a symmetric measure satisfying (6.3) must be a constant multiple of the m-dimensional Hausdorff measure on some m-plane. After this the rectifiability of E follows from Theorem 6.4.

The relation between rectifiability and the L^2-boundedness of the Cauchy and Riesz transforms is more problematic. Let us look at this first for more general kernels. Recall the class $\mathcal{K}_m(\mathbb{R}^n)$ of kernels from Section 4. Theorem 4.1 holds for the operators T_K corresponding to all kernels in $\mathcal{K}_1(\mathbb{R}^2)$. Moreover, for AD$_1$-regular sets F, T_K is bounded on $L^2(F)$ for all kernels $K \in \mathcal{K}_1(\mathbb{R}^2)$ if and only if F is contained in some Ahlfors regular curve, see [DS2]. David and Semmes call such sets F *uniformly rectifiable*. They also develop an extensive theory of m-dimensional uniformly rectifiable sets with relations to singular integrals, Littlewood–Paley-theory and other things, see [DS2] and in particular its long and excellent introduction. We do not give a precise definition here but roughly speaking an AD$_m$-regular set is uniformly rectifiable if it is rectifiable with quantitative estimates. For the proof of the following theorem see [DS1], [DS2] and [MPr].

Theorem 8.2. *Let F be an AD$_m$-regular closed subset of $\mathbb{R}^n$. Then the following are equivalent:*
(1) *F is uniformly rectifiable,*
(2) *T_K is bounded on $L^2(F)$ for all $K \in \mathcal{K}_m(\mathbb{R}^n)$,*
(3) *T_K is bounded on $L^2(F)$ for all kernels $K \in \mathcal{K}_m(\mathbb{R}^n)$ of the form*

$$K(x) = h(|x|)\,x|x|^{-m-1}$$

where h is a C^∞ function $\mathbb{R} \to \mathbb{R}$.

The good part of this theorem is that it guarantees that a large class of singular integral operators are L^2-bounded on uniformly rectifiable sets. The bad part is that in order to conclude the uniform rectifiability from the L^2-boundedness one needs a lot of kernels. It is a fundamental open problem, closely related to the earlier removability questions, to find out whether one single kernel, e.g. $|x|^{-m-1}x$, would be enough. For non-integral m Vihtilä [V] has shown that T_K with $K(x) = |x|^{-m-1}x$ cannot be bounded on $L^2(F)$ for any AD$_m$-regular set F.

Let us return to Conjecture 5.3 for AD_1-regular sets E. According to the above it would be enough to show that if T_K with $K(z) = 1/z$ is bounded on $L^2(E)$, then E must be rectifiable. We would obtain this from Theorem 8.1 if we could show that the L^2-boundedness implies the almost everywhere convergence of the principal values. Unfortunately the usual methods in the theory of the singular integrals do not seem to give any clue how such a proof could go.

Note added in October 1994. Recently Mark Melnikov, Joan Verdera and I showed that in Theorem 8.2 for $n = 2$, $m = 1$ the conditions (1)–(3) are equivalent to the boundedness of the Cauchy integral operator on $L^2(F)$. Thus also Conjecture 5.3 is true for $n = 2$ and AD_1-regular sets.

References

[A] Ahlfors, L. V., Bounded analytic functions. Duke Math. J. 14 (1947), 1–11.

[C] Calderón, A. P., Cauchy integrals on Lipschitz curves and related operators. Proc. Nat. Acad. Sci. U.S.A. 74 (1977), 1324–1327.

[Ca] Carleson, L., Selected Problems on Exceptional Sets. Van Nostrand, Princeton 1967.

[C1] Christ, M., Lectures on Singular Integral Operators. Regional Conference Series in Mathematics 77, Amer. Math. Soc., Providence 1990.

[C2] — A $T(b)$ theorem with remarks on analytic capacity and the Cauchy integral. Colloq. Math. 60/61 (1990), 601–628.

[CMM] Coifman, R. R., McIntosh, A., Meyer, Y., L'integrale de Cauchy définit un opérateur borné sur L^2 pour les courbes lipschitziennes. Ann. of Math. 116 (1982), 361–387.

[CW] Coifman, R. R., Weiss, G., Analyse harmonique non-commutative sur certains espaces homogènes. Lecture Notes in Math. 242, Springer-Verlag, Berlin 1971.

[D1] David, G., Operatéurs intégraux singuliers sur certaines courbes du plan complexe. Ann. Sci. École Norm. Sup. (4) 17 (1984), 157–189.

[D2] — Wavelets and Singular Integrals on Curves and Surfaces. Lecture Notes in Math. 1465, Springer-Verlag, Berlin 1991.

[DS1] David, G., Semmes, S., Singular Integrals and Rectifiable Sets in $\mathbb{R}^n$. Au-delà des graphes lipschitziens. Astérisque 193, 1991.

[DS2] — Analysis of and on Uniformly Rectifiable Sets. Surveys and Monographs 38, Amer. Math. Soc., Providence 1993.

[D] Davie, A. M., Analytic capacity and approximation problems. Trans. Amer. Math. Soc. 171 (1972), 409–444.

[Do] Dolzenko, E. P., The removability of singularities of analytic functions. Uspekhi Mat. Nauk 18 (1963), 135–142 (Russian).

[F1] Fang, X., The Cauchy integral of Calderón and analytic capacity. Ph.D. dissertation, Yale University, 1990.

[F2] — The Cauchy integral, analytic capacity and subsets of quasicircles. Pacific J. Math. (to appear).

[F] Federer, H., Geometric Measure Theory. Springer-Verlag, Berlin 1969.

[G1] Garnett, J., Positive length but zero analytic capacity. Proc. Amer. Math. Soc. 21 (1970), 696–699.

[G2] — Analytic Capacity and Measure. Lecture Notes in Math. 297, Springer-Verlag, Berlin 1972.

[H] Havin, V. P., Boundary properties of integrals of Cauchy type and of conjugate harmonic functions in regions with rectifiable boundary. Mat. Sb. (N.S.) 68 (110) (1965), 499–517 (Russian).

[HH] Havin, V. P., Havinson, S. J., Some estimates of analytic capacity. Soviet Math. Dokl. 2 (1961), 731–734.

[I] Ivanov, L. D., On Denjoy's conjecture. Uspekhi Mat. Nauk 18 (112) (1963), 147–149 (Russian).

[J] Jones, P. W., Square functions, Cauchy integrals, analytic capacity, and harmonic measure. In: Harmonic Analysis and Partial Differential Equations (ed. by J. Garcia-Cuerva), 24–68. Lecture Notes in Math. 1384, Springer-Verlag, 1989.

[JM] Jones, P. W., Murai, T., Positive analytic capacity but zero Buffon needle probability. Pacific J. Math. 133 (1988), 99–114.

[K] Kaufman, R., Hausdorff measure, BMO, and analytic functions. Pacific J. Math. 102 (1982), 369–371.

[Kr] Král, J., Semielliptic singularities. Čas. Pěst. Mat. 109 (1984), 304–322.

[M] Marshall, D. E., Removable sets for bounded analytic functions. In: Linear and Complex Analysis - 199 Research Problems (ed. by V. P. Havin, S. V. Hruščëv, N. K. Nikol'skii), 485–490. Lecture Notes in Math. 1043, Springer-Verlag, 1984.

[M1] Mattila, P., A class of sets with positive length and zero analytic capacity. Ann. Acad. Sci. Fenn. Ser. A I Math. 10 (1985), 387–395.

[M2] — Smooth maps, null-sets for integralgeometric measure and analytic capacity. Ann. of Math. 123 (1986), 303–309.

[M3] — Cauchy singular integrals and rectifiability of measures in the plane. Adv. Math. 115 (1995), 1–34.

[M4] — Geometry of Sets and Measures in Euclidean Spaces. Cambridge University Press, Cambridge 1995.

[MM] Mattila, P., Melnikov, M. S., Existence and weak type inequalities for Cauchy integrals of general measures on rectifiable curves and sets. Proc. Amer. Math. Soc. 120 (1994), 143–149.

[MP] Mattila, P., Paramonov, P. V., On geometric properties of harmonic Lip_1-capacity. Pacific J. Math. (to appear).

[MPr] Mattila, P., Preiss, D., Rectifiable measures in $\mathbb{R}^n$ and existence of principal values for singular integrals. J. London Math. Soc. (to appear).

[Mu1] Murai, T., Construction of H^1 functions concerning the estimate of analytic capacity. Bull. London Math. Soc. 19 (1986), 154–160.

[Mu2] — A Real-Variable Method for the Cauchy Transform, and Analytic Capacity. Lecture Notes in Math. 1307, Springer Verlag, Berlin 1988.

[O] O'Farrell, A. G., Capacities in function theory. In: Proceedings of the International Conference on Potential Theory, Nagoya (Japan), 1990 (ed. by M. Kishi), 93–105. Walter de Gruyter, Berlin 1992.

[P] Paramonov, P. V., On harmonic approximation in the C^1-norm. Math. USSR-Sb. 71 (1992), 183–207.

[Pr] Preiss, D., Geometry of measures in $\mathbb{R}^n$: distribution, rectifiability, and densities. Ann. of Math. 125 (1987), 537–643.

[S] Stein, E. M., Singular Integrals and Differentiability Properties of Functions. Princeton University Press, Princeton 1970.

[U1] Uy, N. X., On Riesz transforms of bounded functions of compact supports. Michigan Math. J. 24 (1977), 169–175.

[U2] — Removable sets of analytic functions satisfying a Lipschitz condition. Ark. Mat. 17 (1979), 19–27.

[U3] — An extremal problem on singular integrals. Amer. J. Math. 102 (1980), 279–290.

[U4] — A removable set for Lipschitz harmonic functions. Michigan Math. J. 37 (1990), 45–51.

[V1] Verdera, J., A weak type inequality for Cauchy transforms of measures. Publ. Mat. 36 (1992), 1029–1034.

[V2] — Removability, capacity and approximation. Complex Potential Theory, NATO ASI Series, Kluwer Academic Publishers (to appear).

[V] Vihtilä, M., The boundedness of Riesz s-transforms of measures in $\mathbb{R}^n$. Preprint.

[Vi1] Vitushkin, A. G., Example of a set of positive length but zero analytic capacity. Dokl. Akad. Nauk. SSSR 127 (1959), 246–249 (Russian).

[Vi2] — Analytic capacity of sets in problems of approximation theory. Russian Math. Surveys 22 (1967), 139–200.

[Z] Zalcman, L., Analytic Capacity and Rational Approximation. Lecture Notes in Math. 50, Springer-Verlag, Berlin 1968.

Bernstein–Walsh type theorems for pluriharmonic functions

Józef Siciak

Abstract. Given a complex-valued function f on a compact subset E of $\mathbb{C}^N$, define $\delta(f, E) := \limsup_{k\to\infty} \sqrt[k]{\inf \|f - p_k\|_E}$, the inf being taken over all pluriharmonic (PH) polynomials p_k of degree $\leq k$. Consider the following conditions:

(1) E has *positive capacity with respect to PH polynomials,* i.e.,

$$|p_k(z)| \leq A^k \|p_k\|_E (1 + |z|)^k, \quad z \in \mathbb{C}^N,\ k \geq 0,\ A = \text{const} > 0;$$

(2) E is determining for $PH(\mathbb{C}^N)$, entire PH functions, and

$$\{f \in C(E); \delta(f, E) = 0\} = \{f|_E; f \in PH(\mathbb{C}^N)\};$$

(3) E is *PH-regular*, i.e., for each $b > 1$ there exist $M > 0$ and a neighborhood U of E such that

$$\|p_k\|_U \leq M b^k \|p_k\|_E, \quad k \geq 0;$$

(4) V_E is continuous in $\mathbb{C}^N$ and for every $b > 1$ there is $M > 0$ such that

$$|p_k(z)| \leq M b^k \|p_k\|_E \exp(k V_E(z)), \quad z \in \mathbb{C}^N, \quad k \geq 0,$$

where V_E is the plurisubharmonic Green function of E;

(5) V_E is continuous in $\mathbb{C}^N$ and $V_E(z) = \lim_{k\to\infty} \chi_k(z)$, where

$$\chi_k(z) := \sup\{\frac{1}{k} \log |p_k(z)|; \|p_k\|_E \leq 1\}$$

and the convergence is locally uniform in $\mathbb{C}^N$;

(6) E is determining for germs of PH functions on E and

$$\{f \in C(E); \delta(f, E) < 1\} = \{f|_E; f \text{ is PH in a neighborhood of } E\}.$$

Theorem I. For every compact set E in $\mathbb{C}^N$ the conditions (1) and (2) are equivalent.

Theorem II. If E is polynomially convex then the conditions (3), (4), (5) and (6) are equivalent.

1991 Mathematics Subject Classification: 31C10, 32E30, 30E10, 41A17, 41A63.

This research was supported by KBN Grant No. 2 1077 91 01.

1. Introduction

Given an open set Ω in $\mathbb{C}^N$ we denote by PSH(Ω) the set of all plurisubharmonic (PSH) functions in Ω. We recall that an upper semicontinuous (usc) function $u\colon \Omega \longrightarrow [-\infty,\infty)$ is called PSH, if for every complex line l its restriction to $\Omega \cap l$ is a subharmonic (SH) function of a complex variable. If $N = 1$ then $\mathrm{SH}(\Omega) = \mathrm{PSH}(\Omega)$. If $N \geq 2$ then $\mathrm{PSH}(\Omega) \subset \mathrm{SH}(\Omega)$ and $\mathrm{PSH}(\Omega) \neq \mathrm{SH}(\Omega)$. PSH functions play a very important role in multidimensional complex analysis. If f is a holomorphic function in Ω ($f \in \mathcal{O}(\Omega)$), then $|f|^\alpha$, $\alpha \log|f| \in \mathrm{PSH}(\Omega)$ for every $\alpha \in \mathbb{R}_+$. By a theorem of Bremermann an usc function $u\colon \Omega \longrightarrow [-\infty,\infty)$ is PSH in a domain of holomorply $\Omega \subset \mathbb{C}^N$ if and only if there exists a sequence $\{f_k\} \subset \mathcal{O}(\Omega)$ such that $\{\frac{1}{k}\log|f_k|\}$ is locally uniformly upper bounded in Ω and

$$u = \left(\limsup_{k\to\infty} \tfrac{1}{k}\log|f_k|\right)^*,$$

where v^* denotes the usc regularization of v.

On the complex plane $\mathbb{C}$ a special role among all SH functions is played by logarithmic potentials. If $N \geq 2$ there is no direct counterpart of logarithmic potential theory. Logarithmic potential theory is strictly related to the theory of the Laplace operator Δ, which is linear and admits a fundamental solution of the Laplace equation $\Delta u = 0$. The pluripotential theory (i.e., the theory of PSH functions) is based on the theory of the Monge–Ampère operator

$$MA[u] := \det\left[\frac{\partial^2 u}{\partial z_j \partial \overline{z_k}}\right]_{j,k=1,\dots,N}$$

which is not linear ($N \geq 2$).

A nice introduction to pluripotential theory in $\mathbb{C}^N$ is given by M. Klimek's recent monograph [4].

In applications of logarithmic potential theory a crucial role is played by the Green function of a nonpolar compact subset of $\mathbb{C}$ with pole at infinity. In particular the Green function is extremely useful in the theory of polynomial or rational approximation on the complex plane (see e.g. [12]).

If $N \geq 2$, one can define a plurisubharmonic Green function with pole at ∞ which plays an analogous role in the theory of approximation by polynomials of N complex variables (see [8], [10]). The definition and fundamental properties of the PSH Green function of a compact set E in $\mathbb{C}^N$ will be recalled in §2.

In §3–§6 we prove a Bernstein–Walsh type inequality for pluriharmonic (PH) polynomials as well as Bernstein–Walsh type theorems on approximation of functions by PH polynomials on compact subsets of $\mathbb{C}^N$.

2. The plurisubharmonic Green function of a compact subset of $\mathbb{C}^N$ and the Bernstein–Walsh type inequality for polynomials of N complex variables

2.1. Put

$$\mathcal{L} := \{u \in \mathrm{PSH}(\mathbb{C}^N);\ u(z) \le \beta + \log(1 + |z|) \text{ on } \mathbb{C}^N\},$$

where $\beta \in \mathbb{R}$ may depend on u. The functions of the class $\mathcal{L}$ play a similar role in $\mathbb{C}^N$ $(N \ge 2)$ to logarithmic potentials on the complex plane. Let $\mathcal{P}(\mathbb{C}^N)$ (resp. $\mathcal{P}_k(\mathbb{C}^N)$) be the vector space of all polynomials (resp. polynomials of degree $\le k$) of N complex variables. It is clear that $\frac{1}{k}\log|p| \in \mathcal{L}$ for every $p \in \mathcal{P}_k(\mathbb{C}^k)$ $(k \ge 1)$, $p \not\equiv 0$.

2.2. An usc function $u\colon \mathbb{C}^N \longrightarrow [-\infty, \infty)$ belongs to $\mathcal{L}$ iff there exists a sequence $\{p_k\}$ such that $p_k \in \mathcal{P}_k(\mathbb{C}^N)$ and

$$u = \left(\limsup_{k\to\infty} \tfrac{1}{k}\log|p_k|\right)^* .$$

2.3. For a subset E of $\mathbb{C}^N$ the following conditions are equivalent:

(1) E is locally pluripolar (plp), i.e., $\forall a \in E\ \exists W \in \mathrm{PSH}(U_a)$ where U_a is an open neighbourhood of a such that $W = -\infty$ on $U_a \cap E$;

(2) E is globally plp, i.e., $\exists W \in \mathrm{PSH}(\mathbb{C}^N)$ such that $W = -\infty$ on E;

(3) E is $\mathcal{L}$-polar, i.e., $\exists W \in \mathcal{L}$ such that $W = -\infty$ on E.

The implication (1) $\Longrightarrow$ (2) is due to B. Josefson [3], and the implication (2) $\Longrightarrow$ (3) was proved in [10].

2.4. If E is a subset of $\mathbb{C}^N$ we define

$$V_E(z) \equiv V(z, E) := \sup\{u(z); u \in \mathcal{L},\ u \le 0 \text{ on } E\},\ z \in \mathbb{C}^N.$$

2.5. If E is a compact subset of $\mathbb{C}^N$ then

$$V_E(z) = \lim_{k\to\infty} V_k(z) = \sup_{k\ge 1} V_k(z), \quad z \in \mathbb{C}^N,$$

where $V_k(z) := \sup\{\frac{1}{k}\log|p(z)|; p \in \mathcal{P}_k(\mathbb{C}^N), \|p\|_E \le 1\}$.

2.6. $V_E^* \equiv +\infty$ iff E is plp.

2.7. E is not plp iff $V_E^* \in \mathcal{L}$.

2.8. For a compact set E in $\mathbb{C}^N$ the following conditions are equivalent:

(1) E is regular with respect to polynomials of N complex variables, i.e., $\forall b > 1$ $\exists M > 0$ $\exists$ an open neighbourhood Ω of E $\forall p \in \mathcal{P}(\mathbb{C}^N)$

$$\|p\|_\Omega \leq M b^{\deg p} \|p\|_E;$$

(2) E is *$\mathcal{L}$-regular*, i.e., V_E is continuous in $\mathbb{C}^N$;

(3) $V_E^* = V_E$ on E (i.e., V_E is continuous at each point of E).

Theorem 2.9. *If E is a nonpluripolar compact subset of $\mathbb{C}^N$ then V_E^* is the unique function $u \in \mathrm{PSH}(\mathbb{C}^N)$ with the following properties:*

(i) $\alpha + \log(1 + |z|) \leq u(z) \leq \beta + \lg(1 + |z|)$, $z \in \mathbb{C}^N$;

(ii) $u = 0$ *on* $E \setminus A$, *where* A *is plp;*

(iii) $\det \left[\dfrac{\partial^2 u}{\partial z_j \partial \overline{z_k}} \right]_{j,k=1,\ldots,N} = 0$ *in* $\mathbb{C}^N \setminus E$.

By definition the condition (iii) *is satisfied, if for every sequence* $u_\nu \in \mathcal{C}^\infty(\mathbb{C}^N) \cap \mathrm{PSH}(\mathbb{C}^N)$ *with* $u_{\nu+1} \leq u_\nu$, $u = \lim_{\nu\to\infty} u_\nu$, *and for every test function* $\phi \in \mathcal{C}_0^\infty(\mathbb{C}^N \setminus E)$

$$\lim_{\nu\to\infty} \int \phi \det \left[\frac{\partial^2 u_\nu}{\partial z_j \partial \overline{z_k}} \right]_{j,k=1,\ldots,N} d\lambda = 0,$$

where λ is Lebesgue measure in $\mathbb{C}^N$.

Proof. If $E \subset \mathbb{C}^N$ is not plp then $u := V_E^*$ satisfies (i) (resp. (ii)) by definition 2.4 (resp. by the Bedford–Taylor Theorem for neglible sets [2]; see [2] or [4] for details). The condition (iii) is then satisfied by [2], Corollary 9.4.

Now let $u_1, u_2 \in \mathrm{PSH}(\mathbb{C}^N)$ satisfy (i), (ii) and (iii). Let W be a function of the class $\mathcal{L}$ with $W \leq 0$ on E and $W = -\infty$ on A. Let D_∞ be the unbounded component of $\mathbb{C}^N \setminus E$. Put $D_R := \{z \in D_\infty; |z| < R\}$. Given $\varepsilon_1 > 0$, $\varepsilon_2 > 0$ put $u := (1 + \varepsilon_2)u_2$ and $v := \varepsilon_1 W + (1 - \varepsilon_1)u_1$. Then there is R_0 such that

$$\liminf_{\substack{z\to w \\ z\in D_R}} (u(z) - v(z)) \geq 0, \quad w \in \partial D_R,\ R > R_0.$$

By Corollary 3.7.6 of [4] we have $u \geq v$ in D_R. By the arbitrary nature of $R > R_0$, $\varepsilon_1 > 0$, $\varepsilon_2 > 0$, we get $u_2 \geq u_1$ in D_∞. By an analogous argument $u_1 \leq u_2$ in D_∞. Therefore $u_1 = u_2$ in D_∞.

If D is a bounded component of $\mathbb{C}^N \setminus E$, Corollary 3.7.6 of [4] applied to $u := u_2$, $v := \varepsilon_1 W_1 + (1 - \varepsilon_1)U_1$ gives $u_2 \geq u_1$ in D. Analogously, one gets $u_2 \leq u_1$ in D. Therefore $u_1 = u_2$ in $\mathbb{C}^N \setminus A$, where A is plp. It follows that $u_1 = u_2$ in $\mathbb{C}^N$. □

2.10. If E is a nonpolar compact subset of the complex plane $\mathbb{C}$ then V_E is the Green function of the unbounded component D_∞ of $\mathbb{C} \setminus E$ with pole at ∞.

2.11. As a direct consequence of 2.4 one gets the following *Bernstein–Walsh type inequality (BWI)* for holomorphic polynomials of N complex variables

$$\log |p_k(z)| \leq \log \|p_k\|_E + kV_E(z), \quad z \in \mathbb{C}^N,$$

where $p_k \in \mathcal{P}_k(\mathbb{C}^N)$, $k = 0, 1, \ldots$.

Later on we shall need the following Lemma.

Lemma 2.12. *If a compact set $E \subset \mathbb{C}^N$ is nonpluripolar then there exists $R_0 > 1$ such that for every $R > R_0$ the open set*

$$E_R := \{z \in \mathbb{C}^N; V_E^*(z) < \log R\}$$

is connected.

If E is $\mathcal{L}$-regular, then the set $F_R := \{z \in \mathbb{C}^N; V_E(z) \leq \log R\}$ is connected for every $R > R_0$.

Proof. Let the ball $B = B(0, \rho)$ be so large that $E \subset B$.
Put $R_0 := \max\{V_E^*(z); z \in \overline{B}\}$. Let $R > R_0$. Suppose E_R is not connected. Let G be the component of E_R disjoined with B. Then the function

$$u(z) := \begin{cases} V_E^*(z), & z \in \mathbb{C}^N \setminus G \\ \log R, & z \in G \end{cases}$$

is an element of the class $\mathcal{L}$ and $u \leq 0$ on $E \setminus A$, where A is a plp subset of E. Therefore $u(z) \leq V_E(z)$ in $\mathbb{C}^N$. But $V_E(z) < \log R = u(z)$ in G. We have got a contradiction which shows that E_R is connected.

Now let E be $\mathcal{L}$-regular. It is clear that if E_R is connected, then $\overline{E_R}$, the closure of E_R, is also connected. If V_E is continuous then $\overline{E_R} \subset F_R$. Suppose, contrary to our claim, that F_R is not connected for some $R > R_0$. Then there are two nonempty compact sets A and B such that $F_R = A \cup B$, $A \cap B = \emptyset$, $\overline{E_R} \subset A$. Hence for each $R' > R$ sufficiently close to R the set $E_{R'}$ is a disconnected neighborhood of F_R, which is impossible (because $E_{R'}$ is connected). □

Later on we shall need the following version of the Polynomial Lemma due to F. Leja [9].

2.13. Polynomial Lemma. *Let $\mathcal{F}$ be a family of polynomials of m complex variables such that*

$$\sup\{|p(x)|; p \in \mathcal{F}\} < \infty$$

for every point x in an open set $G \subset \mathbb{R}^m$. Then for every compact subset E of G and every $b > 1$ there exist $M > 0$ and an open set Ω in $\mathbb{C}^m$ such that $\Omega \supset E$ and

$$\|p\|_\Omega \leq M b^{\log p}, \quad p \in \mathcal{F}.$$

3. Compact sets in $\mathbb{C}^N$ of positive capacity with respect to pluriharmonic polynomials

3.1. Given an open set Ω in $\mathbb{C}^N$ we denote by $PH(\Omega)$ the set of complex–valued pluriharmonic (PH) functions f in Ω of the form $f(z) = g(z) + \overline{h(z)}$, where $g, h \in \mathcal{O}(\Omega)$ (holomorphic functions in Ω).

If Ω is simply connected then a $\mathcal{C}^2$-function f belongs to $PH(\Omega)$ if and only if $\frac{\partial^2 f}{\partial z_j \partial \overline{z}_k} = 0$ $(j, k = 1, \dots, N)$ in Ω. Denote by $PH_k(\mathbb{C}^N)$ the set of all PH polynomials of N complex variables of degree $\leq k$.

3.2. Let $\mathcal{W}$ be a family of complex–valued functions defined on $\mathbb{C}^N$. We say that $E \subset \mathbb{C}^N$ is *determining* for functions of the family $\mathcal{W}$ (shortly: $\mathcal{W}$-*determining*), if the following implication is true:

$$f \in \mathcal{W},\ f = 0 \text{ on } E \implies f = 0 \text{ in } \mathbb{C}^N.$$

3.3. If $d_k := \dim PH_k(\mathbb{C}^N)$ is the dimension of the vector space of PH polynomials of N complex variables of degree $\leq k$, then $d_k = 2\binom{k+N}{N} - 1$.

Let $e_1, \dots, e_{d_k}$ be a fixed basis of $PH_k(\mathbb{C}^N)$ (e. g. the canonical basis consisting of the monomials z^α, $\overline{z}^\beta$, $|\alpha| \leq k$, $|\beta| \leq k$, $\alpha, \beta \in \mathbb{Z}_+^N$). Put

$$V(z_1, \dots, z_s) := \det \left[e_j(z_i)\right]_{j,i=1,\dots,s}, \quad z_1, \dots, z_s \in \mathbb{C}^N,\ s = 1, \dots, d_k.$$

Then E is $PH_k(\mathbb{C}^N)$-determining if and only if there are points $z_1, \dots, z_{d_k} \in E$ with $V(z_1, \dots, z_{d_k}) \neq 0$.

If $V(z_1, \dots, z_{d_k}) \neq 0$, the system of polynomials

$$L_{kj}(z) := V(z_1, \dots, z_{j-1}, z, z_{j+1}, \dots, z_{d_k})/V(z_1, \dots, z_{d_k}),\ \ j = 1, \dots, d_k \qquad (*)$$

is a basis of $PH_k(\mathbb{C}^N)$.

Indeed, $L_{kj}(z_i) = \delta_{ji}$ $(j, i = 1, \dots, d_k)$, which implies that the polynomials (*) are linearly independent elements of $PH_k(\mathbb{C}^N)$. Therefore

$$p(z) = \sum_{j=1}^{d_k} p(z_j) L_{kj}(z), \quad z \in \mathbb{C}^N,\ p \in PH_k(\mathbb{C}^N).$$

Every system of d_k polynomials $\{L_{kj}; j = 1, \dots, d_k\} \subset PH_k(\mathbb{C}^N)$ such that $L_{kj}(z_i) = \delta_{ji}$ $(j, i = 1, \dots, d_k)$, where $\{z_1, \dots, z_{d_k}\}$ is a system of d_k points of $\mathbb{C}^N$, is called a *Lagrange Interpolation Basis* of $PH_k(\mathbb{C}^N)$ with nodes $z_1, \dots, z_{d_k}$.

A subset E of $\mathbb{C}^N$ is $PH_k(\mathbb{C}^N)$-determining if and only if there exists a Lagrange Interpolation Basis $\{L_{kj}; j = 1, \dots, d_k\}$ of $PH_k(\mathbb{C}^N)$ with nodes $z_1, \dots, z_{d_k} \in E$.

3.4. Let E be a compact set in $\mathbb{C}^N$. We say that $\xi_1, \dots, \xi_{d_k} \in E$ are *PH-extremal points of E of order k*, if

$$|V(\xi_1, \dots, \xi_{d_k})| = \max\{|V(z_1, \dots, z_{d_k})|;\ z_j \in E\ (j = 1, \dots, d_k)\}. \qquad (**)$$

As a direct consequence of (*) and (**) one gets the following Proposition.

Proposition 3.5. *Let E be a compact set in $\mathbb{C}^N$ determining for PH polynomials. Let $\{L_{kj}; j=1,\dots,d_k\}$ be a Lagrange Interpolation Basis of $PH_k(\mathbb{C}^N)$ with nodes $\xi_{k1},\dots,\xi_{kd_k}$ – the PH-extremal points of E of order k. Put*

$$L_k f(z) := \sum_{j=1}^{d_k} f(\xi_{kj}) L_{kj}(z), \quad z \in \mathbb{C}^N, \ f \in C(E), \ k \geq 0. \tag{α_k}$$

Then

$$\|L_{kj}\|_E \leq 1, \quad j=1,\dots,d_k; \ k=1,\dots; \tag{1}$$

$$\|L_k f\|_E \leq d_k \|f\|_E, \quad f \in C(E), \ k \geq 0; \tag{2}$$

$$\delta_k(f,E) \leq \|f - L_k f\|_E \leq (1+d_k)\delta_k(f,E), \quad k \geq 0, \ f \in C(E), \tag{3}$$

where

$$\delta_k(f,E) := \inf\{\|f-p\|_E; \ p \in PH_k(\mathbb{C}^N)\}. \tag{4}$$

3.6. Given a compact set E in $\mathbb{C}^N$ we say that E is of *positive capacity with respect to PH polynomials* if $c_{PH}(E) > 0$, where

$$c_{PH}(E) := \liminf_{k\to\infty} \sqrt[k]{c_k(E)},$$

$$c_k(E) \equiv c_k(E, PH) := \inf\{\|p\|_E / \|p\|_B; \ p \in PH_k(\mathbb{C}^N)\}, \ B := \{z \in \mathbb{C}^n; \ |z| \leq 1\}.$$

3.7. Given a compact set E in $\mathbb{C}^N$ we define the *extremal function* χ_E *of* E *with respect to PH polynomials* by the formula

$$\chi_E(z) := \limsup_{k\to\infty} \chi_k(z), \quad z \in \mathbb{C}^N,$$

where

$$\chi_k(z) \equiv \chi_k(z,E) := \sup\{\tfrac{1}{k}\log|p(z)|; \ p \in PH_k(\mathbb{C}^N), \ \|p\|_E \leq 1\}.$$

3.8. We say that a complex-valued function f defined on a compact subset E of $\mathbb{C}^N$ *admits e-rapid approximation by PH polynomials*, if $\lim_{k\to\infty} \sqrt[k]{\delta_k(f,E)} = 0$.

Let $e(E,PH) := \{f \in C(E); \ \lim_{k\to\infty} \sqrt[k]{\delta_k(f,E)} = 0\}$ be the vector space of functions f on E admitting e-rapid uniform approximation on E by PH polynomials.

Theorem 3.9 (Characterization of compact sets with positive PH-capacity). *For a compact set E in $\mathbb{C}^N$ the following conditions are equivalent:*

(i) *E is of positive capacity with respect to PH polynomials.*

(ii) *There exists a positive constant A such that*

$$|p(z)| \leq A^k \|p\|_E (1+|z|)^k, \quad z \in \mathbb{C}^N, \ p \in PH_k(\mathbb{C}^N), \ k \geq 0.$$

(iii) *There exists a positive constant B such that*

$$\chi_E(z) \le B + \log(1+|z|), \quad z \in \mathbb{C}^N.$$

(iv) *E is determining for entire PH functions and*

$$e(E, PH) = \{f|E;\ f \in PH(\mathbb{C}^N)\}.$$

The condition (iv) can equivalently be stated as follows: A function $f\colon E \longrightarrow \mathbb{C}$ admits e-rapid approximation by PH polynomials iff there is a unique entire PH function $\widetilde{f}$ with $\widetilde{f} = f$ on E.

Proof. (i) $\Longrightarrow$ (ii). Regard $\mathbb{C}^N \equiv \mathbb{R}^{2N}$ as a "real" part of $\mathbb{C}^{2N}$. The real unit ball $B := \{z \in \mathbb{R}^{2N};\ |z| \le 1\}$ is a nonpluripolar subset of $\mathbb{C}^{2N}$.
Therefore there is a positive constant A_1 such that

$$|p(z)| \le A_1^k \|p\|_B (1+|z|)^k, \quad p \in \mathcal{P}_k(\mathbb{C}^{2N}),\ z \in \mathbb{C}^{2N}.$$

Hence, by definition 3.6,

$$|p(z)| \le A_1^k \frac{\|p\|_E}{c_k(E)} (1+|z|)^k, \quad p \in PH_k(\mathbb{C}^N),\ k \ge 0,\ z \in \mathbb{C}^N.$$

By (i), $\liminf_{k\to\infty} \sqrt[k]{c_k(E)} = c_{PH}(E) > 0$, so there exists $A > 0$ with $A_1/\sqrt[k]{c_k(E)} \le A$ $(k \ge 0)$, which shows that (i) $\Longrightarrow$ (ii).

(ii) $\Longrightarrow$ (iii). If (ii) is satisfied then (iii) is true with $B := \log A$.

(iii) $\Longrightarrow$ (iv). 1° First we shall show that $\{f|E;\ f \in PH(\mathbb{C}^N)\} \subset e(E, PH)$ for every compact set E in $\mathbb{C}^N$. By the Poisson formula for harmonic functions for every $R > 0$ there exists $M = M(R) > 0$ such that for each $f \in PH(\mathbb{C}^N)$

$$\left| \frac{f^{(\alpha,\beta)}(0)}{\alpha!\beta!} \right| \le M(R) \|f\|_{B(0,R)} R^{-|\alpha|-|\beta|}, \quad \alpha, \beta \in \mathbb{Z}_+^N,$$

where $f^{(\alpha,\beta)} := \left(\frac{\partial}{\partial z}\right)^\alpha \left(\frac{\partial}{\partial \bar z}\right)^\beta f$. Hence

$$\|f - f_k\|_{B(0,r)} \le M(r,R) \|f\|_{B(0,R)} \left(\frac{r}{R}\right)^k, \quad R > r > 0,\ k = 1, 2, \ldots, \tag{*}$$

where

$$f_k(z) := \sum_{|\alpha|+|\beta| \le k} \frac{f^{(\alpha,\beta)}(0)}{\alpha!\beta!} z^\alpha \bar z^\beta.$$

Therefore

$$\lim_{k\to\infty} \sqrt[k]{\delta_k(f, E)} = 0 \tag{**}$$

for each compact set $E \subset \mathbb{C}^N$ and for every $f \in PH(\mathbb{C}^N)$.

2° Now we shall prove that if (iii) is satisfied then E is $PH(\mathbb{C}^N)$-determining. Put $r := \max\{|z|;\ z \in E\}$ and let f be an entire PH function with $f = 0$ on E. Then by (*)

$$\|f_k\|_E \le M(f,R)(\tfrac{r}{R})^k, \quad R > r,\ k = 1, 2, \ldots,$$

where $M(f,R)$ is a positive constant depending only on f and R. Hence by definition 3.7 of χ_k one gets

$$\tfrac{1}{k}\log|f_k(z)| \le \tfrac{1}{k}\log M(f,R) + \log\tfrac{r}{R} + \chi_k(z,E), \quad z\in\mathbb{C}^N,\ k\ge 1,\ R>r,$$

which implies that $f(z) = \lim_{k\to\infty} f_k(z) = 0$ for all $z\in\mathbb{C}^N$.

3° It remains to show that if (iii) is satisfied then $e(E,PH) \subset \{f|E;\, f\in PH(\mathbb{C}^N)\}$. Fix f in $e(E,PH)$. Choose p_k in $PH_k(\mathbb{C}^N)$ such that $\|f-p_k\|_E = \delta_k(f,E)$ and define

$$\widetilde{f} := p_1 + \sum_{k=2}^{\infty}(p_k - p_{k-1}). \tag{***}$$

We claim that the series converges locally uniformly in $\mathbb{C}^N$, so its sum $\widetilde{f}$ is (by Harnack's Theorem) an entire PH function with $\widetilde{f} = f$ on E.

Indeed, $\|p_k - p_{k-1}\|_E \le \delta_k(f,E) + \delta_{k-1}(f,E)$. Therefore

$$\tfrac{1}{k}\log|p_k(z)-p_{k-1}(z)| \le \tfrac{1}{k}\log[2\delta_{k-1}(f,E)] + \chi_k(z,E), \quad z\in\mathbb{C}^N,\ k\ge 1.$$

Hence by (iii) and by (**)

$$\lim_{k\to\infty}\tfrac{1}{k}\log|p_k(z)-p_{k-1}(z)| = -\infty, \quad z\in\mathbb{C}^N,$$

which, by the Polynomial Lemma, shows that the series (***) is convergent locally uniformly in $\mathbb{C}^N$. It is clear that $\widetilde{f} = f$ on E. The proof of the implication (iii) $\Longrightarrow$ (iv) is complete.

(iv) $\Longrightarrow$ (i). The linear space $PH(\mathbb{C}^N)$ endowed with the topology of uniform convergence on compact subsets of $\mathbb{C}^N$ is (by Harnack's Theorem) a Frechet space. For every $b>1$ we define a norm $q_{E,b}$ on $e(E,PH)$ by the formula

$$q_{E,b}(f) := \|f\|_E + \sup_{k\ge 0} b^k\delta_k(f,E), \quad f\in e(E,PH).$$

One can check that $e(E,PH)$ becomes a Frechet space when endowed with the topology determined by the norms $q_{E,b}$, $b>1$ (It is sufficient to take the norms $q_{E,b}$ with $b = 1,2,\dots$).

By (iv) the restriction mapping $\mathcal{R}_E\colon PH(\mathbb{C}^N)\ni f \longrightarrow f|E \in e(E,PH)$ is linear and bijective. We claim that $\mathcal{R}_E$ is continuous. Indeed, given $b>1$, we have by (*)

$$q_{E,b}(f|E) < \|f\|_E + \sup_{k\ge 0} b^k\|f-f_k\|_{B(0,r)} \le \|f\|_{B(0,R)}\,[1+M(r,R)]$$

for all $f\in PH(\mathbb{C}^N)$, where $R>br$, $r:=\max\{|z|;\, z\in E\}$, which shows our claim.

By Banach's open mapping theorem $L := \mathcal{R}_E^{-1}$ is a linear continuous mapping of $e(E,PH)$ onto $PH(\mathbb{C}^N)$ such that $Lf = f$ on E for every $f\in e(E,PH)$. Moreover, $L(p|E) = p$ for each PH polynomial p.

By the continuity of L for every $R>0$ there are $b>1$ and a positive constant M such that

$$\|Lf\|_{B(0,R)} \le M q_{E,b}(f), \quad f\in e(E,PH).$$

In particular

$$\|p\|_{B(0,R)} \le M q_{E,b}(p|E) \le 2M b^{\deg p} \|p\|_E$$

for every PH polynomial p, which completes the proof of Theorem 3.9. □

Corollary 3.10. *A compact set E in $\mathbb{C}^N$ has positive capacity with respect to PH polynomials iff the restriction mapping*

$$\mathcal{R}_E \colon PH(\mathbb{C}^N) \ni f \mapsto f|E \in e(E, PH)$$

is an isomorphism of the Frechet spaces.

3.11. Given $b > 1$ and a compact set E in $\mathbb{C}^N$, put

$$a_b(E, PH) := \{f \in \mathcal{C}(E);\ q_{E,b}(f) < \infty\},$$

where $q_{E,b}(f) := \|f\|_E + \sup_{k \ge 0} b^k \delta_k(f, E)$. Then $a_b(E, PH)$ is a Banach space with the norm $q_{E,b}$. It is clear that $e(E, PH) = \bigcap_{b>1} a_b(E, PH)$.

Lemma 3.12. *Let E be a compact set in $\mathbb{C}^N$ of positive PH-capacity. For each $k \ge 0$ let L_k denote the Lagrange interpolation operator with nodes at PH-extremal points of E defined by (α_k) of Proposition 3.5.*

Then for each $b > 1$ and for every $f \in a_b(E, PH)$ the series

$$Lf := L_1 f + \sum_{k=2}^{\infty} (L_k f - L_{k-1} f) \tag{α}$$

is locally uniformly convergent in $E_b' := \{z \in \mathbb{C}^N;\ \chi_E^*(z) < \log b\}$, its sum Lf is a PH function in E_b', $Lf = f$ on E and $Lf = p$ on $\mathbb{C}^N$ if $f = p|E$, where p is a PH polynomial.

Moreover, for every relatively compact subset U of E_b' there exists $M = M(b, U)$ such that

$$\|Lf\|_U \le M q_{E,b}(f), \quad f \in a_b(E, PH), \tag{β}$$

i.e., $L \colon a_b(E, PH) \ni f \longrightarrow Lf \in PH(E_b')$ is a continuous linear extension operator from $a_b(E, PH)$ to $PH(E_b')$.

Proof. Given $f \in a_b(E, PH)$, we have $\|L_1 f\|_E \le d_1 \|f\|_E \le d_1 q_{E,b}(f)$, $\|L_k f - L_{k-1} f\|_E \le \|L_k f - f\|_E + \|L_{k-1} f - f\|_E \le (1 + d_k)\delta_k(f, E)$ $+(1 + d_{k-1})\delta_{k-1}(f, E) \le 2(1 + d_k)\delta_{k-1}(f, E) \le 2(1 + d_k) b^{-k+1} q_{E,b}(f)$. Therefore

$$|L_k f(z) - L_{k-1} f(z)| \le 2(1 + d_k) b^{-k+1} q_{E,b}(f) \exp[k \chi_k(z, E)]$$

for all $z \in \mathbb{C}^N$ and $k \ge 1$. Given a relatively compact subset U of E_b', put $\log b_1 := \sup_U \chi_E^*$. Then $1 \le b_1 < b$. Choose b_2 with $b_1 < b_2 < b$ and $\varepsilon > 0$ with $b_2 e^\varepsilon < b$. By the Polynomial Lemma there exists $M_1 = M_1(U, \varepsilon) > 0$ such that

$$\|L_k f - L_{k-1} f\|_U \le 2(1 + d_k) b q_{E,b}(f) M_1 \left(\frac{e^\varepsilon b_2}{b}\right)^k, \quad k \ge 1,$$

which implies that the series (α) is uniformly convergent on U and

$$\|Lf\|_U \le q_{E,b}(f)\left[d_1 + 2bM_1\sum_{k=2}^{\infty}(1+d_k)\left(\frac{e^{\varepsilon}b_2}{b}\right)^k\right] = Mq_{E,b}(f)$$

for every $f \in a_b(E, PH)$. □

Corollary 3.13. *The operator L, given by the series (α), defines a continuous linear one-to-one mapping of $e(E, PH)$ onto $PH(\mathbb{C}^N)$ such that $Lf = f$ on E for each $f \in e(E, PH)$ and $L(p|E) = p$ for every PH polynomial p.*

3.14. Remark. It is known [10] that a compact set E in $\mathbb{C}^N$ is of positive capacity with respect to holomorphic polynomials iff E is not pluripolar. It is clear that if $c_{PH}(E) > 0$ then E is not pluripolar (but not vice versa).

One can easily check that all the results of §3 have counterparts for compact subsets of $\mathbb{C}^N$ and holomorphic polynomials of N complex variables.

One can also check that the contents of §3 can be reformulated for compact subsets of $\mathbb{R}^N$ with respect to harmonic polynomials of N real variables.

In particular, if E is a compact set in $\mathbb{R}^N$ then the following conditions are equivalent.

(i) E is of positive capacity with respect to harmonic polynomials, i.e.,

$$\liminf_{k\to\infty}\left(\inf\{\|p\|_E/\|p\|_B;\ p \in H_k(\mathbb{R}^N)\}\right)^{1/k} > 0;$$

(ii) $\{f \in C(E);\ \lim_{k\to\infty}\sqrt[k]{\varepsilon_k(f,E)} = 0\} = \{f|E;\ f \in H(\mathbb{R}^N)\}$, where $H_k(\mathbb{R}^N)$ is the set of harmonic polynomials on $\mathbb{R}^N$ of degree $\le k$, $H(\mathbb{R}^N)$ is the set of entire harmonic functions of N real variables and $\varepsilon_k(f,E) := \inf\{\|f-p\|_E;\ p \in H_k(\mathbb{R}^N)\}$.

4. The Bernstein–Walsh type inequality for pluriharmonic polynomials

4.1. We say that a compact set E in $\mathbb{C}^N$ is *regular with respect to PH polynomials* (shortly *PH-regular*), if for every $b > 1$ there exist $M > 0$ and a neighbourhood Ω of E such that

$$\|p\|_\Omega \le Mb^{\deg p}\|p\|_E$$

for every PH polynomial p.

4.2. It is clear that every PH-regular compact set in $\mathbb{C}^N$ is $\mathcal{L}$-regular (regular with respect to holomorphic polynomials) but not vice versa.

Theorem 4.3. *If a compact set $E \subset \mathbb{C}^N$ is PH-regular then E is $\mathcal{L}$-regular and for each $b > 1$ there is $M > 0$ such that for every PH polynomial p we have the* *following Bernstein–Walsh Type Inequality*

$$|p(z)| \le M\|p\|_E(b\Phi_E(z))^{\deg p}, \quad z \in \mathbb{C}^N, \tag{BWI}$$

where $\Phi_E := \exp V_E$ *and* V_E *is defined by* 2.4.

Proof. 1° First we shall show that (BWI) holds locally in $\mathbb{C}^N$, i.e., $\forall a \in \mathbb{C}^N\ \forall b > 1$ $\exists M > 0\ \exists r > 0$ such that for each PH polynomial p

$$|p(z)| \le M\|p\|_E(b^3\Phi_E(z))^{\deg p}, \quad \text{if } |z-a| < r. \tag{1}$$

By the PH-regularity of E and by the Poisson integral formula for harmonic functions, for each $b > 1$ there is $M_1 > 0$ such that for every PH polynomial p

$$\|dp\|_E \le M_1 b^{\deg p}\|p\|_E,$$

where $\|dp\|_E : = \sum_{j=1}^N \left(\left\|\frac{\partial p}{\partial z_j}\right\|_E + \left\|\frac{\partial p}{\partial \bar z_j}\right\|_E\right)$.

Since the coefficiensts of the $(1,0)$-form $\partial p := \sum_{j=1}^N \frac{\partial p}{\partial z_j} dz_j$ (resp. of the $(0,1)$-form $\bar\partial p := \sum_{j=1}^N \frac{\partial p}{\partial \bar z_j} d\bar z_j$) are holomorphic (resp. antiholomorphic) polynomials, therefore the Bernstein–Walsh inequality for holomorphic polynomials shows that

$$|dp(z)| := \sum_{j=1}^N \left(\left|\frac{\partial p}{\partial z_j}(z)\right| + \left|\frac{\partial p}{\partial \bar z_j}(z)\right|\right) \le 2NM_1\,(b\Phi_E(z))^{\deg p}\,\|p\|_E, \tag{2}$$

for all $z \in \mathbb{C}^N$.

By the PH-regularity of E the inequality (1) is true for every $a \in \mathbb{C}^N$ with $\Phi_E(a) = 1$.

Fix a in $\mathbb{C}^N$ with $\Phi_E(a) =: \rho > 1$. Given $b > 1$ take $r > 0$ so small that $B(a,r) \subset \{b^{-1}\rho < \Phi_E(z) < b\rho\}$. Then

$$b^2\rho \le b^3\Phi_E(z) \quad \text{in } B(a,r). \tag{3}$$

Observe that

$$p(z) = p(z_0) + \int_{\gamma_{z_0 a}} dp + \int_a^z dp, \quad z \in \mathbb{C}^N,$$

where $\gamma_{z_0 a}$ is a path in $E_{b\rho} := \{\Phi_E(z) < b\rho\}$ joining a point z_0 of E with the point a. Hence by (2)

$$|p(z)| \le \|p\|_E + 2NM_1|\gamma_{z_0 a}|(b^2\rho)^{\deg p}\|p\|_E + 2rNM_1(b^2\rho)^{\deg p}\|p\|_E$$

for all z with $|z-a| \le r$. Therefore by (3)

$$|p(z)| \le C\|p\|_E(b^3\Phi_E(z))^{\deg p}\|p\|_E, \quad |z-a| \le r,$$

where $M = M(a,b)$, $r = r(a,b)$ do not depend on p.

2° Now we shall show that (BWI) is true if E is PH-regular and connected. Without loss of generality it is sufficient to check the inequality for real-valued PH polynomials. By the PH regularity of E and by the Poisson integral formula for each $b > 1$ there are $M > 0$ and an open neighbourhood Ω of E such that for every real-valued pluriharmonic polynomial u

$$\|\partial u\|_\Omega \le M b^{\deg u}\|u\|_E.$$

We may assume that Ω is a union of a finite number of balls $B_j = B(a_j, r_j)$ ($j = 1, \dots, s$) with centers $a_j \in E$. Then there is a constant $C > 0$ such that every point z of Ω can be joined in Ω with some a_k by a path $\gamma_{a_k z}$ whose length $|\gamma_{a_k z}| \le C$. Let p_k be a holomorphic polynomial such that $u = \operatorname{Re} p_k$ and $p_k(a_k) = u(a_k)$. Then $p_k(z) = u(a_k) + 2\int_{\gamma_{a_k z}} \partial u$ and $\|p_k\|_E \le \|u\|_E(1 + 2CM)b^{\deg u}$. Hence

$$|u(z)| \le |p_k(z)| \le (1 + 2CM)\|u\|_E \left(b\Phi_E(z)\right)^{\deg u}, \quad z \in \mathbb{C}^N.$$

3° Now let E be an arbitrary PH-regular compact subset of $\mathbb{C}^N$. Take $R_0 > 1$ so large that the set $F := \{\Phi_E(z) \le R_0\}$ is connected (which is possible by Lemma 2.12). By 1° for each $b > 1$ there exists $M_1 > 0$ such that $|p(z)| \le M_1 \left(b\Phi_E(z)\right)^{\deg p} \|p\|_E$ in a neigbourhood of F. By 2°

$$|p(z)| \le M_2 \left(b\Phi_F(z)\right)^{\deg p} \|p\|_F \quad \text{in } \mathbb{C}^N.$$

Hence

$$\begin{aligned} |p(z)| &\le M_2 \left(b\Phi_E(z)/R_0\right)^{\deg p} M_1 \left(bR_0\right)^{\deg p} \|p\|_E \\ &= M_1M_2\|p\|_E \left(b^2\Phi_E(z)\right)^{\deg p}, \quad \text{if } \Phi_E(z) > R_0. \end{aligned}$$

Therefore

$$|p(z)| \le M\|p\|_E \left(b^2\Phi_E(z)\right)^{\deg p}, \quad z \in \mathbb{C}^N,$$

with $M := \max\{M_1, M_1M_2\}$, for every PH polynomial p. By the arbitrary nature of $b > 1$ the proof is complete. □

Corollary 4.4. *If a compact set $E \subset \mathbb{C}^N$ is PH-regular then $\chi_E = V_E$ in $\mathbb{C}^N$.*

Indeed, for every $b > 1$ there exists $M > 0$ such that

$$V_k(z) \le \chi_k(z) \le \tfrac{1}{k}\log M + \log b + V_E(z), \quad z \in \mathbb{C}^N, \ k \ge 1,$$

which implies that $V_E(z) = \chi_E(z) = \lim_{k\to\infty} \chi_k(z)$ and the convergence is locally uniform in $\mathbb{C}^N$.

4.5. Example. Let E be a compact subset of the unit circle

$$\Gamma := \{z \in \mathbb{C};\ |z| = 1\}.$$

Then
(i) E is of positive capacity with respect to harmonic polynomials of two real variables iff the logarithmic capacity of E is positive;
(ii) $\chi_E(z) = 2V_E(z)$, for all $z \in \Gamma$;
(iii) $\chi_E(z) \le 2\sup_{\zeta\in\Gamma} V_E(\zeta) + \log|z|$;
(iv) If $E \ne \Gamma$, then E is not regular with respect to harmonic polynomials.

Proof. If

$$p(z)=\sum_{j=0}^{k}\left(a_j z^j + b_j \bar{z}^j\right), \quad b_0=0, \tag{1}$$

is a harmonic polynomial of degree $\leq k$ then

$$\widetilde{p}(z):=\sum_{j=0}^{k}\left(a_j z^{k+j} + b_j z^{k-j}\right), \quad b_0=0, \tag{2}$$

is a holomorphic polynomial of degree $\leq 2k$ such that $|\widetilde{p}(z)|=|p(z)|$ for all $z\in\Gamma$. On the other hand each holomorphic polynomial $\widetilde{p}$ of degree $\leq k$ can be written in the form (2), while the harmonic polynomial p associated with $\widetilde{p}$ by formula (1) is of degree $\leq k$ and $|p(z)|=|\widetilde{p}(z)|$ on Γ. Hence

$$\chi_k(z,E)=2V_{2k}(z,E), \quad |z|=1,\ k=1,2,\ldots$$

which implies (ii).

By Szegö's inequality [11] (ii) $\Longrightarrow$ (iii). By Theorem 3.9 (iii) $\Longrightarrow$ (i), by Corollary 4.4 (ii) $\Longrightarrow$ (iv). □

4.6. Remark. For H-regular continua on the complex plane Theorem 4.3 was first shown in [9]. For all H-regular compact subsets of $\mathbb{C}$ Theorem 4.3 was proved by Nguyen Thanh Van and B. Djebbar [5]. The idea of reducing the case of an arbitrary PH-regular compact set to the case of a connected one is borrowed from [5].

5. Compact sets regular with respect to PH polynomials and Bernstein–Walsh type theorems for PH functions

5.1. Let E be a compact subset of $\mathbb{C}^N$. We say that a function $f\colon E\longrightarrow\mathbb{C}^N$ *admits a-rapid approximation by PH polynomials,* if $\limsup_{k\to\infty}\sqrt[k]{\delta_k(f,E)}<1$. Denote by

$$a(E,PH):=\{f\in C(E);\ \limsup_{k\to\infty}\sqrt[k]{\delta_k(f,E)}<1\}$$

the vector space of all functions $f\in C(E)$ admitting a-rapid approximation on E by PH polynomials.

Given $b>1$, put

$$a_b(E,PH):=\{f\in C(E);\ q_{E,b}(f)<\infty\},$$

where $q_{E,b}(f):=\|f\|_E+\sup_{k\geq 0}b^k\delta_k(f,K)$. Then $a_b(E,PH)$ is a Banach space with the norm $q_{E,b}$. It is clear that $a(E,PH)=\bigcup_{b>1}a_b(E,PH)$, $a_b(E,PH)\subset a_{b'}(E,PH)$, if $1<b'<b$, and the inclusion is a continuous linear mapping (because $q_{E,b'}(f)\leq q_{E,b}(f)$ for each $f\in a_b(E,PH)$). Therefore $a(E,PH)$ becomes a locally

convex topological vector space of type (β) when endowed with the inductive limit topology

$$a(E, PH) = \liminf_{b \searrow 1} \operatorname{ind} a_b(E, PH).$$

5.2. Put $E^\rho := \{z \in \mathbb{C}^N;\ \operatorname{dist}(z, E) < \rho\}$ and let $PH^\infty(E^\rho)$ be the Banach space of bounded PH functions f on E^ρ with the supremum norm $\|f\|_{E^\rho}$. Then

$$PH(E) := \lim_{\rho \searrow 0} \operatorname{ind} PH^\infty(E^\rho)$$

is a locally convex topological vector space of type (β).

5.3. Direct Bernstein Walsh Theorem (DBWT) for PH functions.
1° *If a compact set $E \subset \mathbb{C}^N$ is polynomially convex then for every PH function f in a neighbourhood of E*

$$\limsup_{k \to \infty} \sqrt[k]{\delta_k(f, E)} < 1.$$

2° *If E is an $\mathcal{L}$-regular compact subset of $\mathbb{C}^N$ and $f \in PH(E_R)$ $(R > 1)$, where $E_R := \{z \in \mathbb{C}^N;\ V_E(z) < \log R\}$, then*

$$\limsup_{k \to \infty} \sqrt[k]{\delta_k(f, E)} \le 1/R.$$

Proof. It is sufficient to show 2°. If $f \in PH(E_R)$, then $f(z) = g(z) + \overline{h(z)}$, where $g, h \in \mathcal{O}(E_R)$. By the direct Bernstein–Walsh type theorem for holomorphic functions [8], there are polynomials $g_k, h_k \in \mathcal{P}_k(\mathbb{C}^N)$ such that

$$\limsup_{k \to \infty} \sqrt[k]{\|g - g_k\|_E} \le 1/R \quad \text{and} \quad \limsup_{k \to \infty} \sqrt[k]{\|h - h_k\|_E} \le 1/R.$$

Hence $\limsup_{k \to \infty} \sqrt[k]{\delta_k(f, E)} \le 1/R$, because $\delta_k(f, E) \le \|g - g_k\|_E + \|h - h_k\|_E$. □

Corollary 5.4. *If a compact set E in $\mathbb{C}^N$ is polynomially convex and PH-regular then E is $PH(E)$-determining, i.e., if f is PH in a neighbourhood of E and $f = 0$ on E, then $f = 0$ in a neighbourhood of E.*

Proof. Let f be a PH function in a neighbourhood Ω of E with $f = 0$ on E. Let $\rho > 1$ be so close to 1 that $\overline{E}_\rho \subset \Omega$. The set $\overline{E}_\rho$ is polynomially convex. By 1° of 5.3 there exist $M_1 > 0$, θ $(0 < \theta < 1)$, and a sequence $p_k \in PH_k(\mathbb{C}^N)$, $k = 1, 2, \ldots$ such that

$$\|p_k\|_E \le \|f - p_k\|_{E_\rho} = \delta_k(f, \overline{E}_\rho) \le M_1 \theta^k, \quad k \ge 1.$$

By the PH-regularity of E for each $b > 1$ there exist M_2 and a neighbourhood U of E such that

$$\|p_k\|_U \le M_2 b^k \|p_k\|_E \le M_1 M_2 (b\theta)^k, \quad k \ge 1.$$

Choose $b > 1$ so close to 1 that $b\theta < 1$. Then $f(z) = \lim_{k\to\infty} p_k(z) = 0$ for all z in $\Omega_1 := E_\rho \cap U$.

5.5. Inverse Bernstein–Walsh Theorem for PH functions. *If a compact set $E \subset \mathbb{C}^N$ is PH-regular, then for every $b > 1$ and for each R with $1 < R < b$ there exists $M > 0$ such that for all f in $a_b(E, PH)$ one has*

$$Lf \in PH(E_R) \quad \text{and } \|Lf\|_{E_R} \le M q_{E,b}(f),$$

where Lf is defined by (α)) of Proposition 3.5.

Moreover, $Lf = f$ on E for each $f \in a_b(E, PH)$, and $L(p|E) = p$ for every PH polynomial p.

Proof. This is a direct consequence of Lemma 3.12 and of Corollary 4.4 $(E'_R = E_R)$. □

Corollary 5.6. *If E is a PH-regular and $\limsup \sqrt[k]{\delta_k(f, E)} \le 1/R$ then there exists $\tilde{f} \in PH(E_R)$ with $\tilde{f} = f$ on E.*

Corollary 5.7. *If E is PH-regular compact subset of $\mathbb{C}^N$ then the mapping L defined by (α) of Proposition 3.5 is a continuous one-to-one linear extension operator from the topological vector space $a(E, PH)$ to the topological vector space $PH(E)$. Moreover, the operator L preserves PH polynomials, i.e., $L(p|E) = p$ for every PH polynomial p.*

Theorem 5.8 (Characterization of PH regular sets). *For a compact set E in $\mathbb{C}^N$ the following conditions are equivalent:*

(1) *E is PH-regular, i.e., for every $b > 1$ there exist $M > 0$ and a neighbourhood Ω of E such that*

$$\|p\|_\Omega \le M b^{\deg p} \|p\|_E$$

for each PH polynomial p;

(2) *E is $\mathcal{L}$-regular (equivalently: regular with respect to holomorphic polynomials of N complex variables) and for every $b > 1$ there exists $M > 0$ such that*

$$|p(z)| \le M \|p\|_E \, (b\Phi_E(z))^{\deg p}, \quad z \in \mathbb{C}^N,$$

for every PH polynomial p, where $\Phi_E := \exp V_E$;

(3) *E is $\mathcal{L}$-regular and $\chi_E(z) = V_E(z)$ in $\mathbb{C}^N$;*

(4) *χ_E is continuous in $\mathbb{C}^N$;*

(5) *$\chi_E^*(z) = 0$, $z \in E$ (i.e., χ_E is continuous at each point z of E).*

If E is polynomially convex then each of the above conditions is equivalent to each of the following ones:

(6) *E is $PH(E)$-determining and for every $R > 1$ and $f \in C(E)$ with $\limsup_{k\to\infty} \sqrt[k]{\delta_k(f, E)} \le 1/R$ there exists $\tilde{f} \in PH(E_R)$ with $\tilde{f} = f$ on E;*

(7) *E is $PH(E)$-determining and for every $f \in C(E)$ with* $\limsup_{k\to\infty} \sqrt[k]{\delta_k(f,E)} < 1$ *there exists an open neighbourhood Ω of E and a PH function $\tilde{f}$ in Ω such that $\tilde{f} = f$ on E;*

(8) *E is $PH(E)$-determining and*

$$a(E, PH) = \{f|E;\ f \in PH(E)\},$$

where $a(E,PH) := \{f \in C(E);\ \limsup_{k\to\infty} \sqrt[k]{\delta_k(f,E)} < 1\}$ *and $PH(E)$ is the space of germs of PH functions on E.*

Proof. (1) $\Longrightarrow$ (2) $\Longrightarrow$ (3) – follow from Theorem 4.3 and Corollary 4.4

(3) $\Longrightarrow$ (4) $\Longrightarrow$ (5) – obvious.

(5) $\Longrightarrow$ (1). Fix $b > 1$ and put $\mathcal{W} := \{pb^{-\frac{1}{2}\deg p};\ p$ is a PH polynomial with $\|p\|_E \le 1\}$, $\Omega' := \{z \in \mathbb{C}^N;\ \chi_E^*(z) < \frac{1}{4}\log b\}$. Let Ω be a relatively compact open subset of Ω' with $E \subset \Omega$. Then by the Polynomial Lemma

$$\|p\|_\Omega \le Mb^{\frac{1}{2}\deg p}, \quad p \in \mathcal{W},$$

which implies (1).

(1) $\Longrightarrow$ (6) – follows from Corollary 5.4 and Theorem 5.5.

(6) $\Longrightarrow$ (7) – obvious.

(7) $\Longrightarrow$ (8) – follows from Theorems 5.3 and 5.5.

(8) $\Longrightarrow$ (1) 1° First we shall show that if (8) is satisfied then the restriction mapping

$$\mathcal{R}_E\colon PH(E) \ni f \mapsto f|E \in a(E, PH)$$

is bijective and continuous. Bijectivity of $\mathcal{R}_E$ is obvious.

To prove the continuity of $\mathcal{R}_E$ we have to show that for every bounded neighbourhood U of E there exist $b > 1$ and $M > 0$ such that

$$q_{E,b}(f|E) \le M\|f\|_U, \quad f \in PH^\infty(U). \tag{*}$$

Without loss of generality we may assume that U is polynomially convex. Fix ρ with $0 < \rho < \operatorname{dist}(E, \mathbb{C}^N \setminus U)$ and put $E^\rho := \{z \in \mathbb{C}^N;\ \operatorname{dist}(z,E) \le \rho\}$. Then $F := \widehat{E^\rho}$, the polynomially convex hull of E^ρ, is PH-regular and $F \subset U$. By Corollary 5.7, applied to F, and by Grothendieck's open mapping theorem [7], the restriction $\mathcal{R}_F = L^{-1}$ is continuous. Therefore there are $b > 1$ and $M > 0$ such that

$$q_{F,b}(f|F) \le M\|f\|_U, \quad f \in PH^\infty(U).$$

But it is clear that $q_{E,b}(f|E) \le q_{F,b}(f|F)$, which implies the required inequality (*).

2° By 1° and Grothendieck's open mapping theorem $L := \mathcal{R}_E^{-1}$ is a continuous linear extension operator from the space $a(E, PH)$ to the space $PH(E)$, i.e., $Lf = f$ on E for all $f \in a(E, PH)$, and for every $b > 1$ there exist $M > 0$ and an open neighbourhoof U of E such that

$$Lf \in PH^\infty(U),\ \|Lf\|_U \le Mq_{E,b}(f),\ \forall f \in a_b(E, PH).$$

If $f = p|E$, where p is a PH polynomial then by $PH(E)$-determinacy of E, we have $L(p|E) = p$ in $\mathbb{C}^N$. Therefore

$$\|p\|_U \leq 2M\|p\|_E b^{\deg p}$$

for each PH polynomial p, which completes the proof of the implication (8) $\Longrightarrow$ (1) as well as of Theorem 5.8. □

5.9. Remark. Let $H_k(\mathbb{R}^N)$ be the vector space of harmonic polynomials on $\mathbb{R}^N$ of degree $\leq k$. Let E be a compact set in $\mathbb{R}^N$. We define the *extremal function H_E of E with respect to harmonic polynomials* by the formula

$$H_E(x) := \limsup_{k\to\infty} H_k(x), \quad x \in \mathbb{R}^N,$$

where

$$H_k(x) \equiv H_k(x, E) := \sup\{\tfrac{1}{k}\log|p(x)|;\ p \in H_k(\mathbb{R}^N), \|p\|_E \leq 1\}.$$

Put $\operatorname{dist}_E(f, H_k(\mathbb{R}^N)) := \inf\{\|f - p\|_E;\ p \in H_k(\mathbb{R}^N)\}$, $f \in C(E)$. Let $H(E)$ be the space of germs of harmonic functions on E.

One can show that the following theorems are true.

I. *E is of positive capacity with respect to harmonic polynomials if and only if $H_E(x) \leq \beta + \log(1 + |x|)$ in $\mathbb{R}^N$ with a positive constant β.*

II. *E is H-regular (i.e., regular with respect to harmonic polynomials) if and only if $H_E^*(x) = 0$ on E.*

III. *If $\mathbb{R}^N \setminus E$ is connected then E is H-regular if and only if E is $H(E)$-determining and*

$$\{f \in C(E);\ \limsup_{k\to\infty} \sqrt[k]{\operatorname{dist}_E(f, H_k(\mathbb{R}^N))} < 1\} = \{f(E);\ f \in H(E)\}.$$

The last result follows from the approximation thorems of [1] through Grothendieck's open mapping theorem.

If $N = 2$, then $H_E \equiv \chi_E$. If $N \geq 3$, we know very litle about the properties of the extremal function H_E. For instance, we do not know if H_E is continuous in $\mathbb{R}^N$ for H-regular compact sets $E \subset \mathbb{R}^N$ $(N \geq 3)$.

6. Examples of PH-regular sets

6.1. A sufficient condition for a compact set E in $\mathbb{C}^N$ to be PH-regular is that E be regular with respect to polynomials of $2N$ real variables (here we identify $\mathbb{C}^{2N}$ with $\mathbb{R}^{2N}$). In particular, if $E = \overline{D}$, where D is a bounded domain with Lipschitz boundary, then E is PH-regular (see e.g. [4], [10]).

6.2. A finite union of PH-regular sets is PH-regular.

6.3. If $E_j \subset \mathbb{C}^{N_j}$ $(j = 1, \ldots, s)$ are PH-regular then $E = E_1 \times \cdots \times E_s$ is PH-regular.

6.4. A compact set E in $\mathbb{C}^N$ is PH-regular, if it has the following property: E is $\mathcal{L}$-regular and there is a sequence of PH-regular sets $\{E_j\}$ such that $E_j \subset E_{j+1}$ and $\lim_{j\to\infty} V_{E_j} = V_E$.

Indeed,

$$V_{E_j} = \chi_{E_j} \geq \chi_E \geq V_E, \quad j = 1, 2, \ldots$$

Hence $\chi_E = V_E$, which by Theorem 5.8 implies thet E is PH-regular.

Corollary 6.5 ([9]). *If $N = 1$ and $E = \overline{D_1} \cup \cdots \cup \overline{D_s}$, where D_j is a bounded simply connected domain in $\mathbb{C}$, then E is regular with respect to harmonic polynomials of two real variables.*

Corollary 6.6. *If E is an $\mathcal{L}$-regular compact set in $\mathbb{C}^N$ and there exists a sequence of PH-regular sets $\{E_j\}$ such that $E_j \subset E_{j+1}$ and $E \setminus \bigcup_{j=1}^{\infty} E_j$ is pluripolar, then E is PH-regular.*

To prove the last Corollary it is sufficient to observe that by [10] $V_{E_j} \searrow V_F$, where $F := \bigcup_1^{\infty} E_j$.

Acknowledgements. A preliminary version of this paper was prepared by the author during his stay at Toulouse University (France), Sept. 11 - Oct. 11, 1992. The author wishes to thank the University for the invitation. He is especially indebted to Professor Nguyen Thanh Van for his great hospitality and stimulating conversations.

Section 4 of this paper was the subject of the author's talk at the "NATO Advanced Workshop on Classical and Modern Potential Theory and Related Fields" held at the Chateau de Bonas, France, July 25-30, 1993. The author is very grateful to the organizers for the invitation.

References

[1] Bagby T., Levenberg N., Bernstein theorems for harmonic functions. In: Methods of Approx. Theory in Complex Analysis and Mathematical Physics 7–18. Lecture Notes in Math. 1550, Springer-Verlag, Berlin 1993.

[2] Bedford E., Taylor B. A., A new capacity for plurisubharmonic functions. Acta Math. 149 (1982), 1–40.

[3] Josefson B., On the equivalence between locally polar and globally polar sets for plurisubharmonc functions on $\mathbb{C}^n$. Arkiv Mat. 16(1) (1978), 109–115.

[4] Klimek M., Pluripotential Theory (London Math. Soc. Monographs). Clarendon Press, Oxford 1991.

[5] Nguyen Thanh Van, Djebbar B., Propriétés asymptotiques d'une suite orthonormale de polynômes harmoniques. Bull. Sci. Math. (2) 113 (1989), 239–251.

[6] Nguyen Thanh Van, Siciak J., Remarques sur l'approximation polynomiale. C.R. Acad. Sci. Paris Sér. I 279 (1974), 95–98.

[7] Robertson A. P., Robertson W., Topological Vector Spaces. Russian translation with Appendix by Raikov, D.A. Mir, Moskva 1967.

[8] Siciak J., On some extremal functions and their applications. Trans. Amer. Math. Soc. 105 (1962), 322–357.

[9] — Asymptotic behaviour of harmonic polynomials bounded on a compact set. Ann. Pol. Math. 20 (1968), 267–278.

[10] — Extremal plurisubharmonic functions in $\mathbb{C}^n$. Ann. Pol. Math. 39 (1981), 175–211.

[11] Szegö G., Über trigonometrische und harmonische Polynome. Math. Ann. 79 (1919), 323–339.

[12] Walsh J. L., Interpolation and approximation by rational functions in the complex domain. Amer. Math. Soc. Coll. Publ. 20, 3rd ed. Amer. Math. Soc., Providence, 1960.

[13] — The approximation of harmonic functions by harmonic polynomials and by harmonic rational functions. Bull. Amer. Math. Soc. 35 (1929), 499–544.

[14] — Maximal convergence of sequences of harmonic polynomials. Ann. of Math. 38 (1927), 321–364.

Potentials for the Dirichlet problem in Lipschitz domains

Gregory C. Verchota

Abstract. Recent results for the Dirichlet problem for higher order elliptic systems in bounded Lipschitz domains of $\mathbb{R}^n$ with data taken in L^p spaces $2 - \epsilon < p \leq \infty$ are discussed.

1991 Mathematics Subject Classification: 35J40, 35J55, 31B10.

0. Introduction

The main purpose of this paper is to survey some recent results for the Dirichlet problem for real constant coefficient elliptic operators homogeneous of degree $2m$ in bounded Lipschitz domains $D \subset \mathbb{R}^n$. The first six sections are mostly devoted to an account of the author's paper with J. Pipher [PV4]. In addition, it is proved that the results of that paper extend to symmetric Legendre–Hadamard systems, and a counter-example is constructed to illustrate the role symmetry plays when dealing with systems that take on boundary data in Lebesgue spaces. The last two sections give an account of a set of papers dealing with the Agmon–Miranda maximum principle in Lipschitz domains. Those results together with theorems on the Lusin area integral lead to the most general result on Lipschitz domain maximum principles, Theorem 8.1. The proofs of the Lusin area estimates and Theorem 8.1 will appear in [DKPV].

I would like to thank the organizers of the 1994 International Conference on Potential Theory for their hospitality and for the opportunity to present these results. I also thank my co-workers in this area whose collaboration I gratefully acknowlege: B. E. J. Dahlberg, C. E. Kenig, J. Pipher, and A. L. Vogel.

1. The Dirichlet problem with L^p data

$D \subset \mathbb{R}^n$ is Lipschitz if there are a finite number of neighborhoods that cover the boundary ∂D so that within each neighbourhood, ∂D may be given as the graph of a Lipschitz function $\phi : \mathbb{R}^{n-1} \to \mathbb{R}$. The maximum of the Lipschitz norms of each ϕ together with the number of neighborhoods essentially describe the

Lipschitz nature of D. See, for example, [D2], [PV4]. Dirichlet data will be taken in $L^p(\partial D)$ spaces with respect to surface measure $d\sigma$ where in general $1 < p \leq \infty$. Thus boundary values will be taken in the sense of nontangential convergence a.e. $d\sigma$. To do this we define nontangential approach regions for each $Q \in \partial D$

$$\Gamma(Q) = \Gamma_\alpha(Q) = \{X \in D : \ |X - Q| \leq (1+\alpha)\,\mathrm{dist}(X, \partial D)\}$$

where α is taken large enough depending on the Lipschitz constant associated with D. Then if $\lim_{\substack{X \to Q \\ X \in \Gamma(Q)}} u(X) = g(Q)$ we say that u has nontangential limit $g(Q)$ at Q.

In order to prove such convergence the Lebesgue dominated convergence theorem is used and thus our object of study becomes the nontangential maximal function

$$N(u)(Q) = \sup_{X \in \Gamma(Q)} |u(X)|.$$

When prescribing data in $L^p(\partial D)$ it is necessary to look for unique solutions u to our equations in the class of solutions for which $N(u) \in L^p(\partial D)$ also. For example, in the unit ball of the complex plane the Poisson kernel function $\frac{1-|z|^2}{|1-z|^2}$ has zero Dirichlet boundary values a.e.. However, its nontangential maximal function is not even in L^1 and thus not in any class of solutions considered here.

Consider systems of equations

$$L(D)u = \sum_{l=1}^{N} L^{kl}(D)u_l = 0, \quad k = 1, \ldots, N \tag{1.1}$$

where $D = (D_1, \ldots, D_n)$, $D_i = \frac{\partial}{\partial X_i}$, $i = 1, \ldots, n$. For $\xi \in \mathbb{R}^n$ and a positive integer m we assume each $L^{kl}(\xi)$ to be homogeneous of degree $2m$ with constant real coefficients and $L^{kl} = L^{lk}$. For all $\xi \in \mathbb{R}^n$ and $\zeta \in \mathbb{C}^N$ and using summation convention we assume the Legendre–Hadamard ellipticity condition

$$Re L^{kl}(\xi)\zeta_k \bar{\zeta}_l \geq E|\xi|^{2m}|\zeta|^2$$

for some ellipticity constant $E > 0$.

Well known approximation techniques reduce solving the L^p-Dirichlet problem for (1.1) to obtaining the a priori estimate

$$\|N(\nabla^{m-1}u)\|_p \leq C\|\nabla^{m-1}u\|_p \tag{1.2}$$

for solutions in smooth domains where C depends only on p, E, m and the Lipschitz nature of the domain. The L^p-norms, as throughout the paper, are taken on ∂D and ∇^m denotes all derivatives in $\mathbb{R}^n$ of order m.

Dirichlet data for each component of a solution is taken in the completion of the array of functions

$$\{g, \nabla g, \ldots, \nabla^{m-1}g : g \in C_0^\infty(\mathbb{R}^n)\}$$

under the norm (on ∂D) $(\sum_{j=0}^{m-1} \|\nabla^j g\|_p^p)^{1/p}$. Call this space of arrays $WA^{m-1,p}(\partial D) = WA^{m-1,p}$.

The Dirichlet problem may now be stated.

Given data in $(WA^{m-1,p})^N$ and the operator L in (1.1) find a unique solution to $Lu = 0$ with

(i) $\|N(\nabla^{m-1}u)\|_p < \infty$ and so that

(ii) $u, \nabla u, \dots, \nabla^{m-1}u$ have nontangential limits a.e. $\quad (D_p)$

to the prescribed data.

As the Lipschitz nature of a domain is scale invariant so is the estimate (1.2). The theme of scale invariance can be considered the main point of solving D_p in Lipschitz domains.

This version of the Dirichlet problem also has a variational formulation in Lipschitz domains. This is because estimate (1.2) implies certain sharp Besov space estimates inside D for solutions. For example, when $p = 2$ (i) of D_2 may be replaced with the requirement that $\nabla^{m-1}u$ belong to $\Lambda^{2,2}_{1/2}(D)$, i.e.,

$$\text{(i}'\text{)}\quad \int_D |\nabla^{m-1}u|^2 + \int_D\int_D \frac{|\nabla^{m-1}u(X) - \nabla^{m-1}u(Y)|^2}{|X-Y|^{n+1}} dX dY < \infty.$$

This follows from the "square function" estimates for solutions

$$\int_D |\nabla^m u(X)|^2 \operatorname{dist}(X) dX \le C\|N(\nabla^{m-1}u)\|_2 \tag{1.3}$$

and if $\nabla^{m-1}u(P)$ vanishes for some fixed point $p \in D$

$$\|N(\nabla^{m-1}u)\|_2 \le C\int_D |\nabla^m u(X)|^2 \operatorname{dist}(X) dX. \tag{1.4}$$

These estimates and their L^p analogs will be established in [DKPV] for the operators (1.1).

2. Some history and a basic argument

For Laplace's equation $\Delta u = 0$ in C^1 domains, D_p for $1 < p < \infty$ was solved independently by Fabes, Jodiet and Riviére [FJR] and Dahlberg [D1], [D2]. The main result of Dahlberg's papers, however, was the solution of D_p for the Laplacian in Lipschitz domains for $2 - \epsilon < p \le \infty$ where ϵ depends on the Lipschitz nature of D. Examples show that this range of p's is best possible. To solve D_p a careful study of harmonic measure was made exploiting the maximum principle and properties of positive solutions such as the Harnack inequality.

For solutions to (1.1) for single equations when $m \ge 2$ or for systems when $m \ge 1$ exploiting positive solutions and maximum principles is in general impossible. Another approach is by the method of layer potentials, leading to the solving of singular integral equations on ∂D. Such potentials may be produced in our case by taking $2m - 1$ derivatives of a fundamental solution matrix $\Gamma(X)$ for L

[J, p. 75–76] and integrating against a density in $(L^p(\partial D))^N$:

$$\mathcal{K}^\alpha f(X) = \int_{\partial D} D^\alpha \Gamma(X-Q) f(Q)\, d\sigma(Q), \qquad X \in D. \tag{2.1}$$

Here $\alpha = (\alpha_1, \dots, \alpha_n)$ is a multi-index of order $|\alpha| = \alpha_1 + \cdots + \alpha_n = 2m-1$. Letting X approach points on ∂D we obtain a matrix of singular integral operators. On C^1 domains (or domains with small Lipschitz constant) these operators can be shown to be bounded in L^p because of Calderon's result [C1]. On general Lipschitz domains boundedness follows from Coifman, McIntosh and Meyer [CMM].

These results together with arguments in [FJR] yield, for us, the important estimate

$$\|N(\mathcal{K}^\alpha f)\|_p \le C\|f\|_p, \qquad 1 < p < \infty \tag{2.2}$$

where C depends only on p, m, E and the Lipschitz nature of D. Recall now the key estimate (1.2). If the derivatives of the solution on the left side of (1.2) can be represented by potentials $\mathcal{K}^\alpha f$ or combinations of such potentials in such a way so that the densities f are bounded by the Dirichlet data, i.e.,

$$\|f\|_p \le C\|\nabla^{m-1}u\|_p, \tag{2.3}$$

then (1.2) and ultimately the solution to D_p follow because of (2.2). One way of attempting such a plan is by the Fredholm method of solving singular integral equations on ∂D. In the case of Laplace's equation we write

$$u(X) = \mathcal{K}f(X) = C_n \int_{\partial D} \frac{\langle N_Q, Q-X\rangle}{|Q-X|^n} f(Q) d\sigma(Q) \tag{2.4}$$

where N_Q denotes the outer unit normal vector and $\langle \cdot, \cdot \rangle$ the scalar product. By (2.2) $\|N(u)\|_p \le C\|f\|_p$. The Dirichlet data for u is

$$(\frac{1}{2}I + K)f(P) = \frac{1}{2}f(P) + \text{p.v.}C_n \int_{\partial D} \frac{\langle N_Q, Q-P\rangle}{|Q-P|^n} f(Q) d\sigma(Q).$$

Thus to solve D_p we need to know

$$\|f\|_p \le C\|(\frac{1}{2}I + K)f\|_p \tag{2.5}$$

and

$$\frac{1}{2}I + K : L^p \to L^p \text{ is onto.} \tag{2.6}$$

The standard methods of establishing (2.5) and (2.6) are to show $\frac{1}{2}I + K$ to be invertible by showing that K is a compact operator and applying the Fredholm theory, or to show that K has small norm. These are the methods used in [FJR] for C^1 domains. The compactness is a consequence of the operator estimates in [C1].

For Lipschitz domains examples in [FJL] show that K is not compact. In addition it can also be shown that $\|K\|$ grows without bound as the Lipschitz nature of D becomes more severe. Nevertheless invertibility of $\frac{1}{2}I+K$ on Lipschitz boundaries has been established by obtaining the equivalence on the adjoint operator

$$\|(\frac{1}{2}I + K^*)f\|_p \approx \|(\frac{1}{2}I - K^*)f\|_p \tag{2.7}$$

for all $f \in L^p(\partial D)$. This was first established for $p = 2$ in [V1] and proved for the optimum range $1 < p < 2+\epsilon$ in Dahlberg–Kenig [DK1] from which (2.5) and (2.6) follow for $2 - \epsilon < p < \infty$. The method (2.7) also has consequences for the Neuman problem and Regularity problem [V1],[DK1] and has been used for Dirichlet and conormal derivative problems including the traction and slip problems of elastostatics and hydrostatics in [DKV2], [FKV], [DK2] and [S]. Another noteworthy method that works for Dirichlet and certain oblique derivative problems for second order symmetric Legendre–Hadamard systems is in [C2] and [G].

There is another way of representing solutions as potentials and carrying out the plan of (2.2) and (2.3) that does not involve inverting systems of integral operators. These systems become very complicated for higher order equations and systems and as far as the author knows are not very well understood on Lipschitz boundaries. The method is the method of [PV4] and may be partially illustrated as follows. For Laplace's equation we use Green's formula

$$u(X) = \mathcal{K}u(X) - S\frac{\partial u}{\partial N}(X) \tag{2.8}$$

where $\mathcal{K}$ is as in (2.4) with the Dirichlet data u in place of f and $S\frac{\partial u}{\partial N}$ is the single layer potential applied to the normal derivative of u. In general

$$Sf(X) = C_n(2-n)^{-1}\int_{\partial D}\frac{f(Q)d\sigma(Q)}{|X-Q|^{n-2}}.$$

Using (2.2) and (2.8) we get

$$\|N(u)\|_p \le C\|u\|_p + \|N(S\frac{\partial u}{\partial N})\|_p \tag{2.9}$$

and we would have (1.2) but for the last term.

In order to take care of the last term we now specialize to the case where D is a domain above the graph of a Lipschitz function $\phi : \mathbb{R}^{n-1} \to \mathbb{R}$, $\|\nabla\phi\|_\infty \le M < \infty$. We will ignore any difficulties at infinity in order to concentrate on the main ideas. These ideas are carried out locally with more technical difficulties in [PV4]. Hence we define Stein-Weiss conjugate harmonic functions to u by letting

$$u_{-1}(x,t) = -\int_t^{+\infty} u(x,s)\, ds \tag{2.10}$$

and $u_j(X) = D_j u_{-1}(X)$, $j = 1, \dots, n-1$ where $X = (x,t)$, $x \in \mathbb{R}^{n-1}$ and $t > \phi(x)$. We next observe that

$$\frac{\partial}{\partial x_j}u_j(x,\phi(x)) = \sqrt{1+|\nabla\phi(x)|^2}\frac{\partial u}{\partial N}(x,\phi(x)).$$

Thus $\frac{\partial u}{\partial N}d\sigma$ in the single layer potential may be replaced with $\frac{\partial}{\partial x_j}u_j\, dx$, the (tangential) derivatives now transferred to the (lower order) kernel and once again we have operators like (2.1) with the $u_j = f$. Consequently (2.9) now becomes

$$\|N(u)\|_p \leq C\|(u_1, \ldots, u_{n-1}, u)\|_p. \tag{2.11}$$

The n-tuple of harmonic functions on the right side of (2.11) is ∇u_{-1}. Therefore (1.2) follows if it is true for harmonic functions h that

$$\|\nabla h\|_p \leq C\|D_n h\|_p \tag{2.12}$$

since $D_n u_{-1} = u$. That (2.12) is true for $p = 2$ for any Lipschitz domain is very easily seen but in general is more difficult for $p > 2$ and is necessarily false for $p < 2$. In fact, as we will see, for certain higher order problems it is necessarily false for intervals of p's containing ∞ also. The necessary falseness for $p < 2$ is a consequence of well-known examples of nonuniqueness for the Dirichlet problem in Lipschitz domains (e.g. see [D2]). The principle is very general:

Given $p < 2, n$, and L as in (1.1) *there is a bounded Lipschitz domain for which D_p is not solvable.*

For $p = 2$ (2.12) is a consequence of the integration by parts formula

$$-\int_{\partial D} |\nabla h|^2 N^n \, d\sigma = -2\int_{\partial D} D_n h \frac{\partial h}{\partial N} \, d\sigma \tag{2.13}$$

where $N = (N^1, \ldots, N^n)$, followed by the Schwarz inequality on the right side. Dahlberg's theorem for $p = 2$, i.e., estimate (1.2) for $m = 1$ and a single equation now follows by (2.11), (2.10) and (2.12).

In effect the Poisson kernel has been decomposed in the following way. Riesz transforms may be defined by

$$\mathcal{R}_j : u|_{\partial D} \to u_j|_{\partial D}, \qquad j = 1, \ldots, n-1.$$

Denote the double-layer kernel of (2.4) by $\mathcal{K}^X(Q)$ and define tangential kernels $\mathcal{K}_j^X(Q) = C_n \frac{N_Q^n(Q_j - X_j) - N_Q^j(Q_n - X_n)}{|Q-X|^n}$. Then harmonic measure at X has been written

$$\left[\mathcal{K}^X - \sum_{j=1}^{n-1} \mathcal{R}_j(\mathcal{K}_j^X)\right] d\sigma;$$

its nontangential L^2 properties established by using Coifman–McIntosh–Meyer [CMM] on the explicit kernels and by using the L^2-boundedness of the Riesz transforms, which follows from (2.13). Various generalizations of (2.13) in the context of boundary value problems have become known in recent years as Rellich formulas. Rellich's interest was in eigenvalue problems for the Laplacian [R]. Payne and Weinberger [PW] used them to study harmonic and biharmonic functions. Nečas extended their use to a variety of higher order problems and systems in his book [N]. It was Jerison and Kenig who introduced them into the study of nontangential limits in their work on harmonic measure and the Neumann problem for the Laplacian [JK1], [JK2].

3. The difficulty with higher order operators

The method described in the last section for decomposing the Poisson kernel and obtaining estimates can be made to work for the systems (1.1). Only one part of the argument remains to be illustrated and we will do that by considering fourth order equations with a single unknown.

Hence let

$$Lu = \sum_{|\alpha|=|\beta|=2} D^\alpha a_{\alpha\beta} D^\beta u = 0$$

and let Γ^X denote the fundamental solution for L with pole at X. Here $m = 2$ so by (1.2) we wish to control $N(\nabla u)$ in terms of ∇u on the boundary. This time in order to use the Green's formula for L we will in analogy to (2.10) take two (i.e., m) primitives of u defining

$$u_{-2}(x,t) = \int_t^{+\infty} \int_{s_1}^{+\infty} u(x, s_2)\, ds_1 ds_2.$$

Thus we want to control $N(\nabla D_n^2 u_{-2})$ by $\nabla D_n^2 u_{-2}$. To go directly to a Green's formula that will yield potentials of the type (2.1) we use integration by parts two different ways on the solid integral

$$\sum_{\alpha,\beta} a_{\alpha\beta} \int_D D_j D^\alpha \Gamma^X D^\beta D_n^2 u_{-2}\, dY$$

to obtain for any j the representation

$$D_j D_n^2 u_{-2}(X) = \sum_{|\gamma|=3} \int_{\partial D} K^\gamma (X - Q) D^\gamma u_{-2}(Q)\, d\sigma(Q).$$

Here the K^γ are certain linear combinations of third derivatives of Γ together with components of the normal. By (2.2) it now suffices to control $\nabla^3 u_{-2}$ by $\nabla D_n^2 u_{-2}$ on the boundary. Since u_{-2} is also a solution we will suppress the subscript and prove

$$\|\nabla^3 u\| \le C \|\nabla D_n^2 u\|. \tag{3.1}$$

The first step would be to control $\nabla^3 u$ by $\nabla^2 D_n u$. To do this we will prove the slightly stronger inequality

$$\|\nabla^2 D_j u\|^2 \le C \|\nabla^3 u\|\, \|\nabla D_j D_n u\|, \qquad j = 1, \ldots, n. \tag{3.2}$$

To see that this is stronger we sum over j and divide by $\|\nabla^3 u\|$ to obtain

$$\|\nabla^3 u\| \le C \|\nabla^2 D_n u\|. \tag{3.3}$$

Then letting $j = n$ in (3.2) and using (3.3) we obtain

$$\|\nabla^3 u\|^2 \le C \|\nabla^3 u\|\, \|\nabla D_n^2 u\|$$

and (3.1) follows.

It remains to prove (3.2). There will be two parts to the proof. In the first part we use a Rellich formula, viz.,

$$
\begin{aligned}
-\sum_{\alpha,\beta} a_{\alpha\beta} \int_{\partial D} & D^\alpha D_j u D^\beta D_j u N^n \, d\sigma \\
&= -\sum_{\alpha,\beta} a_{\alpha,\beta} \int_{\partial D} D_n D^{\alpha''} D_j u \left[D^\beta D_j u N^{\alpha'} - D^\beta D^{\alpha'} u N^j \right] \\
&+ D_n D_j^2 u D^\beta D^{\alpha'} u N^{\alpha''} \, d\sigma \\
&+ \text{ similar term.}
\end{aligned} \tag{3.4}
$$

Here $|\alpha'| = |\alpha''| = 1$ and $\alpha' + \alpha'' = \alpha$ and we recall that N is the downward pointing normal. The Schwarz inequality now yields the right side of (3.2).

For the second part we prove a Gårding inequality on the boundary of a Lipschitz domain, viz.,

$$
\begin{aligned}
-\frac{E}{4} \int_{\partial D} & |\nabla^2 u|^2 N^n \, d\sigma \\
&\leq -\sum_{\alpha,\beta} a_{\alpha\beta} \int_{\partial D} D^\alpha u D^\beta u N^n \, d\sigma + C \int_{\partial D} |\nabla D_n u|^2 \, d\sigma
\end{aligned} \tag{3.5}
$$

for solutions u. Here E is the ellipticity constant and for the remainder of the paper

$$
|\nabla^m u|^2 = \sum_{|\alpha|=m} \frac{m!}{\alpha!} (D^\alpha u)^2.
$$

Since $D_j u$ is a solution, (3.2) follows from (3.5).

The last term on the right of (3.5) is the analog of the lower order term that enters in the standard Gårding inequality [A]. That it is needed may be seen by an example.

Let

$$
Lu = \left[\delta(D_2^4 + D_3^4 + D_4^4) + D_1^4 + D_2^2 D_3^2 + D_3^2 D_4^2 + D_4^2 D_2^2 - 4 D_1 D_2 D_3 D_4\right] u = 0
$$

in $\mathbb{R}^4$ for some $\delta > 0$. This elliptic operator is derived from examples due to Motzkin [M] of semi-positive definite polynomials that cannot be written as the sum of squares of polynomials. For δ small enough, the positive definite $L(\xi)$ shares this property with the semi-positive definite

$$
\xi_1^4 + \xi_2^2 \xi_3^2 + \xi_3^2 \xi_4^2 + \xi_4^2 \xi_2^2 - 4 \xi_1 \xi_2 \xi_3 \xi_4.
$$

Thus there is no way to write the first term on the right of (3.5) in such a way that it can be seen to be positive, let alone something that can dominate the left side of (3.5). These remarks pertain because we are on the boundary of a Lipschitz domain and cannot use the Fourier transform to convert all the derivatives in an integral form $-\sum_{\alpha,\beta} a_{\alpha\beta} \int_{\partial D} D^\alpha u D^\beta u N_n \, d\sigma$ into multipliers and then exploit the positivity of $L(\xi)$ when $|\alpha| = |\beta| \geq 2$. However, we can make limited use of the Parseval formula together with an apparently new theorem about positive definite forms and prove (3.5) and its higher order analogs, which we do in the next section, for all our systems (1.1). Once this is done estimate (1.2) follows for the systems (1.1) by the methods outlined here and in the previous section.

4. The boundary Gårding inequality and remarks on symmetry

If ζ is a function from the multi-indices for $\mathbb{R}^n$ of order $m-1 \geq 0$ into the complex numbers define

$$|\zeta|_{m-1}^2 = \sum_{|\alpha|=m-1} \frac{(m-1)!}{\alpha!} |\zeta(\alpha)|^2. \tag{4.1}$$

If $\zeta = (\zeta_1, \ldots, \zeta_N)$ and each ζ_j is as in (4.1) define

$$\|\zeta\|_{m-1}^2 = \sum_{j=1}^{N} |\zeta_j|_{m-1}^2. \tag{4.2}$$

Denote by e_j the standard basis vectors for $\mathbb{R}^n$. For each $m \geq 2$, $\xi \in \mathbb{R}^n$ and ζ as in (4.2) define

$$\mathcal{Q}(m, \xi, \zeta) = \frac{1}{2} \sum_{k=1}^{N} \sum_{i,j=1}^{n} \sum_{|\alpha|=m-2} |\xi_i \zeta_k(\alpha + e_j) - \xi_j \zeta_k(\alpha + e_i)|^2. \tag{4.3}$$

Let $L = (L^{kl})$ be a strongly elliptic operator of type (1.1). We will write

$$L^{kl}(D) = \sum_{i,j=1}^{n} \sum_{|\alpha|=|\beta|=m-1} D_i D^\alpha a_{\alpha\beta ij}^{kl} D_j D^\beta$$

for $k, l = 1, \ldots, N$. The ellipticity condition then takes the form

$$Re \sum_{k,l=1}^{N} \sum_{i,j=1}^{n} \sum_{|\alpha|=|\beta|=m-1} \xi_i \xi^\alpha a_{\alpha\beta ij}^{kl} \xi_j \xi^\beta \zeta_k \overline{\zeta_l} \geq E|\xi|^{2m} \|\zeta\|_0^2 \tag{4.4}$$

for each $\xi \in \mathbb{R}^n$ and $\zeta \in \mathbb{C}^N$.

Theorem 4.1. *If L satisfies the ellipticity condition* (4.4), *then there exists a C depending only on E, n and m so that for all functions ζ as in* (4.2) *and $\xi \in \mathbb{R}^n$*

$$\frac{1}{2} E|\xi|^2 \|\zeta\|_{m-1}^2$$
$$\leq C\mathcal{Q}(m, \xi, \zeta) + Re \sum_{k,l=1}^{N} \sum_{i,j=1}^{n} \sum_{|\alpha|=|\beta|=m-1} \xi_i \zeta_k(\alpha) a_{\alpha\beta ij}^{kl} \xi_j \overline{\zeta_l(\beta)}. \tag{4.5}$$

We need a lemma.

Lemma 4.2. *For each fixed $\xi \neq 0$ and function ζ as in (4.2), $\mathcal{Q}(m,\xi,\zeta) = 0$ if and only if there exists a fixed $\zeta_0 \in \mathbb{C}^N$ so that $\zeta(\beta) = \xi^\beta \zeta_0$ for all $|\beta| = m-1$.*

Proof. If $\mathcal{Q}(m,\xi,\zeta) = 0$ then for each k and each α we have after summing in i and j in (4.3)

$$|\sum_i \xi_i \zeta_k(\alpha+e_i)|^2 = |\xi|^2 \sum_i |\zeta_k(\alpha+e_i)|^2,$$

i.e., we have equality in the Schwarz inequality, whence there is a complex number which we will again denote by $\zeta_k(\alpha)$ so that for each i

$$\zeta_k(\alpha+e_i) = \zeta_k(\alpha)\xi_i. \tag{4.6}$$

Thus the lemma is true for $m=2$. Assume the lemma true for $m-1$. Then for the ξ and ζ which implies (4.6) when $\mathcal{Q}(m,\xi,\zeta) = 0$ we define $\mathcal{Q}(m-1,\xi,\zeta)$ as in (4.3) where we use now the ζ on the right side of (4.6). By (4.3) and (4.6) $\mathcal{Q}(m-1,\xi,\zeta) = 0$ whence by induction for all $|\alpha| = m-2$, $\zeta_k(\alpha) = (\zeta_0)_k \xi^\alpha$. Thus by (4.6) $\zeta_k(\alpha+e_i) = (\zeta_0)_k \xi^\alpha \xi_i$ for each i and α and the lemma is proved.

Theorem 4.1 now follows as in [PV4] by continuity. I.e., if $\|\zeta\|_{m-1} = 1$ and $\zeta(\alpha) = \zeta_0 \xi^\alpha$ for all $|\alpha| = m-1$ then (4.5) follows from ellipticity (4.4). If $\|\zeta\|_{m-1} = 1$ and $\|\zeta - \zeta_0 \xi^{\cdot}\|_{m-1}$ is small (4.5) is still true with $C = 0$. Otherwise by the lemma, C may be fixed large enough depending only on E, n and m to use the positivity of $\mathcal{Q}$.

We will now prove a boundary Gårding inequality for functions $u \in [C_0^\infty(\mathbb{R}^n)]^N$. Well known approximation techniques suffice to pass to actual solutions.

Theorem 4.3. *Let $\phi : \mathbb{R}^{n-1} \to \mathbb{R}$ be Lipschitz with $\|\nabla\phi\|_\infty \le M$ and let D be the domain above the graph of ϕ. Let L satisfy (4.4). Then there exists a constant $C < \infty$ depending only on E, M, n and m so that for any real $u \in C_0^\infty(\mathbb{R}^n)^N$*

$$-\frac{E}{4}\int_{\partial D} |\nabla^m u|^2 N^n \, d\sigma \le$$

$$-\sum_{k,l=1}^{N}\sum_{i,j=1}^{n}\sum_{|\alpha|=|\beta|=m-1}\int_{\partial D} D_i D^\alpha u_k a^{kl}_{\alpha\beta ij} D_j D^\beta u_l N^n \, d\sigma \tag{4.7}$$

$$+C\int_{\partial D} |\nabla^{m-1} D_n u|^2 \, d\sigma.$$

Proof. For any $i = 1,\ldots,n-1$, D_i may be replaced by $\frac{\partial}{\partial x_i}$ since $\frac{\partial}{\partial x_i} D^\alpha u_k(x,\phi(x)) = D_i D^\alpha u_k(x,\phi(x)) + \frac{\partial \phi}{\partial x_i} D_n D^\alpha u_k(x,\phi(x))$ and Young's inequality may be used to control any terms with $D_n D^\alpha u_k$. Thus by Parseval the integrals in the first term on the right may be written

$$\int_{\mathbb{R}^{n-1}} \xi_i \widehat{D^\alpha u_k(\cdot,\phi(\cdot))}(\xi) a^{kl}_{\alpha\beta ij} \xi_j \overline{\widehat{D^\beta u_l(\cdot,\phi(\cdot))}(\xi)} \, d\xi.$$

For each ξ the integrands are as in Theorem 4.1. Thus (4.7) follows by Theorem 4.1 and Plancherel except for integrals in ξ arising from the Q term in Theorem 4.1. These integrals are of the form

$$\int_{\mathbb{R}^{n-1}} |\xi_i \widehat{D_j D^\gamma u_k}(\xi) - \xi_j \widehat{D_i D^\gamma u_k}(\xi)|^2 \, d\sigma$$

where $|\gamma| = m-2$. By Plancherel these equal integrals of the squares of

$$\frac{\partial}{\partial x_i} D_j D^\gamma u(x,\phi(x)) - \frac{\partial}{\partial x_j} D_i D^\gamma u(x,\phi(x)) =$$
$$\frac{\partial \phi}{\partial x_i} D_n D_j D^\gamma u(x,\phi(x)) - \frac{\partial \phi}{\partial x_j} D_n D_i D^\gamma u(x,\phi(x))$$

which are controlled by the last term on the right of (4.7). This completes the proof.

To obtain estimate (1.2) when $p = 2$ for solutions to (1.1) we form the mth primitive u_{-m} of u as in (2.10) and §3 and use integration by parts on

$$\sum_{|\alpha|=|\beta|=m} a_{\alpha\beta} \int_D D^\gamma D^\alpha \Gamma^X D^\beta D_n^m u_{-m} \, dY, \qquad k = 1, \dots, N \tag{4.8}$$

to obtain a Green's formula

$$D^\gamma u(X) = \sum_{|\delta|=2m-1} \int_{\partial D} K^\delta (X-Q) D^\delta u_{-m}(Q) \, d\sigma(Q)$$

for any $|\gamma| = m-1$. Here Γ is the $N \times N$ fundamental solution matrix for L, the $a_{\alpha\beta}$ are $N \times N$ matrices, and K^δ denotes sums of various $2m-1$ order derivatives of Γ together with multiplication by components of the unit normal vector. By (2.2) and suppressing the primitive notation, we now need the estimate

$$\|\nabla^{2m-1} u\|_2 \le C \|\nabla^{m-1} D_n^m u\|_2 \tag{4.9}$$

for solutions u. This follows by repeated application of

$$\|\nabla^m D^\gamma u\|_2^2 \le C \|\nabla^{m-1} D_n D^\gamma u\|_2 \|\nabla^{2m-1} u\|_2, \qquad \text{for each } |\gamma| = m-1 \tag{4.10}$$

just as in §3. To prove (4.9) we apply Theorem 4.3 to bound the left side of (4.9) by $C\|\nabla^{m-1} D_n D^\gamma u\|_2^2$ plus the integral form involving the coefficients of the operator L, i.e., the first term on the right of (4.7) with $D^\gamma u$ in place of u. Integration by parts in this form leads to the Rellich formula and by the Schwarz inequality the right side of (4.9). Estimate (1.2) is proved for $p = 2$.

It should be remarked that the symmetry of the L^{kl} in (1.1) is used in obtaining the Rellich formula. As an example we give a single second order equation with complex coefficients that is derived from counterexamples due to Kozlov [Ko] to his theorem on the spectrum of certain operator pencils for plane corners. By breaking the equation into real and imaginary parts a system with real coefficients and $-L^{12} = L^{21}$ is obtained. We will argue that there can be no Rellich formula because the Dirichlet problem D_2 cannot be solvable for this equation in the class of bounded Lipschitz domains.

Hence, let $p = -\frac{1}{2} - \frac{\sqrt{3}}{2}i = e^{-i\frac{\pi}{3}}$ and $w = e^{-i\bar{p}\pi}$. Consider the strongly elliptic (Legendre–Hadamard) operator

$$L = (\frac{\partial}{\partial x} - w^{-1}\frac{\partial}{\partial y})(\frac{\partial}{\partial x} - w\frac{\partial}{\partial y}) \tag{4.11}$$

and the solution

$$u(x,y) = (x + wy)^p - (x + w^{-1}y)^p$$

to $Lu = 0$ in the corner $\{(x,y) : x > 0 \text{ and } y > 0\}$. The Dirichlet data is zero a.e. while u behaves like $(x^2+y^2)^{\frac{-1}{4}}$ near the origin. This means that locally $N(u) \notin L^2$ while $N(u) \in L^p$ for all $p < 2$.

In the $\frac{1}{4}$-disc $D = \{x > 0, y > 0,\ x^2+y^2 < 1\}$, u has continuous Dirichlet data. Suppose now that D_2 is solvable in bounded Lipschitz domains. We will show that this assumption leads to a contradiction. Since $L^2(\partial D) \subset L^p(\partial D)$ for $p \leq 2$ we can solve D_2 for the Dirichlet data from u and produce a nontrivial solution v with zero Dirichlet data on all of ∂D with the property that $N(v) \in L^p(\partial D)$ for all $p < 2$ while $N(v) \notin L^2(\partial D)$. Thus there would be no uniqueness and therefore no estimates for the problems D_p, $p < 2$.

But perturbation methods due to Dahlberg and Kenig imply that if D_2 is solvable in bounded Lipschitz domains, then so is D_p for p in an open interval containing 2. This contradicts our assumption. We will discuss these perturbation methods in the next section. This argument on the nonsolvability of D_2 is part of some current research with A. L. Vogel [VV].

5. L^p perturbation techniques and more on symmetry

Once the L^2 theory is established, several things follow by known arguments. Solvability of the Dirichlet problem in L^p for $2 - \epsilon < p < 2 + \epsilon$ is done by real variable methods due to Dahlberg and Kenig. Here $\epsilon > 0$ is in general small depending on the Lipschitz constant M, ellipticity, dimension and order of L. If we had solved integral equations on the boundary, an operator theoretic argument of A. P. Calderon's [C2] would be available to us. But it is not.

Examples of the use of the Dahlberg–Kenig perturbation can be found in [DK2] and [K] for the $2 < p < 2 + \epsilon$ case and in [DKV1] for the $2 - \epsilon < p < 2$ case. In [PV4] it is stated in a general way.

We will illustrate the $2 - \epsilon < p < 2$ case, and indicate how much the perturbation method is independent of any special properties that the equations being studied might have by completing the proof of the nonsolvability in bounded Lipschitz domains of D_2 for the antisymmetric system introduced at the end of the last section. We will do this by establishing that if D_2 is solvable and thus estimate

(1.2) holds (by a uniform boundedness principle) with $p = 2$ and C depending only on the Lipschitz character of the domain, then (1.2) holds for some $p < 2$.

Let u be any solution to $Lu = 0$ in the $\frac{1}{4}$-disc D with Dirichlet data in $L^2(\partial D)$ where L is the operator (4.11). Recall the nontangential maximal function $N = N(u)$ taken over approach regions Γ of §1. Let $\tilde{N} = \tilde{N}(u)$ denote the nontangential maximal function for u taken over uniformly wider approach regions $\tilde{\Gamma}$. Thus $N \leq \tilde{N}$ pointwise. The key point now is that for any $0 < \gamma < 1$ there is a constant C depending only on γ and the Lipschitz nature of D so that

$$\int_{\partial D} N^{2-\gamma} \mathrm{d}\sigma \leq C \int_{\partial D} N^2 \tilde{N}^{-\gamma} \mathrm{d}\sigma.$$

The proof of this rests entirely on the geometry of nontangential approach regions and the real variable theory of Muckenhaupt weights. See [DKV1, p. 129]. Consequently, continuing with the Dahlberg–Kenig technique,

$$\int_{\partial D} N^{2-\gamma} \mathrm{d}\sigma \leq \gamma C \int_0^{+\infty} \lambda^{-1-\gamma} \Big(\int_{\{Q:\tilde{N}(Q) \leq \lambda\}} N^2 \mathrm{d}\sigma(Q) \Big) \mathrm{d}\lambda. \tag{5.1}$$

For each λ the inner integral is taken over a subset of the boundary of the Lipschitz domain $D_\lambda = \bigcup_{\tilde{N}(Q) \leq \lambda} \tilde{\Gamma}(Q)$. The Lipschitz constants associated with the D_λ depend only on D since the Γ and $\tilde{\Gamma}$ are fixed. Consequently by our assumption on the L^2-solvability of the Dirichlet problem in bounded Lipschitz domains

$$\int_{\{\tilde{N} \leq \lambda\}} N^2 \mathrm{d}\sigma \leq C \int_{\{\tilde{N} \leq \lambda\}} u^2 \mathrm{d}\sigma + \lambda^2 \sigma\{\tilde{N} > \lambda\}.$$

So by (5.1)

$$\int_{\partial D} N^{2-\gamma} \mathrm{d}\sigma \leq C \int_{\partial D} u^2 \tilde{N}^{-\gamma} + \gamma C \int_{\partial D} \tilde{N}^{2-\gamma} \mathrm{d}\sigma.$$

In the first integral $\tilde{N}$ may be replaced with the smaller $u|_{\partial D}$. By well-known arguments using the Hardy–Littlewood maximal function (see [FS]) the second integral may be dominated by $\gamma C \int_{\partial D} N^{2-\gamma} \mathrm{d}\sigma$. Thus for γ small enough estimate (1.2) holds for $p = 2 - \gamma$ on ∂D. This is our illustration of the perturbation method.

But having such an estimate contradicts the existence of the function v constructed in the last section. To see this fix a point X_0 in D and define $v_t(X) = v(t(X - X_0) + X_0)$ for $t < 1$. Then each v_t has Dirichlet data in $L^2(\partial D)$ so that by our assumption and the fact that (1.2) now holds for $p = 2 - \gamma$

$$\int_{\partial D} N(v_t)^p \mathrm{d}\sigma \leq C \int_{\partial D} v_t^p \mathrm{d}\sigma.$$

But letting $t \uparrow 1$ and using $N(v) \in L^p$, $p < 2$ we obtain $\int_{\partial D} N(v)^p \mathrm{d}\sigma = 0$ which is not true.

We will close by stating the theorem on the Dirichlet problem obtained by the perturbation techniques illustrated here and the establishment of estimate (1.2) when $p = 2$ from the last section.

Theorem 5.1. *For any bounded Lipschitz domain $D \subset \mathbb{R}^n$, $n \geq 2$ and operator L from* (1.1) *the problem D_p is solvable for $2 - \epsilon < p < 2 + \epsilon$ where $\epsilon > 0$ depends on n, m, E, N and the Lipschitz character of D. Furthermore estimate* (1.2) *holds for $2 - \epsilon < p < 2 + \epsilon$.*

6. The Regularity problem

A problem closely related to the Dirichlet problem and in many ways dual to it is the so-called Regularity problem. Its solution leads to estimates on Green functions and Poisson kernels for the Dirichlet problem.

Data for the Regularity problem for the operators (1.1) is taken to be the functions $g, \ldots, \nabla^{m-1} g$ from N copies of the arrays in $WA^{m,p}(\partial D)$. Thus the data all have tangential derivatives in $L^p(\partial D)$. We will denote the tangential part of the gradient of a function on the boundary by ∇_T. The Regularity problem is the following.

Given data in $(WA^{m,p})^N$ and L in (1.1) find a unique solution to $Lu = 0$ with

(i) $\|N(\nabla^m u)\|_p < \infty$ and so that

(ii) $u_1, \ldots, \nabla^{m-1} u$ have nontangential limits a.e. $\qquad (R_p)$

to the prescribed data.

Theorem 6.1. *For any bounded Lipschitz domain $D \subset \mathbb{R}^n$, $n \geq 2$ and operator L from* (1.1) *the problem R_p is solvable for $2 - \epsilon < p < 2 + \epsilon$ where ϵ depends on n, m, E, N and the Lipschitz character of D. Furthermore*

$$\|N(\nabla^m u)\|_p \leq C \|\nabla_T \nabla^{m-1} u\|_p \tag{6.1}$$

for $2 - \epsilon < p < 2 + \epsilon$ where C depends only on p and the same quantities that determine ϵ.

The proof of this theorem follows the same pattern of argument that leads to Theorem 5.1. Instead of using the mth primitive the $(m-1)$th primitive is used in (4.8) along with $|\gamma| = m$ instead of $m - 1$. This reduces the proof of (6.1) to obtaining extimates on the boundary. For the $p = 2$ case inequality (4.9) becomes

$$\|\nabla^{2m-1} u\|_2 \leq C \|\nabla_T \nabla^{m-1} D_n^{m-1} u\|_2. \tag{6.2}$$

The boundary Gårding inequality is implicitly used to first obtain

$$\|\nabla^{2m-1} u\|_2 \leq C \|\nabla^m D_n^{m-1} u\|_2$$

by invoking the slightly stronger (4.9). Next ellipticity (4.4) yields the point-wise inequality for any $|\gamma| = m$

$$E \sum_{k=1}^{N} |D^\gamma v|^2 \leq \sum_{k,l=1}^{N} \sum_{|\alpha|=|\beta|=m} N^\alpha a_{\alpha\beta}^{kl} N^\beta D^\gamma v_k D^\gamma v_l.$$

As shown in [PV4] the last quantity is equal to

$$N^{2\gamma}\sum_{k,l=1}^{N}\sum_{|\alpha|=|\beta|=m} D^\alpha v_k a^{kl}_{\alpha\beta} D^\beta v_i + \nabla_T \nabla^{m-1} v \text{ terms.} \tag{6.3}$$

Letting $v = D_n^{m-1}u$ we need only a Rellich formula that produces tangential derivatives instead of transverse ones to take care of the first quantity in (6.3) in order to prove (6.2) and complete the L^2-Regularity theory. This is easily done. The distinction may be illustrated by comparing the 4th order Dirichlet–Rellich formula (3.4) with

$$\begin{aligned} -\sum_{\alpha,\beta} a_{\alpha\beta}\int_{\partial D} D^\alpha D_n u D^\beta D_n u N^n d\sigma \\ = -\sum_{\alpha,\beta} a_{\alpha\beta}\int_{\partial D} (N_n D^{\alpha'} - N^{\alpha'} D_n) D^{\alpha''} D_n u D_n D^\beta u \\ + (N_n D^{\alpha''} - N^{\alpha''} D_n) D_n^2 u D^{\alpha'} D^\beta u d\sigma \\ + \text{ similar term.} \end{aligned}$$

The parentheses contain the required tangential derivatives. This establishes (6.1) for $p = 2$. The extension to $2-\epsilon < p < 2+\epsilon$ follows from the Dahlberg–Kenig perturbation methods.

7. Green functions and Poisson kernels

Given a fundamental solution matrix $\Gamma(Y)$ for an operator L from (1.1), the Green function is constructed in the usual way by

$$G^X(Y) = \Gamma(X-Y) - W^X(Y) \tag{7.1}$$

where each row of W^X is a solution of (1.1). Given a bounded Lipschitz domain D and $X \in D$, $W^X(Y)$ is defined by using Theorem 6.1 to solve R_p N times so that

$$\nabla_Q^k G^X(Q) = 0 \text{ for all } Q \in \partial D,\ k = 0, \dots, m-1 \tag{7.2}$$

and

$$N(\nabla^m W^X) \in L^p(\partial D),\ 2-\epsilon < p < 2+\epsilon. \tag{7.3}$$

Let u be a solution to the Dirichlet problem for L. Formally integrating by parts two different ways and using (7.1), (7.2) the Poisson integral formula

$$\begin{aligned} 0 &= \sum_{|\alpha|=|\beta|=m} a_{\alpha\beta}\int_D D^\alpha G^X D^\beta u dY \\ &= \sum_{k=0}^{m-1}\int_{\partial D} D_Q^{m+k} G^X(Q) D_Q^{m-1-k} u(Q) d\sigma(Q) + (-1)^m u(X) \end{aligned} \tag{7.4}$$

may be obtained. Here D^l_Q denotes sums of certain derivatives of order l multiplied by components of the unit normal vector and $a_{\alpha\beta}$ are $N\times N$ matrices of coefficients.

Suppose now we wish to extend the result of Theorem 5.1 on the Dirichlet problem to all $p>2$ including $p=\infty$. By interpolation with L^2 it is enough to do the L^∞ case. In this case the estimate (1.2) becomes

$$|\nabla^{m-1}u(X)| \le C\|\nabla^{m-1}u\|_\infty \text{ for all } X\in D \tag{7.5}$$

with C independent of X. It is the X close to the boundary that are of most interest.

Consider now the $k=0$ term in the Poisson integral formula (7.4). The other terms will be discussed in the next section. Define a Poisson kernel by:

$$K^X_D(Q) = \nabla^{m-1}_X D^m_Q G^X(Q). \tag{7.6}$$

Estimate (7.5) requires that K^X_D have finite $L^1(\partial D)$ norm independent of X. For $t>0$, let $D'=\{X' : tX'\in D\}$ and recall that $\nabla^{2m-1}\Gamma(Y)\sim|Y|^{1-n}$. Then it is easy to see that $K^{tX'}_D(tQ') = K^{X'}_{D'}(Q')t^{1-n}$. Thus $\|K^X_D\|_{L^1(\partial D)} = \|K^{X'}_{D'}\|_{L^1(\partial D')}$ for $X=tX'$. By letting $t=\operatorname{dist}(X,\partial D)$ we see by these observations on rescaling that the required estimate on K^X_D for eventually obtaining (7.5) is

$$\|K^X_D\|_{L^1(\partial D)} \le C \tag{7.7}$$

with C independent of X and $\operatorname{dist}(X,\partial D)=1$. Note that here we are using the fact that D is Lipschitz. The Lipschitz nature of a domain is scale invariant.

Because of the rescaling it is now best to think of D as the domain above the graph of a Lipschitz function $\varphi:\mathbb{R}^{n-1}\to\mathbb{R}$ with $\|\nabla\varphi\|_\infty\le M$ where M is independent of the rescaling. Writing $X=(x,s)$ and $Y=(y,t)$ with $x,y\in\mathbb{R}^{n-1}$, $s,t\in\mathbb{R}$ define

$$\Delta_0 = \{Y\in\overline{D} : \frac{s-t}{|x-y|} > M\} \quad \text{and} \tag{7.8}$$

$$\Delta_k = \{Y\in\overline{D} : \frac{s+2^{k-1}\sqrt{1+M^2}-t}{|x-y|} > M\},\ k=1,2,\ldots.$$

Then $\|K^X_D\|_{L^1(\partial D\cap\Delta_1)}\le C$ follows by (7.6), (7.1), (7.2), Hölder's inequality applied over $\partial D\cap\Delta_1$, and the more precise form of (7.3) that follows from Theorem 6.1, viz.

$$\|N(\nabla^{m-1}_X\nabla^m W^X)\|_{L^p(\Delta_1\cap\partial D)} \le C'\left(\int_{\partial D}|Q-X|^{p(1-n)}d\sigma(Q)\right)^{\frac{1}{p}} \le C \tag{7.9}$$

where C depends only on M,m,n,E,N and $2-\epsilon<p<2+\epsilon$.

To complete the proof of (7.7) we need an estimate on dyadic pieces of ∂D, viz.

$$\|K^X_D\|_{L^1(\partial D\cap\Delta_k\setminus\Delta_{k-1})} \le C\eta^k,\ k=2,3,\ldots \tag{7.10}$$

for some $\eta<1$. When $n=2$ or 3 (7.10) follows for all m and N by the same atomic estimate first used in [DK2] for second order systems (1.1) in dimensions 2 and 3. The estimate was used in these dimensions for the biharmonic operator in [PV1] and the polyharmonic operators in [PV3]. The proof needed here is essentially the same and so will not be repeated. It should be noted, however, that in the proof the solvability of D_p for $2-\epsilon<p<2$ (Theorem 5.1) plays a key role.

The atomic estimate itself is as follows. Suppose $2-\epsilon<p<2+\epsilon$ and v is a solution of R_p with regularity data supported in $\Delta_1\cap\partial D$ (the data is essentially a unit $(1,p)$-atom; see [CW]). Then if $n=2$ or 3 there is a C and $\eta<1$ so that

$$\|N(\nabla^m v)\|_{L^1(\partial D\cap\Delta_k\setminus\Delta_{k-1})}\le C\eta^k\|\nabla_T\nabla^{m-1}v\|_p,\ k=2,3,\dots. \tag{7.11}$$

Now each row of $\nabla_X^{m-1}G^X$ is a solution of (1.1) in the domain $D'=D\setminus\Delta_1$ with regularity data by (7.2) supported in $\partial D'\setminus\partial D$. Furthermore by (7.9) and inspection of Γ this regularity data has norm bounded by a constant as in (7.9). Thus v may be replaced by $\nabla_X^{m-1}G^X$ in (7.11), minor modifications made and (7.10) follows when $n=2$ or 3. We summarize with a theorem.

Theorem 7.1. *Let $D\subset\mathbb{R}^n$, $n=2$ or 3 be a bounded Lipschitz domain. Let u be the unique L^2-solution of* (1.1) *with Dirichlet data $u=\nabla u=\cdots=\nabla^{m-2}u=0$ and $\frac{\partial^{m-1}}{\partial N^{m-1}}u=g\in L^\infty(\partial D)$. Then the Agmon–Miranda maximum principle holds:*

$$\sup_{X\in D}|\nabla^{m-1}u(X)|\le C\|g\|_\infty$$

with C depending ony on the Lipschitz nature of D,m,N, and E.

For $m=1$ and $N=1$ the atomic estimate (7.11) follows by a different proof in [DK1] for all dimensions. For $m=1$ and $N>1$ it is not known if the estimate is true or not in higher dimensions. There are no simple counterexamples in conical domains because of the positive results of Maz'ya, Nazarov and Plamenevskiĭ [MNP]. However when $m>1$, estimate (7.11) must fail in dimensions greater than 3. For $n\ge 5$ this is due to examples in [MP] and [MNP] while for $n=4$ the examples were independently discovered in [PV1] and [KM]. Thus Theorem 7.1 is false for $m>1$ and $n\ge 4$. In fact the examples can be used to show D_p must fail for intervals of p. For example, when $m=2$ and $n=4$, $p>6$ fails. See [PV3].

8. The Agmon–Miranda maximum principle and the Lusin-area integral

We will close with a brief explanation of the proof of the following result on the Agmon–Miranda maximum principle. The full proof will be given in [DKPV].

Theorem 8.1. *Let $D\subset\mathbb{R}^n$, $n=2$ or $n=3$ be a bounded Lipschitz domain. Let L be any system from* (1.1). *Given data in $(WA^{m-1,\infty})^N$, let u be the unique L^2-solution of Theorem* 5.1 *for $Lu=0$ and the given data. Then*

$$\sup_{X\in D}|\nabla^{m-1}u(X)|\le C\|\nabla^{m-1}u\|_\infty$$

where C is as in (1.2) *for $p=\infty$.*

As in §7, rescaling may be employed. Thus assume $\mathrm{dist}(X, \partial D) = 1$ and recall the dyadic subdomains Δ_k of (7.8). Essentially one puts $u = u_1 + u_2$ where u_2 is a solution with data equal to the data of u on $\partial D \setminus \Delta_2$ and data identically zero on Δ_1. Then as with (7.9), the L^2-theory is used locally so that $|\nabla^{m-1}u_1(X)| \le C\|\nabla^{m-1}u\|_\infty$.

For $u_2(X)$ we exploit the Poisson integral formula (7.4), this time analyzing all the terms. This is done by observing again that G^X is a solution in $D' = D \setminus \Delta_1$. Let u' be a solution in D' with data equal to the data of u_2 on $\partial D \setminus \Delta_1$ and with data identically zero on $\partial D' \setminus \partial D$. Then integrating by parts in D' in exactly the same way as in D in (7.4)

$$\begin{aligned}\sum_{|\alpha|=|\beta|=m} a_{\alpha\beta} \int_{D'} D^\alpha G^X D^\beta u' \mathrm{d}Y &= \sum_{k=0}^{m-1} \int_{\partial D \setminus \Delta_1} D^{m+k} G^X D^{m-1-k} u' \mathrm{d}\sigma \\ &= \sum_{k=0}^{m-1} \int_{\partial D} D^{m+k} G^X D^{m-1-k} u_2 \mathrm{d}\sigma \\ &= (-1)^{m-1} u_2(X). \end{aligned} \tag{8.1}$$

But since G^X is a solution in D', primitive solutions G^X_{-k} may be defined as in §2, §3, and §4. Then the left side of (8.1) may also be calculated as

$$\sum_{|\alpha|=|\beta|=m} a_{\alpha\beta} \int_{D'} D^\alpha D_n^{m-1} G^X_{-m+1} D^\beta u' \mathrm{d}Y = \sum_{k=0}^{m-1} \int_{\partial D \setminus \Delta_1} D^{m+k} G^X_{-k} D^{m-1} u_2 \mathrm{d}\sigma. \tag{8.2}$$

Thus Theorem 8.1 follows if we can show each of $\nabla_X^{m-1} D^{m+k} G^X_{-k}$, $k = 0, 1, \dots, m-1$ has $L^1(\partial D)$ norm bounded independently of X whenever $\mathrm{dist}(X, \partial D) = 1$. Since $\nabla_X^{m-1} G^X$ are solutions to the Regularity problem in D' with unit atomic data on $\partial D' \cap \overline{\Delta}_1$, and the $k = 0$ case is estimates (7.7) and (7.11), what is needed is the estimates on solutions to the regularity problem with atomic data, viz.

$$\|N(\nabla^{m+k} u_{-k})\|_1 \le C \|N(\nabla^m u)\|_1, \quad k = 1, 2, \dots \tag{8.3}$$

Estimates (8.3) follow from the L^p analogues of the "square function" estimates (1.3), (1.4). In [DKPV] the Lusin area integrals for solutions u to (1.1)

$$S(u)(Q) = \left(\int_{\Gamma(Q)} \frac{|\nabla^m u(X)|^2}{|X - Q|^{n-2}} \mathrm{d}X \right)^{\frac{1}{2}}$$

will be studied. It will be established that

$$\frac{1}{C} \|N(\nabla^{m-1} u)\|_p \le \|S(u)\|_p \le C \|N(\nabla^{m-1} u)\|_p$$

for $0 < p < \infty$. See [D3]. By a lemma of Stein [St], there is a point-wise estimate

$$S(u)(Q) \le C \left(\int_{\tilde{\Gamma}(Q)} \frac{|\nabla^{m-1} D_n u(X)|^2}{|X - Q|^{n-2}} \mathrm{d}X \right)^{\frac{1}{2}}$$

where $\tilde{\Gamma}(Q)$ is a larger nontangential approach region containing $\Gamma(Q)$. See applications in [V1], [V2] and [PV2]. Consequently, one can prove (8.3) by

$$\begin{aligned}\|N(\nabla^{m+k}u_{-k})\|_p &\leq C\|S(\nabla^k u_{-k})\|_p \\ &\leq C'\|S(\nabla^{k-1}u_{-k+1})\|_p \leq \cdots \leq C''\|N(\nabla^m u)\|_p\end{aligned}$$

and Theorem 8.1 follows because of the representation (8.1), (8.2).

Corollary. *If D is a bounded Lipschitz domain in $\mathbb{R}^n$, $n = 2$ or 3, then D_p is solvable for $2 - \epsilon < p \leq \infty$.*

Supported in part by NSF grant DMS-9401354.

References

[A] Agmon, S., Lectures on Elliptic Boundary Value Problems. D. Van Nostrand, Princeton, New Jersey 1965.

[C1] Calderón, A. P., Cauchy integrals on Lipschitz curves and related operators. Proc. Nat. Acad. Sci. U.S.A 74 (1977), 1324–1327.

[C2] Calderón, A. P., Boundary value problems for the Laplace equation in Lipschitz domains. In: Recent Progress in Fourier Analysis (ed. by I. Peral and J.-L Rubio de Francia), North Holland, Amsterdam 1985, 33–48.

[CMM] Coifman, R. R., McIntosh, A., and Meyer, Y., L'integrale de Cauchy definit un operateur borné sur L^2 pour les courbes lipschitziennes. Ann. of Math. 116 (1982), 361–387.

[CW] Coifman, R. R., Weiss, G., Extensions of Hardy spaces and their uses in analysis. Bull. Amer. Math. Soc. 83 (1977), 569–646.

[D1] Dahlberg, B. E. J., On estimates of harmonic measure. Arch. Rational Mech. Anal. 65 (1977), 272–288.

[D2] Dahlberg, B. E. J., On the Poisson integral for Lipschitz and C^1 domains. Studia Math. 66 (1979), 13–24.

[D3] Dahlberg, B. E. J., Weighted norm inequalities for the Lusin area integral and the non-tangential maximal funcitons for functions harmonic in a Lipschitz domain. Studia Math. 67 (1980), 297–314.

[DK1] Dahlberg, B. E. J., Kenig, C. E., Hardy spaces and the Neumann problem in L^p for Laplace's equation in Lipschitz domains. Ann. of Math. 125 (1987), 437–465.

[DK2] Dahlberg, B. E. J., Kenig, C. E., L^p estimates for the 3-dimensional systems of elastostatics on Lipschitz domains. In: Analysis and partial differential equations, 621–634. Lecture Notes in Pure and Appl. Math. 122, Dekker, New York 1990.

[DKPV] Dahlberg, B. E. J., Kenig, C. E., Pipher, J., Verchota, G. C., Area integral estimates and maximum principles for higher order elliptic equations and systems, (in preparation).

[DKV1] Dahlberg, B. E. J., Kenig, C. E., Verchota, G. C., The Dirichlet problem for the biharmonic equation in a Lipschitz domain. Ann. Inst. Fourier (Grenoble) 36 (1986), 109–135.

[DKV2] Dahlberg, B. E. J., Kenig, C. E., Verchota, G. C., Boundary value problems for the systems of elastostatics on a Lipschitz domain. Duke Math. J. 57 (1988), 795–818.

[FJL] Fabes, E. B., Jodeit, M., Lewis, J. E., Double layer potentials for domains with corners and edges, Indiana Univ. Math. J. 26 (1977), 95–114.

[FJR] Fabes, E. B., Jodeit, M., Riviére, N. M., Potential techniques for boundary value problems on C^1 domains, Acta Math. 141 (1978), 165–186.

[FKV] Fabes, E. B., Kenig, C. E., Verchota, G. C., The Dirichlet problem for the Stokes system on Lipschitz domains, Duke Math. J. 57 (1988), 769–793.

[FS] Fefferman, C., Stein, E., H^p spaces of several variables, Acta Math. 129 (1972), 137–193.

[G] Gao, W., Layer potentials and boundary value problems for elliptic systems in Lipschitz domains. J. Funct. Anal. 95 (1991), 377–399.

[JK1] Jerison, D. S., Kenig, C. E., An identity with applications to harmonic measure. Bull. Amer. Math. Soc. (1980), 447–451.

[JK2] Jerison, D. S., Kenig, C. E., The Neumann problem on Lipschitz domains. Bull. Amer. Math. Soc. 4 (1981), 203–382.

[J] John, F., Plane Waves, Spherical Means. Interscience Publishers Inc., New York 1955.

[K] Kenig, C. E., Elliptic boundary value problems on Lipschitz domains, Ann. Math. Studies 112 (1986), 131–183.

[Ko] Kozlov, V. A., On singularities of solutions of the Dirichlet problem for elliptic equations in the neighborhood of corner points. Leningrad Math. J. 1 (4) (1990), 967–982.

[KM] Kozlov, V. A., Maz'ya, V. G., On the spectrum of the operator pencil generated by the Dirichlet problem in a cone, Math. USSR Sbornik 73 (1) (1992), 27–48.

[MNP] Maz'ya, V. G., Nazarov, S. A., Plamenevskiĭ, B. A., On the singularities of solutions of the Dirichlet problem in the exterior of a slender cone. Math. USSR Sbornik 50 (2) (1985), 415–437.

[MP] Maz'ya, V. G., Plamenevskiĭ, B. A., The maximum principle for biharmonic equations in a region with conical points. Izv. Vyssh. Uceb. Zaved. Mat. 25 (2) (1981), 52–59.

[M] Motzkin, T. S., The arithmetic-geometric inequality. In: Inequalities (ed. by O. Shisha), 205–224. Academic Press, New York 1967.

[N] Nečas, J., Les méthodes directes en theorie des equations elliptiques. Academia, Prague 1967.

[PW] Payne, L., Weinberger, H., New bounds in harmonic and biharmonic problems. J. Math. Phys. 33 (1954), 291–307.

[PV1] Pipher, J., Verchota, G. C., The Dirichlet problem in L^p for biharmonic functions on Lipschitz domains. Amer. J. Math. 114 (1992), 923–972.

[PV2] Pipher, J., Verchota, G. C., The maximum principle for biharmonic functions in Lipschitz and C^1 domains. Comment. Math. Helv. 68 (1993), 385–414.

[PV3] Pipher, J., Verchota, G. C., Maximum principles for the polyharmonic equation on Lipschitz domains. Potential Analysis (to appear).

[PV4] Pipher, J., Verchota, G. C., Dilation invariant estimates and the boundary Gårding inequality for higher order elliptic operators. Ann. of Math. (to appear).

[R] Rellich, F., Darstellung der Eigenwerte von $\Delta u + \lambda u$ durch ein Randintegral. Math. Z. 46 (1940), 635–646.

[S] Shen, Z., A note on the Dirichlet problem for the Stokes system in Lipschitz domains, Proc. Amer. Math. Soc. (to appear).

[St] Stein, E. M., Singular Integrals and Differential Properties of Functions. Princeton University Press, Princeton, NJ, 1970.

[V1] Verchota, G. C., Layer potential and regularity for the Dirichlet problem for Laplace's equation in Lipschitz domains. J. Funct. Anal. 59 (1984), 572–611.

[V2] Verchota, G. C., The Dirichlet problem for the polyharmonic equation in Lipschitz domains. Indiana Univ. Math. J. 39 (1990).

[VV] Verchota, G. C., Vogel, A. L., Nonsymmetric systems in plane corners (to appear).

Dirichlet problem at infinity for harmonic functions on graphs

Wolfgang Woess

Abstract. On a locally finite graph X, harmonic functions are considered with respect to a stochastic transition operator which is adapted to the underlying structure. Given a "natural" compactification $\widehat{X}$ of X with boundary at infinity $\partial X = \widehat{X} \setminus X$, the Dirichlet problem for P-harmonic functions with respect to ∂X consists in the following question: does every continuous function on ∂X have a continuous extension to $\widehat{X}$ which is harmonic on X? After a general discussion of the Dirichlet problem for irreducible Markov chains and the probabilistic approach to its solution, various types of specific graph or group structures are reviewed under this aspect. A few proofs are given which previously, at least in the discrete context, were not available in a "clean" form. Several of the results extend to continuous time and/or space (harmonic functions on Riemannian manifolds).

1991 Mathematics Subject Classification: 31C12, 60J15.

1. Introduction

In these notes, we shall deal with discrete potential theory, i.e., the potential theory associated with denumerable Markov chains. Although perhaps looked upon as "too simple" by some tough potential analysts, this contains all essential features of the theory besides those regarding small scale questions (in the discrete setting there are no arbitrarily small nontrivial neighbourhoods). Interplay with the probabilistic counterpart is immediately at hand without technical obstacles. While classical "continuous" potential theory was well developed before the foundation of modern probability theory and the elaboration of the links between the two fields, discrete potential theory goes back to the fifties (see, e.g., Doob's seminal article [D1]) and was built from the beginning upon the probabilistic concept of Markov chains.

Besides the discretization of typical elliptic partial differential equations and the associated heat equation (compare with Ancona [A2]), which leads to random walks and the associated space-time chain, respectively (I'm not discussing the numerical quality here), the most typical counterpart in the continuous setting is the study of Riemannian manifolds in terms of potential theoretic properties (see, for example, the book by Sario and Nakai [SN]). This analogy is not casual, but precise, as shown by the work done mainly since the eighties by Furstenberg,

Varopoulos, Kanai, Coulhon, Chavel and Feldman, Doyle, Lyons and Sullivan and others, see, e.g., §9 of the survey by the author [W4].

Potential theory on discrete networks (reversible Markov chains; see the recent book by Soardi [So]) evolves along similar lines as that of Laplacians on manifolds. Both the discrete and the continuous theory are encircled by the potential theory associated with Dirichlet forms (Fukushima [Fu]), but in order to get concrete answers to concrete structural questions, one has to leave that level of generality.

The approach adopted here is that of examining the interplay between structural features of the underlying space and potential theoretic and / or probabilistic properties of Markov chains which are adapted to that structure ("random walks"). For a detailed survey of results in this spirit (presented in a probabilistic vein), see [W4]. Also, the lecture notes of Ancona [A2], which deal simultaneously with the discrete and continuous situation, and the "random thoughts" of Lyons [Ly] contain a very high amount of precious information.

Here, we shall only deal with a relatively small portion of this field, namely, the Dirichlet problem at infinity. Given our (infinite, discrete) space and a notion of harmonicity (in terms of a stochastic transition operator), we consider some compactification of the space with associated ideal boundary, the points at infinity. Typically, the space carries some geometrical structure, the transition operator is adapted to this structure in a way which has to be specified, and the compactification is also "natural" in terms of that structure. The Dirichlet problem at infinity then consists in asking whether every continuous function on the boundary admits a continuous extension to the whole compactification which is harmonic on the original space.

In §2, we describe precisely what we mean and explain the equivalent probabilistic formulation of solvability of the Dirichlet problem: (i) the Markov chain governed by the transition operator converges almost surely in the topology of our compactification to a boundary-valued random variable, and (ii) as the starting point approaches the boundary, the corresponding limit distributions converge weakly to the Dirac mass at the limit point. If only (i) holds, then the limit distributions on the boundary (harmonic measure class) can be used to give a probabilistic definition of regularity of a single boundary point. We also discuss further properties of the limit distributions.

The following sections are devoted to the study of the Dirichlet problem for concrete classes of infinite graphs with adapted transition operators and natural compactifications. In most of these considerations, the probabilistic approach will be used.

In §3, we briefly explain why solvability of the Dirichlet problem at infinity is not a reasonable property in a "Euclidean" setup. ("Hyperbolic" features are needed.)

In §4, we consider a particularly simple case: that of nearest neighbour random walks on trees with the end compactification, which is also quite instructive for the subsequent considerations.

§5 is an "intermezzo" regarding the situation where "geometry" and "adaptedness" are linked with the presence of a group of isometries. If a certain convergence property ("contractivity") is satisfied, then a method of Furstenberg [F1], [F2] can be adapted to yield solvability of the Dirichlet problem.

In §6 and §7, we consider two types of compactifications: the end compactification of an arbitrary graph (§6) and the hyperbolic compactification of a graph which is hyperbolic in the sense of Gromov [Gr] (§7). In the group-invariant case, the result of §5 applies. Otherwise, certain geometrical adaptedness conditions yield convergence to the boundary and solvability of the Dirichlet problem (Ancona [A1], Cartwright, Soardi and Woess [CSW], Kaimanovich and Woess [KW]).

The purpose of these notes is to present a flow of results connected with a "structural" approach to some topics of potential and probability theory. Proofs are given in more or less detail, or omitted, according to the author's choice and the available space. Some of them may be new at least in part. It is obvious and unavoidable that this paper has nonempty intersection with the survey [W4]. Finally, the author acknowledges very useful discussions with V. A. Kaimanovich.

2. Dirichlet problem and irreducible Markov chains

Let X be a countable infinite set. *Harmonic functions* on X can be introduced as the invariant functions of a *stochastic transition operator* (matrix) $P = \big(p(x,y)\big)_{x,y\in X}$. "Stochastic" means that P has nonnegative elements and row sums equal to one. It acts on functions $g : X \to \mathbb{R}$ by

$$Pg(x) = \sum_{y\in X} p(x,y)g(y).$$

Thus, a function h is (P-)harmonic if $Ph = h$, i.e., harmonicity is defined in terms of a mean value property (the "Laplacian" is $\Delta = P - I$). A *superharmonic* function is one with $Pg \leq g$ pointwise. We shall always assume *irreducibility:* for every $x, y \in X$, there is n such that $p^{(n)}(x,y) > 0$. Here, $p^{(n)}(x,y)$ is the (x,y)-entry of the matrix power P^n; as usual, $P^0 = I$, the identity matrix. Irreducibility is a kind of "connectedness" or "ellipticity" condition. It is easy to see that it implies validity of the minimum principle for superharmonic functions:

Minimum principle. *If $Pg \leq g$ and $x \in X$ is such that $g(x) \leq g(y)$ for all $y \in X$, then g is constant.*

Indeed,

$$g(x) \geq \sum_y p^{(n)}(x,y)g(y) \geq \sum_y p^{(n)}(x,y)g(x) = g(x),$$

so that it cannot be $g(y) > g(x)$ for any y with $p^{(n)}(x,y) > 0$.

In the subsequent sections, X will always carry a geometrical structure (that of a graph), and P will be adapted to this structure in some way to be specified more precisely in the respective context. To see what we have in mind, consider the following

Basic example. Let X be a graph, i.e., a countable set equipped with a symmetric relation $E \subset X \times X$, the edge set. We think of an edge $[x,y] \in E$ as a segment connecting the endpoints x and y, and say in this case that x and y are neighbours. The degree $\deg(x)$ of a vertex $x \in X$ is its number of neighbours. We shall always consider graphs which are locally finite, i.e., $\deg(x) < \infty$ for all x, and connected. The latter means that for every pair $x, y \in X$ there is a path (sequence of successive neighbours) $x = x_0, x_1, \dots, x_k = y$ from x to y. The minimal k of this type is the *distance* between x and y, denoted $d(x,y)$. This defines an integer-valued metric on X.

The simplest type of a transition matrix P which is adapted to the structure of X is given by

$$p(x,y) = \begin{cases} 1/\deg(x), & \text{if } d(y,x) = 1\,; \\ 0, & \text{otherwise.} \end{cases}$$

Usually, this example is referred to as the *simple random walk*. A function on X is harmonic with respect to this P if its value at each point is the arithmetic mean over the values at the neighbours.

Let us now return to the general setting of an infinite discrete space X together with an irreducible transition operator P which defines harmonic functions on X. Next, we need a notion of infinity, i.e., a *compactification* $\widehat{X}$ of X with ideal *boundary* ∂X, the set of points at infinity. We require that our compactification has all "good" properties: Hausdorff, with countable base of the topology, with X open and dense in $\widehat{X}$, so that ∂X is compact and $\widehat{X}$ is metrizable.

The *Dirichlet problem at infinity,* or more precisely, the Dirichlet problem with respect to P and $\widehat{X}$ then consists of the following question.

Does every continuous function on ∂X admit a continuous extension to $\widehat{X}$ which is P-harmonic on X?

By irreducibility, the solution is unique, if it exists. Indeed, suppose that two harmonic functions h_1 and h_2 give the continuous extension of the same function in $C(\partial X)$. Then $h = h_1 - h_2$ is a continuous extension of the function which is zero on ∂X. By compactness, the extended function assumes a minimum on $\widehat{X}$, and by the minimum principle, this minimum must be attained on the boundary. Hence $h \geq 0$. Analogously, $-h \geq 0$, so that $h_1 = h_2$.

As P is stochastic, the constant functions are always harmonic. Thus, with respect to the one point-compactification, the Dirichlet problem trivially admits solution. This is not interesting. What we want is to distinguish many points at infinity, at least more than one. Hence, we shall always suppose that

$$|\partial X| \geq 2.$$

In this case, solvability of the Dirichlet problem implies the existence of nonconstant bounded harmonic functions: (X,P) does not have the Liouville property. This yields the existence of a finite *Green kernel* $G(x,y) > 0$, $x, y \in X$, such that for every $y \in X$, the function $x \mapsto G(x, y)$ satisfies

$$PG(\cdot, y) = G(\cdot, y) - \delta_y,$$

see below. The *potential* of a function f on X is then $g = Gf$, defined by $Gf(x) = \sum_y G(x, y)f(y)$, whenever this series converges absolutely.

We next turn our attention to the probabilistic formulation of the Dirichlet problem. The matrix P governs a Markov chain with state space X (when X carries a geometrical structure and P is adapted to that structure, we shall call it "random walk"). This is a sequence $(Z_n)_{n \geq 0}$ of X-valued random variables with

$$\Pr[Z_{n+1} = y | Z_n = x] = p(x, y), \quad x, y \in X.$$

Thus, Z_n is the random position at time n, and the position at time $n + 1$ depends only on Z_n and not on the past. More precisely, we also have to specify the starting point Z_0 (or the starting distribution), and we need a probability space on which the Z_n are defined. The simplest construction is that of the *trajectory space* $(\Lambda, \mathcal{A}, \Pr_x)$ $(x \in X)$: we set $\Lambda = X^{\mathbb{N}_0}$, the set of all possible evolutions (trajectories) of the chain. $Z_n : \Lambda \to X$ is the nth projection, and the σ-algebra $\mathcal{A}$ of subsets of Λ is the one generated by the sets (events) $[Z_0 = x_0, \ldots, Z_n = x_n]$, where $n \geq 0$ and $x_0, \ldots, x_n \in X$. The probability measure $\Pr_x$ with respect to starting point x at time 0 is then given by

$$\Pr_x[Z_0 = x_0, \ldots, Z_n = x_n] = \delta_x(x_0)p(x_0, x_1) \ldots p(x_{n-1}, x_n).$$

(By Kolmogorov's extension theorem, this defines $\Pr_x$ uniquely on $\mathcal{A}$.) The corresponding expectation is denoted E_x.

We have

$$p^{(n)}(x, y) = \Pr_x[Z_n = y] = \Pr_{x_0}[Z_{k+n} = y | Z_k = x]$$

whenever x_0 and k are such that $p^{(k)}(x_0, x) > 0$. We write

$$F(x, y) = \Pr_x[\exists\, n \geq 0 : Z_n = y].$$

The Green kernel can be defined as

$$G(x, y) = \sum_{n=0}^{\infty} p^{(n)}(x, y).$$

It satisfies the identity $G(x, y) = F(x, y)G(y, y)$. What we need is its finiteness. The following is well known (recall that we always assume irreducibility), see, e.g., Kemeny, Snell and Knapp [KSK].

Theorem 2.1. *The following statements are equivalent.*

(a) *There are nonconstant positive superharmonic functions.*

(b) *There are $x, y \in X$ such that $G(x, y) < \infty$.*

(c) $G(x, y) < \infty$ *for all* $x, y \in X$.

(d) *there is $x \in X$ such that* $\Pr_x[\exists\, n > 0 : Z_n = x] < 1$.

(e) $\Pr_x[\exists\, n > 0 : Z_n = x] < 1$ *for all* $x \in X$.

(f) *For every finite subset A of X and every ($\iff$ some) $x \in X$,*

$$\Pr_x[\exists n_0 : Z_n \notin A \;\; \forall n > n_0] = 1.$$

In this case, the Markov chain is called *transient;* otherwise, it is called *recurrent.* Statement (d) says that there is a positive probability of never returning to the starting point; statement (f) says that $Z_n \to \infty$ in the sense that it leaves every finite set with probability one after a finite time.

We remark that the problem of transience vs. recurrence is very well studied and that a large number of applicable transience criteria is available, see §4 in [W4]. Henceforth we shall always assume transience. Together with irreducibility, this means that

$$0 < G(x,y) < \infty \quad \forall x, y \in X.$$

As (Z_n) tends to infinity in the sense that it leaves every finite set with probability one, we may ask whether convergence to infinity can be described in more detail. In the presence of our compactification $\widehat{X}$, one way of doing this is to study almost sure convergence of (Z_n) in the topology of $\widehat{X}$. If this holds, then the limit must be a ∂X-valued random variable, denoted Z_∞. In terms of the trajectory space, this means the following. Set

$$\Lambda_\infty = \{\lambda \in \Lambda : \lim_{n\to\infty} Z_n(\lambda) \in \partial X \text{ exists in the topology of } \widehat{X}\}$$

and

$$Z_\infty(\lambda) = \lim_{n\to\infty} Z_n(\lambda).$$

We then say that (Z_n) *converges to the boundary* ∂X, if $\Pr_x(\Lambda_\infty) = 1$ for every $x \in X$. In this case, Z_∞ is measurable with respect to the Borel σ-algebra of ∂X. Write ν_x for the $\Pr_x$-distribution of Z_∞ :

$$\nu_x(B) = \Pr_x[Z_\infty \in B],$$

where B is a Borel set in ∂X (or in $\widehat{X}$ - the part $B \cap X$ is not charged by ν_x). Factoring with respect to the first step and using the Markov property, we get

$$\nu_x = \sum_{y\in X} p(x,y)\nu_y. \tag{1}$$

If we have convergence to the boundary, then $(\nu_x)_{x\in X}$ is a family of Borel probability measures on ∂X, called the *harmonic measure class.* From (1), we get that $\nu_x \geq p^{(n)}(x,y)\nu_y$, and irreducibility yields that all the ν_x are mutually absolutely continuous (this justifies the terminology "measure class") and that a function on ∂X which is ν_x-integrable for some x has this property for all x. The harmonic measure class can be used for "producing" harmonic functions: if φ is a ν_x-integrable function on ∂X, then

$$h_\varphi(x) = \int_{\partial X} \varphi \, d\nu_x \tag{2}$$

defines a P-harmonic function. If φ is continuous, then this is the right candidate for the harmonic function which should solve the Dirichlet problem with boundary function φ. What we want is that

$$\lim_{x\to\xi} h_\varphi(x) = \varphi(\xi) \tag{3}$$

for every $\varphi \in C(\partial X)$ and every $\xi \in \partial X$ (convergence in $\widehat{X}$).

Theorem 2.2. *The Dirichlet problem with respect to P and $\widehat{X}$ is solvable if and only if*
(i) *(Z_n) converges to the boundary, and*
(ii) *for the corresponding harmonic measure class,*

$$\lim_{x\to\xi} \nu_x = \delta_\xi \quad \textit{weakly for every } \xi \in \partial X.$$

This theorem should be well known to specialists; for the case of Brownian motion in a Euclidean domain, see, e.g., Doob [D2]. In lack of a reference in the context of denumerable Markov chains, we give an outline.

Proof. The above discussion shows that (i) and (ii) are sufficient conditions. To prove necessity, assume that the Dirichlet problem admits a solution. For $\varphi \in C(\partial X)$, let h_φ denote the (unique) harmonic function on X which gives the continuous extension of φ to $\widehat{X}$. For $x \in X$, consider the (evaluation) operator

$$V_x : C(\partial X) \to \mathbb{R}, \quad V_x(\varphi) = h_\varphi(x).$$

By uniqueness of the extension, V_x is linear, and by the minimum (maximum) principle, it is bounded. Furthermore, $V_x(\mathbf{1}) = 1$, and if $\varphi \geq 0$ then $V_x(\varphi) \geq 0$ (once more by the minimum principle). Considering the dual of the Banach space $C(\partial X)$ with the max-norm, we see that there must be a unique Borel probability measure ν_x on ∂X such that

$$h_\varphi(x) = \int_{\partial X} \varphi \, d\nu_x.$$

Solvability of the Dirichlet problem now implies (ii) for the measures ν_x, $x \in X$.

To prove (i), first observe that harmonicity of h_φ for all φ yields that (1) holds: $\nu_x = \sum_y p(x,y)\nu_y$. Now choose a countable family $\{\varphi_k : k \in \mathbb{N}\}$ which is dense in $C(\partial X)$. For each φ_k, the sequence

$$h_{\varphi_k}(Z_n) = \int_{\partial X} \varphi_k \, d\nu_{Z_n}, \; n = 0, 1, 2, \ldots,$$

is a bounded martingale. By the martingale limit theorem, it converges almost surely. Hence, using transience, $\Pr_x(\Lambda') = 1$ for every x, where

$$\Lambda' = \left\{ \lambda \in \Lambda : Z_n(\lambda) \to \infty, \; \int_{\partial X} \varphi_k \, d\nu_{Z_n(\lambda)} \text{ converges for every } k \right\}.$$

If $\lambda \in \Lambda'$, then by Helly's principle, there is a Borel probability σ_λ on ∂X such that

$$\lim_{n\to\infty} \nu_{Z_n(\lambda)} = \sigma_\lambda \quad \text{weakly.}$$

By compactness, there must be a subsequence (n') and a point $\xi \in \partial X$ such that $Z_{n'}(\lambda) \to \xi$ in the topology of $\widehat{X}$. By (ii),

$$\lim_{n\to\infty} \nu_{Z_{n'}(\lambda)} = \delta_\xi \quad \text{weakly.}$$

Consequently, $\sigma_\lambda = \delta_\xi$. Now suppose that the sequence $\big(Z_n(\lambda)\big)$ does not converge in $\widehat{X}$. Then it must have another subsequence with a limit $\eta \in \partial X$ different from ξ. Repeating the above argument for that subsequence, we get $\sigma_\lambda = \delta_\eta$. But then $\xi = \eta$. Thus, we have proved that $Z_n(\lambda)$ converges to some point of ∂X, and we get convergence to the boundary.

We still have to show that the resulting limit distributions, i.e., the $\Pr_x$-distributions of Z_∞, coincide with the measures ν_x, $x \in X$. What we have to verify is that

$$\mathrm{E}_x\big(\varphi(Z_\infty)\big) = \int_{\partial X} \varphi \, d\nu_x \quad \text{for every } \varphi \in C(\partial X) \text{ and every } x \in X.$$

To see this, let $\varphi \in C(\partial X)$. Then

$$\varphi(Z_\infty) = \lim_{n\to\infty} h_\varphi(Z_n) \quad \text{almost surely.}$$

By dominated convergence, we may exchange expectation and limit. But by harmonicity of h_φ,

$$\mathrm{E}_x\big(h_\varphi(Z_n)\big) = \sum_y p^{(n)}(x,y)h_\varphi(y) = h_\varphi(x) = \int_{\partial X} \varphi \, d\nu_x$$

for every n. This concludes the proof. □

If we have convergence of (Z_n) to the boundary, then we can "localize" (ii) to get a probabilistic definition of regular points: $\xi \in \partial X$ is *regular*, if

$$\lim_{x\to\xi} \nu_x = \delta_\xi \quad \text{weakly.}$$

It is an instructive exercise to show that regular points constitute a Borel subset of ∂X, compare with Knapp [Kn], who has studied regular points in the case when $\widehat{X}$ is the *Martin compactification*.

Lemma 2.3. *Let ξ be regular.*

(a) *ξ is in the support of ν_x for every x.*

(b) *If, in addition, the support of ν_x contains more than one point, then the Green kernel vanishes at ξ, i.e., $\lim_{y\to\xi} G(y,x) = 0$ for some ($\Longleftrightarrow$ every) x.*

Proof. (a) If U is a neighbourhood of ξ in ∂X, then $\lim_{x \to \xi} \nu_x(U) = 1$. Hence $\nu_x(U) > 0$ for some x. As all the ν_x are mutually absolutely continuous, this must hold for all x.

(b) By the assumption, we may choose a neighbourhood U of ξ such that $\nu_x(U^c) > 0$. We have

$$\nu_y(U^c) \geq F(y,x)\nu_x(U^c). \tag{4}$$

(This inequality can be seen by a simple stopping time argument, or in the following way: $y \mapsto \nu_y(U^c)$ is a positive harmonic function, and $y \mapsto F(y,x)\nu_x(U^c) = Gf(y)$ is a potential with $f(\cdot) = \delta_x(\cdot)\nu_x(U^c)/G(x,x)$. The two functions coincide at x. By the domination principle, the first function is bigger.) If $y \to \xi$ then $\nu_y(U^c) \to 0$, so that $F(y,x) \to 0$ and also $G(y,x) = F(y,x)G(x,x) \to 0$.

The " $\Longleftrightarrow$ every" in statement (b) is an easy exercise. □

We remark here that [Kn] gives an example of an irreducible transition operator P on a space X whose Martin compactification has two points, neither of which is regular. However, in this example there is no reasonable (nonoriented) graph structure on X to which P is "adapted" in the spirit of the present notes.

3. Euclidean lattices

Here, we explain why Euclidean structures are not the right ones for solving the Dirichlet problem at infinity. Let $X = \mathbb{Z}^d$ be the d-dimensional lattice. To view it as a graph, the usual picture is that where every pair of integer points at unit distance is connected by an edge. The lattice carries the structure of an Abelian group acting by (integer) translations. Hence the most reasonable way of saying that a transition operator P is adapted to the structure of $\mathbb{Z}$ is to require (besides irreducibility) that it is translation invariant:

$$p(x,y) = p(0, y-x). \tag{5}$$

Thus, P is determined by its row at 0, which is a probability distribution on $\mathbb{Z}^d$.

Now it is well known since the late fifties by the (as it seems, independent) work of Blackwell [Bl], Choquet and Deny [CD], Dynkin and Malyutov [DM] and Doob, Snell and Williamson [DSW] that all bounded harmonic functions with respect to P satisfying (5) are constant. Thus, the Dirichlet problem at infinity is only solvable with respect to the (trivial) one point-compactification.

Let us mention that this does not mean that there are no reasonable "natural" compactifications of $\mathbb{Z}^d$. A nice example is obtained by embedding $\mathbb{Z}^d$ into the d-dimensional unit disc via the map

$$x \mapsto \frac{x}{1+|x|}$$

($|x|$ denoting the Euclidean length) and taking closure; the boundary $\partial\mathbb{Z}^d$ is then the unit sphere. When P has nonzero finite first moment

$$\bar{\xi} = \sum_x x\, p(0,x) \in \mathbb{R}^d,$$

then the law of large numbers implies that (Z_n) converges to the boundary in this compactification with deterministic limit $\bar{\xi}/|\bar{\xi}|$. In particular, every ν_x is the Dirac mass at this point.

If in addition to nonzero first moment, P has finite range, i.e., $p(0,x) > 0$ only for finitely many x, then it is even known (Ney and Spitzer [NS]) that this compactification is (homeomorphic with) the Martin compactification of $\mathbb{Z}^d$ with respect to P, so that it is without doubt "natural".

4. Nearest neighbour random walks on trees

A *tree* is a connected graph without circuits. In this section, our space will be a locally finite tree, and we shall write T instead of X. Its graph metric will be denoted by d. The geometry of a tree is particularly simple. For every pair of vertices x, y of T, there is a unique path of length $d(x,y)$ connecting the two, the *geodesic segment* $\overline{xy}$. We choose and fix an "origin" (reference point) $o \in T$ and write $|x| = d(o,x)$ for $x \in T$.

There is a natural compactification $\widehat{T}$ of T: a *geodesic ray* is a sequence $[x_0, x_1, \ldots]$ of vertices such that $d(x_i, x_j) = |i-j|$. Two rays are *equivalent*, if they differ only by finite initial pieces. An equivalence class is an *end* of T. The space of all ends is denoted by ∂T. For any $x \in T$ and $\omega \in \partial T$, there is precisely one geodesic ray, denoted $\overline{x\omega}$, which starts at x and represents ω as an equivalence class. Given $\xi, \eta \in \widehat{T} = T \cup \partial T$, we define their *confluent* $c(\xi,\eta)$ to be the last common vertex on $\overline{o\xi}$ and $\overline{o\eta}$ unless $\xi = \eta \in \partial T$, in which case we set $c(\xi,\eta) = \xi$. The topology of $\widehat{T}$ is discrete on T, while for $\omega \in \partial T$, a neighbourhood basis is given by all sets

$$U_x(\xi) = \{\xi \in \widehat{T} : c(\xi,\omega) \in \overline{x\omega}\}$$

where $x \in \overline{o\omega}$. Thus , $\xi_n \to \omega$ if and only of $|c(\xi_n,\omega)| \to \infty$. Set $(\xi|\eta) = |c(\xi,\eta)|$ (for the sake of name-dropping this is nowadays often called the "Gromov product"; it has been used before). Then the topology is induced by the ultrametric

$$\theta(\xi,\eta) = \begin{cases} e^{-(\xi|\eta)}, & \text{if } \xi \neq \eta; \\ 0, & \text{otherwise.} \end{cases}$$

The simplest type of transition operator P on T is one which is of *nearest neighbour* type (this includes the simple random walk):

$$p(x,y) > 0 \iff d(x,y) = 1.$$

In this section, we shall always assume that P is of this form and transient. The seminal paper of Cartier [Ca] contains a detailed study of positive harmonic functions in this setting. The following crucial property for nearest neighbour random walk on a tree is often used: on the way from vertex x to vertex y, (Z_n) must visit every point of $\overline{xy}$. By an easy stopping time argument, this yields

$$F(x,y) = F(x,z)F(z,y), \quad z \in \overline{xy}. \tag{6}$$

In particular, it is quite easy to see that $\widehat{T}$ is the Martin compactification of T with respect to P. This implies convergence to the boundary, but this can be seen more directly.

Lemma 4.1. *If P is nearest neighbour and transient, then the associated random walk (Z_n) converges to ∂T.*

Proof. The set of trajectories λ for which $d\big(Z_n(\lambda), Z_{n+1}(\lambda)\big) = 1$ and $|Z_n(\lambda)| \to \infty$ has $\Pr_x$-measure one. Suppose that such a trajectory has two different accumulation points $\omega_1, \omega_2 \in \partial T$. Let x and y be the neighbours of $c = c(\omega_1, \omega_2)$ on $\overline{c\omega_1}$ and $\overline{c\omega_2}$, respectively. Then there must be infinitely many pairs $m < n$ such that $Z_m(\lambda) \in U_x(\omega_1)$ and $Z_n(\lambda) \in U_y(\omega_2)$. By the nearest neighbour property, between instants m and n, the trajectory must pass through every point on the geodesic segment between $Z_m(\lambda)$ and $Z_n(\lambda)$. The latter contains the point c. Consequently, $Z_k(\lambda) = c$ for infinitely many k, in contradiction with $|Z_k(\lambda)| \to \infty$. □

We now can proceed to regularity, i.e., weak convergence of harmonic measures. Recall our assumption that the boundary has at least two elements. In the following theorem, statements (a) and (c) have been first observed (independently) by Cartwright, Soardi and Woess [CSW] and Benjamini and Peres [BP] under the additional assumption that $|\operatorname{supp} \nu_x| \geq 2$. Statement (b) is a consequence of [BP], Prop. 8, where it is shown that the set of nonregular points has logarithmic capacity zero; here, we give a direct proof of (b).

Theorem 4.2.

(a) *$\omega \in \partial T$ is regular if and only if the Green kernel vanishes at ω.*

(b) *The set of all regular points has ν_x-measure one for every x.*

(c) *The Dirichlet problem is solvable with respect to ∂T if and only if the Green kernel vanishes at infinity, i.e.,*

$$\lim_{|x| \to \infty} G(x,o) = 0.$$

Proof. Our first claim is that

$$\nu_x\big(\{\omega \in \partial T : \lim_{y \to \omega} G(y,x) = 0\}\big) = 1 \tag{7}$$

for some and hence every $x \in T$. (Note that the set itself does not depend on x; also, it is easily seen to be a Borel set.)

Write $v_n(\omega)$ for the nth vertex on $\overline{o\omega}$, $\omega \in \partial T$.

Start with observing that $G(Z_n,o) \to 0$ Pr_o-almost surely (this is well known): indeed, $\big(G(Z_n,o)\big)_n$ is a positive, bounded supermartingale and must converge almost surely. Also,

$$\mathrm{E}_o\big(G(Z_n,o)\big) = \sum_x p^{(n)}(o,x)G(x,o) = G(o,o) - \sum_{k=0}^{n-1} p^{(k)}(o,o) \to 0,$$

so that the a.s. limit must be zero. Next, set

$$\bar{\Lambda} = \big\{\lambda \in \Lambda : Z_0(\lambda) = o,\ d\big(Z_n(\lambda), Z_{n+1}(\lambda)\big) = 1,\\ G\big(Z_n(\lambda),o\big) \to 0,\ Z_n(\lambda) \to Z_\infty(\lambda) \in \partial T\big\}.$$

Then $\mathrm{Pr}_o(\bar{\Lambda}) = 1$. Each trajectory $\lambda \in \bar{\Lambda}$ must visit every point on $\overline{o\,Z_\infty(\lambda)}$. For $k \in \mathbb{N}$, let $n_k = n_k(\lambda) = \min\{n : Z_n(\lambda) = v_k\big(Z_\infty(\lambda)\big)\}$. Then $n_k \to \infty$, and we get that $G\Big(v_k\big(Z_\infty(\lambda)\big),o\Big) = G\big(Z_{n_k}(\lambda),o\big) \to 0$ almost surely. Consequently,

$$\mathrm{Pr}_o\big[Z_\infty \in \{\omega : G\big(v_k(\omega),o\big) \to 0\}\big] = 1.$$

But if $G\big(v_k(\omega),o\big) \to 0$ and $y \to \omega$, then by (6)

$$G(y,o) = F(y,o)G(o,o) = F\big(y,c(y,\omega)\big)F\big(c(y,\omega),o\big)G(o,o)\\ \le F\big(c(y,\omega),o\big)G(o,o) \to 0,$$

and we get the statement (7) for $x = o$.

To prove (a), let $U = U_x(\omega)$ be a basic neighbourhood of $\omega \in \partial T$. Formula (4) sharpens to

$$\nu_y(U^c) = F(y,x)\nu_x(U^c), \quad x \in U \cap T$$

(as the random walk must pass through x on the way from U to U^c). If the Green kernel vanishes at ω then also $F(y,x) \to 0$, when $y \to \omega$. Thus $\nu_y(U^c) \to 0$, and it must be $\nu_y(U) \to 1$ for every basic neighbourhood of ω, so that ω is regular. Conversely, the "only if"-part has been proved in Lemma 2.3, unless $\mathrm{supp}\,\nu_x$ consists of a single $\omega \in \partial T$. But in this case, $\nu_x = \delta_\omega$, and we know from (7) that the Green kernel must vanish at ω.

As we now know that the regular points are precisely those where the Green kernel vanishes, (b) follows from (7). Statement (c) follows from (a) by compactness. ⊓

We remark that, even though in the case of nearest neighbour random walks on trees the set of regular points has harmonic measure one, there are examples where $\mathrm{supp}\,\nu_x$ contains nonregular points. See Kaimanovich and Woess [KW] for an example; in [BP] one even finds examples where the nonregular points are dense in $\mathrm{supp}\,\nu_x$.

5. Group invariance and contractive compactifications

When a group Γ acts by isometries (automorphisms) on the graph X with respect to the graph metric d, then a natural type of adaptedness condition for P will be *group invariance:*

$$p(\gamma x, \gamma y) = p(x, y) \quad \text{for all } x, y \in X, \gamma \in \Gamma. \tag{8}$$

This is of interest in particular when Γ acts *transitively* or *quasi transitively* (with finitely many orbits) on X. The group $\mathrm{Aut}(X)$ of automorphisms (self-isometries) of a locally finite, connected graph X carries the topology of pointwise convergence:

$$\gamma_n \to \gamma \iff \text{for every } x \in X \text{ there is } n_x \text{ with } \gamma_n x = \gamma x \; \forall\, n \geq n_x.$$

A neighbourhood base of the identity ι is given by the pointwise stabilizers of finite subsets of X. Each of these is an open and compact subgroup of $\mathrm{Aut}(X)$, so that $\mathrm{Aut}(X)$ is a locally compact, totally disconnected group. See Trofimov [Tr] and Woess [W2] for more details concerning this topology.

Property (8) remains valid for the closure of Γ. Assume that Γ is a closed subgroup of $\mathrm{Aut}(X)$ which acts transitively on X and that P satisfies (8). In this case, one can lift the random walk (Z_n) from X to Γ. Write Γ_x for the stabilizer of vertex x in the group Γ, and choose a reference vertex o. As Γ_o is open and compact, we may normalize left Haar measure $d\gamma$ on Γ such that Γ_o has measure one. We now define a Borel probability measure μ on Γ by

$$\mu(d\gamma) = p(o, \gamma o)\, d\gamma. \tag{9}$$

Observe that μ is Γ_o-invariant on the left, i.e., $\mu(\alpha B) = \mu(B)$ for every Borel set $B \subset \Gamma$ and $\alpha \in \Gamma_o$. Irreducibility of P (on X) yields that μ is irreducible (on Γ), i.e., the support of μ generates Γ as a closed semigroup. If h is a function on X, then we lift it to Γ by defining $\tilde{h}(\gamma) = h(\gamma o)$. Vice versa, if $\tilde{h}$ is a function on Γ which is Γ_o*-invariant on the right* (that is, $\tilde{h}(\gamma\alpha) = \tilde{h}(\gamma)$ for every $\alpha \in \Gamma_o$), then we project $\tilde{h}$ onto a function on X by $h(x) = \tilde{h}(\gamma o)$, where $\gamma \in \Gamma$ maps o to x (by transitivity, such γ exists). This inverts the lifting $h \mapsto \tilde{h}$. One sees that h is P-harmonic on X if and only if $\tilde{h}$ is μ-harmonic on Γ, i.e.,

$$\tilde{h}(\gamma) = \int_\Gamma \tilde{h}(\gamma\alpha)\, \mu(d\alpha) \quad \text{for every } \gamma \in \Gamma. \tag{10}$$

In the sequel, let us forget for a while that μ has arisen from P via (9). Just assume that μ is an irreducible Borel probability measure on Γ. Also, our considerations will make sense for more general types of locally compact groups with countable base of the topology.

Following the lines of Woess [W3], we now introduce the type of (not necessarily discrete) groups, spaces and compactifications for which we are going to study the Dirichlet problem.

Let (X, d) be a *proper metric space*. Proper means that closed balls in X are compact. $\mathrm{Aut}(X)$ is the group of all self-isometries of X, it carries the topology of uniform convergence on compact sets. Consider a closed subgroup Γ of $\mathrm{Aut}(X)$ (not necessarily assumed to act transitively on X). By a compactification $\widehat{X}$ of X we shall always mean one which has all "good" properties, as above. Again, we write ∂X for the boundary.

We say that $\widehat{X}$ is a *Γ-compactification* if every $\gamma \in \Gamma$ extends to a homeomorphism of $\widehat{X}$ onto itself. If this holds, then we say that $\widehat{X}$ is a *contractive* Γ-compactification, provided that the following two properties are satisfied.

(A) If (x_n), (y_n) are sequences in X, then

$$x_n \to \xi \in \partial X \quad \text{and} \quad \sup_n d(y_n, x_n) < \infty \quad \text{imply} \quad y_n \to \xi.$$

(B) Whenever (γ_n) is a sequence in Γ such that for some $x \in X$,

$$\gamma_n x \to \xi \in \partial X \quad \text{and} \quad \gamma_n^{-1} x \to \eta \in \partial X,$$

then $\gamma_n \zeta \to \xi$ uniformly for $\zeta \in \partial X$ outside of every neighbourhood of η.

Condition (B) means that for every choice of neighbourhoods U of ξ and V of η in ∂X, there is an index $n(U,V)$ such that $\gamma_n(\widehat{X} \setminus V) \subset U$ for all $n \geq n(U,V)$. This is certainly not a "Euclidean" property; it bears a strong flavour of hyperbolicity. Indeed, the typical example is the hyperbolic plane $\mathbb{H}$ (consider the unit disc-model with hyperbolic metric). The group of Möbius transformations leaving $\mathbb{H}$ invariant acts transitively as a group of isometries. If we take $\widehat{\mathbb{H}}$ to be the standard compactification (Euclidean closure of the unit disc), then properties (A) and (B) hold. Γ might be, for example, a finitely generated Fuchsian group.

Another typical example where (A) and (B) hold is the homogenous tree $T = T_q$, where all vertices have degree $q+1 \geq 3$, together with its end compactification and the group of all automorphism. Indeed, for every locally finite tree T, its end compactification is a contractive $\mathrm{Aut}(T)$-compactification. We shall see extensions of these examples in §§6–7.

The above definition is due to [W3], but "contractivity" is already hidden in two important papers by Furstenberg [F1], [F2], where—among many other things—convergence to the boundary of certain Markov processes is studied. If one wants to apply Furstenberg's techniques to trees (Cartwright and Soardi [CS]) or other classes of graphs [W1], then one comes across properties (A) and (B) more or less automatically. A very similar property has appeared in the study of quasiconformal groups acting on Euclidean space, see Gehring and Martin [GM], who called their groups "convergence groups".

If (A) and (B) hold, then the action of Γ on $\widehat{X}$ shares many common features with the action of Möbius transformations on $\mathbb{H}$, see [W3]: one can classify the elements of Γ into *elliptic*, *parabolic* or *hyperbolic* according to whether they fix a compact subset of X, or (otherwise) fix one or two points of ∂X, respectively. The *limit set* $L(\Gamma)$ is defined as the set of accumulation points in ∂X of an orbit Γx. By (A), it is independent of $x \in X$, and $|L(\Gamma)| \in \{0, 1, 2, \infty\}$. As we assume Γ to be closed, $L(\Gamma)$ is empty if and only if Γ is compact. Otherwise, we may consider $L(\Gamma)$ as the boundary of a compactification $\widehat{\Gamma}$ of Γ itself: a sequence (γ_n)

in Γ converges to $\xi \in L(\Gamma) \subset \partial X$ if $\gamma_n x \to \xi$ for some ($\iff$ every) $x \in X$. Finally, we have the following alternative [W3].

Theorem 5.1. *If $\widehat{X}$ is a contractive Γ-compactification of X then either*

(a) *Γ contains two hyperbolic elements which freely generate a discrete free subgroup of Γ, and the orbits of Γ on $L(\Gamma)$ are infinite, or*

(b) *Γ has the* fixed set property: *it fixes a nonempty compact subset of X, a unique element of ∂X, or a unique pair of elements of ∂X.*

In case (a), the limit set is infinite, but this may also happen in the second subcase of (b) (while it is empty in the first subcase and consists of that pair in the third). Case (b) is in some sense "degenerate"; the case interesting for us is (a).

Theorem 5.2. *Let Γ be a closed group of automorphisms of the proper metric space X, and let $\widehat{X}$ be a contractive Γ-compactification of X. Suppose that Γ does not have the fixed set property.*

(a) *For every irreducible probability measure μ on Γ, the Dirichlet problem for μ-harmonic functions on Γ is solvable with respect to $L(\Gamma)$.*

(b) *If, in particular, X is a locally finite, connected graph, Γ acts transitively and P is a Γ-invariant, irreducible transition operator on X, then the Dirichlet problem for P-harmonic functions on X is solvable with respect to ∂X.*

This can be proved, with some intermediate work, by reading results in [F1], [F2] in the right order, compare with [CS] and [W1]. Here, we prefer to give a complete, ordered proof — which turns out to be rather simple.

Proof. (a) As Γ acts on $L(\Gamma)$ on the left by homeomorphisms, we can consider the convolution between μ and a probability ν on $L(\Gamma)$. This is the probability on $L(\Gamma)$ defined by

$$\int_{L(\Gamma)} f(\xi)\, \mu * \nu(d\xi) = \int_{L(\Gamma)} \int_{\Gamma} f(\gamma\xi)\, \mu(d\gamma)\nu(d\xi).$$

By compactness of $L(\Gamma)$, there must be a *μ-stationary* probability measure ν on $L(\Gamma)$, i.e., one that satisfies $\mu * \nu = \nu$. (To see this, take an arbitrary probability σ on $L(\Gamma)$, for example a point mass, and consider the sequence of measures $\frac{1}{n}(\sigma + \mu * \sigma + \cdots + \mu^{(n-1)} * \sigma)$. It must have a weak accumulation point in the space of Borel probabilities on $L(\Gamma)$, and the latter is stationary.)

The assumption that Γ does not have the fixed set property is now used to show that ν is continuous (carries no point mass), compare with [CS] or [W1]. Let $M = \max\{\nu(\{\xi\}) : \xi \in L(\Gamma)\}$. Suppose that $M > 0$. Then

$$S = \{\xi \in L(\Gamma) : \nu(\{\xi\}) = M\}$$

is finite. As $\mu * \nu = \nu$, we have $\gamma S = S$ for every $\gamma \in \operatorname{supp} \mu$. But $\operatorname{supp} \mu$ generates Γ, so that Γ must fix S. This contradicts Theorem 5.1(a).

Next, let $f \in C(L(\Gamma))$. We claim that the solution of the Dirichlet problem with boundary function f is given by

$$h(\gamma) = \int_{L(\Gamma)} f(\gamma\xi)\,\nu(d\xi), \quad \gamma \in \Gamma. \tag{11}$$

We have

$$\int_\Gamma h(\gamma\alpha)\,\mu(d\alpha) = \int_{L(\Gamma)} f(\gamma\xi)\,\mu * \nu(d\xi) = h(\gamma),$$

and h is μ-harmonic. Let $\zeta \in L(\Gamma)$, and let (γ_n) be a sequence such that $\gamma_n \to \zeta$ in the topology of $\widehat{\Gamma}$, i.e., $\gamma_n o \to \zeta$ in $\widehat{X}$ (such a sequence exists for every $\zeta \in L(\Gamma)$ by the definition of the limit set). By compactness, we may assume without loss of generality that $\gamma_n^{-1} o \to \eta \in L(\Gamma)$. But then contractivity yields that $\gamma_n \xi \to \zeta$ for every $\xi \in L(\Gamma) \setminus \{\eta\}$. Now, $\nu(\{\eta\}) = 0$, so that

$$f(\gamma_n\xi) \to f(\zeta) \quad \text{for } \nu\text{-almost every } \xi \in L(\Gamma).$$

By dominated convergence, we get

$$h(\gamma_n) = \int_{L(\Gamma)} f(\gamma_n\xi)\,\nu(d\xi) \to \int_{L(\Gamma)} f(\zeta)\,\nu(d\xi) = f(\zeta).$$

To see that (a) implies (b), suppose that μ arises from P as in (9). Then μ is Γ_o-invariant on the left, so that we also have

$$\delta_\alpha * \nu = \delta_\alpha * \mu * \nu = \mu * \nu = \nu, \quad \alpha \in \Gamma_o.$$

Consequently, the function h defined by (11) is Γ_o-invariant on the right (we should have written $\tilde{h}$ instead of h), and as in (10) it projects onto a function on X which is P-harmonic and solves the Dirichlet problem on $\widehat{X}$ with boundary function $f \in C(L(\Gamma)) = C(\partial X)$. □

Remark 5.3. The assumption of non validity of the fixed set property is used to show that there is a stationary measure ν on $L(\Gamma)$ which is continuous. Thus, the result also holds if one only assumes existence of a continuous stationary measure. Indeed, this can occur even when Γ fixes a unique point in $L(\Gamma)$, while $L(\Gamma)$ is infinite. See [CKW] for such a situation.

Remark 5.4. In (b), with some minor effort, one can replace "transitive" with "almost transitive". (One method to work this out is to use the induced random walk on one of the orbits.)

Random walks on groups. We now add a few observations of probabilistic nature in the context of a locally compact group Γ with a countable base of the topology, together with an irreducible probability measure μ on Γ.

Instead of the trajectory space, we consider the infinite product space $(\Gamma^{\mathbb{N}}, \mathcal{A}, \Pr)$, where $\mathcal{A}$ is the product σ-algebra and $\Pr = \mu^{\mathbb{N}}$. The nth projections onto Γ constitute a sequence of i.i.d. random variables X_n, each one with distribution μ. The *right random walk on* Γ *with law* μ is the sequence of Γ-valued random variables

$$S_n = \iota \cdot X_1 \dots X_n,\ n \geq 0.$$

The distribution of S_n is $\mu^{(n)}$, the nth convolution power of μ.

If μ arises as in (9) by lifting a random walk on X to Γ, then it is an easy exercise to show that the sequence $(S_n o)$ is a model of the original random walk (Z_n) on X with starting point o, see Woess [W1]. By "model" we mean that $(S_n o)$ is a Markov chain on X with transition operator P; the probability space on which it is defined is of course not the standard trajectory space. Furthermore, if $\gamma \in \Gamma$, then $(\gamma S_n o)$ "models" (Z_n) starting at $x = \gamma o$.

As in the proof of Theorem 2.2, from statement (a) of Theorem 5.2 one can now deduce convergence of (S_n) to the boundary. Indeed, the existence of nonconstant bounded μ-harmonic functions on Γ yields transience, i.e.,

$$d(S_n o, o) \to \infty \quad \text{almost surely.}$$

If $f \in C(L(\Gamma))$ then the sequence of random variables

$$\delta_{S_n} * \nu(f) = \int_{L(\Gamma)} f(S_n \xi)\, \nu(d\xi)$$

is a bounded martingale with respect to the filtration generated by (S_n). Hence, it converges almost surely. Writing $\Lambda = \Gamma^{\mathbb{N}}$ (by abuse of the notation used in the other sections), we proceed as in the proof of Theorem 2.2. Using a countable dense family of functions in $C(L(\Gamma))$, we get that for Pr-almost every $\lambda \in \Lambda$, there is a probability measure σ_λ on $L(\Gamma)$, such that

$$\delta_{S_n(\lambda)} * \nu \to \sigma_\lambda \quad \text{weakly.}$$

Repeating the argument of Theorem 2.2, we now get that σ_λ is a point mass and that $(S_n(\lambda))$ converges to that point. A posteriori on gets as well that ν is the limit distribution of (S_n) for starting point $S_0 = \iota$, so that for arbitrary starting point γ, the limit distribution is $\delta_\gamma * \nu$.

Note that in the whole argument, the hypothesis of irreducibility is not fully used. It is enough to assume that $\operatorname{supp}\mu$ generates Γ as a closed group. In this case, it is not necessarily true that ν is supported by the whole of $L(\Gamma)$, and one cannot use the minimum principle for showing that the solution of the Dirichlet problem is unique.

In the discrete setting, Theorem 5.2 may become useful in two ways: when X is nondiscrete, but Γ is discrete, or vice versa. Let us now explain an application of the first type (cases of the second type will be considered in §§6–7).

Harmonic functions on Fuchsian groups. Let Γ be a finitely generated *Fuchsian group,* i.e, a discrete group of Möbius transformations $z \mapsto \frac{az+b}{cz+d}$ of the unit disc. If we consider the latter as (a model of) the hyperbolic plane $\mathbb{H}$, then Γ is a closed subgroup of $\operatorname{Aut}(\mathbb{H})$. As mentioned above, the Euclidean closure $\widehat{\mathbb{H}}$ is easily seen to be a contractive $\operatorname{Aut}(\mathbb{H})$-compactification of $\mathbb{H}$, see also §6. The limit set of Γ is a well-studied object, see, e.g., Beardon [Be]. In particular, unless $|L(\Gamma)| = 2$

(in which case Γ is called *elementary*), the limit set is always uncountable and Γ does not have the fixed set property.

Corollary 5.5. *Let Γ be a non-elementary Fuchsian group and μ an irreducible probability measure on Γ. Then the Dirichlet problem is solvable for μ-harmonic functions with respect to the limit set of Γ, and the random walk with law μ converges almost surely to a $L(\Gamma)$-valued random variable whose distribution is continuous and supported by the whole of $L(\Gamma)$.*

This extends previous results of Series [Se] and Ancona [A1], who considered only the case when $\operatorname{supp}\mu$ is finite. At the same time, the proof is considerably simpler.

Cayley graphs. One might want to view results on random walks on finitely generated groups in the context of graphs, as in the rest of these notes. Let Γ be any finitely generated group, and let S be a symmetric ($S^{-1} = S$) set of generators. Then the *Cayley graph* of Γ with respect to S has vertex set Γ, while the edges are given by all unordered pairs $[\gamma, , \gamma\alpha]$, where $\gamma \in \Gamma$ and $\alpha \in S$. This is a connected graph, where all vertices have the same finite degree, and Γ acts by left multiplication as a discrete subgroup of the automorphism group. (In many cases, it is the whole automorphism group.) The Cayley graphs of Γ with respect two different (finite, symmetric) generating sets S_1, S_2 have equivalent metrics d_1, d_2, i.e., $C^{-1}d_1 \le d_2 \le Cd_1$ for some $C > 0$.

If μ is a probability measure on Γ, then the random walk with law μ has transition operator given by

$$p(\alpha, \gamma) = \mu(\{\alpha^{-1}\gamma\}).$$

It is clear that this alone makes sure that P is well adapted to the underlying (group) structure; the random walk will be of nearest neighbour type on the Cayley graph, when $\operatorname{supp}\mu = S$. However, the above results apply even when $\operatorname{supp}\mu$ is infinite.

6. Hyperbolic graphs

A graph X is called *hyperbolic* if the metric space (X, d) is hyperbolic in the sense of Gromov [Cr], where d is the discrete metric of X.

More generally, let (X, d) be a proper metric space which is also *geodesic:* any pair of points $x, y \in X$ is connected by at least one *geodesic segment* $\overline{xy}$, i.e., an isometric imbedding of the interval $[0, d(x, y)]$ into X which sends 0 to x and $d(x, y)$ to y. (A locally finite, connected graph has these properties, when we regard each edge as a copy of the unit interval and extend the metric to all points on the edges in the obvious way.) Once again, we choose and fix a reference point $o \in X$

and write $|x| = d(x, o)$. As in §4, the "Gromov product" is defined by

$$(x|y) = \frac{1}{2}\big(|x| + |y| - d(x,y)\big).$$

The space X is said to be *hyperbolic* if there is a $\delta \geq 0$ (possibly large) such that

$$(x|y) \geq \min\{(x|z), (z|y)\} - \delta \quad \text{for all } x, y, z \in X.$$

In a hyperbolic space, the geometric significance of the quantity $(x|y)$ is the following: if $\overline{xy}$ is a geodesic segment between x and y (note that in general it is not necessarily unique), then $(x|y)$ is the distance between o and $\overline{xy}$ up to an error of at most $\pm const \cdot \delta$, where *const* is a fixed number. Another, equivalent, definition of hyperbolicity is that there exists a constant $\delta' \geq 0$, such that for every geodesic triangle $\overline{xyz}$ (i.e., three points x, y, z together with a geodesic segment = side between any two of them) the following holds: for any point on one side there is a point on one of the other two sides at distance $\leq \delta'$.

Details on hyperbolic spaces can be collected together from Gromov [Gr], Ghys and de la Harpe [GH], Coornaert, Delzant and Papadopoulos [CDP], Short [Sh] and Bowditch [Bo] (it may be useful to consult more than one of these sources).

Examples. Locally finite trees are hyperbolic with $\delta = \delta' = 0$.

In the continuous setting, the typical examples are the upper half spaces $\mathbb{H}_n$ with hyperbolic metric (equivalently, one may use the open unit ball model) in arbitrary dimension n. More generally, every complete Riemannian manifold with bounded geometry and curvature between two negative bounds is Gromov hyperbolic.

If X is a graph which is *nicely embedded* into $\mathbb{H}_n$ then it is hyperbolic. By nicely embedded we mean that (i) $X \subset \mathbb{H}_n$, and (ii) there is $L > 0$ such that $L^{-1}d_X(x,y) \leq d_{\mathbb{H}_n}(x,y) \leq L d_X(x,y)$. Here, $\mathbb{H}_n$ might be replaced by any other Gromov hyperbolic space.

In particular, (the Cayley graphs of) finitely generated Fuchsian groups are hyperbolic. A typical example of a graph which does not arise from a group and is nicely embedded into $\mathbb{H} = \mathbb{H}_2$ is the one whose picture can be found in the note by Cartwright and Woess [CW].

Besides these, also (the Cayley graphs of) certain small cancellation groups are Gromov hyperbolic. Free products with amalgamation and HNN extensions of hyperbolic groups over finite subgroups are again hyperbolic. However, the Cartesian product of two hyperbolic spaces is not hyperbolic (unless one of the two is compact).

A hyperbolic space X has a "natural" *hyperbolic compactification* $\widehat{X}$. One way of introducing it is the following. If $a > 0$ is sufficiently small (depending on δ), then

$$\rho_a(x,y) = \begin{cases} e^{-a\cdot(x|y)}, & \text{if } x \neq y, \\ 0, & \text{otherwise,} \end{cases}$$

is "almost" a metric, i.e.,

$$\theta_a(x,y) = \inf\left\{\sum_{i=1}^{n} \rho_a(x_{i-1},x_i) : n \geq 0,\ x = x_0, x_1, \ldots, x_{n-1}, x_n = y \in X\right\}$$

is a metric on X such that $C \cdot \rho_a \leq \theta_a \leq \rho_a$ for some $C > 0$ (depending on δ). Now take the completion of X in this new metric. The set of new points added to X in this way is the *hyperbolic boundary* ∂X. If X is a graph, then the topology of $\widehat{X} = X \cup \partial X$ is just the one defined by θ_a. However, even if (X,d) is nondiscrete, (X,θ_a) becomes discrete, as one easily verifies. Consequently, in general, the topology on $\widehat{X}$ is the original one on X (a neighbourhood base at $x \in X$ consists of the open d-balls centered in x), while it is the one induced by θ_a on ∂X (a neighbourhood base at $\xi \in \partial X$ consists of the open θ_a-balls centered in ξ). This compactification is independent of the choice of a (sufficiently small) and has all "good" properties in our sense.

From the relation between ρ_a and θ_a, one sees that a sequence (x_n) in X converges to some boundary point if and only if

$$\lim_{m,n\to\infty} (x_m|x_n) = \infty, \tag{12}$$

and another sequence (y_n) converges to the same boundary point if and only if

$$\lim_{n}(x_n|y_n) = \infty. \tag{13}$$

In fact, another way of introducing the boundary is to consider equivalence classes of sequences satisfying (12), where equivalence is defined by (13). If $\gamma \in \mathrm{Aut}(X)$, then (γx_n) also satisfies (12). In this way, one works out that γ extends to a homeomorphism of $\widehat{X}$.

If a countable set X is equipped with two different (locally finite, connected) graph structures giving rise to equivalent graph metrics, then hyperbolicity of one implies hyperbolicity of the other, and the hyperbolic compactifications coincide. In particular, for a finitely generated group, hyperbolicity is independent of the choice of the set of generators and corresponding Cayley graph metric, so that one speaks of *hyperbolic groups.*

Examples. For locally finite trees, the hyperbolic compactification coincides with the end compactification described in §4. The hyperbolic compactification of $\mathbb{H}_n$, using the open unit disc model, is the Euclidean closure in $\mathbb{R}^n$. For a complete Riemannian manifold with bounded geometry and pinched negative curvature, the hyperbolic boundary coincides with the visibility sphere at infinity.

If X is a locally finite, connected graph which is nicely imbedded into $\mathbb{H}_n$, then (again using the unit disc model) $\widehat{X}$ is Euclidean closure. Thus, ∂X is the limit set of X, i.e., the set of accumulation points of $X \subset \mathbb{H}_n$ in $\partial\mathbb{H}_n$.

If X is the Cayley graph of a Fuchsian group Γ, then one has to be more careful. Indeed, the hyperbolic compactification is the same for isometric spaces and, in particular, for isomorphic groups. In other words, the compactification depends only on the abstract group Γ, while it is well possible to have two different realizations of Γ (i.e., groups consisting of concrete Möbius transformations) which have significantly different limit sets. For example, the Cayley graph of a *free group* with respect to the free generators is a homogeneous tree T, and the hyperbolic boundary is the space of ends ∂T of that tree: a Cantor set. Now, one may realize the free group as a Fuchsian group in such a way that the limit set is a Cantor set homeomorphic with ∂T, but one may also define a realization such that the limit set is the whole unit circle. The point is that a concrete Fuchsian group Γ may have parabolic elements (elements with a single fixed point in $\partial\mathbb{H}$), in which case none of its Cayley graphs is nicely imbedded. This occurs when the standard fundamental domain of the group has cusps, i.e., it comes from a punctured surface. It is obvious that one can enlarge the puncture to become a small hole such that the fundamental groups remain isomorphic. In other words, moving slightly apart any pair of sides of the fundamental domain which form a cusp, one passes to the fundamental domain of an isomorphic Fuchsian group Γ_1 which has no parabolic elements, and any of its Cayley graphs is nicely embedded. The limit set of Γ_1 is (homeomorphic with) the hyperbolic boundary of Γ and of Γ_1. Furthermore, the isomorphism $\Gamma_1 \to \Gamma$ extends to a continuos surjection $\Gamma_1 \cup L(\Gamma_1) \to \Gamma \cup L(\Gamma)$ which is one-to-one except at most at a countable number of points in $L(\Gamma)$ where it is two-to-one. Thus, the difference between the hyperbolic boundary and the limit set of a Fuchsian group is "not very big", and in most cases, the two coincide.

In general, an infinite, finitely generated hyperbolic group has either two boundary points, in which case it has a cyclic subgroup with finite index, or the boundary is infinite and the group contains a free subgroup on two generators. In the latter case, the group fixes no point of its boundary. Besides the proofs to be found in the above references for hyperbolic spaces, another way to see this is Prop. 4 in [W3].

The following is proved in [W3].

Proposition 6.1. *If (X,d) is a proper, geodesic space which is hyperbolic, then its hyperbolic compactification $\widehat{X}$ is a contractive* $\mathrm{Aut}(X)$*-compactification.*

Thus, we can apply Theorem 5.2.

Corollary 6.2. *Let X be a locally finite, connected graph which is hyperbolic. Suppose that its hyperbolic boundary ∂X has more than two points, and let Γ be a subgroup of* $\mathrm{Aut}(X)$ *which acts quasi transitively and does not fix any boundary point. If P is an arbitrary irreducible, Γ-invariant transition operator on X, then the Dirichlet problem for P-harmonic functions is solvable with respect to ∂X.*

In the above situation, the boundary must be infinite. In particular, Corollary 6.2 applies to random walks on hyperbolic groups with infinite boundary.

Let us now turn our attention to situations when the transition operator is not invariant under a group which acts transitively or quasi transitively. In this case, one has to find geometrical "substitutes" for the algebraic adaptedness condition of group invariance. In the sequel, X will always be a connected, locally finite graph.

We say that P is *uniformly irreducible* if there are $K < \infty$ and $\varepsilon_0 > 0$ such that the following holds:

$$\text{For every } x, y \in X \text{ with } d(x,y) = 1 \text{ there is } k \leq K \text{ such that } p^{(k)}(x,y) \geq \varepsilon_0. \tag{14}$$

Note that this implies $\deg(x) \leq K/\varepsilon_0$ for every x. If P is invariant under a quasi transitive group then irreducibility is automatically uniform.

We say that P has *bounded range* if

$$\sup\{d(x,y) : x, y \in X, p(x,y) > 0\} < \infty. \tag{15}$$

For an arbitrary irreducible transition operator P, the number

$$\rho(P) = \limsup_{n\to\infty} (p^{(n)}(x,y))^{1/n} \tag{16}$$

is independent of x, y and usually called its *spectral radius*. (This terminology may be somewhat misleading in general.) Pruitt [Pr] has shown that $\rho(P)$ is the smallest $t > 0$ for which there is a positive function u on X satisfying $Pu \leq t \cdot u$. Besides the problem of transience / recurrence, it is one of the basic questions to decide whether $\rho(P) = 1$ or $\rho(P) < 1$. The latter is a kind of "coercitivity" property. It is well studied for random walks on groups. When the group is nonamenable, $\rho(\mu) < 1$, and the converse is true for symmetric μ, see Kesten [K1], [K2], Day [Da]. In absence of a group, the following variant of symmetry is used.

We say that P is *reversible* if there is a positive function (measure) $c(\cdot)$ on X such that

$$c(x)p(x,y) = c(y)p(y,x) \quad \text{for all } x, y \in X, \tag{17}$$

and P is called *strongly reversible* if in addition

$$M^{-1} \leq c(\cdot) \leq M \quad \text{for some } M > 0. \tag{18}$$

For strongly reversible P, the property $\rho(P) < 1$ is linked with the graph structure as follows.

Theorem 6.3. *If P is strongly reversible, uniformly irreducible and has bounded range, then $\rho(P) < 1$ if and only if X satisfies the strong isoperimetric inequality*

$$\inf\left\{\frac{\text{Volume(A)}}{\text{Surface(A)}} : A \subset X,\ A \text{ finite}\right\} > 0.$$

Here, Volume(A) is the sum of the degrees of all vertices in A, and Surface(A) is the number of edges with one endpoint in A and the other in $X \setminus A$. For a simple random walk, this theorem is due to Dodziuk [Do], see also Dodziuk and Kendall [DK]. The version stated here is due to Gerl [Ge]. A further extension is due to

Kaimanovich [Ka], who shows that bounded range can be replaced with "tight step length distributions", thus also generalizing the group-invariant case.

The following significant result is due to Ancona [A1].

Theorem 6.4. *Suppose that X is a hyperbolic graph and that P is uniformly irreducible with bounded range and satisfies $\rho(P) < 1$. Then the hyperbolic compactification $\widehat{X}$ coincides with the Martin compactification of X with respect to P. In particular, (Z_n) converges to the boundary.*

Furthermore, if the Green kernel vanishes at infinity, then the Dirichlet problem is solvable with respect to ∂X. This holds, in particular, if in addition to the above hypotheses, P is strongly reversible.

Note that it is a rather difficult theorem in Martin boundary theory that the Markov chain converges to the Martin boundary; Dynkin [Dy] is the best readable source. On the other hand, in all cases where the Martin compactification is explicitly known, simpler "geometric" proofs of convergence to the boundary are at hand. This does not become transparent from [A1], but has been worked out in [KW, Th. 1], see below.

After Theorem 6.4, in absence of group invariance, a next step is to replace bounded range by a more general property. The *step length distribution* at $x \in X$ is the probability on $\mathbb{N}_0$ given by

$$\sigma_x(n) = \sum_{y:d(x,y)=n} p(x,y).$$

We say that P satisfies the *uniform first moment condition* if

$$\sum_{n=1}^{\infty} \sup_{x\in X} \sigma_x\big([n,\infty)\big) < \infty. \tag{19}$$

This condition enables one to compare (Z_n) with the random walk on the nonnegative integers whose law has tail $\sup_x \sigma_x\big([n,\infty)\big)$.

Theorem 6.5. *Suppose that P is uniformly irreducible and satisfies the uniform first moment condition.*

(a) *If $\rho(P) < 1$, then (Z_n) converges to the hyperbolic boundary ∂X.*

(b) *If there are $\lambda < 1$ and $M < \infty$ such that*

$$p^{(n)}(x,y) \le M\,\lambda^n \quad \text{for all } x,y \in X,\ n \in \mathbb{N},$$

then the Dirichlet problem is solvable with respect to ∂X. This holds, in particular, when P is strongly reversible and $\rho(P) < 1$.

We remark that this theorem, as well as several of the other results presented in these notes, also apply to Markov processes with discrete / continuous time in the continuous setting. In particular, Theorem 6.5(b) applies - with obvious reformulation - to Brownian motion on complete Riemannian manifolds with bounded geometry and pinched negative curvature. This is explained by Kaimanovich in the last section of [KW] and furnishes a simplification of previous proofs by Anderson [An] and Sullivan [S2].

7. Ends of graphs

While in the last section we considered a compactification which is only defined for a special class of graphs, here we shall study one that can be associated with any locally finite connected graph X. The concept of *ends* in the discrete setting goes back to Freudenthal [Fr]; a lot of work has later been done by Halin [H1], [H2] and others. For the vertex-transitive case, see, e.g., Stallings [St], Dunwoody [Du] or Thomassen and Woess [TW].

As in §4, a *geodesic ray* is a sequence $\pi = [x_0, x_1, \dots]$ of vertices such that $d(x_i, x_j) = |i - j|$. If A is a finite subgraph of X, then the graph $X \setminus A$ falls apart into (at most) finitely many connected components. Among these, precisely one must contain all but finitely many points of π, and we say that π ends up in this component. Two rays π, π' are *equivalent,* if for every finite $A \subset X$, they end up in the same component of $X \setminus A$. (Note that for trees, this is the same as in §4.) An *end* is an equivalence class of rays. In this section, we write ∂X for the set of all ends of X. For A as above, we augment every component of $X \setminus A$ by those $\omega \in \partial X$ whose representative rays end up in that component. We write $C(A, \omega)$ for the augmented component of $X \setminus A$ which contains ω.

Let $\widehat{X} = X \cup \partial X$. The topology on $\widehat{X}$ is discrete on X, while a neighbourhood base at $o \in \partial X$ is given by all $C(A, \omega)$, where $A \subset X$ is finite. This defines a compactification with all required properties. Once again, if X is equipped with two different graph structures giving rise to equivalent metrics, the two have the same end compactification. Therefore, for finitely generated groups, the end compactification is independent of the choice of the generating set.

If X is a hyperbolic graph, then the space of ends of X consists precisely of the connected components of the hyperbolic boundary ("connected" refers to the topology of the latter). More precisely, the identity on X extends to a continuous surjection of the hyperbolic onto the end compactification, which is one-to-one between the components of $\partial_{\mathsf{hyp}} X$ and $\partial_{\mathsf{end}} X$. However, let me repeat again that the end structure of a non hyperbolic graph may be very rich and complicated. (It seems that only few people have been willing to expose themselves both to the graph theoretical subtleties of ends and to the related probability and potential theory, unless remaining at the level of trees.) The following is a simple exercise.

Lemma 7.1. *Let X be a locally finite, connected graph. Then the end compactification of X is a contractive* Aut(X)*-compactification.*

The case which is of particular interest is that when Aut(X) acts transitively or quasi transitively on X. In this case, X is finite, or it has one, two or infinitely many ends. The case interesting for us is the one when X has infinitely many ends. From Theorem 5.2 we get

Corollary 7.2. *Let X be a locally finite, connected graph with infinitely many ends, and let Γ be a subgroup of $\mathrm{Aut}(X)$ which acts quasi transitively and does not fix any end. If P is an arbitrary irreducible, Γ-invariant transition operator on X, then the Dirichlet problem for P-harmonic functions is solvable with respect to ∂X.*

This applies, in particular, to random walks on finitely generated groups with infinitely many ends. Indeed, in this case the group cannot fix one of its ends. See [W1], §9 for more details. If Γ is a closed subgroup of $\mathrm{Aut}(X)$ which acts quasi transitively and does fix one of the infinitely many ends, then the group has to be nonunimodular. Even in this case, although Corollary 7.2 does not apply, the Dirichlet problem is solvable for the simple random walk and other related transition operators [W1].

Let us now turn our attention to the situation when no quasi transitive group is around. First of all, we remark that Theorem 4.2 remains valid when instead of nearest neighbour random walks, one considers transient, uniformly irreducible transition operators with bounded range on trees. Equivalently, one may consider such transition operators on graphs X whose metric is equivalent with the metric of a tree defined on the same vertex set. In general, when X is not necessarily tree-like, points (b) and (c) of the following have been proved in [CSW], while (a) is a straightforward extension of Lemma 4.1.

Theorem 7.3. *Suppose that X has at least two ends, and let P be a transient, uniformly irreducible transition operator on X with bounded range.*
(a) *The random walk associated with P converges to the (end) boundary.*
(b) *If the Green kernel vanishes at $o \in \partial X$, then ω is regular.*
(c) *The Dirichlet problem is solvable with respect to the end compactification if and only if the Green kernel vanishes at infinity.*

In comparing this result with Theorem 6.4, note that it is not needed that $\rho(P) < 1$. In general, the Martin boundary will be bigger than the space of ends. Without the hypothesis of bounded range, a result analogous to Theorem 6.5 is proved in [KW].

Theorem 7.4. *Suppose that X has at least two ends, and that P is uniformly irreducible and satisfies the uniform first moment condition.*
(a) *If $\rho(P) < 1$ then (Z_n) converges to the boundary in the end compactification.*
(b) *If there are $\lambda < 1$ and $M < \infty$ such that*

$$p^{(n)}(x,y) \le M\,\lambda^n \quad \text{for all } x, y \in X,\ n \in \mathbb{N},$$

then the Dirichlet problem is solvable with respect to the space of ends. This holds, in particular, when P is strongly reversible and $\rho(P) < 1$.

8. Final remarks

In principle, the results reviewed so far are all that is known to the author regarding the Dirichlet problem at infinity in the context of discrete potential theory. In particular, the author is not aware of any reference in the context of Markov chains considering the relation between the potential theoretic property of *resolutivity* of the boundary and the probabilistic property of almost sure convergence to the boundary. (The author acknowledges a discussion with J. Bliedtner on this subject.)

Another question of interest (to which the author intends to come back in future work) is that of studying the Dirichlet problem at infinity *relative to a fixed positive harmonic function.* In our setting of a discrete space X with boundary at infinity ∂X and transition operator P, this means the following: fix a positive harmonic function h. Given a continuous function φ on ∂X, is there a harmonic function g on $\widehat{X}$ such that

$$\lim_{x \to \xi} g(x)/h(x) = \varphi(\xi)$$

for every $\xi \in \partial X$ (or for every ξ in a reasonable subset of ∂X)? This is the same as studying the (usual) Dirichlet problem for the *h-process,* whose transition probabilities are given by

$$p_h(x, y) = p(x, y)h(y)/h(x).$$

This is of particular interest when ∂X is the Martin boundary and the support of the harmonic measure class is considerably smaller than ∂X, so that the "Poisson integral" (2) leaves many boundary points uncharged. In this case, one will look for a "natural" positive harmonic function, whose representing measure has support containig the whole minimal boundary. In the discrete setting, to the author's knowledge only Picardello and Sjögren [PS] have considered this problem. (However, see also Sullivan [S1, Cor. 30].) They study Fatou convergence with respect to the minimal Martin boundary for the simple random walk on the Cartesian product of two homogeneous trees. As a preliminary step, with a natural choice of reference function h, they prove regularity of all minimal boundary points relative to that function.

Note added in proof. Recently, in a beautiful paper, Benjamini and Schramm have studied the Dirichlet problem at infinity for transient planar graphs.

References

[A1] Ancona, A., Positive harmonic functions and hyperbolicity. In: Potential Theory, Surveys and Problems. Lecture Notes in Math. 1344 (ed. by J. Král et al.), 1–23. Springer-Verlag, Berlin 1988.

[A2] — Théorie du potentiel sur les graphes et les variétés. In: École d'Été de Probabilités de St. Flour XVIII–1988. Lecture Notes in Math. 1427 (ed. by P. L. Hennequin), 4–112. Springer-Verlag, Berlin 1990.

[An] Anderson, M. T., The Dirichlet problem at infinity for manifolds of negative curvature. J. Diff. Geometry 18 (1983), 701–721.

[Be] Beardon, A. F., The Geometry of Discrete Groups. Springer-Verlag, New York 1983.

[BP] Benjamini, I., Peres, Y., Random walks on a tree and capacity in the interval. Ann. Inst. H. Poincaré Prob. Stat. 28 (1992), 557–592.

[Bl] Blackwell, D., On transient Markov processes with a countable number of states and stationary transition probabilities. Ann. Math. Statist. 26 (1955), 654–658.

[Bo] Bowditch, B., Notes on Gromov's hyperbolicity criterion for path-metric spaces. In: Group Theory from a Geometrical Viewpoint (ed. by E. Ghys et al.), 64–167. ICTP, World Scientific, Singapore 1991.

[Ca] Cartier, P., Fonctions harmoniques sur un arbre. Symposia Math. 9 (1972), 203–270.

[CKW] Cartwright, D., Kaimanovich, V. A., Woess, W., Random walks on the affine group of local fields and of homogeneous trees. Ann. Inst. Fourier (Grenoble) 44 (1994), 1–46.

[CS] Cartwright, D., Soardi, P. M., Convergence to ends for random walks on the automorphism group of a tree. Proc. Amer. Math. Soc. 107 (1989), 817–823.

[CSW] Cartwright, D., Soardi, P. M., Woess, W., Martin and end compactifications of non locally finite graphs. Trans. Amer. Math. Soc. 338 (1993), 679–693.

[CW] Cartwright, D., Woess, W., Infinite graphs with nonconstant Dirichlet finite harmonic functions. SIAM J. Discrete Math. 5 (1992), 380–385.

[CD] Choquet, G., Deny, J., Sur l'équation de convolution $\mu = \mu * \sigma$. C. R. Acad. Sci. Paris 250 (1960), 799–801.

[CDP] Coornaert, M., Delzant, T., Papadopoulos, A., Geométrie et Théorie des groupes: Les groupes hyperboliques de Gromov. Lecture Notes in Math. 1441, Springer-Verlag, Berlin 1990.

[Da] Day, M. M., Convolutions, means and spectra. Illinois J. Math. 8 (1964), 100–111.

[Do] Dodziuk, J., Difference equations, isoperimetric inequality, and transience of certain random walks. Trans. Amer. Math. Soc. 284 (1984), 787–794.

[DK] Dodziuk, J., Kendall, W. S., Combinatorial Laplacians and isoperimetric inequality. In: From Local Times to Global Geometry, Control and Physics, Pitman Res. Notes Math. Ser. 150 (ed. by K. D. Elworthy), 68–74. Longman Sci. Tech., Harlow 1986.

[D1] Doob, J. L., Discrete potential theory and boundaries. J. Math. Mech. 8 (1959), 433–458.

[D2] — Classical Potential Theory and its Probabilistic Counterpart. Springer-Verlag, New York 1984.

[DSW] Doob, J. L., Snell, J. L., Williamson, R. E., Application of boundary theory to sums of independent random variables. In: Contributions to Probability and Statisitics, 182–197. Stanford Univ. Press, Stanford 1960.

[Du] Dunwoody, M. J., Cutting up graphs. Combinatorica 2 (1982), 15–23.

[Dy] Dynkin, E. B., Boundary theory of Markov processes (the discrete case). Russian Math. Surveys 24 (1969), 1–42.

[DM] Dynkin, E. B., Malyutov, M. B., Random walks on groups with a finite number of generators. Soviet Mat. Doklady 2 (1961), 399–402.

[Fr] Freudenthal, H., Über die Enden diskreter Räume und Gruppen. Comment. Math. Helv. 17 (1944), 1–38.

[Fu] Fukushima, M., Dirichlet Forms and Markov Processes. North Holland, Amsterdam 1980.

[F1] Furstenberg, H., Non commuting random products. Trans. Amer. Math. Soc. 108 (1963), 377–428.

[F2] -- Random walks and discrete subgroups of Lie groups. In: Advances in Probability and Related Topics 1 (ed. by P. Ney), 1–63. Dekker, New York 1971.

[Ge] Gerl, P., Random walks on graphs with a strong isoperimetric inequality. J. Theoret. Probab. 1 (1988), 171–187.

[GH] Ghys, E., de la Harpe, P. (eds.), Sur les Groupes Hyperboliques d'apres Mikhael Gromov. Birkhäuser, Basel 1990.

[GM] Gehring, F. W., Martin, G. J., Discrete quasiconformal groups I. Proc. London Math. Soc. 55 (1987), 331–358.

[Gr] Gromov, M., Hyperbolic groups. In: Essays in Group Theory (ed. by S. M. Gersten), 75–263. Springer-Verlag, New York 1987.

[H1] Halin, R., Über unendliche Wege in Graphen. Math. Ann. 157 (1964), 125–137.

[H2] — Automorphisms and endomorphisms of infinite locally finite graphs. Abh. Math. Sem. Univ. Hamburg 39 (1973), 251–283.

[Ka] Kaimanovich, V. A., Dirichlet norms, capacities, and generalized isoperimetric inequalities for Markov operators. Potential Anal. 1 (1992), 61–82.

[KW] Kaimanovich, V. A., Woess, W., The Dirichlet problem at infinity for random walks on graphs with a strong isoperimetric inequality. Probab. Theory Rel. Fields 91 (1992), 445–466.

[KSK] Kemeny, J. G., Snell, J. L., Knapp, A. W., Denumerable Markov Chains. 2nd edn., Springer-Verlag, New York 1976.

[K1] Kesten, H., Symmetric random walks on groups. Trans. Amer. Math. Soc. 92 (1959), 336–354.

[K2] — Full Banach mean values on countable groups. Math. Scand. 7 (1959), 146–156.

[Kn] Knapp, A. W., Regular boundary points in Markov chains. Proc. Amer. Mat. Soc. 17 (1966), 435–440.

[Ly] Lyons, T., Random thoughts on reversible potential theory. In: Summer School on Potential Theory, Joensuu 1990 (ed. by Ilpo Laine), 71–114. Joensuu Publ. in Sciences 26, Joensuu 1992.

[NS] Ney, P., Spitzer, F., The Martin boundary for random walk. Trans. Amer. Math. Soc. 121 (1966), 116–132.

[PS] Picardello, M. A., Sjögren, P., Boundary behaviour of eigenfunctions of the Laplacian in a bi-tree. J. Reine Angew. Math. 424 (1992), 137–148.

[Pr] Pruitt, W. E., Eigenvalues of non-negative matrices. Annals Math. Statist. 35 (1964), 1797–1800.

[SN] Sario, L., Nakai, M., Classification Theory of Riemann Surfaces. Springer-Verlag, Berlin 1970.

[Se] Series, C., The infinite word problem and limit sets in Fuchsian groups. Ergod. Th. Dynam. Sys. 1 (1981), 337–360.

[Sh] Short, H. (ed.), Notes on word hyperbolic groups. In: Group Theory from a Geometrical Viewpoint (ed. by E. Ghys et al.), 3–63. ICTP, World Scientific, Singapore 1991.

[So] Soardi, P. M., Potential Theory on Infinite Networks. Lecture Notes in Math. 1590. Springer-Verlag, Berlin 1994.

[St] Stallings, J., Group Theory and Three-dimensional Manifolds. Yale Univ. Press, New Haven 1971.

[S1] Sullivan, D., The density at infinity of a discrete group of hyperbolic motions. Publ. Math. IHES 50 (1979), 171–202.

[S2] — The Dirichlet problem at infinity for a negatively curved manifold. J. Diff. Geometry 18 (1983), 723–732.

[TW] Thomassen, C., Woess, W., Vertex-transitive graphs and accessibility. J. Combin. Th., Ser. B 58 (1993), 248–268.

[Tr] Trofimov, V. I., Automorphism groups of graphs as topological groups. Math. Notes (Transl. Mat. Zametki) 38 (1986), 717–720.

[W1] Woess, W., Boundaries of random walks on graphs and groups with infinitely many ends. Israel J. Math. 68 (1989), 271–301.

[W2] — Topological groups and infinite graphs. Discrete Math. 95 (1991), 373–384.

[W3] — Fixed sets and free subgroups of groups acting on metric spaces. Math. Z. 214 (1993), 425–440.

[W4] — Random walks on infinite graphs and groups – a survey on selected topics. Bull. London Math. Soc. 26 (1994), 1–60.

Quasiadditivity and measure property of capacity and the tangential boundary behavior of harmonic functions

Hiroaki Aikawa and Alexandr A. Borichev

Dedicated to Professor F-Y. Maeda on the occasion of his 60th birthday

Abstract. A certain quasiadditive property of capacity is shown. Namely, it is proved that if a set E is dispersely decomposed into subsets, then the capacity of E is comparable to the sum of the capacities of the subsets. From the quasiadditivity it is derived that the Lebesgue measure of a certain expanded set is estimated by the capacity of the original set. The estimation has an application to the boundary behavior of harmonic functions.

1991 Mathematics Subject Classification: Primary 31B15, 31B25.

1. Introduction

Throughout this article we denote by A a positive constant whose value is unimportant and may change from line to line. We say that two positive quantities f and g are comparable, written $f \approx g$, if there exists a constant A such that $A^{-1}g \le f \le Ag$. We say that a capacity C is quasiadditive if

$$C(E) \approx \sum C(E_j)$$

for some decomposition $E = \bigcup E_j$. In [3] and [4] the first author considered the quasiadditivity for certain capacities with respect to the Whitney decomposition. Here we give a different type of quasiadditivity.

Let $K(r) \not\equiv 0$ be a nonnegative nonincreasing lower semicontinuous (l. s. c.) function for $r > 0$. We assume that

$$\lim_{r \to 0} K(r) = \infty \quad \text{and} \quad \lim_{r \to \infty} K(r) = 0.$$

For $x \in \mathbb{R}^N$ we define $K(x) = K(|x|)$, and assume that $K(x)$ is integrable over $\mathbb{R}^N$. We define the capacity C_K by

$$C_K(E) = \inf\{\|\mu\| : K * \mu \ge 1 \text{ on } E\}.$$

Define a positive function $\eta(r)$ by $|B(0, \eta(r))| = C_K(B(0, r))$ and put $\eta^*(r) = \max\{\eta(r), 2r\}$.

Theorem 1. *Suppose* $\{B(x_j, \eta^*(r_j))\}$ *is disjoint and* E *is an analytic subset of* $\bigcup B(x_j, r_j)$. *Then*

$$C_K(E) \approx \sum C_K(E \cap B(x_j, r_j)).$$

Theorem 1 has a counterpart in L^p-capacity and in energy capacity. Let $1 < p < \infty$. We define

$$C_{K,p}(E) = \inf\{\|f\|_p^p : K * f \geq 1 \text{ on } E,\ f \geq 0\}.$$

Define a positive function $\eta_p(r)$ by $|B(0, \eta_p(r))| = C_{K,p}(B(0, r))$ and put $\eta_p^*(r) = \max\{\eta_p(r), 2r\}$.

Theorem 2. *Suppose* $\{B(x_j, \eta_p^*(r_j))\}$ *is disjoint and* E *is an analytic subset of* $\bigcup B(x_j, r_j)$. *Then*

$$C_{K,p}(E) \approx \sum C_{K,p}(E \cap B(x_j, r_j)).$$

The energy capacity is defined by

$$e_K(E) = \sup\{\|\mu\|^2 :\ \mu \text{ is concentrated on } E,\ \int K * \mu d\mu \leq 1\}.$$

We define $\eta_e(r)$ by $|B(0, \eta_e(r))| = e_K(B(0, r))$ and put $\eta_e^*(r) = \max\{\eta_e(r), 2r\}$.

Theorem 3. *Suppose* $\{B(x_j, \eta_e^*(r_j))\}$ *is disjoint and* E *is an analytic subset of* $\bigcup B(x_j, r_j)$. *Then*

$$e_K(E) \approx \sum e_K(E \cap B(x_j, r_j)).$$

From Theorems 1–3 we can deduce the following measure property of C_K, $C_{K,p}$ and e_K. For notational convenience we extend $C_{K,p}$, η_p and η_p^* for $p \geq 1$. Thus, $C_{K,1}$, η_1 and η_1^* mean C_K, η and η^*, respectively. For a proper subset E we put $\delta_E(x) = \operatorname{dist}(x, E^c)$ and

$$\widetilde{E}_{K,p} = \bigcup_{x \in E} B(x, \eta_p^*(\delta_E(x))), \quad \widetilde{E}_{K,e} = \bigcup_{x \in E} B(x, \eta_e^*(\delta_E(x))).$$

Theorem 4. *Let* $p \geq 1$. *There is a positive constant* A *depending only on* N, K *and* p *such that*

$$|\widetilde{E}_{K,p}| \leq A C_{K,p}(E),$$
$$|\widetilde{E}_{K,e}| \leq A e_K(E).$$

Theorem 4 has an application to the tangential boundary behavior of harmonic functions. We shall later give Theorem 5, a generalization of Theorem 4, in connection with Nagel–Stein approach regions ([10]). We shall introduce a notion of "thin sets" to obtain precise description of the tangential boundary behavior of harmonic functions given as the Poisson integral of certain potentials. We shall combine it with Theorem 5 and observe that [9, Theorem 2.9] follows.

The plan of this article is as follows: We prove Theorems 1–3 in the next section. Since the proofs are similar, we shall give a complete proof only for Theorem 3. For Theorems 1 and 2 we refer the reader to [6]. In Section 3 we shall prove Theorem 4 by using the usual covering lemma. We shall also indicate that if we invoke the covering lemma due to Nagel–Stein [10], then we obtain Theorem 5, a generalization of Theorem 4. In Section 4 we shall give some applications of Theorem 5 to the boundary behavior of harmonic functions.

2. Proof of Theorems 1–3

We prepare an elementary lemma.

Lemma 1. *Let* $0 < 2r \le R$. *Suppose* $x \notin B(x_0, R)$ *and let* $\rho = \operatorname{dist}(x, B(x_0, r))$. *Then*

$$|B(x,\rho) \cap B(x_0,R)| \ge A|B(x_0,R)|,$$

where A *depends only on the dimension.*

Proof. Let x_1 be the point on the line segment connecting x_0 and x such that $|x_1 - x_0| = \frac{3}{4}R$. It is easy to see that

$$\rho = |x - x_0| - r = |x - x_1| + \frac{3}{4}R - r \ge |x - x_1| + \frac{1}{4}R.$$

We observe that

$$B(x_1, \frac{1}{4}R) \subset B(x,\rho) \cap B(x_0,R),$$

since if $y \in (x_1, \frac{1}{4}R)$, then

$$|x - y| < |x - x_1| + \frac{1}{4}R \le \rho \text{ and } |x_0 - y| < |x_0 - x_1| + \frac{1}{4}R = R.$$

Thus the required inequality follows. □

Proof of Theorem 3. For simplicity we write E_j for $E \cap B(x_j, r_j)$. It is sufficient to show that $\sum e_K(E_j) \le Ae_K(E)$. Let μ_j be the e_K-equilibrium measure for E_j, i.e.,

$$\mu_j \text{ is concentrated on } E_j,$$
$$\int K * \mu_j d\mu_j = 1,$$
$$\|\mu_j\|^2 = e_K(E_j).$$

Let $\mu_j^* = e_K(E_j)^{1/2}\mu_j$ and let $\mu^* = \sum \mu_j^*$. We observe that

$$\int K * \mu_j^* d\mu_j^* = \|\mu_j^*\| = e_K(E_j) \text{ and } \|\mu^*\| = \sum e_K(E_j). \tag{1}$$

We put

$$f_j = \frac{\|\mu_j^*\|}{|B(x_j, \eta_e^*(r_j))|}\chi_{B(x_j,\eta_e^*(r_j))},$$

$d\mu_j' = f_j dx$ and $\mu' = \sum \mu_j'$. Then $\|\mu_j'\| = \|\mu_j^*\| = e_K(E_j)$ and $f_j \le \chi_{B(x_j,\eta_e^*(r_j))}$, since $|B(x_j,\eta_e^*(r_j))| \ge |B(x_j,\eta_e(r_j))| = e_K(B(x_j,r_j)) \ge e_K(E_j) = \|\mu_j^*\|$. We observe from the disjointness of $B(x_j,\eta_e^*(r_j))$ that $\sum f_j \le \sum \chi_{B(x_j,\eta_e^*(r_j))} \le 1$. Hence

$$K * \mu' = K * \left(\sum f_j\right) \le K * 1 = \int K dx < \infty. \tag{2}$$

Let us compare $K * \mu^*$ and $K * \mu'$. Suppose first $x \notin B(x_j,\eta_e^*(r_j))$. We apply Lemma 1 with $x_0 = x_j$, $R = \eta_e^*(r_j)$ and $r = r_j$. Let $\rho_j = \operatorname{dist}(x, B(x_j, r_j))$. We have

$$\begin{aligned} K * \mu_j'(x) &\ge \int_{B(x,\rho_j)\cap B(x_j,\eta_e^*(r_j))} K(x-y)d\mu_j'(y) \\ &\ge K(\rho_j)\mu_j'(B(x,\rho_j)\cap B(x_j,\eta_e^*(r_j))) \\ &= K(\rho_j)\|\mu_j^*\|\frac{|B(x,\rho_j)\cap B(x_j,\eta_e^*(r_j))|}{|B(x_j,\eta_e^*(r_j))|} \\ &\ge AK(\rho_j)\|\mu_j^*\|. \end{aligned}$$

Obviously, $K * \mu_j^*(x) \le K(\rho_j)\|\mu_j^*\|$, whence

$$K * \mu_j^*(x) \le AK * \mu_j'(x). \tag{3}$$

Now suppose $x \notin \bigcup B(x_j,\eta_e^*(r_j))$. Then (3) holds for all j, so that by (2)

$$K * \mu^*(x) = \sum K * \mu_j^*(x) \le A \sum K * \mu_j'(x) = AK * \mu'(x) \le A.$$

Suppose $x \in B(x_j,\eta_e^*(r_j))$. Then, by the disjointness of $\{B(x_j,\eta_e^*(r_j))\}$, we have $x \notin \bigcup_{i\ne j} B(x_i,\eta_e^*(r_i))$. Hence (2) and (3) yield

$$K * \mu^*(x) = K * \mu_j^*(x) + \sum_{i\ne j} K * \mu_i^*(x) \le K * \mu_j^*(x) + A.$$

Therefore

$$\begin{aligned}\int K*\mu^* d\mu^* &= \int_{\mathbb{R}^N\setminus\cup B(x_j,\eta_\varepsilon^*(r_j))} K*\mu^* d\mu^* + \sum \int_{B(x_j,\eta_\varepsilon^*(r_j))} K*\mu^* d\mu^* \\ &\le A\|\mu^*\| + \sum \int_{B(x_j,\eta_\varepsilon^*(r_j))} K*\mu_j^* d\mu^* \\ &= A\|\mu^*\| + \sum \int_{B(x_j,\eta_\varepsilon^*(r_j))} K*\mu_j^* d\mu_j^* \\ &= (A+1)\sum e_K(E_j),\end{aligned}$$

where the last equality follows from (1).

Now the proof is easy. Let

$$\widetilde{\mu} = \left(\sum e_K(E_j)\right)^{-1/2} \mu^*.$$

We have

$$\int K*\widetilde{\mu} d\widetilde{\mu} \le A,$$

$$\|\widetilde{\mu}\| = \left(\sum e_K(E_j)\right)^{1/2}.$$

Obviously $\widetilde{\mu}$ is concentrated on E. Hence by definition

$$e_K(E) \ge A\|\widetilde{\mu}\|^2 = A\sum e_K(E_j).$$

Thus the required inequality follows. The theorem is proved. □

The proofs of Theorems 1 and 2 can be carried out in a similar way with the help of Lemma 1 and the following dual definition of C_K and $C_{K,p}$ (cf. [8, Theorem 14]). For details we refer to [6].

Theorem A. *Let E be an analytic set. Then*

$$C_K(E) = \sup\{\|\mu\| : \mu \textit{ is concentrated on } E, K*\mu \le 1 \text{ on } \mathbb{R}^N\}.$$

Theorem B. *Let E be an analytic set. Then*

$$C_{K,p}(E) = \sup\{\|\mu\|^p : \mu \textit{ is concentrated on } E, \|K*\mu\|_q \le 1\},$$

where $\dfrac{1}{p} + \dfrac{1}{q} = 1$.

3. Proof of Theorem 4

Since the proofs are similar, we shall prove only the first inequality. Take an arbitrary compact subset F of $\widetilde{E}_{K,p}$. By the usual covering lemma we can find $x_j \in E$ such that

$$F \subset \bigcup B(x_j, 5\eta_p^*(r_j)),$$
$$\{B(x_j, \eta_p^*(r_j))\} \text{ is disjoint},$$
$$r_j = \delta_E(x_j).$$

Let $E' = \bigcup B(x_j, r_j)$. By definition this is a subset of E. We apply Theorems 1 and 2 for $B(x_j, r_j)$ and E'. We obtain

$$\sum C_{K,p}(B(x_j, r_j)) \le AC_{K,p}(E') \le AC_{K,p}(E).$$

On the other hand we have

$$|F| \le \sum |B(x_j, 5\eta_p^*(r_j))| = A\sum |B(x_j, \eta_p^*(r_j))|.$$

It is easy to see that

$$|B(0, \eta_p^*(r))| \approx |B(0, \eta_p(r))| = C_{K,p}(B(0,r)).$$

Hence

$$|F| \le AC_{K,p}(E).$$

Since F is an arbitrary compact subset of $\widetilde{E}_{K,p}$, the required inequality follows. The theorem is proved. □

In Theorem 4 we have considered the enlargement based on balls. We can replace balls by the so-called Nagel–Stein regions. Let Ω be a set in $\mathbb{R}_+^{N+1}$ with $\overline{\Omega} \cap \partial\mathbb{R}_+^{N+1} = \{0\}$. Put $\Omega(y) = \{x : (x,y) \in \Omega\}$. We say that Ω satisfies the Nagel–Stein condition (abbreviated to (NS)), if

(i) $|\Omega(y)| \le Ay^N$ with $A = A(\Omega)$;
(ii) there is $\alpha > 0$ such that

$$(x_1, y_1) \in \Omega \text{ and } |x - x_1| < \alpha(y - y_1) \implies (x,y) \in \Omega.$$

Obviously, the nontangential cone $\Gamma = \{(x,y) : |x| < y\}$ satisfies (NS). The section $\Gamma(y)$ is the open ball with center at 0 and radius y. So, $\Omega(y)$ may be regarded as an extension of a ball. For E we put

$$\widetilde{E}_{K,p;\Omega} = \bigcup_{x \in E} \left(x - \Omega(\eta_p^*(\delta_E(x)))\right).$$

This is a generalization of $\widetilde{E}_{K,p}$.

Theorem 5. *Let Ω satisfy* (NS). *Then*

$$|\widetilde{E}_{K,p;\Omega}| \leq AC_{K,p}(E),$$

where $A > 0$ depends only on N, K, p and Ω.

Theorem 5 can be proved in a similar way with the help of the covering lemma due to Nagel–Stein [10, pp. 90–92]. For details we refer to [6].

4. Boundary behavior of harmonic functions

In what follows we are interested in the boundary behavior of harmonic functions in $\mathbb{R}_+^{N+1}$. Hereafter we let $1 < p < \infty$. Following the idea in [2] and [7] we introduce the notion of thinness at the boundary. For a set $E \subset \mathbb{R}_+^{N+1}$ we put $E_t = \{(x,y) \in E : 0 < y < t\}$ and $E^* = \bigcup_{(x,y)\in E} B(x,y)$. We recall that $B(x,y)$ is the N-dimensional ball with center at x and radius y, so that the set E^* is a set on the boundary $\mathbb{R}^N = \partial\mathbb{R}_+^{N+1}$. We shall combine the above notation and write

$$E_t^* = \bigcup_{(x,y)\in E, 0<y<t} B(x,y).$$

Definition. Let $E \subset \mathbb{R}_+^{N+1}$. We say that E is $C_{K,p}$-thin at $\partial\mathbb{R}_+^{N+1}$ if

$$\lim_{t\to 0} C_{K,p}(E_t^*) = 0.$$

Remark. If E is $C_{K,p}$-thin at $\partial\mathbb{R}_+^{N+1}$, then the essential projection of E

$$\{x : \text{ for any } t > 0 \text{ there is a positive number } y < t \text{ such that } (x,y) \in E\ \}$$

is of $C_{K,p}$-capacity 0, and hence of measure 0.

From Theorem 5 we have

Theorem 6. *Suppose Ω satisfies* (NS). *Let $\Omega_{K,p} = \{(x,y) : x \in \Omega(\eta_p^*(y))\}$. If E is $C_{K,p}$-thin at $\partial\mathbb{R}_+^{N+1}$, then*

$$\left|\bigcap_{t>0}\{x : (x + \Omega_{K,p}) \cap E_t \neq \emptyset\}\right| = 0.$$

In other words, for almost all $x \in \partial\mathbb{R}_+^{N+1}$, $x + \Omega_{K,p}$ lies eventually outside E, i.e., there is $t = t_x > 0$ such that $E_t \cap (x + \Omega_{K,p}) = \emptyset$.

Proof. It is not so difficult to see that

$$\{x \in \mathbb{R}^N : (x + \Omega_{K,p}) \cap E \neq \emptyset\} \subset \bigcup_{x\in E^*} (x - \Omega_{K,p}(\delta_{E^*}(x))) = \bigcup_{x\in E^*} (x - \Omega(\eta_p^*(\delta_{E^*}(x))))$$

(see [6, Lemma 2]). Hence Theorem 5 yields

$$|\{x \in \mathbb{R}^N : (x + \Omega_{K,p}) \cap E \neq \emptyset\}| \leq AC_{K,p}(E^*).$$

Apply this inequality to E_t replacing E. Then the definition of thinness implies that

$$|\{x : (x + \Omega_{K,p}) \cap E_t \neq \emptyset\}| \leq AC_{K,p}(E_t^*) \to 0 \text{ as } t \to 0.$$

Thus the theorem follows. □

Remark. It is not so difficult to see that $\eta_p^*(r)/r \to \infty$ as $r \to 0$ (cf. [1]). Hence $\Omega_{K,p}$ is a tangential region.

For a function f on $\partial\mathbb{R}_+^{N+1}$ we denote by $PI(f)$ its Poisson integral. In [6] we have proved

Theorem 7. *Let $\Omega \subset \mathbb{R}_+^{N+1}$ and suppose $\overline{\Omega} \cap \partial\mathbb{R}_+^{N+1} = \{0\}$. Suppose $f \in L^p(\mathbb{R}^N)$. Then there is a set $E \subset \mathbb{R}_+^{N+1}$ such that E is $C_{K,p}$-thin at $\partial\mathbb{R}_+^{N+1}$ and that*

$$\lim_{P \to x, P \in (x+\Omega)\setminus E} PI(K * f)(P) = K * f(x) \tag{4}$$

for $C_{K,p}$-a.e. $x \in \partial\mathbb{R}_+^{N+1}$, i.e., there is a set $F \subset \partial\mathbb{R}_+^{N+1}$ such that $C_{K,p}(F) = 0$ and (4) holds for $x \in \partial\mathbb{R}_+^{N+1} \setminus F$.

As a corollary to Theorems 6 and 7 we have the following theorem. This is a generalization of [9, Theorem 2.9].

Corollary. *Let $\Omega \subset \mathbb{R}_+^{N+1}$ and suppose Ω satisfies (NS). Suppose $f \in L^p(\mathbb{R}^N)$. Then*

$$\lim_{P \to x, P \in x + \Omega_{K,p}} PI(K * f)(P) = K * f(x)$$

for almost all $x \in \partial\mathbb{R}_+^{N+1}$.

Remark. In the proof of Theorem 7 we use the nontangential maximal function, which is of type (p,p) for $p > 1$ but not of type $(1,1)$. Hence the assumption $1 < p < \infty$ is necessary. However, we can show similar results even for $p = 1$ under an additional assumption on the kernel K. In fact, if

$$K(r) \approx r^{-N} \int_0^r K(t) t^{N-1} dt \text{ for small } r > 0,$$

then the same conclusions of Theorem 7 and Corollary hold. For details we refer to [6].

Remark. The approach region $\Omega_{K,p}$ in Theorem 7 is in some sense most tangential. If one considers less tangential approach regions, then one may obtain boundary limit theorems with smaller boundary exceptional sets (corresponding to F in Theorem 7) which can be measured by the Hausdorff measure. This problem was considered in [5].

References

[1] Aikawa, H., Comparison of L^p-capacity and Hausdorff measure. Complex Variables 15 (1990), 223–232.

[2] — Thin sets at the boundary. Proc. London Math. Soc. (3) 65 (1992), 357–382.

[3] — Quasiadditivity of Riesz capacity. Math. Scand. 69 (1991), 15–30.

[4] — Quasiadditivity of capacity and minimal thinness. Ann. Acad. Sci. Fenn. Ser. A. I. Mathematica 18 (1993), 65–75.

[5] — Bessel capacity, Hausdorff content and the tangential boundary behavior of harmonic functions, preprint.

[6] Aikawa, H., Borichev, A. A., Quasiadditivity and measure property of capacity and the tangential boundary behavior of harmonic functions, preprint.

[7] Borichev, A. A., Boundary behavior in Dirichlet-type spaces. Uppsala University Department of Mathematics Report 3 (1994).

[8] Meyers, N. G., A theory of capacities for potentials of functions in Lebesgue classes. Math. Scand. 26 (1970), 255–292.

[9] Nagel, A., Rudin W., Shapiro, J. H., Tangential boundary behavior of functions in Dirichlet-type spaces. Ann. of Math. 116 (1982), 331–360.

[10] Nagel, A., Stein, E. M., On certain maximal functions and approach regions. Adv. in Math. 54 (1984), 83–106.

The Riesz–Herglotz representation for positive harmonic functions via Choquet's theorem

David H. Armitage

Abstract. It is proved by elementary methods that a positive harmonic function on the unit ball B of $\mathbb{R}^N$ is extremal if and only if it has limit 0 at every point save one of ∂B. This allows the Riesz–Herglotz (Poisson integral) representation of positive harmonic functions on B to be deduced from Choquet's theorem.

1991 Mathematics Subject Classification: 31B05.

1. Introduction

Let $\mathcal{H}^+$ denote the cone of non-negative harmonic functions on the open unit ball B of $\mathbb{R}^N$, where $N \geq 2$. The Riesz–Herglotz theorem states that if $h \in \mathcal{H}^+$, then there exists a unique measure μ on ∂B such that

$$h(x) = \int_{\partial B} K(x,y)d\mu(y) \qquad (x \in B),$$

where K is the Poisson kernel of B, given by

$$K(x,y) = \frac{1 - ||x||^2}{||x-y||^N} \qquad (x \in B, y \in \partial B) \tag{1}$$

with $||\cdot||$ denoting the Euclidean norm on $\mathbb{R}^N$ (see, e.g., [4; 1.II.14]).

The Riesz–Herglotz theorem illustrates Choquet's theorem [2] (or see [8]), as we now explain. Let $\mathcal{H}$ be the vector space $\{h_1 - h_2 : h_1, h_2 \in \mathcal{H}^+\}$, equipped with the topology of local uniform convergence. Then $\mathcal{H}$ is metrizable, and the closed convex cone $\mathcal{H}^+$ induces a lattice ordering on $\mathcal{H}$. Define

$$X = \{h \in \mathcal{H}^+ : h(0) = 1\},$$

and let E denote the set of extreme points of X. Then X is a metrizable compact convex base for $\mathcal{H}^+$, and therefore by Choquet's theorem, for each $h \in \mathcal{H}^+$ there exists a unique measure μ on E such that

$$h(x) = \int_E H(x)d\mu(H).$$

In view of the following well-known proposition, we may regard μ as a measure on ∂B and thus obtain the Riesz–Herglotz theorem.

Proposition. *The set of extreme points of X is*

$$E = \{K(\cdot, z) : z \in \partial B\}. \tag{2}$$

Although the Riesz–Herglotz theorem illustrates Choquet's theorem, it is not usually presented as a consequence of Choquet's theorem, because the Riesz–Herglotz theorem itself is invoked in the proof of the proposition. Phelps [8, p. 118] asks for an elementary proof of (2) which would allow the Riesz–Herglotz theorem to be deduced from Choquet's theorem. The purpose of this note is to give such a proof.

I am very grateful to Ivan Netuka for bringing Phelps' question to my attention. He also kindly indicated several elementary proofs of (2) for the case $N = 2$ ([3], [5], [6], [7]).

2. Proof of the proposition

We require some well-known and easily proved facts about Poisson integrals of continuous functions. Let $B(r)$ denote the open ball of centre 0 and radius r in $\mathbb{R}^N$ (so that $B(1) = B$). The Poisson kernel K_r of $B(r)$ is given by

$$K_r(x, y) = \frac{r^{N-2}(r^2 - ||x||^2)}{||x - y||^N} \qquad (x \in B(r), y \in \partial B(r)).$$

Let σ_r denote $(N-1)$-dimensional measure on $\partial B(r)$, normalized so that $\sigma_r(\partial B(r)) = 1$. If f is a real-valued function defined and continuous at least on $\partial B(r)$, then the Poisson integral of f is defined by

$$I_{f,r}(x) = \int_{\partial B(r)} f(y) K_r(x, y) d\sigma_r(y) \qquad (x \in B(r)).$$

The function $I_{f,r}$ is harmonic on $B(r)$. Further, if h is harmonic on some neighbourhood of the closed ball $\overline{B(r)}$, then $h = I_{h,r}$ on $B(r)$. (See, e.g., [4, 1.II.1].)

The main step in our proof of (2) is to show that if $h \in E$, then there exists $z \in \partial B$ such that

$$\lim_{x \to w} h(x) = 0 \qquad (w \in \partial B \backslash \{z\}). \tag{3}$$

Suppose for the moment that this has been established, and suppose that $h \in E$. By making an appropriate choice of axes, we may assume that (3) holds with $z = (0, \ldots, 0, -1)$. Let h^* be the image of h under the Kelvin transform relative to the sphere of centre z and radius 2, that is

$$h^*(x) = \left(\frac{2}{||x - z||}\right)^{N-2} h(x^*), \tag{4}$$

where

$$x^* = \frac{4(x-z)}{||x-z||^2} + z.$$

Then h^* is positive and harmonic on the open half-space $D = \{x = (x_1, \dots, x_N) : x_N > 1\}$ (see, e.g. [4; 1.II.12]) and h^* vanishes continuously on the finite boundary ∂D. It follows from a classical uniqueness theorem of Bouligand [1] that $h^*(x)$ is proportional to $x_N - 1$. The relation (4) and the definition (1) of K then show that h is proportional to $K(\cdot, z)$, and since $h(0) = 1 = K(0, z)$, we have $h = K(\cdot, z)$. Thus $E \subseteq \{K(\cdot, z) : z \in \partial B\}$. For the reverse inclusion, note that since E is non-empty it must contain some element $K(\cdot, z)$ and hence, since harmonicity is preserved by rotations of axes, all such elements. We remark that Bouligand's proof, like the proof we are about to give, is elementary and depends on a study of Poisson integrals of continuous functions.

It remains to show that if $h \in E$, then (3) holds for some $z \in \partial B$. We prove this by contradiction. Suppose then that $h \in E$ and that there exist distinct points $z_1, z_2 \in \partial B$ such that

$$\limsup_{x \to z_j} h(x) > 0 \qquad (j = 1, 2). \tag{5}$$

Let S be an $(N-1)$-dimensional subspace of $\mathbb{R}^N$ chosen so that $z_j \in \Omega_j$, where Ω_1, Ω_2 are the connected components of $\partial B \backslash S$. Let $(f_n), (g_n)$ be sequences of non-negative continuous functions on ∂B such that

$$f_n = 0 \text{ on } \{y \in \Omega_2 : \operatorname{dist}(y, S) > n^{-1}\}, \tag{6}$$

$$g_n = 0 \text{ on } \{y \in \Omega_1 : \operatorname{dist}(y, S) > n^{-1}\}$$

and

$$f_n(y) + g_n(y) = h(r_n y) \qquad (y \in \partial B),$$

where $r_n = (n-1)/n$. Write $F_n = I_{f_n,1}, G_n = I_{g_n,1}$. Then $F_n, G_n \in \mathcal{H}^+$ and

$$F_n(x) + G_n(x) = I_{h,r_n}(r_n x) = h(r_n x) \qquad (x \in B).$$

Hence

$$\lim_{n\to\infty} (F_n(x) + G_n(x)) = h(x) \qquad (x \in B).$$

The sequence (F_n) is locally uniformly bounded on B and therefore has a subsequence (F_{n_j}) which converges to some function $F \in \mathcal{H}^+$. (These remarks depend only on elementary results which can be proved using the Harnack inequalities and hence can ultimately be deduced from the simple properties of Poisson integrals of continuous functions mentioned above; see, e.g., [4; 1.II, 2-3].) It follows that (G_{n_j}) converges on B to the function $G \in \mathcal{H}^+$ such that $F + G = h$. Since $h \in E$, either $F = h$ or $G = h$; suppose without loss of generality that $F = h$. Let $\delta = \frac{1}{3} \operatorname{dist}(z_2, \Omega_1)$. If $n^{-1} < \delta$ and x is a point of B such that $||x - z_2|| < \delta$, then in view of (6), $||x-y|| > \delta$ whenever $y \in \operatorname{supp} f_n$, so that

$$\begin{aligned} F_n(x) &= \int_{\partial B} (1-||x||^2)||x-y||^{-N} f_n(y)d\sigma_1(y) \\ &\le \delta^{-N}(1-||x||^2)\int_{\partial B} h(r_n y)d\sigma_1(y) \\ &= \delta^{-N}(1-||x||^2)h(0). \end{aligned}$$

Hence $h(x) = F(x) \to 0$ as $x \to z_2$, contrary to (5).

References

[1] Bouligand, G., Démonstration élémentaire d'un théorème de détermination à un facteur constant près d'une fonction harmonique. Mathematica 6 (1932), 80–85.

[2] Choquet, G., Existence et unicité des représentations intégrales au moyen des points extrémaux dans les cônes convexes. Séminaire Bourbaki 139 (1956), 15pp.

[3] Clunie, J. G., Some remarks on extreme points in function theory. In: Aspects of Contemporary Complex Analysis (ed. by D. A. Brannan and J. G. Clunie), 137–146. Academic Press, New York 1980.

[4] Doob, J. L., Classical Potential Theory and its Probabilistic Counterpart. Springer-Verlag, New York–Berlin–Heidelberg 1983.

[5] Edgar, G. A., Two integral representations. Lecture Notes in Math. 1033, 193–198. Springer-Verlag, Berlin 1983.

[6] Hallenbeck, D. J., A note on the Riesz–Herglotz representation. Ann. Univ. Mariae Curie-Skłodowska 39 (1985), 61–64.

[7] Holland, F., Extreme points of a class of functions with positive real part. Math. Ann. 202 (1973), 85–87.

[8] Phelps, R. R., Lectures on Choquet's Theorem. Van Nostrand, Princeton 1966.

Spherical harmonics and applications associated with the Weinstein operator

Zouhir Ben Nahia and Nejib Ben Salem

Abstract. In this paper we are concerned with the Weinstein operator defined on the upper half space $\mathbb{R}_n^+ = \{x \in \mathbb{R}^n ; x = (x_1, \ldots, x_n), x_n > 0\}$ by

$$\Delta_{n,\alpha} = \sum_{i=1}^{n} \frac{\partial^2}{\partial x_i^2} + \frac{2\alpha+1}{x_n} \frac{\partial}{\partial x_n}, \quad \alpha > -\frac{1}{2}.$$

We define the spherical harmonics associated with this operator, compute them and resolve a Dirichlet problem.

1991 Mathematics Subject Classification: 42C15.

1. Introduction

We consider the Weinstein operator defined on $\mathbb{R}^n \times]0, +\infty[$ by

$$\begin{aligned} \Delta_{n,\alpha} &= \sum_{i=1}^{n} \frac{\partial^2}{\partial x_i^2} + \frac{2\alpha+1}{x_n} \frac{\partial}{\partial x_n}, \quad \alpha > -\frac{1}{2}, \\ &= \Delta_{n-1} + l_{\alpha,n}; \end{aligned}$$

here Δ_{n-1} is the Laplacian in $\mathbb{R}^{n-1}$ and $l_{\alpha,n}$ is the Bessel operator with respect to the last variable x_n, i.e.,

$$l_{\alpha,n} = \frac{\partial^2}{\partial x_n^2} + \frac{2\alpha+1}{x_n} \frac{\partial}{\partial x_n}.$$

We construct an isomorphism R_α satisfying

$$\Delta_{n,\alpha} R_\alpha = R_\alpha \Delta_n$$

where Δ_n is the Laplacian on $\mathbb{R}^n$. (This operator R_α is called the transmutation operator.)

Next, we introduce the spherical harmonics associated with $\Delta_{n,\alpha}$. With the help of R_α, we give a tool for computing spherical harmonics associated with $\Delta_{n,\alpha}$. Using zonal associated with $\Delta_{n,\alpha}$, we construct the Poisson kernel associated with this operator $\Delta_{n,\alpha}$ on the ball $B(0,r)$. Its expression is given by

$$p(u,v) = c_{n,\alpha}\left(2\langle u_n, v_n\rangle \mid u \mid\right)^{-\frac{n+2\alpha-1}{2}} e^{i\pi\left(\frac{n}{2}-\frac{3}{2}\right)} 2^{\alpha+\frac{1}{2}} \frac{\Gamma(\alpha+1)}{\Gamma\left(\frac{2\alpha+3-n}{2}\right)}$$
$$\times \left(\frac{1-2\langle u',v'\rangle \mid u \mid + \mid u \mid^2}{2\langle u_n, v_n\rangle \mid u \mid} - 1\right)^{2-\frac{n}{4}+\frac{3}{4}} Q^{\frac{2-n}{2}}_{\frac{2\alpha-1}{2}}\left(\frac{1-2u'v' \mid u \mid + \mid u \mid^2}{2\langle u_n, v_n\rangle}\right),$$

where

$$u = (u_1, \dots, u_{n-1}, u_n) \in B(0,R),$$
$$u' = (u_1, \dots, u_{n-1}),$$
$$\langle u', v'\rangle = \sum_{i=1}^{n-1} u_i v_i,$$

$c_{n,\alpha}$ is a suitable real constant and Q^{β}_{α} is the Legendre function of the second kind (see [1]).

Next, we resolve the Dirichlet problem associated with the Weinstein operator: given any continuous function f on S^{n-1} (where S^{n-1} denotes the unit sphere), even in the last variable, equal to zero on $S^{n-1} \cap \mathbb{R}^{n-1}$, we are looking for a function F equal to f on S^{n-1} and W-harmonic (i.e., $\Delta_{n,\alpha}F = 0$) on the unit ball.

2. Spherical harmonics and zonal associated with Weinstein operator

We denote by

- $x = (x_1, \dots, x_n)$, an element of $\mathbb{R}^n$; we let $\mid x \mid = \left(x_1^2 + \cdots + x_n^2\right)^{\frac{1}{2}}$;
- Δ_n, the laplacian on $\mathbb{R}^n$, i.e., $\Delta_n = \sum\limits_{i=1}^{n} \frac{\partial^2}{\partial x_i^2}$;
- $\Delta_{n,\alpha}$, the operator on $\mathbb{R}^{n-1} \times]0,+\infty[$ — the Weinstein operator — defined by

$$\Delta_{n,\alpha} = \sum_{i=1}^{n} \frac{\partial^2}{\partial x_i^2} + \frac{2\alpha+1}{x_n}\frac{\partial}{\partial x_n}, \quad \alpha > -\frac{1}{2};$$

- $C^{\infty}_{*}(\mathbb{R}^n)$, the space of C^{∞} functions on $\mathbb{R}^n$, even in the last variable;
- E, the subspace of functions in $C^{\infty}_{*}(\mathbb{R}^n)$ such that $\Delta_n f(x_1, \dots, x_n) = 0$, for all $x \in \mathbb{R}^{n-1} \times]0,+\infty[$;
- E_{α}, the subspace of functions in $C^{\infty}_{*}(\mathbb{R}^n)$ such that $\Delta_{n,\alpha} f(x_1, \dots, x_n) = 0$, for all $x \in \mathbb{R}^{n-1} \times]0,+\infty[$;
- $P_{m,*}$, the space of homogeneous polynomials of degree m, even in the last variable;

- HS_m, the space of spherical harmonics of degree m;
- $PH_{m,*}$, the subspace of harmonic polynomials in $P_{m,*}$;
- $PH^{\alpha}_{m,*}$, the subspace of polynomials P in $P_{m,*}$ such that $\Delta_{n,\alpha}P = 0$.

2.1 Operators of transmutation associated with the operators Δ_n and $\Delta_{n,\alpha}$

Theorem 2.1. *The operator R_α defined on $C_*^\infty(\mathbb{R}^n)$ by*

$$R_\alpha f(x_1, \ldots, x_n) = \begin{cases} \frac{2\Gamma(\alpha+1)}{\sqrt{\pi}\Gamma(\frac{\alpha+1}{2})} x_n^{-2\alpha} \int\limits_0^{x_n} \left(x_n^2 - y^2\right)^{\frac{\alpha-1}{2}} f(x_1, \ldots, x_{n-1}, y)\, dy & \text{if } x_n > 0, \\ f(x_1, \ldots, x_{n-1}, 0) & \text{if } x_n = 0, \end{cases}$$

is an isomorphism onto itself.

Proof. Let $R_\alpha = I_{n-1} \otimes \chi_\alpha^{-1}$, where χ_α^{-1} is the Riemann–Liouville operator — see [3] — and I_{n-1} is the identity on $\mathbb{R}^{n-1}$. Our result is deduced from a theorem in [3]. □

Remark 2.1. The same theorem given in [3] may also give an expression of R_α^{-1}.

Proposition 2.1. *The operator R_α satisfies*

$$\Delta_{n,\alpha} R_\alpha f = R_\alpha \Delta_n f,$$

for all $f \in C_*^\infty(\mathbb{R}^n)$.

Proof. The result is deduced from the property $\chi_\alpha^{-1} \circ D^2 = l_\alpha \circ \chi_\alpha^{-1}$, where l_α is the Bessel operator defined on $]0, +\infty[$ by

$$l_\alpha u = u'' + \frac{2\alpha + 1}{r} \frac{d}{dr}$$

and $D^2 = \frac{d^2}{dr^2}$ — see Theorem 1, [3], p. 40. □

Proposition 2.2. *R_α is an isomorphism of $P_{m,*}$ onto itself.*

Proof. From the definition of the operator R_α, we have that the image of any polynomial of $P_{m,*}$ by R_α is a polynomial of $P_{m,*}$; also, we have the same result with the operator R_α^{-1}.

Our result is then a consequence of Theorem 2.1. □

Corollary 2.1. *R_α is an isomorphism of $PH_{m,*}$ onto $PH^\alpha_{m,*}$.*

Proof. This result is an immediate consequence of the two above propositions. □

2.2. Spherical harmonics associated with Weinstein operator

Let us introduce in $\mathbb{R}^n$ the spherical coordinates $r, \theta_1, \dots, \theta_{n-1}$ which are related to Cartesian coordinates by the formulas

$$\begin{cases} x_1 = r\sin\theta_{n-1}\dots\sin\theta_2\sin\theta_1, \\ x_2 = r\sin\theta_{n-1}\dots\sin\theta_2\cos\theta_1, \\ \vdots \\ x_{n-1} = r\sin\theta_{n-1}\cos\theta_{n-2}, \\ x_n = r\cos\theta_{n-1}, \end{cases}$$

where $0 \le r < \infty$, $0 \le \theta_1 < 2\pi$ and $0 \le \theta_k < \pi$, $k \ne 1$.

So the expressions of the variables $r, \theta_1, \dots, \theta_{n-1}$ as functions of the Cartesian coordinates are

$$\cos\theta_k = \frac{x_{k+1}}{r_{k+1}}, \quad \sin\theta_k = \frac{r_k}{r_{k+1}}, \quad r = r_n,$$

where we have $r_k^2 = x_1^2 + \dots + x_k^2$ with $r_1 = x_1$.

Therefore the expression of the Laplacian in spherical coordinates is

$$\Delta_n = \frac{\partial^2}{\partial r^2} + \frac{n-1}{r}\frac{\partial}{\partial r} + \frac{1}{r^2}\Delta_n(\theta),$$

with

$$\Delta_n(\theta) = \frac{1}{\sin^{n-2}\theta_{n-1}}\frac{\partial}{\partial\theta_{n-1}}\sin^{n-2}\theta_{n-1}\frac{\partial}{\partial\theta_{n-1}} + \dots + \frac{1}{\sin^2\theta_{n-1}\dots\sin^2\theta_2}\frac{\partial^2}{\partial\theta_1^2}.$$

The operator $\Delta_n(\theta)$ is the angular part of the Laplacian Δ_n or also the Laplacian on the unit sphere S^{n-1}.

Lemma 2.1. *The expression of the operator $\Delta_{n,\alpha}$ in spherical coordinates is given by*

$$\Delta_{n,\alpha} = \frac{\partial^2}{\partial r^2} + \frac{n+2\alpha}{r}\frac{\partial}{\partial r} + \frac{1}{r^2}\Delta_{n,\alpha}(\theta),$$

where $\Delta_{n,\alpha}(\theta) = \Delta_n(\theta) - (2\alpha+1)\tan\theta_{n-1}\frac{\partial}{\partial\theta_{n-1}}$.

Proof. Let

$$f(x_1, \dots, x_n) = f(r\sin\theta_{n-1}\dots\sin\theta_2\sin\theta_1, \dots, r\cos\theta_{n-1}) =: g(r, \theta_{n-1}, \dots, \theta_1).$$

So, we get

$$\begin{aligned}\frac{\partial f}{\partial x_n} &= \frac{\partial g}{\partial r}\frac{\partial r}{\partial x_n} + \frac{\partial g}{\partial \theta_{n-1}}\frac{\partial \theta_{n-1}}{\partial x_n} \\ &= \frac{x_n}{r}\frac{\partial g}{\partial r} + \frac{\partial g}{\partial \theta_{n-1}}\frac{-1}{\sqrt{1-\frac{x_n^2}{r^2}}}\frac{r-\frac{x_n^2}{r}x}{r^2} \\ &= \frac{x_n}{r}\frac{\partial g}{\partial r} - \frac{1}{r}\frac{\partial g}{\partial \theta_{n-1}}\frac{1}{\sin\theta_{n-1}}\frac{\sin^2\theta_{n-1}}{r}.\end{aligned}$$

From this we conclude

$$\frac{1}{x_n}\frac{\partial f}{\partial x_n} = \frac{1}{r}\frac{\partial g}{\partial r} - \frac{1}{r^2}\frac{\partial g}{\partial \theta_{n-1}}\tan\theta_{n-1}.$$

Consequently, the expression of $\Delta_{n,\alpha}$ in spherical coordinates is given by

$$\Delta_{n,\alpha} = \frac{\partial^2}{\partial r^2} + \frac{n+2\alpha}{r}\frac{\partial}{\partial r} + \frac{1}{r^2}\Delta_{n,\alpha}(\theta)$$

where $\Delta_{n,\alpha}(\theta) = \Delta_n(\theta) - (2\alpha+1)\tan\theta_{n-1}\frac{\partial}{\partial\theta_{n-1}}$. □

Definition. We call a *W-spherical harmonic of degree* m, a function on S^{n-1} which is a restriction to the unit sphere of a homogeneous polynomial P of degree m in n variables, even in the last variable, such that $\Delta_{n,\alpha}P = 0$, i.e., of a polynomial of $PH^{\alpha}_{m,*}$.

Let us put $Y^{\alpha}_m(\theta) = \frac{P_{m,\alpha}(x)}{|x|^m}$, where $\theta = \frac{x}{|x|}$ and $P_{m,\alpha} \in PH^{\alpha}_{m,*}$.

Proposition 2.3. *$Y^{\alpha}_m(\theta)$ is an eigenfunction of $\Delta_{n,\alpha}(\theta)$.*

Proof. For all $P_{m,\alpha} \in PH^{\alpha}_{m,*}$, we have

$$\begin{aligned}0 &= \Delta_{n,\alpha}P_{m,\alpha}(x) \\ &= \frac{\partial^2}{\partial r^2}\left(r^m Y^{\alpha}_m(\theta)\right) + \frac{n+2\alpha}{r}\frac{\partial}{\partial r}\left(r^m Y^{\alpha}_m(\theta)\right) + \frac{1}{r^2}\Delta_{n,\alpha}(\theta)\left(r^m Y^{\alpha}_m(\theta)\right) \\ &= \left[m(m-1)Y^{\alpha}_m(\theta) + (n+2\alpha)mY^{\alpha}_m(\theta) + \Delta_{n,\alpha}(\theta)\right]r^{m-2}.\end{aligned}$$

Hence

$$\begin{aligned}\Delta_{n,\alpha}(\theta)Y^{\alpha}_m(\theta) &= \left[-m(m-1) - m(m+2\alpha)\right]Y^{\alpha}_m(\theta) \\ &= -m(m+n+2\alpha-1)Y^{\alpha}_m(\theta).\end{aligned}$$

Therefore, the numbers $\lambda_m = -m(m+n+2\alpha-1)$ are the eigenvalues of $\Delta_{n,\alpha}(\theta)$ of multiplicity $\sigma(m) = \dim PH^{\alpha}_{m,*}$.

The W-spherical harmonics $Y^{\alpha}_{m,k}(\theta)$, $1 \le k \le \sigma(m)$, are the eigenfunctions of $\Delta_{n,\alpha}(\theta)$ associated with the eigenvalue λ_m. □

Example. $n = 2$. $PH_{m,*}$ is of dimension 1 and is generated by the polynomial

$$(x_1 + ix_2)^m + (x_1 - ix_2)^m.$$

From the above expansion, $PH^{\alpha}_{m,*}$ is of dimension 1 and is generated by

$$\begin{aligned} &R_\alpha\{(\xi_1 + i\xi_2)^n + (\xi_1 - i\xi_2)^n\}(x_1, x_2) \\ &= \frac{\Gamma(\alpha+1)}{\sqrt{\pi}\Gamma(\alpha+\frac{1}{2})}[\int_0^\pi (x_1 + ix_2\cos\varphi)^n \sin^{2\alpha}\varphi d\varphi + \int_0^\pi (x_1 - ix_2\cos\varphi)^n \sin^{2\alpha}\varphi d\varphi \\ &= 2\frac{\Gamma(\alpha+1)}{\sqrt{\pi}\Gamma\left(\alpha+\frac{1}{2}\right)}\int_0^\pi (x_1 + ix_2\cos\varphi)^n \sin^{2\alpha}\varphi d\varphi. \end{aligned}$$

So

$$\begin{aligned} Y^{\alpha}_m(\theta) &= R_\alpha\{(\xi_1 + i\xi_2)^n + (\xi_1 - i\xi_2)^n\}(\cos\theta, \sin\theta) \\ &= 2\frac{\Gamma(\alpha+1)}{\sqrt{\pi}\Gamma\left(\alpha+\frac{1}{2}\right)}\int_0^\pi (\cos\theta + i\sin\theta\cos\varphi)^n \sin^{2\alpha}\varphi d\varphi. \end{aligned}$$

Now, the integral representations of Gegenbauer's polynomials are given by

$$C^p_l(\cos\theta) = \frac{\Gamma(2p+l)}{2^{2p-1}\Gamma(2p)\, l!}\int_0^\pi (\cos\theta - i\sin\theta\cos\varphi)^l \sin^{2p-1}\varphi d\varphi;$$

therefore

$$Y^{\alpha}_m(\theta) = cC_m^{\frac{\alpha+1}{2}}(\cos\theta),$$

where $c \in \mathbb{R}$.

Notations.

- $S^{n-1}_+ = \{x \in \mathbb{R}^n; \mid x \mid = 1, x_n > 0\}$.
- σ is the measure on S^{n-1} defined by $d\sigma = \sin^{n-2}\theta_{n-1}\ldots\sin\theta_2 d\theta_1\ldots d\theta_{n-1}$.
- $L^2\left(S^{n-1}_+, \cos^{2\alpha+1}\theta_{n-1}d\sigma\right)$
$= \{f; S^{n-1}_+ \to \mathbb{C}, f \text{measurable and} \int \mid f \mid^2 \cos^{2\alpha+1}\theta_{n-1}d\sigma < \infty\}$.

Theorem 2.2. $\int_{S^{n-1}_+} Y^{\alpha}_m(\theta)\, Y^{\alpha}_k(\theta)\cos^{2\alpha+1}\theta_{n-1}d\sigma = 0$, *for* $m, k \in \mathbb{N}, m \neq k$.

Proof. It suffices to apply the Green–Riemann formula. □

2.3. Zonal associated with the Weinstein operator

We denote by $HS_{k,\alpha} = \{Y; Y$ is the restriction of an element of $PH^{\alpha}_{m,*}$ to $S^{n-1}\}$. The elements of this space are called spherical harmonics associated with the Weinstein operator.

As the map on $HS_{k,\alpha}$, defined by $Y \longrightarrow Y(u)$, where u is a fixed point on S_+^{n-1}, is a linear form on the Hilbert space $HS_{k,\alpha}$ of finite dimension (for the scalar product induced by $L^2\left(S_+^{n-1}, \cos^{2\alpha+1}\theta_{n-1}d\sigma\right)$), we have from the Riesz representation theorem the following proposition:

Proposition 2.4. *For all $k \in \mathbb{N}$ and for all $u \in S_+^{n-1}$, there exists a unique $Z_u^k \in HS_{k,\alpha}$ such that*

$$Y(u) = \langle Y, Z_u^k \rangle = \int_{S_+^{n-1}} Y(v)\, Z_u^k(v) \cos^{2\alpha+1}\theta_{n-1} d\sigma$$

for all $Y \in HS_{k,\alpha}$.

Definition. Z_u^k is called a *zonal of degree k and pole u* associated with the Weinstein operator $\Delta_{n,\alpha}$.

Theorem 2.3. *For all $u, v \in S_+^{n-1}$, we have*

$$Z_v^k(u) = c_{n,\alpha} \int_0^\pi C_k^{\frac{n+2\alpha-1}{2}}\left(\langle u', v'\rangle + u_n v_n \cos\theta\right) \sin^{2\alpha}\theta d\theta,$$

with $c_{n,\alpha}$ being the constant defined by

$$c_{n,\alpha} = \frac{n+2\alpha+2k-1}{k!\pi^{n/2}} \frac{\Gamma(n+2\alpha+k-1)}{\Gamma(n+2\alpha)} \Gamma\left(\frac{2\alpha+1+n}{2}\right) \frac{1}{\Gamma(\alpha+1/2)}.$$

Proof. In the first step, we prove the result for α half integer, i.e., a multiple of 1/2. In the next step, we use a theorem of Carlson (see [3], p. 186), of an analytic continuation type, to obtain the result. □

3. Dirichlet problem

For $0 \le r < 1, u, v \in S_+^{n-1}$, we put

$$p(ru, v) = \sum_{k=0}^{\infty} r^k Z_v^k(u).$$

Proposition 3.1. *For v fixed in S_+^{n-1}, the map $x \longrightarrow p(x, v)$, $x \in B_n^+$ is a harmonic function associated with the Weinstein operator (B_n^+ is the upper unit ball).*

Let us compute the expression of $p(ru, v)$.

We have

$$p(ru,v) = c_{n,\alpha}\sum_{k=0}^{\infty} r^k \int_0^{\pi} C_k^{\frac{n+2\alpha-1}{2}}\left(u'v' + u_n v_n \cos\theta\right)\sin^{2\alpha}\theta d\theta$$

$$= c_{n,\alpha}\int_0^{\pi}\sum_{k=0}^{\infty} r^k C_k^{\frac{n+2\alpha-1}{2}}\left(u'v' + u_n v_n \cos\theta\right)\sin^{2\alpha}\theta d\theta$$

$$= c_{n,\alpha}\int_0^{\pi}\left[1 - 2r\left(u'v' + u_n v_n\cos\theta\right) + r^2\right]^{\frac{-n+2\alpha-1}{2}}\sin^{2\alpha}\theta d\theta.$$

If $u_n v_n > 0$, using an integral formula (see [3]), we obtain the following expression:

$$p(u,v) = c_{n,\alpha}\left(2u_n v_n r\right)^{\frac{-n+2\alpha-1}{2}} e^{i\pi\left(\frac{n}{2}-\frac{3}{2}\right)} 2^{\frac{\alpha+1}{2}}\frac{\Gamma(\alpha+1)}{\Gamma(\frac{2\alpha+3-n}{2})}$$
$$\times\left(\frac{1-2\langle u',v'\rangle r + r^2}{2u_n v_n r} - 1\right)^{2-\frac{n}{4}+\frac{3}{4}} Q_{\frac{2\alpha-1}{2}}^{\frac{2-n}{2}}\left(\frac{1-2\langle u',v'\rangle r + r^2}{2u_n v_n}\right),$$

with $n < 3 + 2\alpha$; here $Q_\nu^\mu(x)$ is the Legendre function of the second type defined by

$$Q_\nu^\mu(x) = \frac{\Gamma(1+\nu+\mu)\Gamma(-\mu)}{2\Gamma(1+\nu-\mu)}\left(\frac{1-x}{1+x}\right)^{\frac{1}{2\mu}} F\left(-\nu,\nu+1;\mu+1;\frac{1}{2}-\frac{x}{2}\right)$$
$$+\frac{1}{2}\Gamma(\mu)\cos(\mu\pi)\left(\frac{1+x}{1-x}\right)^{\frac{1}{2\mu}} F\left(-\nu,\nu+1;1-\mu;\frac{1}{2}-\frac{x}{2}\right)$$

and F is the hypergeometric function defined by

$$F(a,b;c;z) = \sum_{n=0}^{\infty}\frac{(a)_n (b)_n}{(c)_n}\frac{z^n}{n!},$$

where $(a)_n = \frac{\Gamma(a+n)}{\Gamma(a)}$ (see [1]).

Proposition 3.2. *The kernel* $p(x,v)$, *with* $v \in S_+^{n-1}$ *and* $x \in B_n^+$ *has the following properties:*

- $p(x,v) \geq 0$;
- $\int_{S_+^{n-1}} p(x,v)\, v_n^{2\alpha+1} d\sigma(v) = 1$;
- *For* $\omega \in S_+^{n-1}$, *let* $x = r\omega$ $(0 \leq r < 1)$; *then, for each* $\delta > 0$,

$$\int_{S_+^{n-1}\cap\{|v-\omega|>\delta\}} p(v,r\omega)\, v_n^{2\alpha+1} d\sigma(v) \to 0$$

as $r \to 1$, *uniformly in* ω.

Theorem 3.1. *Suppose $n < 3 + 2\alpha$. Given an arbitrary continuous function f defined on S^{n-1}, even in the last variable and vanishing on $S^{n-1} \cap \mathbb{R}^{n-1}$, the function*

$$F(x) = \int_{S_+^{n-1}} f(v)\, p(x, v)\, v_n^{2\alpha+1} d\sigma(v),$$

where $x \in B_n^+$, is continuous for $|x| \leq 1$, harmonic with respect to the Weinstein operator and attains the boundary values $f(x)$.

References

[1] Erdélyi, A., Magnus, W., Oberhettinger, F., Tricomi, F. G., Higher transcendental functions, Vol. 1. Bateman Manuscript Project. McGraw-Hill Book Company, Inc., New York 1953.

[2] Titchmarsh, E. C., The theory of functions. Oxford University Press, Oxford 1939.

[3] Trimèche, K., Transformation intégrale de Weyl et théorème de Paley-Wiener associés à un opérateur différentiel singulier sur $(0, \infty)$. J. Math. Pures Appl. 60 (1981), 51–98.

On a mean value property associated with the Weinstein operator

Zouhir Ben Nahia and Nejib Ben Salem

Abstract. In this paper we are concerned with the Weinstein operator defined on the upper half space $\mathbb{R}_n^+ = \{x \in \mathbb{R}^n : x = (x_1, \dots, x_n), x_n > 0\}$ by

$$\Delta_W = \sum_{i=1}^{n} \frac{\partial^2}{\partial x_i^2} + \frac{2\alpha+1}{x_n}\frac{\partial}{\partial x_n}, \quad \alpha > -\frac{1}{2}.$$

We define a mean value property associated with Δ_W. We prove that a function defined on $\mathbb{R}^n$, which is even in the last variable, satisfies the equation $\Delta_W = 0$ if and only if it satisfies a mean value property associated with the operator Δ_W.

1991 Mathematics Subject Classification: 35B05 31B05.

1. Introduction

The mean value of a function u taken over the surface of a sphere of radius $r > 0$ about the point x in $\mathbb{R}^n$ is defined by

$$\begin{aligned} M_{x,r}(u) &= \frac{1}{C_{n-1}} \int_S u(x + r\omega)d\sigma(\omega) \\ &= \frac{1}{C_{n-1}} \int_S \tau_x u(r\omega)d\sigma(\omega), \end{aligned}$$

where S is the unit sphere in $\mathbb{R}^n$, σ is the surface measure on S and $C_{n-1} = \sigma(S)$. τ_x denotes the classical translation related to the Laplacian Δ on $\mathbb{R}^n$ defined by

$$\tau_x u(y) = u(x+y).$$

Also, it is well known that harmonic functions u on $\mathbb{R}^n$ can be characterized by the so called mean value property $u(x) = M_{x,r}(u)$.

In this paper we are concerned with the Weinstein operator defined on $\mathbb{R}^{n-1} \times]0, +\infty[$ by

$$\Delta_W = \sum_{i=1}^{n} \frac{\partial^2}{\partial x_i^2} + \frac{2\alpha+1}{x_n}\frac{\partial}{\partial x_n}, \quad \alpha > -\frac{1}{2}.$$

This operator intervenes in fluid mechanics and has interested many authors; see, for instance, [1], [2], [6] and [10].

A natural question is whether or not solutions u of the Weinstein equation $\Delta_W u = 0$ satisfy some sort of mean value property.

Let us mention that mean value properties of second order differential operators have been treated by many mathematicians (see [4] and references cited in this paper).

The main purpose of this work is to establish a mean value property associated with the Weinstein operator. In order to do this, we shall attach to the operator Δ_W a generalized translation.

Obviously we have $\Delta_W = \Delta_{n-1} + l_{\alpha,n}$, where Δ_{n-1} is the Laplacian on $\mathbb{R}^{n-1}$ and $l_{\alpha,n}$ is the Bessel operator defined on $]0, +\infty[$ with respect to the variable x_n:

$$l_{\alpha,n} = \frac{\partial^2}{\partial x_n^2} + \frac{2\alpha+1}{x_n}\frac{\partial}{\partial x_n}.$$

It is well known (see [7]) that with the Bessel operator for $\alpha > -\frac{1}{2}$ we can associate a generalized translation given by

$$T_r^\alpha f(s) = \frac{\Gamma(\alpha+1)}{\sqrt{\pi}\Gamma(\alpha+\frac{1}{2})}\int_0^\pi f(\sqrt{r^2+s^2+2rs\cos\theta})\sin^{2\alpha}\theta d\theta,$$

where $r, s \in \mathbb{R}^+$, and f is an even continuous function on $\mathbb{R}$. Hence with the Weinstein operator Δ_W for $\alpha > -\frac{1}{2}$ we can associate a generalized translation given by

$$T_x f(y) = \frac{\Gamma(\alpha+1)}{\sqrt{\pi}\Gamma(\alpha+\frac{1}{2})}$$
$$\times \int_0^\pi f(x_1+y_1, \ldots, x_{n-1}+y_{n-1}, \sqrt{x_n^2+y_n^2+2x_ny_n\cos\theta})\sin^{2\alpha}\theta d\theta,$$

where f is a continuous function on $\mathbb{R}^n$, even in the last variable and $x \in \mathbb{R}^{n-1} \times [0, +\infty[$.

Using this generalized translation, we define the mean value property of a continuous function u on $\mathbb{R}^n$ even in the last variable associated with the Weinstein operator by

$$M_{x,r}^W(u) = \frac{1}{C_{n,\alpha}}\int_{S^+} T_x u(r\omega)\omega_n^{2\alpha+1}d\sigma(\omega),$$

where S^+ is the upper hemisphere, that is $S^+ = \{x \in S; x = (x_1, \ldots, x_n), x_n > 0\}$ and $C_{n,\alpha} = \int_{S^+} \omega_n^{2\alpha+1}d\sigma(\omega)$; here $d\sigma(\omega) = d\sigma(\omega_1, \ldots, \omega_n)$ denotes the element of surface area on S^+.

We prove that every solution u of the Weinstein equation $\Delta_W u = 0$, even in the last variable, satisfies the mean value property for the Weinstein operator ; that is for $x = (x_1, \ldots, x_n) \in \mathbb{R}^n, x_n > 0$ and for $r > 0$, we have $u(x) = M_{x,r}^W(u)$.

Conversely, we prove that every continuous function, even in the last variable and satisfying the formula $u(x) = M^W_{x,r}(u)$ for all $x \in \mathbb{R}^n, x_n > 0$ and for $r > 0$, is a solution of the Weinstein equation $\Delta_W u = 0$.

We conclude this introduction with a summary of the contents of this paper. In the *second* section, we define the generalized translation operators associated with the Weinstein operator, the convolution product and the Fourier transform associated with the Weinstein operator. In the *third* section we recall the definition of a fundamental solution of the operator Δ_W and we compute a fundamental solution by using the Fourier transform associated with Δ_W. The *fourth* section contains the main result of the paper: every solution of the equation $\Delta_W u = 0$, even in the last variable, satisfies a mean value property and conversely every continuous function satisfying the mean value property for the Weinstein operator is a solution of $\Delta_W u = 0$.

2. Generalized translation and Fourier transform associated with the Weinstein operator

2.1. Generalized translation operators associated with the Weinstein operator

We consider the Weinstein operator Δ_W defined on $\mathbb{R}^{n-1} \times]0, +\infty[$ by

$$\Delta_W = \sum_{i=1}^{n} \frac{\partial^2}{\partial x_i^2} + \frac{2\alpha+1}{x_n}\frac{\partial}{\partial x_n}, \quad \alpha > -\frac{1}{2}.$$

Definition. The *generalized translation* operators associated with Δ_W are defined by

$$T_x f(y) = \frac{\Gamma(\alpha+1)}{\sqrt{\pi}\Gamma(\alpha+\frac{1}{2})} \times \int_0^{\pi} f(x_1+y_1, \ldots, x_{n-1}+y_{n-1}, \sqrt{x_n^2+y_n^2+2x_ny_n\cos\theta}) \sin^{2\alpha}\theta d\theta. \tag{2.1}$$

Here $x = (x_1, \ldots, x_n)$, $y = (y_1, \ldots, y_n) \in \mathbb{R}^{n-1} \times [0, +\infty[$ and f is a continuous function on $\mathbb{R}^{n-1} \times [0, +\infty[$.

Proposition 2.1. *The operators T_x satisfy the following properties for all continuous functions on $\mathbb{R}^{n-1} \times [0, +\infty[$:*
(i) $T_x f(y) = T_y f(x)$, *for all* $x, y \in \mathbb{R}^{n-1} \times [0, +\infty[$;
(ii) $T_0 f(x) = f(x)$, *for all* $x \in \mathbb{R}^{n-1} \times [0, +\infty[$;
(iii) $\Delta_W(T_x f) = T_x(\Delta_W f)$.

Remark 2.1. The previous generalized translation associated with the Weinstein operator Δ_W provides the space $\mathbb{R}^{n-1} \times [0,+\infty[$ with a structure of a hypergroup (see [5]) which is the direct product of the group $\mathbb{R}^{n-1}$ and the Bessel–Kingman hypergroup $[0,+\infty[$ (see [3] and [5]; in this last paper the author uses the name convo for hypergroup).

Notations.

- $x = (x_1, \dots, x_{n-1}, x_n) \in \mathbb{R}^{n-1} \times [0,+\infty[$;
- $x' = (x_1, \dots, x_{n-1}) \in \mathbb{R}^{n-1}$;
- $| x | = (\sum_{i=1}^{n} x_i^2)^{\frac{1}{2}}$;
- $\langle x', \xi' \rangle$: the Euclidean inner product on $\mathbb{R}^{n-1}$ of x' and ξ';
- dx: the Lebesgue measure on $\mathbb{R}^n$;
- $L^1(\mathbb{R}^{n-1} \times [0,+\infty[, x_n^{2\alpha+1}dx)$: the space of measurable functions on $\mathbb{R}^{n-1} \times [0,+\infty[$ which are integrable with respect to the Haar measure $x_n^{2\alpha+1}dx$ (of the hypergroup $\mathbb{R}^{n-1} \times [0,+\infty[$);
- $D_*(\mathbb{R}^{n-1} \times [0,+\infty[)$: the space of C^∞ functions on $\mathbb{R}^n$, even in the last variable x_n and with compact support;
- $S_*(\mathbb{R}^n)$: the Schwartz space consisting of functions which are even in the last variable;
- $S'_*(\mathbb{R}^n)$: the topological dual of $S_*(\mathbb{R}^n)$.

2.2. Convolution product associated with the Weinstein operator

Definition. The *convolution product* associated with Δ_W of two functions f in $L^1(\mathbb{R}^{n-1} \times [0,+\infty[, x_n^{2\alpha+1}dx)$ and g in $L^p(\mathbb{R}^{n-1} \times [0,+\infty[)$, $1 \le p \le +\infty$, is the function defined by

$$f * g(x) = \int_{\mathbb{R}^{n-1} \times [0,\infty[} T_x f(y) g(y) y_n^{2\alpha+1} dy, \tag{2.2}$$

where $T_x f(y)$ is given by the relation (2.1).

Proposition 2.2. *We have the following properties:*

(i) *For all f in $L^1(\mathbb{R}^{n-1} \times [0,+\infty[, x_n^{2\alpha+1}dx)$ and g in $L^p(\mathbb{R}^{n-1} \times [0,+\infty[, x_n^{2\alpha+1}dx)$, the function $f*g$ is in $L^p(\mathbb{R}^{n-1} \times [0,+\infty[, x_n^{2\alpha+1}dx)$ and $\|f*g\|_p \le \|f\|_1 \|g\|_p$. (Here $\|f\|_p^p = \int_{\mathbb{R}^{n-1} \times [0,+\infty[} | f(x) |^p x_n^{2\alpha+1} dx$ if $1 \le p < +\infty$ and $\|f\|_\infty = supess_{x \in \mathbb{R}^{n-1} \times [0,+\infty[} | f(x) |$.)*

(ii) *If f is continuous and g is C^∞ with compact support, then the function $f * g$ is C^∞.*

Definition. Let f be in $L^p(\mathbb{R}^{n-1} \times [0, +\infty[, x_n^{2\alpha+1}dx)$ and μ a Borel measure on $\mathbb{R}^{n-1} \times [0, +\infty[$. We define the function $\mu * f$ by

$$\mu * f(x) = \int_{\mathbb{R}^{n-1} \times [0,+\infty[} T_x f(y) d\mu(y).$$

2.3. Generalized Fourier transform associated with the operator Δ_W

Definition. Let f be in $L^1(\mathbb{R}^{n-1} \times [0, +\infty[, x_n^{2\alpha+1}dx)$. The *Fourier transform associated with* Δ_W of the function f is the function $F_W(f)$ defined by

$$F_W(f)(\xi) = \int_{\mathbb{R}^{n-1} \times [0,+\infty[} f(x) \exp(-i\langle x', \xi' \rangle) j_\alpha(x_n \xi_n) x_n^{2\alpha+1} dx,$$

where j_α is the function defined by

$$j_\alpha(t) = \begin{cases} \frac{2^\alpha \Gamma(\alpha+1) J_\alpha(t)}{t^\alpha} & \text{if } t \neq 0, \\ 1 & \text{if } t = 0, \end{cases}$$

J_α being the Bessel function of the first kind and of order α (see [9]).

We observe that $j_\alpha(\lambda t)$ is the unique solution of the system

$$\begin{cases} l_\alpha u(t) &= -\lambda^2 u(t), \quad \lambda \in \mathbb{N}, \\ u(0) &= 1, \\ u'(0) &= 0, \end{cases}$$

where l_α is the Bessel operator defined on $]0, +\infty[$ by

$$l_\alpha u(t) = u''(t) + \frac{2\alpha + 1}{t} u'(t).$$

Definition. The *Fourier transform associated with* Δ_W *of a distribution* u in $S'_*(\mathbb{R}^n)$ is defined by

$$\langle F_W(u), \varphi \rangle = \langle u, F_W(\varphi) \rangle.$$

Here, for a distribution defined by a function u in $S_*(\mathbb{R}^n)$, $\langle u, \varphi \rangle$ is defined by

$$\langle u, \varphi \rangle = \int_{\mathbb{R}^{n-1} \times [0,+\infty[} u(x) \varphi(x) x_n^{2\alpha+1} dx.$$

3. Fundamental solution of the operator Δ_W

We recall the following definition:

Definition. A *fundamental solution* of the operator Δ_W is a distribution K such that $\Delta_W K = -\delta$, where δ denotes the point mass at the origin.

In the following we shall compute a fundamental solution of the Weinstein operator by using the Fourier transform associated with Δ_W. Precisely, we have the following theorem :

Theorem 3.1. *For $n \geq 2$, the distribution defined by the function*

$$K(x) = \frac{\Gamma(\alpha + \frac{n}{2} - \frac{1}{2})}{2\pi^{\frac{n-1}{2}}\Gamma(\alpha+1)} \mid x \mid^{-2\alpha-n+1}$$

is a fundamental solution of the operator Δ_W.

Proof. For a suitable integer k, we denote by U_{F_k} the distribution defined by

$$\langle U_{F_k}, \varphi \rangle = \int_{\mathbb{R}^{n-1}\times[0,+\infty[} \mid x \mid^{-k} \varphi(x) x_n^{2\alpha+1} dx.$$

Then we have for φ in $S_*(\mathbb{R}^n)$

$$\langle F_W(U_{F_k}), \varphi \rangle = \int_{\mathbb{R}^{n-1}\times[0,+\infty[} F_W(\varphi)(x) \mid x \mid^{-k} x_n^{2\alpha+1} dx.$$

Now, by observing that for every positive integer k we have

$$\mid x \mid^{-k} = \frac{1}{\Gamma(\frac{k}{2})} \int_0^{+\infty} \exp(-t \mid x \mid^2) t^{\frac{k}{2}-1} dt,$$

we obtain

$$\langle F_W(U_{F_k}), \varphi \rangle$$
$$= \frac{1}{\Gamma(\frac{k}{2})} \int_{\mathbb{R}^{n-1}\times[0,+\infty[} \left(\int_0^{+\infty} \left(\int_{\mathbb{R}^{n-1}\times[0,+\infty[} \omega(x,t) dx \right) t^{\frac{k}{2}-1} dt \right) \varphi(\xi) \xi_n^{2\alpha+1} d\xi,$$

where $\omega(x,t) = \exp(-t \mid x \mid^2) \exp(-i\langle x', \xi' \rangle) j_\alpha(x_n \xi_n) x_n^{2\alpha+1}$.

A computation gives the following integral

$$\frac{1}{\Gamma(\frac{k}{2})} \int_{\mathbb{R}^{n-1}\times[0,+\infty[} \exp(-t \mid x \mid^2) \exp(-i\langle x', \xi' \rangle) j_\alpha(x_n \xi_n) x_n^{2\alpha+1} dx$$
$$= \frac{1}{2^{2\alpha+\frac{n+1}{2}}\Gamma(\alpha+1)\Gamma(\frac{k}{2})} t^{-\alpha-\frac{n}{2}-\frac{1}{2}} \exp(-\frac{\mid \xi \mid^2}{4t}).$$

Hence,

$$\langle F_W(U_{F_k}), \varphi\rangle = \frac{\Gamma(\alpha+1)\pi^{\frac{n-1}{2}}}{2\Gamma(\frac{k}{2})} \times \frac{\int_{\mathbb{R}^{n-1}\times[0,+\infty[}(\int\limits_0^{+\infty} t^{-\alpha-\frac{n}{2}-\frac{3}{2}+\frac{k}{2}} \exp(-\mid\xi\mid^2}{4t)dt)\varphi(\xi)\xi_n^{2\alpha+1}d\xi}$$

by setting $s = \frac{|\xi|^2}{4t}$ in the integral $\int\limits_0^{\infty} t^{-\alpha-\frac{n}{2}-\frac{3}{2}+\frac{k}{2}} \exp(-\frac{|\xi|^2}{4t})dt$.

We obtain

$$\langle F_W(U_{F_k}), \varphi\rangle = \frac{\Gamma(\alpha+1)2^{2\alpha+n-k}\Gamma(\alpha+\frac{n}{2}-\frac{k}{2}+\frac{1}{2})\pi^{\frac{n-1}{2}}}{\Gamma(\frac{k}{2})} \times \int_{\mathbb{R}^{n-1}\times[0,+\infty[} \mid\xi\mid^{-2\alpha-n+k-1} \varphi(\xi)\xi_n^{2\alpha+1}d\xi.$$

We deduce that

$$F_W(\Delta_W U_{F_k}) = -2^{2\alpha+n-k}\pi^{\frac{n-1}{2}}\frac{\Gamma(\alpha+\frac{n}{2}-\frac{k}{2}+\frac{1}{2})\Gamma(\alpha+1)}{\Gamma(\frac{k}{2})}U_{F_{2\alpha+n-k-1}}.$$

By setting $k = 2\alpha+n-1$, we find

$$F_W(\Delta_W U_{F_k}) = -\frac{2\pi^{\frac{n-1}{2}}\Gamma(\alpha+1)}{\Gamma(\alpha+\frac{n}{2}-\frac{1}{2})}U_1,$$

where U_1 denotes the distribution associated with the constant function 1.

Finally, by using the fact that $F_W(\delta) = U_1$, we deduce the result. □

4. Mean value property for the Weinstein operator

With the help of the generalized translation operators, we shall define the mean value of a function for the Weinstein operator.

Definition. Let u be a continuous function on $\mathbb{R}^n$ even in the last variable. We say that u satisfies the *mean value property* associated with the Weinstein operator Δ_W, if for every $r > 0$ and x in $\mathbb{R}^n$, we have

$$u(x) = \frac{1}{C_{n,\alpha}}\int\limits_{S^+} T_x u(r\omega)\omega_n^{2\alpha+1}d\sigma(\omega),$$

where T_x denotes the generalized translation operator associated with Δ_W given by the relation (2.1), S^+ is the upper hemisphere: $S^+ = \{x\in\mathbb{R}^n, \mid x\mid = 1, x_n > 0\}$,

σ is the surface measure on S^+ and

$$C_{n,\alpha} := \int\limits_{S^+} \omega_n^{2\alpha+1} d\sigma(\omega) = \frac{\pi^{\frac{n-1}{2}}\Gamma(\alpha+1)}{\Gamma\left(2\alpha+\frac{n+1}{2}\right)}.$$

Definition. A function u is said to be *W-harmonic* in a domain D if it is twice differentiable and satisfies the Weinstein equation $\Delta_W u = 0$ at every point of the domain.

Theorem 4.1. Mean value theorem. *If u is a W-harmonic function on $\mathbb{R}^n$ even in the last variable, then u satisfies the mean value property associated with the Weinstein operator.*

For the proof of this theorem, we shall need some results given in the following lemmas.

Lemma 4.1. *W-harmonicity is preserved under generalized translation. That is, if u is W-harmonic, then for all $x \in \mathbb{R}^{n-1} \times [0, +\infty[$, $T_x u$ is also W-harmonic.*

Proof. It is a consequence of the property (iii) given in Proposition 2.1. By taking into account the previous lemma, we can assume $x = 0$. □

Lemma 4.2. *Let u and v be functions even in the last variable and twice differentiable in a neighborhood of the ball $B(0,r)$. Then we have*

$$\int\limits_{B^+(0,r)} (u(x)\Delta_W v(x) - v(x)\Delta_W u(x))x_n^{2\alpha+1}dx$$

$$= \int\limits_{S^+(0,r)} (u\frac{\partial v}{\partial \nu} - v\frac{\partial u}{\partial \nu})\omega_n^{2\alpha+1} d\sigma(\omega),$$

where $B^+(0,r) = \{x \in \mathbb{R}^n, |x| < r, x_n > 0\}$, $S^+(0,r) = \{x \in \mathbb{R}^n, |x| = r, x_n > 0\}$, $\frac{\partial}{\partial \nu}$ is the differentiation in the direction of the outward directed normal to $S^+(0,r)$.

Proof. The proof of this lemma is deduced from the classical Green formula associated with the Laplacian Δ_n on $\mathbb{R}^n$. □

Lemma 4.3. (Gauss lemma). *For any W-harmonic function u in a neighborhood of $B(0,r), r > 0$, even in the last variable, we have*

$$\int\limits_{S^+(0,r)} \frac{\partial u}{\partial \nu}\omega_n^{2\alpha+1} d\sigma(\omega) = 0.$$

Proof. It is a consequence of the previous lemma by letting $v = 1$. □

Having set down these lemmas, the proof of Theorem 4.1 is immediate by applying Lemma 4.2 to the functions u and $v(x) = K(x)$ (K is the fundamental solution of Δ_W given in 3) and by taking into account Lemma 4.3. Hence we obtain

$$u(0) = \frac{1}{C_{n,\alpha}} \int\limits_{S+} u(r\omega)\omega_n^{2\alpha+1} d\sigma(\omega).$$

By applying this relation to the function $T_x u$, we find

$$T_x u(0) = \frac{1}{C_{n,\alpha}} \int\limits_{S+} T_x u(r\omega)\omega_n^{2\alpha+1} d\sigma(\omega).$$

So,

$$u(x) = \frac{1}{C_{n,\alpha}} \int\limits_{S+} T_x u(r\omega)\omega_n^{2\alpha+1} d\sigma(\omega).$$

□

In the rest of this section we shall prove a converse of Theorem 4.1. Before the statement of this converse we give the following results:

Lemma 4.4. *Let u be an even C^{2p+2} function on $\mathbb{R}$. Then u satisfies the following formula*

$$u(r) = \sum_{k=0}^{p} b_k(r) l_\alpha^k u(0) + O(r^{2p+2}),$$

where $b_k(r) = \frac{\Gamma(\alpha+1)}{k!\Gamma(\alpha+k+1)}(\frac{r}{2})^{2k}$ and l_α is the Bessel operator defined in 2.3.

This formula is called Delsarte generalized Taylor formula. For the proof, see [8].

Proposition 4.1. *Let u be a twice differentiable function on $\mathbb{R}^n$, even in the last variable. Then we have*

$$\Delta_W u(0) = \lim_{r \to 0} \frac{2(n+2\alpha+1)}{r^2} \left[\frac{1}{C_{n,\alpha}} \int\limits_{S+} u(r\omega)\omega_n^{2\alpha+1} d\sigma(\omega) - u(0) \right].$$

Proof. The result is obtained by combination of the Taylor formula and the Delsarte generalized Taylor formula associated with the Bessel operator $l_{\alpha,n}$ given in the previous lemma. □

Theorem 4.2. *Let u be a continuous function even in the last variable and satisfying the mean value property associated with the Weinstein operator Δ_W at every point $x \in \mathbb{R}^n$. Then*

(i) *u is a C^∞ function;*

(ii) *u is W-harmonic.*

Proof. (i) We choose a radial function φ in $D_*(\mathbb{R}^n)$ such that

$$\int_{\mathbb{R}^{n-1}\times[0,+\infty[} u(x)\varphi(x)x_n^{2\alpha+1}dx = 1$$

and having its support in the ball $B(0,1)$. Put $\varphi_\epsilon(x) = \epsilon^{-n-2\alpha-1}\varphi(\frac{x}{\epsilon})$ for $\epsilon > 0$. Then we let $u_\epsilon = u * \varphi_\epsilon$, the convolution product associated with Δ_W of u and φ_ϵ defined by the relation (2.2). By Proposition 2.2-ii, we have $u_\epsilon \in D_*(\mathbb{R}^n)$. Using polar coordinates, we write

$$\begin{aligned} u_\epsilon(x) &= \int_{\mathbb{R}^{n-1}\times[0,+\infty[} T_x u(y)\varphi_\epsilon(y)y_n^{2\alpha+1}dy \\ &= \int_0^\epsilon \varphi_\epsilon(r)\Big(\int_{S+} T_x u(r\omega)\omega_n^{2\alpha+1}d\sigma(\omega)\Big)r^{n+2\alpha}dr \\ &= \int_0^\epsilon \varphi_\epsilon(r)C_{n,\alpha}M^W_{x,u}(r)r^{n+2\alpha}dr. \end{aligned}$$

Now for $0 \le r < \epsilon$, we have $M^W_{x,u}(r) = u(x)$. Thus $u_\epsilon(x) = u(x)$, for ϵ small enough and, consequently, u coincides with a C^∞ function in a neighborhood of x.

(ii) From Proposition 4.1, we have

$$\Delta_W u(0) = \lim_{r\to 0}\frac{2(n+2\alpha+1)}{r^2}\left[\frac{1}{C_{n,\alpha}}\int_{S+} u(r\omega)\omega_n^{2\alpha+1}d\sigma(\omega) - u(0)\right].$$

By applying this relation to the function $T_x u$, we find

$$\Delta_W T_x u(0) = \lim_{r\to 0}\frac{2(n+2\alpha+1)}{r^2}\left[\frac{1}{C_{n,\alpha}}\int_{S+} T_x u(r\omega)\omega_n^{2\alpha+1}d\sigma(\omega) - T_x u(0)\right],$$

$$\Delta_W u(x) = \lim_{r\to 0}\frac{2(n+2\alpha+1)}{r^2}\left[\frac{1}{C_{n,\alpha}}\int_{S+} T_x u(r\omega)\omega_n^{2\alpha+1}d\sigma(\omega) - u(x)\right].$$

We conclude that if u satisfies the mean value property associated with Δ_W, then $\Delta_W u = 0$ (i.e., u is W-harmonic). □

References

[1] Brelot-Collin, B., Brelot, M., Représentation intégrale des solutions positives de l'équation $L_k(u) = \sum_1^n \frac{\partial^2 u}{\partial x_i^2} + \frac{k}{x_n}\frac{\partial u}{\partial x_n} = 0$, ($k$: constante réelle) dans le demi-espace $E(x_n > 0)$ de $\mathbb{R}^n$. Bull. Acad. Royale de Belgique 58 (1972), 317–326.

[2] — Allure à la frontière des solutions positives de l'équation de Weinstein : $L_k(u) = \Delta u + \frac{k}{x_n}\frac{\partial u}{\partial x_n} = 0$, dans le demi-espace $E(x_n > 0)$ de $\mathbb{R}^n$ ($n \ge 2$). Acad. Royale de Belgique 59 (1973), 1100–1117.

[3] Heyer, H., Probability theory on hypergroups : A survey. Lecture Notes in Math. 1064, 481–545. Springer-Verlag, Berlin 1984.

[4] Hoh, W., Jacob, N., Remarks on mean value properties of solutions of second order differential operators. Exposition. Math. 9 (1991), 367–377.

[5] Jewett, R.I., Spaces with an abstract convolution of measures. Adv. Math. 18 (1975), 1–101.

[6] Leutwiler, H., Best constants in the Harnack inequality for the Weinstein equation. Aequation. Math. 34 (1987), 304–315.

[7] Kingman, J. F. C, Random walks with spherical symmetry. Acta Math 109 (1963), 11–53.

[8] Trimèche, K., Convergences des séries de Taylor généralisées au sens de Delsarte. C. R. Acad. Sci. Paris Sér. I Math. 286 (1978), 63–66.

[9] Watson, G. N., A treatise on the theory of Bessel functions. Cambridge Univ. Press, London, New York, New Rochelle, Melbourne, Sydney 1962.

[10] Weinstein, A., Discontinues integrals and generalized potential theory. Trans. Amer. Math. Soc. 63 (1948).

Once more about the semipolar sets and regular excessive functions

Lucian Beznea and Nicu Boboc

Abstract. If $\mathcal{U}$ is a proper submarkovian resolvent of kernels on a Lusin mesurable space $(X, \mathcal{X})$ and λ is a finite measure on X then we show that an analytic subset A of X is λ-semipolar (i.e., the union of a semipolar set and a λ-polar set) iff the function $W(\chi_A)$ is λ-negligible for any proper submarkovian resolvent $\mathcal{W}$ on $(X, \mathcal{X})$ having the same excessive functions as $\mathcal{U}$.

1991 Mathematics Subject Classification: 31D05, 60J45.

Introduction

In this paper we consider a proper submarkovian resolvent of kernels $\mathcal{U} = (U_\alpha)_{\alpha>0}$ on a Lusin measurable space $(X, \mathcal{X})$ such that that the set $\mathcal{E}_\mathcal{U}$ of all $\mathcal{U}$-excessive functions on X which are $\mathcal{U}$-a.e. finite is min–stable, contains the positive constant functions and generates the σ-algebra $\mathcal{X}$. In this framework we show (Theorem 4.3) that if λ is a finite measure on X then the following assertions are equivalent for an analytic subset A of X:

- the set A is λ-semipolar (i.e., a union of a semipolar set and a λ-polar set);
- the function $W(\chi_A)$ is λ-negligible for any proper submarkovian resolvent $\mathcal{W} = (W_\alpha)_{\alpha>0}$ on $(X, \mathcal{X})$ with $\mathcal{E}_\mathcal{W} = \mathcal{E}_\mathcal{U}$;
- the set A is μ-negligible for any finite measure μ on X which does not charge the λ-semipolar sets.

Our proof is purely analytical. The same result has been obtained under hypothesis (L) (existence of a reference measure) in [16] and [6]. In the case when X is a locally compact space with countable base and the initial kernel U of $\mathcal{U}$ is a Hunt kernel, important contributions are given by Feyel in [14]. On the other hand for the above result there are many probabilistic approaches. For the case with hypothesis (L), see [1] and for the general case of right processes, see the survey article [10] of Dellacherie.

The first results from this paper are related to the fine carrier of a regular excessive function using the standard technique concerning the Choquet boundary. We show (Theorem 2.6) that any bounded $\mathcal{U}$-excessive function s, for which there exists a Ray topology on X such that s is upper semi-continuous and such that there exists a compact subset K of X with $B^K s = s$, is regular. (Note that in the case considered in [14] one has "upper semi-continuity $\Longrightarrow$ regularity", assertion which fails in our situation.) Also a bounded $\mathcal{U}$-excessive function s on X will be regular iff there exists a bounded kernel W on X satisfying the complete maximum principle and such that $Wf \in \mathcal{E}_{\mathcal{U}}$ for any bounded positive $\mathcal{X}$-measurable function f on X and $W1 = s$ (see Theorem 3.4). If s is a bounded regular $\mathcal{U}$-excessive function on X and $F := \mathrm{carr}_{\mathrm{f}}\, s$ is its fine carrier then F is the smallest fine closed subset of X such that $R^F s = s$, it is a Ray G_δ-set and a basic set. Moreover, we have

$$\begin{aligned} B^F t &= \sup\{s' \in \mathcal{R} \mid (\exists)\alpha > 0 \text{ such that } s' \prec \alpha s,\ s' \leq t\} \\ &= \sup\{s' \in \mathcal{E}_{\mathcal{U}} \mid s' \text{ is bounded and regular, } \mathrm{carr}_{\mathrm{f}}\, s' \subset F,\ s' \leq t\} \end{aligned}$$

where $t \in \mathcal{R}$ and $\mathcal{R}$ is an arbitrary Ray cone containing s. We prove that an analytic subset A of X will be λ-semipolar iff $\lambda(q) = 0$ for any bounded regular $\mathcal{U}$-excessive function q with $\mathrm{carr}_{\mathrm{f}}\, q \subset A$. In particular, if A is a basic set then for any finite measure λ there exists a bounded regular $\mathcal{U}$-excessive function s with $\mathrm{carr}_{\mathrm{f}}\, s \subset A$ and such that $A \setminus \mathrm{carr}_{\mathrm{f}}\, s$ is λ-polar and λ-negligible (Theorem 4.4). This is an analytical version of a recent result of Fitzsimmons and Getoor; Theorem 1.6 from [15].

The starting point for our approach was the article [14] of Feyel. Some arguments are inspired by this paper but our proofs are essentially different. An important tool was the procedure of Dellacherie concerning the existence of a Cantor scheme associated to a set which is not λ-mince (cf. [8]).

Finally we note the profitable discussions with R. K. Getoor on this subject.

1. Preliminaries on the réduite

Throughout the paper, $\mathcal{U} = (U_\alpha)_{\alpha>0}$ will be a submarkovian resolvent on $(X, \mathcal{X})$ as in Introduction.

Recall now some considerations concerning the Ray compactification (see, e.g., [12]). We suppose that the initial kernel $U = U_0$ of $\mathcal{U}$ is bounded. A *Ray cone* associated with $\mathcal{U}$ will be a subcone $\mathcal{R}$ of the bounded $\mathcal{U}$-excessive functions which is min-stable, separable in the uniform norm, generates the σ-algebra $\mathcal{X}$ and moreover $1 \in \mathcal{R}$, $U((\mathcal{R}-\mathcal{R})_+) \subset \mathcal{R}$, $U_\alpha(\mathcal{R}) \subset \mathcal{R}$, $\alpha > 0$. The topology on X generated by a Ray cone is called *Ray topology*. We consider the Ray compactification Y of X with respect to $\mathcal{R}$. For any $p \in \mathcal{R}$ we denote by $\widetilde{p}$ the real continuous extension of p to Y. The set $\widetilde{\mathcal{R}} := \{\widetilde{p} \mid p \in \mathcal{R}\}$ separates the points of Y, $1 \in \widetilde{\mathcal{R}}$ and it is min-stable. Since $\widetilde{\mathcal{R}} - \widetilde{\mathcal{R}}$ is a vector sublattice of $\mathcal{C}(Y)$ which contains the constant functions it follows that any positive linear functional on $\widetilde{\mathcal{R}} - \widetilde{\mathcal{R}}$ is given by a positive measure on Y. We denote by $\widetilde{\mathcal{U}} = (\widetilde{U}_\alpha)_{\alpha>0}$ the bounded submarkovian resolvent on Y such that

$$\widetilde{U}_\alpha(\widetilde{s}) = \widetilde{U_\alpha(s)}, \qquad (\forall)\, s \in \mathcal{R},\ \alpha \geq 0.$$

Obviously $\widetilde{\mathcal{U}}$ is a Ray resolvent on Y. Since $(X, \mathcal{X})$ is a Lusin measurable space it follows that X is a Borel subset of Y and $\mathcal{B}(Y)|_X = \mathcal{X}$, where $\mathcal{B}(Y)$ denotes the σ-algebra of all Borel subsets of Y.

Let us denote by $\mathcal{E}^u_{\mathcal{U}}$ the convex cone of all universally measurable $\mathcal{U}$-excessive functions which are finite $\mathcal{U}$-a.e. If $A \in \mathcal{X}$ and $s \in \mathcal{E}^u_{\mathcal{U}}$ then the réduite $R^A s$ of s on A, $R^A s := \inf\{t \in \mathcal{E}_{\mathcal{U}} \mid t \geq s \text{ on } A\} = \inf\{t \in \mathcal{E}^u_{\mathcal{U}} \mid t \geq s \text{ on } A\}$, is universally measurable and $\lambda(R^A s) = R^A_s(\lambda) := \inf\{\lambda(t) \mid t \in \mathcal{E}_{\mathcal{U}},\ t \geq s \text{ on } A\}$ for any finite measure λ on X with $\lambda(s) < \infty$. Moreover, if $p \in \mathcal{E}_{\mathcal{U}}$ is bounded then the map

$$A \longmapsto R^A_p(\lambda) =: c^p_\lambda(A)$$

is a capacity on X endowed with the Ray topology generated by a Ray cone containing p (cf. [3]). If $A \in \mathcal{X}$ and $s \in \mathcal{E}^u_{\mathcal{U}}$ we denote by $B^A s$ the $\mathcal{U}$-excessive regularization of the $\mathcal{U}$-supermedian function $R^A s$, i.e., $B^A s := \widehat{R^A s}$.

A bounded $\mathcal{U}$-excessive function $p \in \mathcal{E}_{\mathcal{U}}$ will be called *generator of* $\mathcal{E}_{\mathcal{U}}$ if for any $s \in \mathcal{E}_{\mathcal{U}}$ there exists an increasing sequence $(s_n)_{n\in\mathbb{N}} \subset \mathcal{E}_{\mathcal{U}}$ and $(\alpha_n)_{n\in\mathbb{N}} \subset \mathbb{R}_+$ such that $\sup_{n\in\mathbb{N}} s_n = s$ and $s_n \prec \alpha_n p$ for all $n \in \mathbb{N}$ ($\prec$ denotes the specific order on $\mathcal{E}_{\mathcal{U}}$). For instance if $p \in \mathcal{E}_{\mathcal{U}}$ is a bounded function of the form $p = Uq$ with $0 < q < 1$ then it is a generator of $\mathcal{E}_{\mathcal{U}}$.

Let p be generator of $\mathcal{E}_{\mathcal{U}}$. As usual a subset M of X is called: *thin at the point* $x \in X$ if there exists $A \in \mathcal{X}$ such that $M \subset A$ and $B^A p(x) < p(x)$; *totally thin* if it is thin at any point of X; *semipolar* if it is a countable union of totally thin sets. A set $A \in \mathcal{X}$ is called *polar* (resp. λ-*polar* where λ is a measure on X) if $B^A 1 = 0$ (resp. $\lambda(B^A 1) = 0$). An arbitrary subset M of X is called polar (resp. λ-polar) if there exists a set $A \in \mathcal{X}$ such that $M \subset A$ and A is polar (resp. λ-polar). A λ-*semipolar* set is the union of a semipolar set with a λ-polar set.

Remark 1.1. In the above definition of λ-polarity for a set $M \subset X$, the $\mathcal{X}$-measurable set A may be choosen to be in addition fine clopen, if M is fine open (the fine topology is the topology on X generated by $\mathcal{E}_{\mathcal{U}}$).

Proposition 1.2 (Mokobodzki). *For any analytic subset A of X, any bounded element $p \in \mathcal{E}_{\mathcal{U}}$ and any finite measure λ on X we have*

$$R^A_p(\lambda) = \sup\{\nu(\chi_A p) \mid \nu \leq_{\mathcal{E}_{\mathcal{U}}} \lambda\}.$$

Proof. We consider a Ray cone $\mathcal{R}$ such that $p \in \mathcal{R}$. If A is Ray compact then the above formula is well known. Since the map $A \longmapsto R^A_p(\lambda)$ is a capacity it follows that for any universally capacitable subset A of X we have

$$\begin{aligned} R^A_p(\lambda) &= \sup\{R^K_p(\lambda) \mid K \subset A,\ K \text{ Ray compact }\} \\ &= \sup\{\nu(\chi_A p) \mid \nu \leq_{\mathcal{E}_{\mathcal{U}}} \lambda\} \leq R^A_p(\lambda) \end{aligned}$$

and therefore the required relation holds for any analytic subset A of X. □

Lemma 1.3. *Let Y be a compact metrisable space and let $\leq$ be an order relation on Y such that the set $\{(x,y) \in Y \times Y \mid x \leq y\}$ is analytic. Then for any analytic function $f : Y \longrightarrow \overline{\mathbb{R}}_+$ the function $f^\vee : Y \longrightarrow \overline{\mathbb{R}}_+$ defined by $f^\vee(x) := \sup\{f(y) \mid y \leq x\}$ is analytic.*

Proof. For any subset A of Y let us define the set $A^\vee := \{x \in Y \mid (\exists)\ y \in A,\ y \leq x\}$. Since $[f^\vee > \alpha] = [f > \alpha]^\vee$, $\alpha > 0$, to prove that $f^\vee$ is analytic it is sufficient to show that $A^\vee$ is analytic whenever A is analytic. Let us define $\mathcal{O} := \{(x,y) \in Y \times Y \mid y \leq x\}$ and let π be the second projection from $Y \times Y$ into Y; $\pi((y,x)) = x$. Since $A^\vee = \pi((A \times Y) \cap \mathcal{O})$ and since $(A \times Y) \cap \mathcal{O}$ is an analytic set in $Y \times Y$ we conclude that $A^\vee$ is also analytic, completing the proof. □

Theorem 1.4. *For any analytic subset A of X and any $p \in \mathcal{E}_\mathcal{U}$ the function $R^A p$ is analytic.*

Proof. Obviously we may suppose that p is bounded. In this case we consider a Ray cone $\mathcal{R}$ such that $p \in \mathcal{R}$ and let Y be the Ray compactification of X. From [12] it follows that the function

$$\nu \longmapsto \nu(\chi_A p)$$

is analytic on $\mathcal{M}_1(Y) := \{\nu \in \mathcal{M}(Y) \mid \nu(1) \leq 1\}$. For any finite measure λ on X we have

$$\sup\{\nu(\chi_A p) \mid \nu \leq_{\widetilde{\mathcal{R}}} \lambda\} = \sup\{\nu(\chi_A p) \mid \nu(Y \setminus X) = 0,\ \nu \leq_{\mathcal{E}_\mathcal{U}} \lambda\}$$

and therefore, by Lemma 1.3 and Proposition 1.2, we conclude that the function $x \longmapsto R_p^A(x)$ is analytic. □

Remark. From Theorem 1.4 it follows that if A is an analytic subset of X and λ is a finite measure on X then the following assertions are equivalent:
- the set A is λ-polar;
- the set A is $\lambda \circ U$-polar;
- $\lambda(B^A 1) = 0$;
- any compact subset of A is λ-polar.

Recall that a *$\mathcal{U}$-excessive measure* is a σ-finite measure m on X such that $m(\alpha U_\alpha) \leq m$ for all $\alpha > 0$. We denote by $\mathrm{Exc}_\mathcal{U}$ the convex cone of all $\mathcal{U}$-excessive measures on X. A *potential* is a $\mathcal{U}$-excessive measure of the form $\mu \circ U$ where μ is a positive measure on X.

Lemma 1.5. *Let G be a fine open subset of X and m a $\mathcal{U}$-excessive measure on X. Then the set G is m-polar iff it is m-negligible.*

Proof. If A is a $\mathcal{X}$-measurable set such that $G \subset A$ then we have $B^A 1 = 1$ on G and

$$G \subset [U(\chi_A) > 0].$$

If $m(A) = 0$ then $m(U_\alpha(\chi_A)) = 0$ for all $\alpha > 0$ and therefore $U(\chi_A) = 0$ m-a.e. Consequently, G is m-polar. □

An analytic fine closed subset A of X is called *basic* if $B^A s(x) = s(x)$ for any $s \in \mathcal{E}_\mathcal{U}$ and $x \in A$.

Recall that the set X is called *semi-saturated* (cf. [4]) if any excessive measure dominated by a potential is a potential. Note that always one can enlarge the base space X up to a semi-saturated set, preserving the excessive measures and functions. Also if X is semi-saturated then it is a Baire space with respect to the fine topology.

Corollary 1.6. *Suppose that X is semi-saturated and let A be an analytic subset of X which is a basic set and λ a finite measure on X. Then there exists an analytic basic subset A_1 of A such that the set $A \setminus A_1$ is λ-polar and λ-negligible and such that A_1 is a Baire space with the fine topology. In particular, if A is an analytic basic set which is λ-semipolar then it is λ-polar.*

Proof. Let p be a generator of $\mathcal{E}_\mathcal{U}$. Then $A = [B^A p = p]$ and there exists $s \in \mathcal{E}_\mathcal{U}$, $s \le B^A p$ such that the set $[s < B^A p]$ is $\lambda \circ U$ negligible. From Lemma 1.5 we get that $[s < B^A p]$ is a λ-polar set and by Remark 1.1 there exists a $\mathcal{X}$-measurable fine clopen set M which is λ-polar and $[s < B^A p] \subset M$. If we define $A_1 := A \setminus M$ then obviously A_1 is a basic subset of A and $A \setminus A_1$ is a λ-polar set. Note that $A_1 = [s = p] \setminus M$ and the set $[s = p]$ is a G_δ-set with respect to the Ray topology on X induced by a Ray cone $\mathcal{R}$ such that $s, p \in \mathcal{R}$. Since X is semi-saturated we deduce (see [4]) that A_1 is a Baire space with the fine topology.

Let now A be in addition λ-semipolar. From the above considerations one can suppose that A is a Baire space. If $M_0 \in \mathcal{X}$ is a fine clopen λ-polar set such that $A \setminus M_0$ is semipolar then, since $A \setminus M_0$ is a basic set which is a Baire space, we get that $A \setminus M_0 = \emptyset$ and we conclude that A is λ-polar and λ-negligible. □

Lemma 1.7. *For any increasing (resp. decreasing) family $(s_i)_{i\in I}$ in $\mathcal{E}^u_\mathcal{U}$ and any finite measure λ on X there exists an increasing (resp. decreasing) sequence $(s_{i_n})_{n\in\mathbb{N}}$ such that for any $i \in I$ the set*

$$\Big[s_i > \sup_{n\in\mathbb{N}} s_{i_n}\Big] \quad \Big(\text{resp. } \Big[s_i < \widehat{\inf_{n\in\mathbb{N}} s_{i_n}}\Big]\Big)$$

is λ-polar and λ-negligible.

The assertion follows from Lemma 1.5, using standard arguments.

Remark. The above result is the main tool in obtaining quasi-Lindelöf properties for the fine topology (see [13] and [15]).

Theorem 1.8 (Feyel). *For any analytic subset A of X the set*

$$A' := \{x \in X \mid A \setminus \{x\} \textit{ is not thin at } x\}$$

is an analytic set and the set $\{x \in X \mid \{x\}$ *is semipolar*$\}$ *is a* $G_{\delta\sigma}$*-set in any Ray topology on* X.

Proof. We use essentially a technique from [14]. One may suppose that X is semi-saturated. Let $\mathcal{R}$ be a Ray cone on X and $\mathcal{W} = (W_\alpha)_{\alpha>0}$ the resolvent on $X \times X$ given by

$$W_\alpha f(x,y) := U_\alpha f(\cdot, y)(x).$$

It is easy to see that $\mathcal{W}$ is a submarkovian resolvent on $X \times X$ and a positive $\mathcal{X} \otimes \mathcal{X}$-measurable function σ on $X \times X$ will be $\mathcal{W}$-excessive iff for any $y \in X$ the function $x \longmapsto \sigma(x,y)$ is $\mathcal{U}$-excessive on X. If p is a generator of $\mathcal{E}_{\mathcal{U}}$, $p \in \mathcal{R}$, then for any $x, y \in X$ we have

$$R^{A\setminus\{x\}}p(y) = {}^{\mathcal{W}}R^{A\times X\setminus\Delta}p(y,x),$$

where ${}^{\mathcal{W}}R$ denotes the réduite with respect to $\mathcal{E}_{\mathcal{W}}$. Indeed, if $s \in \mathcal{E}_{\mathcal{U}}$ is such that $s \geq p$ on $A \setminus \{x\}$ then the function σ on $X \times X$ defined by

$$\sigma(y,z) := \begin{cases} s(y), & z \neq x, \\ p(y), & z = x \end{cases}$$

is $\mathcal{W}$-excessive and

$$\sigma(y,z) \geq p(y), \qquad (\forall)\ (y,z) \in A \times X \setminus \Delta.$$

Hence ${}^{\mathcal{W}}R^{A\times X\setminus\Delta}p(y,x) \leq R^{A\setminus\{x\}}p(y)$. Conversely, if $\sigma \in \mathcal{E}_{\mathcal{W}}$ is such that $\sigma(y,z) \geq p(y)$ for all $(y,z) \in A \times X \setminus \Delta$ then the function $y \longmapsto \sigma(y,z)$ is $\mathcal{U}$-excessive and

$$\sigma(y,z) \geq p(y), \qquad (\forall)\ y \in A \setminus \{x\}.$$

We conclude that $R^{A\setminus\{x\}}p(y) \leq {}^{\mathcal{W}}R^{A\times X\setminus\Delta}p(y,x)$. On the other hand the set $A \setminus \{x\}$ is analytic in X and $A \times X \setminus \Delta$ is analytic in $X \times X$. By Theorem 1.4 it follows that the function $(x,y) \longmapsto {}^{\mathcal{W}}R^{A\times X\setminus\Delta}p(y,x)$ is analytic on $X \times X$. From the above considerations and from the equality

$$\{x \in X \mid R^{A\setminus\{x\}}p(x) = p(x)\} = \{x \in X \mid {}^{\mathcal{W}}R^{A\times X\setminus\Delta}p(x,x) = p(x)\}$$

it follows that the set A' is analytic and a fine G_δ-set.

Let us now define $X_0 := \{x \in X \mid \{x\}$ is thin at $x\}$. Using the above notations and arguments we get

$$R^{\{x\}}p(y) = {}^{\mathcal{W}}R^{\Delta}p(y,x)$$

and, consequently, $\widehat{R^{\{x\}}p}(y) = \widehat{{}^{\mathcal{W}}R^{\Delta}p}(y,x)$. Hence we have $X_0 = \{x \in X \mid \widehat{{}^{\mathcal{W}}R^{\Delta}p}(x,x) < p(x)\}$. On the other hand since X is a Lusin measurable space such that any excessive measure dominated by $\varepsilon_x \circ U$ is a potential it follows that $X \times X$ possesses the same property with respect to $\mathcal{W}$. From this fact we deduce that for any Ray closed subset F of $X \times X$ the réduit ${}^{\mathcal{W}}R^F p$ is upper semi-continuous with respect to the Ray topology. Hence ${}^{\mathcal{W}}R^{\Delta}p$ is upper semi-continuous and therefore its excessive regularization $\widehat{{}^{\mathcal{W}}R^{\Delta}p}$ is the limit of an increasing sequence of upper semi-continuous functions. We deduce that the set $M := [\widehat{{}^{\mathcal{W}}R^{\Delta}p} < p]$ is a $G_{\delta\sigma}$-set in the given Ray topology and therefore X_0 is a $G_{\delta\sigma}$-set in X since it is identified with $\Delta \cap M$ which is a $G_{\delta\sigma}$-set in Δ. □

Remark. From Theorem 1.8 it follows that if A is an analytic subset of X then the set $b(A) := \{x \in X \mid A \text{ is not thin at } x\}$ is also analytic.

Proposition 1.9. *Let A be an analytic subset of X and $(A_\xi)_{\xi \in \Omega}$ (where Ω is the first uncountable ordinal) the family of subsets of X defined inductively by:*

$$A_0 := A, \qquad A_\xi := (\bigcap_{\eta<\xi} A_\eta) \cap b(\bigcap_{\eta<\xi} A_\eta).$$

Then for any finite measure λ on X there exists $\xi_\lambda \in \Omega$ such that the set $A_{\xi_\lambda} \setminus b(A_{\xi_\lambda})$ is λ-polar and λ-negligible.

Proof. Let p be a generator of $\mathcal{E}_\mathcal{U}$. By Lemma 1.7 there exists a decreasing sequence $(\xi_n)_{n\in\mathbf{N}}$ in Ω such that for any $\xi \in \Omega$ the set $\left[B^{A_\xi}p < \bigwedge_{n\in\mathbf{N}} B^{A_{\xi_n}}p\right]$ is λ-polar and λ-negligible. Since $B^{A_\xi}p \geq R^{A_{\xi+1}}p \geq B^{A_{\xi+1}}p$ for any $\xi \in \Omega$ it follows that if we take $\xi_\lambda := \bigwedge_{n\in\mathbf{N}} \xi_n + 1$ then

$$A_{\xi_\lambda} \setminus A_{\xi_\lambda+1} \subset \left[R^{A_{\xi_\lambda}}p > R^{A_{\xi_\lambda+1}}p\right] \subset \left[\bigwedge_{n\in\mathbf{N}} B^{A_{\xi_n}}p > B^{A_{\xi_\lambda+1}}p\right]$$

and therefore the set $A_{\xi_\lambda} \setminus b(A_{\xi_\lambda})$ is λ-polar and λ-negligible. □

Let λ be a finite measure on X. A set $A \in \mathcal{X}$ is called λ-*mince* (cf. [8]; the French terminology!) if the capacity on X given by $c_\lambda^1(F) := R_1^F(\lambda)$ has the following property: for any family $(K_i)_{i\in I}$ of mutually disjoint $\mathcal{X}$-measurable subsets of A the set $\{i \in I \mid c_\lambda^1(K_i) > 0\}$ is at most countable. (In this definition the sets K_i may be considered Ray compact).

An arbitrary subset of X is called λ-*mince* if it is contained in a $\mathcal{X}$-measurable set which is λ-mince.

Remark. From [8], [9] and [11] it follows that an analytic subset A of X is λ-mince iff any Ray compact subset of A is λ-mince.

Lemma 1.10 (Dellacherie–Feyel). *Let A be an analytic subset of X such that A is a Baire space with the fine topology and let λ be a finite measure on X. If the set A is λ-mince and $A \setminus A'$ is λ-polar and λ-negligible then the set A is also λ-polar and λ-negligible.*

Proof. Let p be a generator of $\mathcal{E}_\mathcal{U}$ and μ be a measure on A such that $\mu \leq_{\mathcal{E}_\mathcal{U}} \lambda$ and such that

$$\mu(M) = 0 \Longleftrightarrow c_\lambda^p(M) = 0$$

for any subset $M \in \mathcal{X}|_A$ of A (cf. [8], using also Proposition 1.2). From [11] we get for any $k \in \mathbb{N}$ an open subset D_k in A maximal with respect to the relation

$$c_\lambda^p(D_k) \geq k\mu(D_k),$$

and therefore $\gamma_\mu(M) := c_\lambda^p(D_k \cup M) - c_\lambda^p(D_k) \leq k\mu(M)$, for any $M \in \mathcal{X}|_A$. Since the submeasure on A given by

$$M \longmapsto \gamma_\mu(M) = \int (R^{D_k \cup M}p(x) - R^{D_k}p(x))d\mu(x)$$

is strongly subadditive and with bounded variation, we deduce that the submeasure

$$M \longmapsto \gamma_x(M) := R^{D_k \cup M}p(x) - R^{D_k}p(x)$$

is also with bounded variation for any $x \in A$ except for a μ-negligible set. If $x \in A \cap A' \setminus \overline{D_k}^f$ is such a point then we have

$$R^{D_k \cup M}p(x) = R^{D_k}p(x) + \gamma_x(M) < p(x)$$

where $M := V \cap A \setminus \{x\}$ for a small enough fine neighbourhood V of x. (For a subset D of X we have denoted by $\overline{D}^f$ its fine closure.) This leads to a contradiction. Hence $A \cap A' \setminus \overline{D_k}^f$ is μ-negligible and

$$\mu(D_k) \leq \frac{1}{k} c_\lambda^p(D_k) < \frac{1}{k} c_\lambda^p(A).$$

If we put $F_k := (\overline{D_k}^f \setminus D_k) \cap A$, $E_k := A \setminus \overline{D_k}^f$ then it follows that F_k is fine meagre in A, E_k is μ-negligible and

$$A = D_k \cup F_k \cup E_k\ ,\ \mu(D_k) \leq \frac{1}{k} c_\lambda^p(A).$$

We get $A = \bigcup_{k=1}^{\infty} F_k \cup E$ where $E := \left(\bigcup_{k=1}^{\infty} E_k\right) \cup \bigcap_{k=1}^{\infty} D_k$. Obviously the set E is μ-negligible and therefore $A \cap \overline{E}^f$ is also μ-negligible. Therefore the set $A \setminus \overline{E}^f$ is of the form $A \setminus \overline{E}^f \subset \bigcup_{k=1}^{\infty} F_k$ where F_k is fine meagre. Since $A \setminus \overline{E}^f$ is a Baire space with the fine topology we deduce that $A \setminus \overline{E}^f = \emptyset$ and therefore A is μ-negligible or equivalently it is λ-polar and λ-negligible. □

Remark. If A is not a Baire space with the fine topology then the assertion from Lemma 1.10 fails. For instance take $X := [0,1]$, $A := \mathbb{Q} \cap X$ and let $\mathcal{U}$ be such that $\mathcal{E}_{\mathcal{U}}$ coincides with the set of all positive concave functions on X.

2. Quasicontinuity and regularity for excessive functions

Recall that a $\mathcal{U}$-excessive function $s \in \mathcal{E}_{\mathcal{U}}$ is termed *regular* (cf. [18]) if for any increasing sequence $(s_n)_{n\in\mathbb{N}} \subset \mathcal{E}_{\mathcal{U}}$ with $\sup_{n\in\mathbb{N}} s_n = s$ we have

$$\bigwedge_{n\in\mathbb{N}} R(s - s_n) = 0$$

($\bigwedge$ is the infimum operation in $\mathcal{E}_{\mathcal{U}}$).

We denote by $(\mathcal{E}_{\mathcal{U}})_r$ the set of all bounded regular $\mathcal{U}$-excessive functions.

Remark. 1. A $\mathcal{U}$-excessive function of the form Uf where $f \geq 0$ is regular.

2. The set of all regular $\mathcal{U}$-excessive functions is a convex subcone of $\mathcal{E}_{\mathcal{U}}$ which is solid in the specific order.

3. Any regular $\mathcal{U}$-excessive function is the sum of a sequence from $(\mathcal{E}_{\mathcal{U}})_r$.

Proposition 2.1. *Let $s \in \mathcal{E}_{\mathcal{U}}$ and $(s_n)_{n\in\mathbb{N}}$ be an increasing sequence of regular elements from $\mathcal{E}_{\mathcal{U}}$ such that $\sup_{n\in\mathbb{N}} s_n = s$ and such that $\bigwedge_{n\in\mathbb{N}} R(s - s_n) = 0$. Then s is regular.*

Proof. Let $(t_k)_{k\in\mathbb{N}}$ be an increasing sequence in $\mathcal{E}_{\mathcal{U}}$ such that $\sup_{k\in\mathbb{N}} t_k = s$. Since the sequence $(t_k \wedge s_n)_k$ increases to s_n and since s_n is regular we get $\bigwedge_{k\in\mathbb{N}} R(s_n - t_k \wedge s_n) = 0$. From $R(s - t_k) \leq R(s - s_n) + R(s_n - t_k \wedge s_n)$ it follows that

$$\bigwedge_{k\in\mathbb{N}} R(s - t_k) \leq R(s - s_n) + \bigwedge_{k\in\mathbb{N}} R(s_n - t_k \wedge s_n) = R(s - s_n),$$

$$\bigwedge_{k\in\mathbb{N}} R(s - t_k) \leq \bigwedge_{n\in\mathbb{N}} R(s - s_n) = 0$$

□

If C is an H-cone we recall that an element $s \in C$ is called *quasicontinuous* (cf. [3]) if for any increasing family $(s_i)_{i\in I}$ in C with $\bigvee_{i\in I} s_i = s$ we have

$$\bigwedge_{i\in I} R(s - s_i) = 0.$$

Proposition 2.2. *Let $s \in C$ and $(s_i)_{i\in I}$ be an increasing family of quasicontinuous elements in C with $\bigvee_{i\in I} s_i = s$ and such that $\bigwedge_{i\in I} R(s - s_i) = 0$. Then s is quasicontinuous.*

Proof. It is similar with the proof of Proposition 2.1. □

In [2] we have considered the H-cone $\overline{\overline{\mathcal{E}_{\mathcal{U}}}}$ which is the dual of the H-cone $\mathrm{Exc}_{\mathcal{U}}$. In fact an element $\varphi \in \overline{\overline{\mathcal{E}_{\mathcal{U}}}}$ is a map $\varphi : \mathrm{Exc}_{\mathcal{U}} \longrightarrow \overline{\mathbb{R}}_+$ which is additive, increasing, continuous in order from below (with respect to the natural order on $\mathrm{Exc}_{\mathcal{U}}$) and such that the set $\{m \in \mathrm{Exc}_{\mathcal{U}} \mid \varphi(m) < \infty\}$ is increasingly dense in $\mathrm{Exc}_{\mathcal{U}}$. We have $\mathcal{E}_{\mathcal{U}}^u \subset \overline{\overline{\mathcal{E}_{\mathcal{U}}}}$ where an element $s \in \mathcal{E}_{\mathcal{U}}^u$ is identified with the functional on $\mathrm{Exc}_{\mathcal{U}}$ given by $m \longmapsto L(m, s)$, L being the energy functional associated with $\mathcal{U}$. For any $s, t \in \mathcal{E}_{\mathcal{U}}$ we have $R(s-t) = \overline{\overline{R}}(s-t)$ where $\overline{\overline{R}}$ is the réduit in $\overline{\overline{\mathcal{E}_{\mathcal{U}}}}$. Also the lattice operations $\bigvee$ and $\bigwedge$ are the same in $\mathcal{E}_{\mathcal{U}}$ and $\overline{\overline{\mathcal{E}_{\mathcal{U}}}}$ for countable families from $\mathcal{E}_{\mathcal{U}}$. Moreover, for any $\varphi \in \overline{\overline{\mathcal{E}_{\mathcal{U}}}}$ and any $m \in \mathrm{Exc}_{\mathcal{U}}$ there exists a $\mathcal{U}$-excessive function s_m on X such that

$$\varphi(\theta) = L(\theta, s_m), \qquad (\forall)\, \theta \in \mathrm{Exc}_{\mathcal{U}},\ \theta \ll m$$

and the function s_m is uniquely determined up to a m-polar set. This result, which is essentially well known (cf. [7]), becomes in the particular case considered in [13] the representation of the supermedian functional φ as an excessive function on X. We note also that any element from $\overline{\overline{\mathcal{E}_{\mathcal{U}}}}$ is the supremum of an increasing family $(s_i)_{i \in I} \subset \overline{\overline{\mathcal{E}_{\mathcal{U}}}}$ such that for any $i \in I$, s_i is a specific minorant in $(\overline{\overline{\mathcal{E}_{\mathcal{U}}}})$ of an element from $\mathcal{E}_{\mathcal{U}}$.

Theorem 2.3. *An excessive function $s \in \mathcal{E}_{\mathcal{U}}$ is regular iff it is quasicontinuous in $\overline{\overline{\mathcal{E}_{\mathcal{U}}}}$.*

Proof. Obviously if s is quasicontinuous in $\overline{\overline{\mathcal{E}_{\mathcal{U}}}}$ then it is regular. Conversely, suppose that s is regular in $\mathcal{E}_{\mathcal{U}}$ and let $(f_n)_{n \in \mathbb{N}}$ be a sequence of bounded positive $\mathcal{X}$-measurable functions on X such that $(Uf_n)_{n \in \mathbb{N}}$ increases to s. Since s is regular we get $\bigwedge_{n \in \mathbb{N}} R(s - Uf_n) = 0$. On the other hand it is known (cf. [2]) that Uf_n is quasicontinuous in $\overline{\overline{\mathcal{E}_{\mathcal{U}}}}$ for all $n \in \mathbb{N}$. From the above considerations and from Proposition 2.2 we conclude that s is also quasicontinuous. □

The following result is a version of Theorem 2.3 from [3]. Further we suppose that X is semi-saturated.

Proposition 2.4. *Let $\mathcal{P}$ be a convex subcone of* $\mathrm{Exc}_{\mathcal{U}}$ *such that $\mathcal{P}$ is solid in* $\mathrm{Exc}_{\mathcal{U}}$ *with respect to the natural order and such that $\mathcal{P}$ is increasingly dense in* $\mathrm{Exc}_{\mathcal{U}}$. *Then an element $s \in \mathcal{E}_{\mathcal{U}}$ will be regular iff for any decreasing family $(m_i)_{i \in I}$ in $\mathcal{P}$ such that $\inf_{i \in I} L(m_i, s) < \infty$ we have*

$$L\Big(\bigwedge_{i \in I} m_i, s\Big) = \inf_{i \in I} L(m_i, s).$$

Lemma 2.5. *Let $(s_n)_{n \in \mathbb{N}}$ be a decreasing sequence in $\mathcal{E}_{\mathcal{U}}$ such that $\inf_{n \in \mathbb{N}} s_n \in \mathcal{E}_{\mathcal{U}}$ and such that s_n is regular for any $n \in \mathbb{N}$. Then $\inf_{n \in \mathbb{N}} s_n$ is regular.*

Proof. We consider the convex cone $\mathcal{P}$ of all $\mathcal{U}$-excessive measures of the form $\mu \circ U$ where μ is a finite measure and $\mu(s_1) < \infty$. Let $(\mu_i \circ U)_{i \in I}$ be a decreasing family in $\mathcal{P}$ and $\mu \circ U$ the element of $\mathcal{P}$ given by

$$\mu \circ U(f) = \inf_{i \in I} \mu_i \circ U(f)$$

for any positive bounded $\mathcal{X}$-measurable function f on X such that Uf is bounded. We have obviously

$$\mu_i(\inf_{n \in \mathbb{N}} s_n) = \inf_{n \in \mathbb{N}} \mu_i(s_n), \ \mu(\inf_{n \in \mathbb{N}} s_n) = \inf_{n \in \mathbb{N}} \mu(s_n).$$

Since s_n is regular we get from Proposition 2.4 that

$$\mu(s_n) = \inf_{i \in I} \mu_i(s_n), \qquad (\forall)\ n \in \mathbb{N}.$$

It follows that

$$\mu(\inf_{n \in \mathbb{N}} s_n) = \inf_{n \in \mathbb{N}} \inf_{i \in I} \mu_i(s_n) = \inf_{i \in I} \inf_{n \in \mathbb{N}} \mu_i(s_n) = \inf_{i \in I} \mu_i(\inf_{n \in \mathbb{N}} s_n).$$

Again from Proposition 2.4 we conclude that $\inf_{n \in \mathbb{N}} s_n$ is a regular $\mathcal{U}$-excessive function. □

Theorem 2.6. *Let $s \in \mathcal{E}_{\mathcal{U}}$ be a bounded and Ray upper semi-continuous function on X such that there exists a Ray compact set K with $B^K s = s$. Then s is regular.*

Proof. Let $\mathcal{R}$ be the Ray cone which generates the topology on X. We may suppose that $R(u-v) \in \mathcal{R}$ for any $u, v \in \mathcal{R}$. We remark that if F is a Ray compact subset of X and μ, ν are two measures on F such that $\mu \leq_{\mathcal{R}} \nu$ then $\mu(s) \leq \nu(s)$. Therefore, s being upper semi-continuous, there exists a decreasing sequence $(r_n)_{n \in \mathbb{N}}$ in $\mathcal{R}$ such that $\inf_{n \in \mathbb{N}} r_n = s$ on F. On the other hand from $B^K s = s$ it follows that $B^{K \cup F} s = s$ and, as a consequence, using the preceding considerations we get

$$s = \inf \{r \mid r \in \mathcal{R},\ r \geq s\} \text{ on } X.$$

Since $\mathcal{R}$ is separable it follows that there exists a decreasing sequence $(u_n)_{n \in \mathbb{N}}$ in $\mathcal{R}$ such that $\inf_{n \in \mathbb{N}} u_n = s$.

We construct now inductively a sequence $(Uf_m)_{m \in \mathbb{N}}$ such that $Uf_m \in \mathcal{R}$ and such that

$$s < Uf_1 \text{ on } K,\ Uf_1 \leq u_1 + 1,\ s < Uf_{m+1} \text{ on } K,$$

$$Uf_{m+1} \leq \min\left(Uf_m\ , u_{m+1} + \frac{1}{m+1}\right).$$

Since $mU_m u_1 \nearrow u_1$ and since $mU_m u_1 \in \mathcal{R}$, there exists $\alpha_1 \in \mathbb{N}$ such that $\alpha_1 U_{\alpha_1} u_1 + \alpha_1 U_{\alpha_1} 1 > s$ on K. Obviously if we take $Uf_1 := \alpha_1 U_{\alpha_1} u_1 + \alpha_1 U_{\alpha_1} 1$ then we have: $Uf_1 \in \mathcal{R}$, $s < Uf_1$ on K, $Uf_1 \leq u_1 + 1$. Suppose that the system

$(Uf_i)_{1\le i\le m}$ has already been constructed. Since

$$s < \sup_{\alpha>0} \alpha U_\alpha \left(\min\left(Uf_m\,, u_{m+1} + \frac{1}{m+1}\right)\right) \quad \text{on } K$$

it follows that there exists $\alpha_{m+1} \in \mathbb{N}$ such that

$$s < \alpha_{m+1} U_{\alpha_{m+1}} \left(\min\left(Uf_m\,, u_{m+1} + \frac{1}{m+1}\right)\right) \quad \text{on } K.$$

If we take

$$Uf_{m+1} := \alpha_{m+1} U_{\alpha_{m+1}} \left(\min\left(Uf_m\,, u_{m+1} + \frac{1}{m+1}\right)\right)$$

then we get $Uf_{m+1} \in \mathcal{R}$, $Uf_{m+1} \le \min\left(Uf_m\,, u_{m+1} + \frac{1}{m+1}\right)$ and $s < Uf_{m+1}$ on K.

By construction and using the hypothesis $B^K s = s$ it follows that

$$s \le Uf_{n+1} \le Uf_n \le u_n + \frac{1}{n}, \qquad (\forall)\, n \in \mathbb{N}.$$

Hence the sequence $(Uf_n)_{n\in\mathbb{N}}$ is decreasing and $\inf_{n\in\mathbb{N}} Uf_n = s$. Since Uf_n is regular, using Lemma 2.5 we conclude that s is also a regular $\mathcal{U}$-excessive function. □

3. Fine carrier for regular excessive functions

In this section we suppose that the set X is semi-saturated with respect to $\mathcal{U}$.

Definition. Let $s \in \mathcal{E}_\mathcal{U}$ be bounded. The set

$$\{x \in X \mid \mu \le_{\mathcal{E}_\mathcal{U}} \varepsilon_x,\ \mu(s) = s(x) \Longrightarrow \mu = \varepsilon_x\}$$

is called the *fine carrier of* s and will be denoted by $\mathrm{carr}_\mathfrak{f}\, s$.

Remark. 1. Since X is semi-saturated it follows that $\mathrm{carr}_\mathfrak{f}\, s \subset [s > 0]$.

2. If $s, t \in \mathcal{E}_\mathcal{U}$ are bounded and $s \prec t$ then $\mathrm{carr}_\mathfrak{f}\, s \subset \mathrm{carr}_\mathfrak{f}\, t$.

We consider now a submarkovian resolvent $\mathcal{V} = (V_\alpha)_{\alpha>0}$ on X such that V is bounded and $\mathcal{E}_\mathcal{U} = \mathcal{E}_\mathcal{V}$. For any bounded function $s \in \mathcal{E}_\mathcal{U}$ we take a Ray cone $\mathcal{R}$ on X such that $s \in \mathcal{R}$ and $R(u-v) \in \mathcal{R}$ for any $u, v \in \mathcal{R}$ and we denote by Y the Ray compactification with respect to $\mathcal{R}$. As in Section 1, for any $u \in \mathcal{R}$ we denote by $\widetilde{u}$ the continuous extension of u to Y and we put $\widetilde{\mathcal{R}} := \{\widetilde{u} \mid u \in \mathcal{R}\}$. There exists a submarkovian resolvent $\widetilde{\mathcal{V}} = (\widetilde{V}_\alpha)_{\alpha>0}$ on Y such that $\widetilde{V}_\alpha(\widetilde{u}) = \widetilde{V_\alpha(u)}$ for all $u \in \mathcal{R}$. Since $(X, \mathcal{X})$ is Lusin measurable space we deduce that X is a Borel subset of Y.

Let us denote by $\mathcal{S}_s$ the convex cone of real continuous functions on Y given by

$$\mathcal{S}_s := \widetilde{\mathcal{R}} - \mathbb{R}_+ \widetilde{s}$$

and by $\mathrm{Ch}_s Y$ the Choquet boundary of Y with respect to this convex cone (i.e., $\mathrm{Ch}_s Y := \{y \in Y \mid \mu \leq_{\mathcal{S}_s} \varepsilon_y \Longrightarrow \mu = \varepsilon_y\}$).

Theorem 3.1. *If $s \in (\mathcal{E}_\mathcal{U})_r$ then $\mathrm{carr}_\mathfrak{f}\, s = \mathrm{Ch}_s Y \cap X$. In particular, $\mathrm{carr}_\mathfrak{f}\, s$ is a G_δ-set with respect to the Ray topology on X generated by $\mathcal{R}$. Also we have:*
(a) *$A := \mathrm{carr}_\mathfrak{f}\, s$ is a basic set.*
(b) *For any $t \in \mathcal{E}_\mathcal{U}$ we have $B^A t \in \mathcal{E}_\mathcal{U}$ and*

$$B^A t = \sup\{s' \in \mathcal{R} \mid (\exists)\alpha > 0 \text{ such that } s' \prec \alpha s,\ s' \leq t\}\ ,\ (\forall) t \in \mathcal{R}.$$

(c) *$B^A s = s$.*

Proof. The inclusion $\mathrm{Ch}_s Y \cap X \subset \mathrm{carr}_\mathfrak{f}\, s$ is obvious. Let now $x \in \mathrm{carr}_\mathfrak{f}\, s$ and let μ be a measure on Y such that $\mu \leq_{\mathcal{S}_s} \varepsilon_x$. We have $\mu(\widetilde{t}) \leq \widetilde{t}(x)$ for all $t \in \mathcal{R}$ and $\mu(\widetilde{s}) = \widetilde{s}(x)$. The map $f \longmapsto l(f) := \mu(\widetilde{Vf})$ defined on the bounded positive $\mathcal{X}$-measurable functions on X is a $\mathcal{V}$-excessive measure on X such that $l(f) \leq Vf(x)$ for any positive $\mathcal{X}$-measurable function f on X. Since X is semi-saturated with respect to $\mathcal{U}$ then X is semi-saturated with respect to $\mathcal{V}$ and therefore there exists a measure ν on X such that $l(f) = \nu(Vf)$ for all f. Hence $\nu(Vf) = l(f) = \mu(\widetilde{Vf}) \leq \widetilde{V}f(x) = Vf(x)$ and consequently $\nu(nV_n t) = \mu(\widetilde{nV_n t})$ for all $t \in \mathcal{R}$. On the other hand since s is regular we have $R(s - nV_n s)(x) \searrow 0$ and therefore from $\mu(\widetilde{s} - \widetilde{nV_n s}) \leq \mu(R(\widetilde{s - nV_n s})) \leq R(\widetilde{s - nV_n s})(x) = R(s - nV_n s)(x)$ we deduce that

$$\sup_{n \in \mathbb{N}} \mu(\widetilde{nV_n s}) = \mu(\widetilde{s}),\ \nu(s) = \sup_{n \in \mathbb{N}} \nu(nV_n s) = \sup_{n \in \mathbb{N}} \mu(\widetilde{nV_n s}) = \mu(\widetilde{s}) = \widetilde{s}(x).$$

From the above considerations we get $\nu = \varepsilon_x$. As a consequence

$$\mu(\widetilde{t}) \geq \sup_{n \in \mathbb{N}} \mu(\widetilde{nV_n t}) = \sup_{n \in \mathbb{N}} \nu(nV_n t) = \sup_{n \in \mathbb{N}} nV_n t(x) = t(x) = \widetilde{t}(x)$$

for any $t \in \mathcal{R}$. Hence $\mu = \varepsilon_x$.

It is known that (cf. [5]) $\mathrm{Ch}_s Y = \bigcap_{t \subset \mathcal{R}} [Q(\widetilde{s} - \widetilde{t}) = \widetilde{s} - \widetilde{t}\,]$, where Q is the Bauer functional given by

$$Qf := \inf\{\widetilde{t} - \alpha\widetilde{s} \mid t \in \mathcal{R},\ \alpha > 0,\ \widetilde{t} - \alpha\widetilde{s} \geq f\}$$

for any finite continuous function f on Y. It follows that $\mathrm{Ch}_s Y$ is a G_δ-set in Y and therefore the set $A := \mathrm{carr}_\mathfrak{f}\, s$ is a G_δ-set in X with respect to the Ray topology on X generated by $\mathcal{R}$. We want to show that

$$-Q(-\widetilde{t})|_X = \sup\{s' \in \mathcal{R} \mid (\exists)\alpha > 0 \text{ such that } s' \prec \alpha s,\ s' \leq t\}.$$

Indeed, we have:

$$\begin{aligned} Q(-\tilde{t})|_X &= \inf\{t' - \alpha s \mid t' \in \mathcal{R},\ \alpha > 0, t' - \alpha s \geq -t\} \\ &= \inf_{\alpha>0}(\inf\{t' - \alpha s \mid t' \in \mathcal{R}, t' - \alpha s \geq -t\}) \\ &= \inf_{\alpha>0}(\inf\{t' - \alpha s \mid t' \in \mathcal{R}, t' \geq \alpha s - t\}) = \inf_{\alpha>0}(R(\alpha s - t) - \alpha s), \end{aligned}$$

$$\begin{aligned} -Q(-\tilde{t})|_X &= \sup_{\alpha>0}(\alpha s - R(\alpha s - t)) \\ &= \sup\{s' \in \mathcal{R} \mid (\exists)\alpha > 0 \text{ such that } s' \prec \alpha s, s' \leq t\}. \end{aligned}$$

Since $\alpha < \beta \Longrightarrow \alpha s - R(\alpha s - t) \leq \beta s - R(\beta s - t)$ it follows that $-Q(-\tilde{t})|_X \in \mathcal{E}_{\mathcal{U}}$. On the other hand $Q(\tilde{s} - \tilde{t}) = Q(\tilde{s}) + Q(-\tilde{t}) = \tilde{s} + Q(-\tilde{t})$ for any $t \in \mathcal{R}$ and therefore

$$\bigcap_{t\in\mathcal{R}}[Q(\tilde{s} - \tilde{t}) = \tilde{s} - \tilde{t}] = \bigcap_{t\in\mathcal{R}}[-Q(-\tilde{t}) = \tilde{t}], \quad A = X \cap \bigcap_{t\in\mathcal{R}}[-Q(-\tilde{t}) = \tilde{t}].$$

Hence A is a fine closed subset of X. To finish the proof it is sufficient to show that $-Q(-\tilde{t})|_X = R^A t$, for all $t \in \mathcal{R}$. Let $x \in X$ and $t \in \mathcal{R}$. There exists a measure μ_x on Y such that $\mu_x \leq_{S_*} \varepsilon_x$ and such that μ_x is minimal with respect to the order relation $\leq_{S_*}$. We have $\mu_x(-\tilde{t}) = Q(-\tilde{t})(x)$. From the first part of the proof there exists a measure μ'_x on X such that $\mu'_x \leq_{S_*} \mu_x$ and therefore, μ_x being minimal, we have $\mu'_x = \mu_x$. Hence μ_x is carried by $A = \mathrm{Ch}_s Y \cap X$. Consequently we get

$$u \in \mathcal{E}_{\mathcal{U}},\ u \geq t \text{ on } A \Longrightarrow u(x) \geq \mu_x(u) \geq \mu_x(t) = -Q(-\tilde{t})(x),$$

$$R^A t(x) \geq -Q(-\tilde{t})(x).$$

Since $-Q(-\tilde{t})|_X \in \mathcal{E}_{\mathcal{U}}$ and $-Q(-\tilde{t})|_A = \tilde{t}|_A = t|_A$ we have also $-Q(-\tilde{t})|_X \geq R^A t$, completing the proof. □

Corollary 3.2. *For any $s \in (\mathcal{E}_{\mathcal{U}})_r$ the set* $\mathrm{carr_f}\, s$ *is the smallest fine closed subset F of X such that $R^F s = s$.*

Proof. From Theorem 3.1 the set $A := \mathrm{carr_f}\, s$ is fine closed and $R^A s = s$. Let now F be an arbitrary fine closed subset of X such that $R^F s = s$. Since X is semi-saturated it follows that for any $x \in X$ there exists a measure μ_x on X such that $R^F t(x) = \mu_x(t)$ for all $t \in \mathcal{E}_{\mathcal{U}}$. From $R^F s(x) = s(x)$ we get $\mu_x \leq_{S_x} \varepsilon_x$ and consequently

$$x \in A \Longrightarrow \mu_x = \varepsilon_x \Longrightarrow R^F t(x) = t(x), \qquad (\forall)\, t \in \mathcal{E}_{\mathcal{U}}.$$

The set F being fine closed we conclude that $A \subset F$. □

Remark. The fine carrier for regular excessive functions was considered also by Feyel in [14] where the above result has been obtained by different methods.

Corollary 3.3. *Let $s \in (\mathcal{E}_{\mathcal{U}})_r$ and $A := \operatorname{carr}_{\mathrm{f}} s$. Then for any $t \in \mathcal{E}_{\mathcal{U}}$ we have*

$$R^A t = \sup\{s' \in (\mathcal{E}_{\mathcal{U}})_r \mid s' \leq t,\ \operatorname{carr}_{\mathrm{f}} s' \subset A\}.$$

Theorem 3.4. *For any $s \in (\mathcal{E}_{\mathcal{U}})_r$ there exists a unique kernel W on X such that $W1 = s$ and $Wf \in \mathcal{E}_{\mathcal{U}}$ for any positive bounded $\mathcal{X}$-measurable function f on X and such that*

$$t \in \mathcal{E}_{\mathcal{U}},\ t \geq Wf \text{ on } [f > 0] \Longrightarrow t \geq Wf.$$

Proof. We may suppose that the Ray cone $\mathcal{R}$ is closed with respect to the uniform convergence. Therefore $\widetilde{\mathcal{R}}$ is a cone of potentials on the Ray compactification Y. Hence (cf. [17]) there exists a kernel $\widetilde{W}$ on Y such that $\widetilde{W}1 = \widetilde{s}$, $\widetilde{W}g \in \mathcal{R}$ and $\delta(\widetilde{W}g) \subset \operatorname{supp} g$ for any finite continuous function g on Y, where for any $u \in \widetilde{\mathcal{R}}$, $\delta(u)$ denotes the Choquet boundary of Y with respect to the convex cone $\mathcal{S}_u := \widetilde{\mathcal{R}} - \mathbb{R}_+ u$. Note that for any bounded positive upper semi-continuous function g on Y we have $\delta(\widetilde{W}g) \subset \operatorname{supp} g$.

We show first that if g is a positive bounded upper semi-continuous function on Y such that $g = 0$ on X then $\widetilde{W}g|_X = 0$. Obviously it will be sufficient to consider the case when g is the characteristic function of a closed subset F of Y such that $F \cap X = \emptyset$. Since there exists a decreasing sequence $(g_n)_{n\in\mathbb{N}}$ of finite continuous functions on Y, $g_n \leq 1$ such that $\inf_{n\in\mathbb{N}} g_n = \chi_F$ and since $\widetilde{W}g_n \in \widetilde{\mathcal{R}}$, $\widetilde{W}g_n \prec \widetilde{W}1 = \widetilde{s}$ it follows that $\widetilde{W}g_n|_X \prec_{\mathcal{R}} s$ and therefore $\widetilde{W}(\chi_F)|_X \prec_{\mathcal{E}_{\mathcal{U}}} s$. Hence $\widetilde{W}g_n|_X$ and $\widetilde{W}(\chi_F)|_X$ are elements from $(\mathcal{E}_{\mathcal{U}})_r$. Theorem 3.1 implies that $\operatorname{carr}_{\mathrm{f}} \widetilde{W}g_n|_X = \delta(\widetilde{W}g_n) \cap X$ for any $n \in \mathbb{N}$. From the preceding considerations we get

$$\operatorname{carr}_{\mathrm{f}} \widetilde{W}(\chi_F)|_X \subset \bigcap_{n\in\mathbb{N}} \operatorname{carr}_{\mathrm{f}} \widetilde{W}g_n|_X \subset F \cap X = \emptyset$$

and therefore from Theorem 3.1 c) we deduce that $\widetilde{W}(\chi_F)|_X = \emptyset$. Consequently $\widetilde{W}f|_X = \widetilde{W}(f\chi_X)|_X$ for any positive bounded Borel measurable function f on Y and therefore there exists a kernel W on X such that $Wf = \widetilde{W}(\widetilde{f})|_X$ where f is a positive bounded $\mathcal{X}$-measurable function on X and $\widetilde{f}$ denotes a measurable extension of f to Y.

Let now f be a positive bounded $\mathcal{X}$-measurable function X and $t \in \mathcal{E}_{\mathcal{U}}$ be such that $t \geq Wf$ on $[f > 0]$. We take an upper semi-continuous function g on Y such that $[g > 0] \subset X$, $g \leq f$. As in the preceding considerations we get $Wg \in (\mathcal{E}_{\mathcal{U}})_r$ and $\operatorname{carr}_{\mathrm{f}} Wg \subset \overline{[g > 0]}^{f}$. From Theorem 3.1 c) we get $t \geq B^{\operatorname{carr}_{\mathrm{f}} Wg} Wg = Wg$ and therefore $t \geq Wf$ on X.

The uniqueness follows by standard arguments. □

4. Semipolar sets and regular excessive functions

In this section we suppose that the set X is semi-saturated.

Definition. An element $s \in (\mathcal{E}_{\mathcal{U}})_r$ is called *diffuse* if there are no points $x \in X$ and s' a specific minorant of s with $\operatorname{carr}_{\mathfrak{f}} s' = \{x\}$. The set of all diffuse elements of $(\mathcal{E}_{\mathcal{U}})_r$ is denoted by $(\mathcal{E}_{\mathcal{U}})_{rd}$.

If W is the kernel associated with s as in Theorem 3.4 then s will be diffuse iff $W(\chi_{\{x\}}) = 0$ for any $x \in X$.

Proposition 4.1. *For any $s \in (\mathcal{E}_{\mathcal{U}})_{rd}$ we have*

$$\operatorname{carr}_{\mathfrak{f}} s = (\operatorname{carr}_{\mathfrak{f}} s)'.$$

Proof. Suppose that $s \in (\mathcal{E}_{\mathcal{U}})_{rd}$. Since $\operatorname{carr}_{\mathfrak{f}} s$ is fine closed we have obviously $(\operatorname{carr}_{\mathfrak{f}} s)' \subset \operatorname{carr}_{\mathfrak{f}} s$. Let now $x \in \operatorname{carr}_{\mathfrak{f}} s$ and suppose that $x \notin (\operatorname{carr}_{\mathfrak{f}} s)'$, i.e., the set $\operatorname{carr}_{\mathfrak{f}} s \setminus \{x\}$ is thin at x. Then the set $\operatorname{carr}_{\mathfrak{f}} s \setminus \{x\}$ is fine closed and since s is diffuse we get $W(\chi_{\{x\}}) = 0$. Hence

$$s = W(\chi_{\operatorname{carr}_{\mathfrak{f}} s \setminus \{x\}}) = B^{\operatorname{carr}_{\mathfrak{f}} s \setminus \{x\}} s$$

and therefore from Corollary 3.2 we get $\operatorname{carr}_{\mathfrak{f}} s \subset \operatorname{carr}_{\mathfrak{f}} s \setminus \{x\}$ which is a contradiction. □

Theorem 4.2. *Let A be an analytic subset of X and λ a finite measure on X. Then the following assertions are equivalent:*
(1) *The set A is λ-mince.*
(2) *For any $q \in (\mathcal{E}_{\mathcal{U}})_{rd}$ with $\operatorname{carr}_{\mathfrak{f}} q \subset A$ we have $\lambda(q) = 0$.*

Proof. (1) $\Longrightarrow$ (2). Let $q \in (\mathcal{E}_{\mathcal{U}})_{rd}$ be such that $\operatorname{carr}_{\mathfrak{f}} q \subset A$. Since from Proposition 4.1 we have $\operatorname{carr}_{\mathfrak{f}} s = (\operatorname{carr}_{\mathfrak{f}} s)'$ it follows that the set $\operatorname{carr}_{\mathfrak{f}} q$ is λ-mince, it is a Baire space with the fine topology and $\operatorname{carr}_{\mathfrak{f}} q \setminus (\operatorname{carr}_{\mathfrak{f}} q)' = \emptyset$. From Lemma 1.10 we deduce that the set $\operatorname{carr}_{\mathfrak{f}} q$ is λ-polar and λ-negligible and therefore

$$\lambda(q) = \lambda(B^{\operatorname{carr}_{\mathfrak{f}} q} q) = 0.$$

(2) $\Longrightarrow$ (1). Let p be a generator of $\mathcal{E}_{\mathcal{U}}$ and $\mathcal{R}$ a Ray cone such that $p \in \mathcal{R}$. Suppose that A is not λ-mince. Then there exist (cf. [8]; see also [11]) a Ray compact subset K of A, $\alpha > 0$ and a continuous and surjective map $\Psi : K \longrightarrow T := \{0,1\}^{\mathbb{N}}$ such that $K_t := \Psi^{-1}(\{t\})$ is compact and $\lambda(R^{K_t} p) > \alpha$. If we denote by $\mathcal{W} = (W_\alpha)_{\alpha > 0}$ the family of kernels on $X \times T$ defined by $W_\alpha f(x,t) := U_\alpha(f(\cdot,t))(x)$ we deduce that $\mathcal{W}$ is a submarkovian resolvent on $X \times T$ such that a positive measurable function s on $X \times T$ will be $\mathcal{W}$-supermedian (resp. $\mathcal{W}$-excessive) iff for any $t \in T$ the function $s(\cdot,t)$ is $\mathcal{U}$-supermedian (resp. $\mathcal{U}$-excessive). Moreover we have

$$^{\mathcal{W}}R^{\operatorname{graf}\Psi} p(x,t) = R^{K_t} p(x)\,, \qquad (\forall)\ (x,t) \in X \times T.$$

We deduce that the function $(x,t) \mapsto {}^{\mathcal{W}}R^{\mathrm{graf}\,\Psi}p(x,t) = R^{K_t}p(x)$ is upper semi-continuous. If we put (as in [14]) $q(x):=\int_T R^{K_t}p(x)dt$ where dt denotes the Cantor diffuse measure on T then we get that q is upper semi-continuous, $\mathcal{U}$-supermedian and $q \le p$. Since $\alpha U_\alpha q(x)=\int_T \alpha U_\alpha R^{K_t}p(x)dt$ and $R^{K_t}p=B^{K_t}p=\sup\limits_{\alpha>0} \alpha U_\alpha R^{K_t}p$ on $X\setminus K_t$, using the fact that $(K_t)_{t\in T}$ is a partition of K and dt is a diffuse measure it follows that $\sup\limits_{\alpha>0} \alpha U_\alpha q(x) = \int_T R^{K_t}p(x)dt = q(x)$ and therefore q is $\mathcal{U}$-excessive. Further we want to show that $R^K q=q$. Indeed, for any $x \in X$ let $(G_n)_{n\in\mathbb{N}}$ be a decreasing sequence of fine open sets such that $G_n \supset K$ and $\inf\limits_{n\in\mathbb{N}} R^{G_n}q(x)=R^K q(x)$. For any $n \in \mathbb{N}$ we denote by $\varepsilon_x^{G_n}$ the measure on X such that $\varepsilon_x^{G_n}(s)=R^{G_n}s(x)$ for all $s \in \mathcal{E}_{\mathcal{U}}$. We prove that for any $t \in T$ we have $\varepsilon_x^{G_n}(R^{K_t}p) = R^{K_t}p(x)$. We consider a decreasing sequence $(\Gamma_m)_{m\in\mathbb{N}}$ of fine open sets such that $K_t \subset \Gamma_m \subset G_n$ and such that $\inf\limits_{m\in\mathbb{N}} R^{\Gamma_m}p(x)= R^{K_t}p(x)$, $\inf\limits_{m\in\mathbb{N}} \varepsilon_x^{G_n}(R^{\Gamma_m}p)=\varepsilon_x^{G_n}(R^{K_t}p)$. We have $\varepsilon_x^{G_n}(R^{\Gamma_m}p)=R^{G_n}(R^{\Gamma_m}p)(x)=R^{\Gamma_m}p(x)$ and therefore $\varepsilon_x^{G_n}(R^{K_t}p) = R^{K_t}p(x)$. Using the preceding considerations we get:

$$\begin{aligned} R^K q(x) &= \inf_{n\in\mathbb{N}} R^{G_n}q(x) = \inf_{n\in\mathbb{N}} \varepsilon_x^{G_n}(q) = \inf_{n\in\mathbb{N}} \int_T \varepsilon_x^{G_n}(R^{K_t}p)dt \\ &= \inf_{n\in\mathbb{N}} \int_T R^{K_t}p(x)dt = q(x). \end{aligned}$$

By Theorem 2.6 it follows that q is regular. From Corollary 3.2 we deduce that $\mathrm{carr_f}\, q \subset K \subset A$ and obviously we have $\lambda(q) > \alpha$. To obtain a contradiction it remains to show that q is diffuse. If W is the kernel on X associated with q as in Theorem 3.4 then we have $W(\chi_{\{x\}}) \prec q = \sum\limits_{k=1}^{2^n} q_k$ where $q_k := \int_{T_k} R^{K_t}pdt$ and $(T_k)_{1\le k\le 2^n}$ is a partition of T such that $\int_{T_k} dt < \frac{1}{2^n}$. Obviously q_k is also regular and $\mathrm{carr_f}\, q_k$ is contained in the compact set $\bigcup\limits_{t\in T_k} K_t$. If k_0 is such that $x \in \bigcup\{K_t \mid t \in T_{k_0}\}$ then we have $W(\chi_{\{x\}}) \prec q_{k_0}$ and therefore $W(\chi_{\{x\}}) \le p/2^n$, $W(\chi_{\{x\}}) = 0$, completing the proof. □

Theorem 4.3. *Let A be an analytic subset of X and λ a finite measure on X. Then the following assertions are equivalent:*

(1) *The set A is λ-semipolar.*
(2) *For any $q \in (\mathcal{E}_{\mathcal{U}})_r$ with $\mathrm{carr_f}\, q \subset A$ we have $\lambda(q) = 0$.*
(3) *Any Ray compact subset of A is λ-semipolar.*
(4) *The function $W(\chi_A)$ is λ-negligible for any proper submarkovian resolvent $\mathcal{W} = (W_\alpha)_{\alpha>0}$ on X such that $\mathcal{E}_{\mathcal{W}} = \mathcal{E}_{\mathcal{U}}$.*
(5) *The set A is μ-negligible for any measure μ on X which does not charge the λ-semipolar sets.*

Proof. (3) $\Longrightarrow$ (2). Let $q \in (\mathcal{E}_{\mathcal{U}})_r$ be such that $\mathrm{carr}_f\, q \subset A$. Since

$$\lambda(q) = \lambda(W1) = \sup\{\lambda(W(\chi_K))| \ K \text{ Ray compact}, \ K \subset \mathrm{carr}_f\, q)\}$$

to show that $\lambda(q) = 0$ it will be sufficient to consider the case when $\mathrm{carr}_f\, q \subset K$ where K is a Ray compact subset of A and W is the kernel on X associated with q as in Theorem 3.4. By hypothesis $\mathrm{carr}_f\, q$ is λ-semipolar and therefore, being also a basic set, from Corollary 1.6 it follows that it is λ-polar. Hence $\lambda(q) = \lambda(B^{\mathrm{carr}_f\, q} q) = 0$.

(3) $\Longrightarrow$ (1). From (3) $\Longrightarrow$ (2) and Theorem 4.2 we get that A is λ-mince. Let now μ be a finite measure on X such that for any $M \in \mathcal{X}|_A$ we have

$$\mu(M) = 0 \Longleftrightarrow c^1_\lambda(M) = 0.$$

Then there exists an increasing sequence $(K_n)_{n\in\mathbb{N}}$ of compact subsets of A with $\mu(A \setminus \bigcup_{n\in\mathbb{N}} K_n) = 0$. Therefore $A \setminus \bigcup_{n\in\mathbb{N}} K_n$ is λ-polar and consequently A is λ-semipolar.

(2) $\Longrightarrow$ (3). Let us denote by Z the set of all points $x \in X$ such that $\{x\}$ is not thin at x. Theorem 1.8 implies that $Z \in \mathcal{X}$. If K is a compact subset of $A \cap Z$ then the $\mathcal{U}$-excessive function $q := B^K 1$ is Ray upper semi-continuous and $B^K q = q$. From Theorem 2.6 we get $q \in (\mathcal{E}_{\mathcal{U}})_r$ and $\mathrm{carr}_f\, q \subset K$. By hypothesis it follows that

$$0 = \lambda(q) = \lambda(B^K q)$$

and therefore K is λ-polar. Hence $A \cap Z$ is λ-polar and λ-negligible.

Let now K be a compact subset of A. By Proposition 1.9 there exists an analytic fine G_δ-subset K_0 of K such that $K \setminus K_0$ is semipolar and the set $K_0 \setminus b(K_0)$ is λ-polar and λ-negligible. From

$$K_0 \setminus K_0' \subset (K \cap Z) \cup (K_0 \setminus b(K_0))$$

we deduce that the set $K_0 \setminus K_0'$ is λ-polar and λ-negligible. Theorem 4.2 implies that K_0 is λ-mince. Since K_0 is a Baire space with the fine topology, by Lemma 1.10 we conclude that K_0 is λ-polar and λ-negligible. Hence K is λ-semipolar.

The implications (1) $\Longrightarrow$ (5) $\Longrightarrow$ (4) $\Longrightarrow$ (2) are immediate, using also Theorem 3.4. □

Remark. In Theorem 4.2 and Theorem 4.3 the assumption that X is semi-saturated is not necessary. If $q \in (\mathcal{E}_{\mathcal{U}})_r$ then $\mathrm{carr}_f\, q$ denotes the fine carrier of q considered as a regular excessive function on a semi-saturated extension of X.

Theorem 4.4. *Let A be an analytic subset of X and λ a finite measure on X. Then there exists a sequence $(s_n)_{n\in\mathbb{N}}$, $s_n \in (\mathcal{E}_{\mathcal{U}})_r$ with $\mathrm{carr}_f\, s_n \subset A$ for all $n \in \mathbb{N}$ and such that the set*

$$A \setminus \overline{\bigcup_{n\in\mathbb{N}} \mathrm{carr}_f\, s_n}^{f}$$

is λ-semipolar. If moreover A is a basic set then there exists $s \in (\mathcal{E}_{\mathcal{U}})_r$ with $\mathrm{carr}_f\, s \subset A$ and such that the set $A \setminus \mathrm{carr}_f\, s$ is λ-polar and λ-negligible.

Proof. Using Lemma 1.7 we get that there exists a sequence $(s_n)_{n\in\mathbb{N}}$ in $(\mathcal{E}_{\mathcal{U}})_r$ such that $\mathrm{carr_f}\, s_n \subset A$ for all $n \in \mathbb{N}$ and such that for any $s \in (\mathcal{E}_{\mathcal{U}})_r$ the set

$$[B^{\mathrm{carr_f}\, s} p > B^{\bigcup_{n\in\mathbb{N}} \mathrm{carr_f}\, s_n} p]$$

is λ-polar and λ-negligible, where p is a generator of $\mathcal{E}_{\mathcal{U}}$. Let now $q \in (\mathcal{E}_{\mathcal{U}})_r$ with $\mathrm{carr_f}\, q \subset A \setminus \overline{\bigcup_{n\in\mathbb{N}} \mathrm{carr_f}\, s_n}^f$. Since $\mathrm{carr_f}\, q \subset [B^{\mathrm{carr_f} q} p > B^{\bigcup_{n\in\mathbb{N}} \mathrm{carr_f}\, s_n} p]$ it follows that the set $\mathrm{carr_f}\, q$ is λ-polar and consequently

$$\lambda(q) = \lambda(B^{\mathrm{carr_f}\, q} q) = 0.$$

From Theorem 4.3 we get that the set $A \setminus \overline{\bigcup_{n\in\mathbb{N}} \mathrm{carr_f}\, s_n}^f$ is λ-semipolar.

Suppose now that A is a basic set. By Corollary 1.6 we may suppose that A is a Baire space with the fine topology. From the first part of the theorem there exists $s \in (\mathcal{E}_{\mathcal{U}})_r$ with $\mathrm{carr_f}\, s \subset A$ and such that the set $A \setminus \mathrm{carr_f}\, s$ is λ-semipolar and therefore λ-mince. On the other hand $A \setminus \mathrm{carr_f}\, s$ is a Baire space and

$$(A \setminus \mathrm{carr_f}\, s) \setminus (A \setminus \mathrm{carr_f}\, s)' \subset (A \setminus \mathrm{carr_f}\, s) \cap Z$$

where Z is the set of all points x such that $\{x\}$ is not thin at x (see Theorem 1.8). Since any compact subset of $(A \setminus \mathrm{carr_f}\, s) \cap Z$ is λ-semipolar and a basic set, from Corollary 1.6 it follows that it is λ-polar and λ-negligible. Hence the set $(A \setminus \mathrm{carr_f}\, s) \setminus (A \setminus \mathrm{carr_f}\, s)'$ is λ-polar and λ-negligible. From Lemma 1.10 we conclude that the set $A \setminus \mathrm{carr_f}\, s$ is λ-polar and λ-negligible. □

Remark. If A is an analytic fine closed subset of X and λ is a finite measure on X then there exists $q \in (\mathcal{E}_{\mathcal{U}})_r$ such that $\mathrm{carr_f}\, q \subset A$ and such that the set $A \setminus \mathrm{carr_f}\, q$ is λ-semipolar.

The assertion was also obtained by probabilistic methods by Dellacherie in [10]. From this result one can deduce using an analytical argumentation (see [15]) the existence of the λ-quasi-fine support for any measure which does not charge the λ-semipolar sets.

Corollary 4.5. *If F is an analytic basic subset of X and λ is a finite measure on X then there exists a bounded kernel W on X such that W satisfies the complete maximum principle, $Wf \in \mathcal{E}_{\mathcal{U}}$ for any bounded positive $\mathcal{X}$-measurable function f on X, $W(\chi_{X\setminus F}) = 0$ and such that the set $F \setminus \mathrm{carr_f}\, W1$ is λ-polar and λ-negligible.*

Remark. 1. The above corollary is the analytical version of Theorem 1.6 from [15] since the existence of the kernel W corresponds in the framework of Markov processes to the existence of a continuous additive functional having $\mathrm{carr_f}\, W1$ as its fine support.

2. The measure $\mu := \lambda \circ W$ (with the notations from Corollary 4.5) has the following properties: μ does not charge the λ-semipolar sets; $\mu(X \setminus F_\mu) = 0$, where $F_\mu := \mathrm{carr_f}\, W1$; if D is a fine closed $\mathcal{X}$-measurable subset of X with $\mu(X \setminus D) = 0$ then the set $F_\mu \setminus D$ is λ-polar and λ-negligible. In this way, in the terminology of

Fitzsimmons–Getoor, the set F_μ is the λ-quasi-fine support of the measure μ. So we derive from the above corollary Theorem 1.5 in [15].

References

[1] Azema, J., Une remarque sur les temps de retour. Trois applications. In: Séminaire de Probabilités VI, Lecture Notes in Math. 258, 35–50. Springer-Verlag, Berlin 1972.

[2] Beznea, L., Boboc, N., Duality and biduality for excessive measures. Rev. Roumaine Math. Pures Appl. 39 (1994), 419–438.

[3] — Excessive functions and excessive measures: Hunt's theorem on balayages, quasi-continuity. In: Classical and Modern Potential Theory and Applications, NATO ASI Series C 430, 77–92. Kluwer, Dordrecht 1994.

[4] — On the integral representation for excessive measures. Rev. Roumaine Math. Pures Appl. (to appear).

[5] Boboc, N., Bucur, Gh., Conuri convexe de funcţii continue pe spaţii compacte. Ed. Academiei, Bucureşti 1974.

[6] Boboc, N., Bucur, Gh., Cornea, A., Order and convexity in potential theory: *H*-cones. Lecture Notes in Math. 853. Springer-Verlag, Berlin 1981.

[7] Boboc N., Cornea A., Cônes convexes ordonnés. *H*-cônes et biadjoints des *H*-cônes. C.R. Acad. Sci. Paris, série A, 270 (1970), 1679–1682.

[8] Dellacherie, C., Capacités et processus stochastiques. Springer-Verlag, Berlin 1972.

[9] — Appendice a l'exposé precedent. In: Séminaire de Probabilités XVI, Lecture Notes in Math. 920, 29-40. Springer-Verlag, Berlin 1982.

[10] — Autour des ensembles semi-polaire. In: Sem. on Stoch. Processes 1987, 65–92. Birkhäuser, Boston 1988.

[11] Dellacherie, C., Feyel, D., Mokobodzki, G., Integrales de capacites fortement sous-additives. In: Séminaire de Probabilités XVI, Lecture Notes in Math. 920, 8–28. Springer-Verlag, Berlin 1982.

[12] Dellacherie, C., Meyer, P. A., Probabilités et Potentiel, Ch. I–IV, XII–XVI. Hermann, Paris 1975, 1987.

[13] Feyel, D., Représentation des fonctionnelles surmédianes. Z. Wahrscheinlichkeitstheorie verw. Gebiete 58 (1981), 183–198.

[14] — Sur la théorie fine du potentiel. Bull. Soc. Math. France 111 (1983), 41–57.

[15] Fitzsimmons, P. J., Getoor, R. K., A weak Quasi-Lindelöf Property and Quasi-fine Supports of Measures. Preprint, 1993.

[16] Hansen, W., Semi-polar sets and quasi-balayages. Math. Ann. 257 (1981), 495–517.

[17] Mokobodzki, G., Structure des cônes de potentiels. In: Séminaire Bourbaki 377, 1969/1970. Lecture Notes in Math. 180, 239–252. Springer-Verlag, Berlin 1971.

[18] — Pseudo-quotient de deux measures, application à la dualité. In: Séminaire de Probabilités VII, Lecture Notes in Math. 321, 318–321. Springer-Verlag, Berlin 1973.

Duality for semi-dynamical systems

Mounir Bezzarga and Gheorghe Bucur

Abstract. A remarkable measure associated with a semi-dynamical system, a dual resolvent of kernels and energy form are constructed here. An algebraic and topological characterization is given for a sufficiently large class of coexcessive functions. We answer a question from [5].

1991 Mathematics Subject Classification: 31D05, 58F99.

Introduction

Starting with a semi-dynamical system $(X, \mathcal{B}, \Phi, \omega)$ we construct so called "Lebesgue measure" $\wedge$ on $\mathcal{B}_0$ which charges any trajectory of X. We associate to Φ the canonical resolvent $\mathcal{V} = (V_\alpha)_\alpha$ on X_0 and a dual resolvent $\mathcal{V}^* = (V^*_\alpha)_{\alpha \geq 0}$ w.r. to the Lebesgue measure. The excessive functions with respect to $\mathcal{V}^*$, i.e., the coexcessive functions with respect to $\mathcal{V}$ are characterized as positive real $\mathcal{B}_0$-measurable functions on $X_0 = X \setminus \{\omega\}$ which vanish outside a countable union of trajectories and are totally increasing and totally continuous on the left. A natural energy form between $\mathcal{E}(\mathcal{V})$ and $\mathcal{E}(\mathcal{V}^*)$ is given, namely

$$[s, s^*] := \sup\{\int fs^* d\wedge \mid f \geq 0;\ V_0 f \leq s\} = \sup\{\int sg d\wedge \mid g \geq 0, V_0^* g \leq s^*\}.$$

It is shown that $\mathcal{E}(\mathcal{V}^*)$ is a naturally solid and closed in order from below convex subcone of the H-cone Exc $=$ Exc$(\mathcal{V})$ of all $\mathcal{V}$-excessive measures on $(X_0, \mathcal{B}_0)$. The above energy form coincides with the restriction to $\mathcal{E}(\mathcal{V}) \times \mathcal{E}(\mathcal{V}^*)$ of the well known energy form $E : \mathcal{E}(\mathcal{V}) \times \text{Exc} \longrightarrow \overline{\mathbb{R}}_+$ given by

$$E(s; \mu) = \sup\{\mu(f) \mid f \geq 0,\ V_0 f \leq s\}.$$

The cone $\mathcal{E}(\mathcal{V}^*)$ separates $\mathcal{E}(\mathcal{V})$ but generally $\mathcal{E}(\mathcal{V}^*) \neq$ Exc as it is proved by Theorem 2.7. This is an answer to a question from [5].

1. Lebesgue measure associated with a semi-dynamical system

In [3] we gave the notion of *semi-dynamical system* as being a separable measurable space $(X, \mathcal{B})$ with a distinguished point ω and a measurable map

$$\Phi : \mathbb{R}_+ \times X \longrightarrow X$$

having the following properties:

a) For any point $x \in X$ there exists an element $\rho(x)$ of $\overline{\mathbb{R}}_+$ such that $\Phi(t,x) \neq \omega$ for all $t \in [0, \rho(x))$ and $\Phi(t,x) = \omega$ for all $t \in \mathbb{R}_+$, $t \geq \rho(x)$. Moreover, $\rho(\omega) = 0$ and $\rho(x) > 0$ for every $x \in X$, $x \neq \omega$.

b) For any $s, t \in \mathbb{R}_+$ and any $x \in X$ we have

$$\Phi(s, \Phi(t,x)) = \Phi(s+t, x).$$

c) $\Phi(0,x) = x$ for all $x \in X$.

d) If $x, y \in X$ are such that $\Phi(t,x) = \Phi(t,y)$ for all $t \in \mathbb{R}_+$, $t > 0$, then $x = y$. The above semi-dynamical system is termed *transient* if there exists a sequence $(A_n)_{n \in \mathbb{N}}$ of measurable subsets of X such that $X \setminus \{\omega\} = \bigcup_n A_n$ and for any $x \in X \setminus \{\omega\}$ we have

$$\lambda(\{t \in \mathbb{R}_+ \mid \Phi(t,x) \in A_n\}) < \infty \;\; (\forall)\, n \in \mathbb{N}$$

where λ is the Lebesgue measure on $\mathbb{R}$.

It is proved that: *for any semi-dynamical system $(X, \mathcal{B}, \Phi, \omega)$ the family $\mathcal{V} = (V_\alpha)_{\alpha \geq 0}$ of kernels on the measurable space $(X_0, \mathcal{B}_0)$, where $X_0 := X \setminus \omega$, $\mathcal{B}_0 := \{A \in \mathcal{B} \mid A \subset X_0\}$, defined by*

$$V_\alpha f(x) = \int_{\mathbb{R}_+} e^{-\alpha t} \bar{f}(\Phi(t,x))dt \;\; (\forall)\, x \in X_0, \; (\forall)\, \alpha \in \mathbb{R}_+$$

is a submarkovian resolvent and it is proper if and only if the semi-dynamical system is transient. (For any positive $\mathcal{B}_0$-measurable function $f : X_0 \longrightarrow \mathbb{R}_+$ we have denoted by $\bar{f}$ the function on X equal f on X_0 and equal zero at ω.) For any point $x \in X_0$ we denote also by Γ_x the trajectory of x, i.e., $\Gamma_x = \{\Phi(t,x) \mid t \in [0, \rho(x))\}$. It is proved that $\Gamma_x \in \mathcal{B}_0$ *and the map* $\Phi_x : [0, \rho(x)) \longrightarrow \Gamma_x$ *is a measurable isomorphism between* $[0, \rho(x))$ *and* Γ_x *endowed with the trace measurable structures. The function* Φ_x *is defined by* $\Phi_x(t) = \Phi(t,x)$.

If $(X, \mathcal{B}, \Phi, \omega)$ is a semi-dynamical system then for any $x, y \in X_0$ we put $x \leq y$ if $y \in \Gamma_x$; it is proved that the above relation "$\leq$" is an order relation if the semi-dynamical system is transient. In this case the subset Γ_x is a totally ordered subset of X_0.

Moreover, the sets

$$\mathcal{O} := \{(x,y) \in X_0 \times X_0 \mid x \leq y\}, \; \mathcal{O}^o := \{(x,y) \in \mathcal{O} \mid x \neq y\}$$

are measurable if we endow $X_0 \times X_0$ with the product measurable structure. Also for any element A of $\mathcal{B}_0$ the set $\tilde{A}$ defined by $\tilde{A} := \{x \in X_0, \mid (\exists)\ a \ \in A;\ x \le a\}$ belongs to $\mathcal{B}_0$.

In the sequel, we shall suppose that $(X, \mathcal{B}, \Phi, \omega)$ is a transient semi-dynamical system.

Definition. A subset Γ of X_0 is called a *maximal trajectory* if Γ is totally ordered subset of X_0 with respect to the above order relation such that there is no minorant x_0 of Γ, $x_0 \in X_0 \setminus \Gamma$, and for any $x \in \Gamma$ we have $\Gamma_x \subset \Gamma$.

Proposition 1.1.
(a) *For any point $x \in X_0$ there is a maximal trajectory Γ such that $x \in \Gamma$.*
(b) *Any maximal trajectory Γ is a countable union of trajectories of X_0.*

Proof. (a) Let x be an arbitrary element of X_0 and let us consider the set $\mathcal{T}$ of all totally ordered non-empty subsets of X_0. We endow $\mathcal{T}$ with the following order relation

$$T', T'' \in \mathcal{T}, \qquad T' \le T'' \text{ if } T' \subset T''.$$

Obviously, the set $\mathcal{T}$ is inductively ordered by the above order relation. Hence for any element T of $\mathcal{T}$ there exists a maximal element Γ of $\mathcal{T}$ such that $T \geq \Gamma$. It is obvious that a maximal trajectory of X_0 is nothing else than a maximal element of $\mathcal{T}$. Hence Γ is a maximal trajectory. In particular, taking $T = \Gamma_x$, we get a maximal trajectory Γ such that $\Gamma_x \subset \Gamma$.

(b) Let Γ be a maximal trajectory. If Γ has a minimal element x_0 then $\Gamma = \Gamma_{x_0}$. In the contrary case we fix an element x_0 of Γ and for any $x \in \Gamma$, $x \in \Gamma$, $x \le x_0$, $x \neq x_0$ we denote by t_x the element of $[0, \rho(x))$ for which $\Phi(t_x, x) = x_0$. It is obvious that if $x', x'' \in \Gamma$ and $x' \le x_0$, $x'' \le x_0$ we have

$$x' \le x'' \Longleftrightarrow t_{x''} \le t_{x'}.$$

Let now $(x_n)_n$ be a sequence of Γ such that $x_n \le x_0$ for all $n \in \mathbb{N}$ and such that

$$\sup_n t_{x_n} = \sup\ \{t_x \mid x \in \Gamma, x \le x_0\}.$$

One can easily verify that $\Gamma = \bigcup\limits_n \Gamma_{x_n}$. □

Corollary 1.2. *Any maximal trajectory of X_0 is measurable.*

For any $x \in X_0$ we consider the measure λ_x on its trajectory Γ_x given by $\lambda_x(A) = \lambda(\Phi_x^{-1}(A))$ where $A \in \mathcal{B}_0$, $A \subset \Gamma_x$ and λ is the Lebesgue measure on the real interval $[0, \rho(x))$.

Proposition 1.3. *If x, y are two arbitrary elements of X_0 then for any $A \in \mathcal{B}_0$, $A \subset \Gamma_x \cap \Gamma_y$ we have $\lambda_x(A) = \lambda_y(A)$.*

Proof. If $A = \emptyset$ the assertion is obviously true. Suppose now that $\Gamma_x \cap \Gamma_y \neq \emptyset$ and for any $z \in \Gamma_x \cap \Gamma_y$ we denote by $t(x,z)$ (resp. $t(y,z)$) the element of $[0,\rho(x))$ (resp. $[0,\rho(y))$ such that $\Phi(t(x,z),x) = z$ (resp. $\Phi(t(y,z),y) = z$).

Since the set $\Gamma_x \cap \Gamma_y$ is closed w.r. to the topology τ_Φ on X_0 associated with the semi-dynamical system $(X, \mathcal{B}, \Phi, \omega)$ (see [3]) we deduce that the element z_0 of X_0 given by $z_0 = \Phi(\alpha, x)$, where $\alpha := \inf \{t(x,z) \mid z \in \Gamma_x \cap \Gamma_y\}$, is an element of $\Gamma_x \cap \Gamma_y$ and moreover $\Gamma_x \cap \Gamma_y = \Gamma_{z_0}$. If we put $\beta := t(y, z_0)$ then for any $r, s \in [0, \rho(z_0))$, $r < s$ we have

$$\Phi_{z_0}([r,s)) = \Phi_x([\alpha + r, \alpha + s)) = \Phi_y([\beta + r, \beta + s)).$$

Hence

$$\lambda_x(\Phi_{z_0}([r,s)) = s - r = \lambda_y(\Phi_{z_0}([r,s)))$$

and since the Borel structure on $[0,\rho(z_0))$ is generated by the set of all intervals $[r,s)$ with $0 \leq r < s < \rho(z_0)$ we deduce $\lambda_x(\Phi_{z_0}(B)) = \lambda_y(\Phi_{z_0}(B))$ for any Borel subset B of $[0,\rho(z_0))$. Since Φ_{z_0} is a measurable isomorphism between $[0,\rho(z_0))$ and Γ_{z_0} we get $\lambda_x(A) = \lambda_y(A)$ for any $A \in \mathcal{B}_0$, $A \subset \Gamma_x \cap \Gamma_y$. □

Corollary 1.4. *For any countable set* $(\Gamma_{x_n})_{n \in \mathbb{N}}$ *of trajectories of* X_0 *there exists a unique measure* λ_M *on the set* $M := \bigcup_n \Gamma_{x_n}$ *such that* $\lambda_M(A) = \lambda_{x_n}(A)$ *for any* $A \in \mathcal{B}_0$, $A \subset M \cap \Gamma_{x_n}$.

Definition. For any Borel subset A of X_0 we put

$$\wedge(A) = \sup_M \lambda_M(A \cap M)$$

where M runs the set of all countable union of trajectories of X_0. The map

$$A \longrightarrow \wedge(A)$$

is a measure on $\mathcal{B}_0$ which will be termed the *Lebesgue measure associated with the semi-dynamical system* $(X, \mathcal{B}, \Phi, \omega)$.

2. Duality and the energy form associated with a semi-dynamical system

Starting with a transient semi-dynamical system $(X, \mathcal{B}, \Phi, \omega)$, in [3] we have associated a suitable topology τ_Φ and we have characterized the set $\mathcal{E}$ of all excessive functions on X_0 w.r. to the submarkovian resolvent

$$\mathcal{V} = (V_\alpha)_{\alpha \geq 0}, \quad V_\alpha f(x) := \int_0^\infty e^{-\alpha t} \bar{f}(\Phi(t,x))dt$$

as being the set of all $\mathcal{B}_0$-measurable positive, real functions on X_0 which are decreasing (w.r. to the order relation previously defined) and continuous or only lower semi-continuous w.r. to this topology.

In this section we shall give a characterization for a remarkable part $\mathcal{E}^{*\sigma}$ of elements of the σ-dual of $\mathcal{E}$ and we introduce a convenient energy form between $\mathcal{E}$ and $\mathcal{E}^{*\sigma}$ using the above Lebesgue measure Λ associated with this semi-dynamical system.

Let $\mathcal{V} = (V_\alpha)_\alpha$ be the resolvent family of kernels associated with the semi-dynamical system $(X, \mathcal{B}, \Phi, \omega)$. We recall ([1], [6], [7], [8]) that a σ-finite measure μ on $\mathcal{B}_0$ is called *excessive* with respect to $\mathcal{V}$ if we have $\alpha V_\alpha \mu \leq \mu$ for all $\alpha \in \mathbb{R}_+$ and $\mu = \lim\limits_{\alpha \to \infty} \alpha V_\alpha \mu = \sup\limits_{\alpha \in \mathbb{R}_+} \alpha V_\alpha \mu$. We know that in our situation a σ-finite measure μ is $\mathcal{V}$-excessive iff $\alpha V_\alpha \mu \leq \mu$ for all $\alpha \in \mathbb{R}_+$ and, moreover (see [1], [6], [7], [8]), *a σ-finite measure μ is $\mathcal{V}$-excessive iff there exists a sequence $(\lambda_n)_{n \in \mathbb{N}}$ of positive measures on $\mathcal{B}_0$ such that the sequence $(V_0 \lambda_n)_n$ increases to μ.*

We recall also (see [6]) that a map $\theta : \mathcal{E} \to \overline{\mathbb{R}}_+$ is termed a *σ-H-integral* or a *coexcessive functional* if

a) $\theta(u+v) = \theta(u) + \theta(v)$, $\theta(\alpha u) = \alpha\theta(u)$ $(\forall)\ u, v \in \mathcal{E}$, $(\forall)\ \alpha \in \mathbb{R}_+$;

b) $u \leq v \Longrightarrow \theta(u) \leq \theta(v)$, $u, v \in \mathcal{E}$;

c) if $(u_n)_n$ increases to u we have $\theta(u) = \sup_n \theta(u_n)$; $u, u_n \in \mathcal{E}$;

d) for any $u \in \mathcal{E}$ there exists a sequence $(u_n)_n$ in $\mathcal{E}$, increasing to u such that $\theta(u_n) < \infty$ $(\forall)\ n \in \mathbb{N}$.

The family $\mathcal{E}^\sigma$ of all σ-H-integrals is an H-cone as it is shown in [6]. Recall that a σ-finite measure μ on $(X_0, \mathcal{B}_0)$ is $\mathcal{V}$-excessive if and only if the following relation holds

$$V_0 f \leq V_0 g \Longrightarrow \mu(f) \leq \mu(g)$$

for all $\mathcal{B}_0$-measurable positive real functions f, g (see [1], [6], [7], [8]). Hence for any $\mathcal{V}$-excessive measure μ the map $\theta_\mu : \mathcal{E} \longrightarrow \overline{\mathbb{R}}_+$ defined by $\theta_\mu(V_0 f) = \mu(f)$, for any positive $\mathcal{B}_0$-measurable function f, is a σ-H-integral and, conversely, for any σ-H-integral θ' on $\mathcal{E}$ the map $m_{\theta'}$ on $\mathcal{B}_0$ defined by $m_{\theta'}(A) = \theta'(V_0(1_A))$ is a $\mathcal{V}$-excessive measure for which $\theta_{m_{\theta'}} = \theta'$. In this way the map $\theta \longrightarrow \mu_\theta$ from $\mathcal{E}^\sigma$ to $\mathrm{Exc}(\mathcal{V})$ — the set of all excessive measures, becomes, as it was shown in [6], an isomorphism of H-cones. Moreover, the map $(s, \mu) \longrightarrow E(s, \mu)$ from $\mathcal{E} \times \mathrm{Exc}(\mathcal{V})$ into $\overline{\mathbb{R}}_+$ given by

$$E(s, \mu) := \sup \{\mu(f) \mid Vf \leq s;\ f : X_0 \to \overline{\mathbb{R}}_+\ ,\ f\mathcal{B}\text{-measurable function}\}$$

realizes a duality between $\mathrm{Exc}(\mathcal{V})$ and $\mathcal{E}$.

In the sequel, we shall denote by $\mathcal{B}_0^\sigma$ the family of all subsets M of $\mathcal{B}_0$ which are countable unions of maximal trajectories of X_0. We consider also the function Ψ from $X_0 \times X_0$ into $\mathbb{R}_+$ given by

$$\Psi(x, y) = \begin{cases} 0 & \text{if } x \not\leq y, \\ t & \text{if } x \leq y \text{ and } \Phi(t, x) = y. \end{cases}$$

Using the properties of a semi-dynamical system one can show that for any $M \in \mathcal{B}_0^\sigma$ the restriction of Ψ to $M \times M$ is measurable if we endow $M \times M$ with the trace Borel structure. Let $\mathcal{O}$ be the subset of all elements (x, y) from $X_0 \times X_0$ such that $x \leq y$. As we know, this set is measurable and therefore the function $G : X_0 \times X_0 \longrightarrow \mathbb{R}_+$ given by

$$G(x,y) = \begin{cases} 0 & \text{if } x \not\leq y, \\ 1 & \text{if } x \leq y, \end{cases}$$

is measurable. For any $M \in \mathcal{B}_0^\sigma$ and any positive $\mathcal{B}_0$-measurable function f on X_0 such that $f = 0$ on $X_0 \setminus M$ we shall denote

$$V_\alpha^* f(y) = \int_M e^{-\alpha\Psi(x,y)} G(x,y) f(x) d\wedge(x) = \int e^{-\alpha\Psi(x,y)} G(x,y) f(x) d\wedge(x)$$

for any $\alpha \in \mathbb{R}_+$ and any $x \in X_0$. Obviously $V_\alpha^* f$ is a positive $\mathcal{B}_0$-measurable function on X_0 and $V_\alpha^* f(x) = 0$ if $x \notin M$.

Definition [10]. Let f be a positive function defined on X_0. We say that f is totally increasing if for any $x \in X_0$ and any finite subset A of pairwise uncomparable elements of X_0 such that $a \leq x$ for any $a \in A$ we have

$$\sum_{a \in A} f(a) \leq f(x).$$

Notation. Let Γ be a maximal trajectory of X_0 and let x be an element of Γ. We shall denote

$$\Gamma_x := \{y \in \Gamma \mid y \geq x\} = \{\Phi(t,x) \mid t \in [0, \rho(x))\},$$
$$\Gamma_x^- := \{y \in \Gamma \mid y \leq x,\ y \neq x\} = \Gamma \setminus \Gamma_x.$$

Remark 2.1. If Γ', Γ'' are two trajectories (or two maximal trajectories) of X_0 then we have either $\Gamma' \cap \Gamma'' = \emptyset$ or there exists an $x \in X_0$ such that $\Gamma'_x = \Gamma''_x = \Gamma_x$ and $\Gamma_x^{'-} \cap \Gamma_x^{''-} = \emptyset$.

Definition. A totally increasing function $f = X_0 \longrightarrow \overline{\mathbb{R}}_+$ is called *totally continuous* on the left if for any $x \in X_0$ we have

$$f(x_0) = \sup_{\mathcal{T}} \left\{ \sum_{\Gamma \in \mathcal{T}} \sup_{y \in \Gamma_{x_0}^-} f(y) \right\}$$

where $\mathcal{T}$ runs over the set of all finite families of maximal trajectories of X_0 such that $\Gamma_x = \bigcap_{\Gamma \in \mathcal{T}} \Gamma$ and $\Gamma_x^{'-} \cap \Gamma_x^{''-} = \emptyset$ for all Γ', Γ'' of $\mathcal{T}$.

Remark 2.2. It follows from the definition of the Lebesgue measure $\wedge$ on $\mathcal{B}_0$ that for any positive $\mathcal{B}_0$-measurable function f on X_0 such that $f = 0$ on $X_0 \setminus M$ for some $M \in \mathcal{B}_0^\sigma$ the function $V_\alpha^* f$ is totally continuous on the left for any $\alpha \in \mathbb{R}_+$. Moreover $V_0^* f$ is totally increasing.

Theorem 2.3. *Let M be an element of $\mathcal{B}_0^g$ and let $\mathcal{F}(M)$ be the set of all positive, $\mathcal{B}_0$-measurable functions on X_0 vanishing on $X_0 \setminus M$. We have:*
(a) $V_\alpha^*(af+bg) = a\, V_\alpha^* f + bV_\alpha^* g \quad (\forall)\ a,b \in \mathbb{R}_+;\ f,g \in \mathcal{F}(M)$;
(b) $(V_\alpha^* f_n)_n \uparrow V_\alpha^* f$ *for any sequence* $(f_n)_n \subset \mathcal{F}(M)$, $(f_n)_n \uparrow f$;
(c) $V_\alpha^* f = V_\beta^* f + (\beta-\alpha)V_\alpha^* V_\beta^* f \quad (\forall)\ \alpha,\beta \in \mathbb{R}_+,\ f \in \mathcal{F}(M)$;
(d) $V_\alpha^* V_\beta^* f = V_\beta^* V_\alpha^* f \quad (\forall)\ \alpha,\beta \in \mathbb{R}_+,\ \alpha<\beta,\ f \in \mathcal{F}(M)$;
(e) $\int f V_\alpha^* g d\wedge = \int g V_\alpha f d\wedge \quad (\forall)\ f,g \in \mathcal{F}(M),\ \alpha \in \mathbb{R}_+$.

Proof. The properties (a), (b) follow directly from the definitions.
(e) We consider $f,g \in \mathcal{F}(M)$ and a positive real number α. We have

$$\int g(u)V_\alpha f(u) d\wedge(u) = \int g(u)\left(\int_{\mathbb{R}_+} e^{-\alpha t}\bar{f}(\Phi(t,u))dt\right) d\wedge(u)$$

$$= \int g(u)\int_0^{\rho(u)} e^{-\alpha t} f(\Phi(t,u))dtd\wedge(u)$$

$$= \int g(u)\int e^{-\alpha\Phi_u^{-1}(\Phi(t,u))} f(\Phi(t,u))dtd\wedge(u)$$

$$= \int g(u)\left(\int e^{-\alpha\Psi(u,v)}G(u,v)f(v)d\wedge(v)\right)d\wedge(u)$$

$$= \int f(v)\left(\int e^{-\alpha\Psi(u,v)}G(u,v)g(u)d\wedge(u)\right)d\wedge(v) = \int f(v)V_\alpha^* g(v)d\wedge(v).$$

As for the assertions (c) and (d), we remark that from the property (e) it follows that for any $f \in \mathcal{F}(M)$ and any $\alpha,\beta \in \mathbb{R}_+$, $\alpha<\beta$ we have

$$V_\alpha^* f = V_\beta^* f + (\beta-\alpha)V_\alpha^* V_\beta^* f = V_\beta^* f + (\beta-\alpha)V_\beta^* V_\alpha^* f \quad \wedge\text{-a.s. on } X_0.$$

On the other hand, the functions $V_\alpha^* f$, $V_\beta^* f + (\beta-\alpha)V_\alpha^* V_\beta^* f$, $V_\beta^* f + (\beta-\alpha)V_\beta^* V_\alpha^* f$ being totally continuous on the left, it follows that they coincide everywhere on X_0. □

Corollary 2.4. *The set $\mathcal{E}^{*\sigma}$ of all positive $\mathcal{B}_0$-measurable functions s^* on X_0, $s^*<\infty$, $s^*=0$ on $X_0\setminus M$ for some element M from $\mathcal{B}_0^g$ and such that s is excessive with respect to the resolvent $\mathcal{V}^* = (V_\alpha^*)_{\alpha\geq 0}$, i.e., $\alpha V_\alpha^* s^* \leq s^*$ for all $\alpha \in \mathbb{R}_+$ and $\sup\limits_{\alpha\in\mathbb{R}_+} \alpha V_\alpha^* s^* = s^*$, is an H-cone with respect to the pointwise algebraic operations and order relation. Moreover, the map $(s,s^*) \longrightarrow [s,s^*]$ from $\mathcal{E}\times\mathcal{E}^{*\sigma}$ into $\overline{\mathbb{R}}_+$ is a duality between $\mathcal{E}$ and $\mathcal{E}^{*\sigma}$ where*

$$[s,s^*] = \sup\left\{\int fs^* d\wedge \mid f\geq 0,\ \mathcal{B}_0\text{-measurable},\ V_0 f \leq s\right\} =$$

$$= \sup\left\{\int gsd\wedge \mid g \in \mathcal{F}(M),\ V_0^* g \leq s^*\right\},$$

i.e., we have

(a) *For any* $s^* \in \mathcal{E}''^{\sigma}$ *the map* $s \longrightarrow [s, s^*]$ *is an* H*-integral on* $\mathcal{E}$ *and for any* $s \in \mathcal{E}$ *the map* $s^* \longrightarrow [s, s^*]$ *is an* H*-integral on* $\mathcal{E}''^{\sigma}$ *(see [2]).*
(b) *If* $s \in \mathcal{E}$, $s \neq 0$ *then there exists* $s^* \in \mathcal{E}^{*\sigma}$ *such that* $[s, s^*] > 0$.

The proof is standard and is based on the fact that for any $s^* \in \mathcal{E}^{*\sigma}$ there exists a sequence $(f_n)_n$ from $\mathcal{F}(M)$, for some $M \in \mathcal{B}_0^{\sigma}$, such that $(V_0^* f_n)_n \uparrow s^*$. Also we use the fact that for any increasing family $(s_i^*)_{i\in I}$ of $\mathcal{E}^{*\sigma}$ which is dominated by an element of $\mathcal{E}^{*\sigma}$ there exists an increasing sequence $(s_{i_n}^*)_{n\in N}$ such that $\sup\limits_{i\in I} s_i^* = \sup\limits_{n\in \mathbb{N}^*} s_{i_n}^*$.

Theorem 2.5. *Let* $M \in \mathcal{B}_0^{\sigma}$ *and* $f \in \mathcal{F}(M)$ *be such that* $f < \infty$ *on* X_0. *Then* $f \in \mathcal{E}^{*\sigma}$ *if and only if* f *is totally increasing and totally continuous on the left on* X_0.

Proof. We have already remarked that any element of $\mathcal{E}^{*\sigma}$ is totally increasing and totally continuous on the left on X_0. Suppose now that $f \in \mathcal{F}(M)$ for some $M \in \mathcal{B}_0^{\sigma}$, $f < \infty$ is a totally increasing and totally continuous on the left on X_0.

We shall prove that $\alpha V_\alpha^* f \leq f$ for any $\alpha \in \mathbb{R}_+$. Let $(\Gamma_n)_{n\in\mathbb{N}}$ be a countable family of maximal trajectories on X_0 such that $M = \bigcup\limits_n \Gamma_n$ and let x be a point of M. Obviously we have

$$\alpha V_\alpha^* f(x) = \sup_{k\in\mathbb{N}} \int_{\bigcup\limits_{n\leq k} \Gamma_n} \alpha e^{-\alpha\Psi(y,x)} G(y,x) f(y) d\wedge(y)$$

and we shall prove inductively that

$$\int_{M'} \alpha e^{\alpha\Psi(y,x)} G(y,x) f(y) d\wedge(y) \leq f(x)$$

where $M' \subset M$, $M = \bigcup\limits_{i\in J} \Gamma_i$, $J \subset \mathbb{N}$, card $J = k$.

Indeed, if $k = 1$ then we have

$$\int_{\Gamma} \alpha e^{-\alpha\Psi(y,x)} G(y,x) f(y) d\wedge(y) = \int_{\Gamma_x^-} \alpha e^{-\alpha\Psi(y,x)} f(y) d\wedge(y)$$

$$\leq \int_{\Gamma_x^-} \alpha e^{-\alpha\Psi(y,x)} f(x) d\wedge(y) = f(x) \int_{\Gamma_x^-} \alpha e^{-\alpha\Psi(y,x)} d\wedge(y)$$

$$= f(x) \sup_n \int_{\Gamma_{x_n}\setminus\Gamma_x} \alpha e^{-\alpha\Psi(y,x)} d\wedge(y) \leq f(x) \sup_n \int_{\Gamma_{x_n}\setminus\Gamma_x} \alpha e^{-\alpha[\Phi_{x_n}^{-1}(x)-\Phi_{x_n}^{-1}(y)]} d\wedge(y)$$

$$= f(x) \sup_n \int_0^{\Phi_{x_n}^{-1}(x)} \alpha e^{-\alpha[\Phi_{x_n}^{-1}(x)-t]} dt = f(x) \sup_n \int_0^{\Phi_{x_n}(x)} \alpha e^{-\alpha u} du \leq f(x)$$

where $(x_n)_n$ is a decreasing sequence on Γ such that $\Gamma = \bigcup\limits_n \Gamma_{x_n}$.

We suppose now that

$$\int_{M'} \alpha e^{-\alpha\Psi(y,x)} G(y,x) f(y) d\wedge(y) \leq f(x)$$

for any $M' \in \mathcal{B}_0^\sigma$, $M' \subset M$, $M' = \bigcup_{i\in J'} \Gamma_i$ and card $J' \leq k$. Let $M'' = \bigcup_{i\in J''} \Gamma_i$ with card $J'' = k+1$ and let us suppose that $x \in \bigcap_{i\in J''} \Gamma_i$. Two posibilities arise:

a) There exists a point $y \in M''$, $y \leq x$, $y \neq x$ and there exist at least two trajectories Γ_{i_1}, Γ_{i_2}; $i_1, i_2 \in J''$ such that $\Gamma_{i_1} \cap \Gamma_{i_2} = \Gamma_y$.

b) For any $i_1, i_2 \in J''$ we have $\Gamma_{i_1} \cap \Gamma_{i_2} = \Gamma_x$.

In the second situation we have, for any $x_i \in \Gamma_i$, $x_i \neq x$, $i \in J''$, $f(x) \geq \sum_{i\in J''} f(x_i)$ and using the preceding considerations

$$f(x_i) \geq \int_{\Gamma_i} \alpha e^{-\alpha\Psi(y,x_i)} G(y,x_i) f(y) d\wedge(y) \qquad (\forall)\ i \in J''.$$

Hence we get

$$f(x) \geq \sum_{i\in J''} \sup_{x_i\in\Gamma_i\setminus\Gamma_x} f(x_i) = \sum_{i\in J''} \int_{\Gamma_i} \alpha e^{-\alpha\Psi(y,x)} G(y,x) f(y) d\wedge(y),$$

$$f(x) \geq \int \alpha e^{-\alpha\Psi(y,x)} G(y,x) f(y) d\wedge(y).$$

In the case a) we denote by x_0 the smallest element of the set $\bigcap_{i\in J''} \Gamma_i$ and we suppose that there exist at least two trajectories Γ_{i_1}, Γ_{i_2} and $z \leq x_0$, $z \neq x_0$ such that $z \in \Gamma_{i_1} \cap \Gamma_{i_2}$. So the set J'' may be decomposed in the form $J'' = J_1 \cup J_2 \cup \cdots \cup J_p$ where $J_i \cap J_j = \emptyset$ if $i \neq j$, card $J_i \geq 1$ for any $i = 1, 2, \cdots, p$, $p < k+1$. There exist pairwise uncomparable y_i, $y_i \in \bigcap_{j\in J_i} \Gamma_j$, $i = 1, \ldots, p$. We may also suppose that

$$(\Gamma_{x_0} \setminus \Gamma_{y_i}) \cap (\Gamma_{x_0} \setminus \Gamma_{y_j}) = \emptyset \quad (\forall)\ i,j \leq p,\ i \neq j,$$

and we have, using the induction hypothesis,

$$\int_{M''} \alpha e^{-\alpha\Psi(y,x_0)} G(y,x) f(y) d\wedge(y)$$

$$= \sum_{i=1}^{p} \sup\left\{ \int_{\bigcup_{j\in J_i} (\Gamma_j \setminus \Gamma_{y_i'})} \alpha e^{-\alpha\Psi(y,y_i')} f(y) d\wedge(y);\ y_i \leq y_i' \leq x_0,\ y_i' \neq x_0 \right\}$$

$$\leq \sum_{i=1}^{p} \sup\{f(y_i') \mid y_i \leq y_i' \leq x_0,\ y_i' \neq x_0\} \leq f(x_0).$$

In the same way we may show that

$$\int_{M''} \alpha e^{-\alpha\Psi(y,x)} G(y,x) f(y) d\wedge(y) \leq f(x).$$

Hence the given function f is supermedian with respect to the resolvent $\mathcal{V}^* = (V_\alpha^*)_{\alpha\geq 0}$ and therefore we have $f(x) = \lim_{\alpha\to\infty} \alpha V_\alpha^* f(x)$ $\mathcal{V}$-a.s. on X_0. Since for any subset $A \in \mathcal{B}_0^g$, $A \subset M$ we have $V_\alpha^*(1_A) = 0$ if and only if $\wedge(A) = 0$ we deduce that the function f coincides $\wedge$-a.s. on X_0 with the excessive function $\hat{f}$ on X_0 given by $\hat{f}(x) = \lim_{\alpha\to\infty} \alpha V_\alpha f(x)$. Since f and $\hat{f}$ are totally increasing and totally continuous on the left on X_0 we get $f = \hat{f}$ on X_0. □

Theorem 2.6.

(a) *For any $s^* \in \mathcal{E}^{*\sigma}$ the measure $s^*\wedge$ is excessive w.r. to the resolvent $\mathcal{V} = (V_\alpha)_{\alpha\geq 0}$. For $s_1^*, s_2^* \in \mathcal{E}^{*\sigma}$ we have $s_1^* \leq s_2^* \Longleftrightarrow s_1^*\wedge \leq s_2^*\wedge$.*

(b) *If $f : X_0 \longrightarrow \overline{\mathbb{R}}_+$ is $\mathcal{B}_0$-measurable and the measure $f \cdot \wedge$ is excessive w.r. to $\mathcal{V}$ then there exists $s^* \in \mathcal{E}^{*\sigma}$ such that $f = s^*$ $\wedge$-a.s. on X_0.*

(c) *If we denote by* Exc$(\mathcal{V})$ = Exc *the H-cone of all excessive measures on $(X_0, \mathcal{B}_0)$ with respect to $\mathcal{V}$ then the set $\{s^*\wedge \mid s^* \in \mathcal{E}^{*\sigma}\}$, denoted also by $\mathcal{E}^{*\sigma}$, is naturally solid and closed in order from below convex subcone of* Exc.

Proof. (a) Let s^* be an element of $\mathcal{E}^{*\sigma}$ and let f be a $\mathcal{B}_0$-measurable and positive function on X_0. We have

$$\int (\alpha V_\alpha f) s^* d\wedge = \int f \alpha V_\alpha^* s^* d\wedge \leq \int f s^* d\wedge \qquad (\forall)\ \alpha > 0$$

and therefore the σ-finite measure $s^*\wedge$ is excessive.

If s_1^*, s_2^* are two elements of $\mathcal{E}^{*\sigma}$ we have obviously $s_1^* \leq s_2^* \Longrightarrow s_1^*\wedge \leq s_2^*\wedge$. Conversely, if $s_1^*\wedge \leq s_2^*\wedge$ then $s_1^* \leq s_2^*$ $\wedge$-a.s. on X_0 and therefore $\alpha V_\alpha^* s_1^* \leq \alpha V_\alpha^* s_2^*$ on X_0 for all $\alpha \in \mathbb{R}_+$. Hence $s_1^* \leq s_2^*$ on X_0.

(b) Let $f : X_0 \longrightarrow \overline{\mathbb{R}}_+$ be a $\mathcal{B}_0$-measurable function such that the measure $f\wedge$ is excessive. Hence the measure $f\wedge$ is σ-finite and therefore $f = 0$ $\wedge$-a.s. on $X_0 \setminus M$ for some subset $M \in \mathcal{B}_0^g$ and $f < \infty$ $\wedge$-a.s. on M. If $g : X_0 \longrightarrow \mathbb{R}_+$ is $\mathcal{B}_0$-measurable we have

$$\int g \alpha V_\alpha^* f d\wedge = \int (\alpha V_\alpha g) f d\wedge \leq \int g f d\wedge \qquad (\forall)\ \alpha > 0$$

and therefore, g being arbitrary, $\alpha V_\alpha^* f \leq f$ $\wedge$-a.s. on M.

Using standard arguments one can show that there exists $s^* \in \mathcal{E}^{*\sigma}$ such that $f = s^*$ $\wedge$-a.s. on X_0.

(c) Let $(s_i^*)_i$ be an increasing family of elements from $\mathcal{E}^{*\sigma}$ such that the measure $\mu : \mathcal{B}_0 \longrightarrow \overline{\mathbb{R}}_+$ given by

$$\mu(A) = \sup_{i\in I} (s_i^*\wedge)(A)$$

is excessive. Then there exists a countable subset $(i_n)_{n\in N}$ of I such that the sequence $(s_{i_n}^*\wedge)_n$ increases to μ. Using (a) and (b) we deduce that the sequence

$(s^*_{i_n})_n$ is increasing to an element s^* of $\mathcal{E}^{*\sigma}$. Hence $\mu = s^*\wedge$ and therefore $\mathcal{E}^{*\sigma}$ is a closed in order from below convex subcone of Exc. Let now $\mu \in$ Exc and $s^* \in \mathcal{E}^{*\sigma}$ be such that $\mu \leq s^*\wedge$. We deduce that the measure μ is carried by a subset $M \in \mathcal{B}_0^\sigma$ and therefore there exists $f : X_0 \longrightarrow \overline{\mathbb{R}}_+$, $\mathcal{B}_0$-measurable such that $\mu = f\wedge$. Using the point (b) we get $\mu = t^*\wedge$ for some $t^* \in \mathcal{E}^{*\sigma}$. Hence $\mathcal{E}^{*\sigma}$ is naturally solid convex subcone of Exc. □

Theorem 2.7. *The following assertions are equivalent:*
(a) *$\mathcal{E}$ is a standard H-cone.*
(b) *$\mathcal{E}^{*\sigma}$ is a standard H-cone.*
(c) *The H-cone $\mathcal{E}^{*\sigma}$ has a unit.*
(d) *$X_0 \in \mathcal{B}_0^\sigma$ or, equivalently, the measure $\wedge$ is σ-finite.*

Proof. The relation (b) $\Longrightarrow$ (c) is obvious. For the assertion (c) $\Longrightarrow$ (a) we consider a unit element u^* of the H-cone $\mathcal{E}^{*\sigma}$. We deduce that $u^* > 0$ on X_0. Indeed, let x_0 be an arbitrary element of X_0. If x_0 is minimal with respect to the order relation on X_0 then the function s^* equal to 1 on Γ_{x_0} and equal to 0 on $X_0 \setminus \Gamma_{x_0}$ is an element of $\mathcal{E}^{*\sigma}$ and therefore $u^*(x_0) > 0$. If the point x_0 is not minimal then we consider $y_0 \in X_0$, $y_0 \leq x_0$, $y_0 \neq x_0$. The function s^* equal to 1 on $\Gamma'_{y_0} \setminus \{y_0\}$ and equal to 0 outside this set is also an element of $\mathcal{E}^{*\sigma}$ and again $s^*(x_0) = 1$, i.e., $u^*(x_0) > 0$.

Hence $u^* > 0$ on X_0. Since $u^* = 0$ on $X_0 \setminus M$ for some subset $M \in \mathcal{B}_0^\sigma$ we deduce that $X_0 \in \mathcal{B}_0^\sigma$ and therefore $\wedge$ is a σ-finite measure.

Since the resolvent $\mathcal{V} = (V_\alpha)_\alpha$ is absolutely continuous with respect to $\wedge$ we deduce that $\mathcal{E}$ is a standard H-cone.

(a) $\Longrightarrow$ (b). If $\mathcal{E}$ is standard then, using [5] and the fact that the subcone $\mathcal{E}^{*\sigma}$ of Exc separates $\mathcal{E}$ we deduce from the previous theorem, the equality Exc $= \mathcal{E}^{*\sigma}$ and therefore $\mathcal{E}^{*\sigma}$ is a standard H-cone.
The relation (a) $\Longleftrightarrow$ (d) is obvious. □

References

[1] Beznea, L., Boboc, N., Duality and biduality for excessive measures. Preprint Series of Inst. of Math. of the Romanian Academy, No. 15/1992.

[2] Boboc, N., Bucur, Gh., Cornea A., Order and Convexity in Potential Theory: H-cones Lecture Notes in Math. 853, Springer-Verlag, Berlin 1981.

[3] Bezzarga, M., Bucur, Gh., Théorie du Potentiel pour les Systemes semi-dynamiques. Rev. Roum. Math. Pures Appl. 39 (1994), 439–456.

[4] Bhatia, N. P., Hajek, O., Local Semi-Dynamical Systems. Lecture Notes in Math. 90, Springer-Verlag, Berlin 1969.

[5] Bucur, Gh., Grecea, V., Zeddinni N., Inductive and Projective limits of H-cones; Duality. Preprint Series of Inst. of Math. of the Romanian Academy, 1994.

[6] Cornea, A., Licea, G., Order and Potential Resolvent Families of Kernels. Lecture Notes in Math. 494, Springer-Verlag, Berlin 1975.

[7] Dellacherie, C., Meyer, P.A., Probabilités et Potentiel, vol. XII–XIV. Hermann, Paris 1987.

[8] Getoor, R. K., Excessive Measures. Birkhäuser, Boston 1990.

[9] Hmissi, M., Cones de Potentiels stables par produit et systémes semi-dynamiques. Exposition. Math 7 (1989), 265–273.

Removable singularities for Hardy spaces in subdomains of $\mathbb{C}$

Anders Björn

Abstract. In this paper we give some new results concerning removable singularities for Hardy spaces of analytic functions. A new result for $p < 1$ concerning relations with Riesz capacities (and Hausdorff dimension) is given. New results for singularities lying on rectifiable chord-arc curves and for singularities being Cantor sets are also given.

1991 Mathematics Subject Classification: Primary: 30D55, Secondary: 30B40, 30C85, 32A35, 32D20.

1. Introduction

The problem was first posed by Michel Parreau, [10] and [11], in 1950. For a more detailed discussion, and full proofs of all new results, we refer the interested reader to Björn [1]. See also Björn [2] for some open problems concerning relations with Hausdorff measures.

2. Hardy spaces

Let $\mathbb{S} = \mathbb{C} \cup \{\infty\}$ be the Riemann sphere. Let $\Omega \subset \mathbb{S}$ or $\Omega \subset \mathbb{C}^n$ be an open connected set (a domain) and $0 < p < \infty$.

Definition 1. Let

$$\begin{aligned} A(\Omega) &= \{f : f \text{ is analytic in } \Omega\},\\ H^p(\Omega) &= \{f \in A(\Omega) : |f|^p \text{ has a harmonic majorant in } \Omega\},\\ H^\infty(\Omega) &= \Big\{f \in A(\Omega) : \sup_{z\in\Omega} |f(z)| < \infty\Big\}. \end{aligned}$$

Let $a \in \Omega$, the norming-point, be fixed and let

$$\|f\|^p_{H^p(\Omega)} = (\text{least harmonic majorant of } |f|^p)(a), \quad f \in H^p(\Omega).$$

Remarks.

(i) There always exists a pointwise least harmonic majorant of $|f|^p$ if $f \in H^p(\Omega)$.

(ii) For $1 \le p \le \infty$, $H^p(\Omega)$ is a Banach space. The space $H^2(\Omega)$ is a Hilbert space.
(iii) Different a give equivalent norms.
(iv) For the unit disc, $\mathbb{D} \subset \mathbb{C}$, we have, with $a = 0$,

$$\|f\|^p_{H^p(\mathbb{D})} = \sup_{0<r<1} \frac{1}{2\pi} \int_0^{2\pi} |f(re^{i\theta})|^p \, d\theta = \lim_{r \to 1-} \frac{1}{2\pi} \int_0^{2\pi} |f(re^{i\theta})|^p \, d\theta,$$

the classical definition.
(v) If $n = 1$, these spaces are conformally invariant, i.e., invariant under conformal mappings.
(vi) For the upper half plane these spaces are conformally equivalent with the ones in $\mathbb{D}$, and thus are *not* the standard ones.
(vii) Let $\omega(E; \Omega, a)$ denote the harmonic measure for E with respect to the domain Ω, evaluated at $a \in \Omega$. If $\overline{\Omega}_k \subset \Omega$, $\Omega_1 \subset \Omega_2 \subset \cdots$, $\Omega = \bigcup_{k=1}^{\infty} \Omega_k$ and $\partial\Omega_k$ is regular for the Dirichlet problem associated with the Laplace operator, then

$$\|f\|^p_{H^p(\Omega)} = \lim_{k \to \infty} \int_{\partial\Omega_k} |f(\zeta)|^p \omega(d\zeta; \Omega_k, a).$$

(viii) By applying Jensen's inequality to (vii), we see that, for every f, the function $p \mapsto \|f\|_{H^p(\Omega)}$ is non-decreasing.
(ix) The inclusions $H^p(\Omega) \supset H^q(\Omega) \supset H^\infty(\Omega)$ hold for $0 < p < q < \infty$.

3. Removable singularities

Definition 2. Let $E \subset \Omega$ be relatively closed and such that $\Omega \setminus E$ is non-empty and connected. In the literature there are three definitions of when E is a *removable singularity* for $H^p(\Omega \setminus E)$.
I If $H^p(\Omega \setminus E) \subset A(\Omega)$.
II If $H^p(\Omega \setminus E) = H^p(\Omega)$ (as sets).
III If $\|f\|_{H^p(\Omega \setminus E)} = \|f\|_{H^p(\Omega)}$ for all $f \in H^p(\Omega \setminus E)$.

Remarks.
(i) In the first case we mean that every $f \in H^p(\Omega \setminus E)$ can be analytically continued to the whole of Ω.
(ii) In the last case, where we use the same norming-point, E is called isometrically removable.
(iii) For H^∞ the three definitions coincide.
(iv) In the plane case removable singularities have to be totally disconnected.

Definition 3. $H^p(\Omega)$ is *trivial* if $H^p(\Omega) = \{f : f \text{ constant}\}$.

Theorem 4 (Conway–Dudziak–Straube, 1987 [3]). *Let $0 < p < \infty$ and $\Omega \subset \mathbb{C}^n$ be a domain. Let E be a relatively closed subset of Ω such that $\Omega \setminus E$ is non-empty* *and connected. Then E is an isometrically removable singularity for $H^p(\Omega \setminus E)$ if and only if* $\operatorname{cap}(E) = 0$ *or $H^p(\Omega \setminus E)$ is trivial.*

Remarks.

(i) By cap we denote the usual capacity for harmonic functions in $\mathbb{R}^{2n} (= \mathbb{C}^n)$, i.e., the logarithmic capacity if $n = 1$ and the Newtonian capacity if $n > 1$.

(ii) The condition is almost independent of p and Ω, the dependence lying only in the triviality of $H^p(\Omega \setminus E)$.

(iii) If $H^p(\Omega \setminus E)$ is trivial, then II-removability is the same as isometrical removability.

(iv) Michel Parreau [11] was the first to prove a version of this theorem in 1952 when he proved that closed polar subsets of a Riemann surface are removable for H^p. For further historical remarks, see Björn [1].

Theorem 5 (Hartogs). *Let Ω be an open set in $\mathbb{C}^n$, $n > 1$, and let K be a compact subset of Ω such that $\Omega \setminus K$ is connected. For every $f \in A(\Omega \setminus K)$, one can then find $F \in A(\Omega)$ so that $f = F$ in $\Omega \setminus K$.*

Remarks.

(i) Thus, all compact subsets are I-removable, if $n > 1$.

(ii) It is not obvious, but can be proved, that if $f \in H^p(\Omega \setminus K)$, then $F \in H^p(\Omega)$, and hence all compact subsets are also II-removable, if $n > 1$.

4. The plane case

For the rest of this paper we study the plane case for compact sets. We will assume that K is a totally disconnected compact set and hence that $\Omega \setminus K$ is a domain if Ω is a domain. Dennis Hejhal proved the following important theorem.

Theorem 6 (Hejhal, 1971 [7] and 1973 [8]). *Let $0 < p < \infty$. Then the following are equivalent:*

(a) *K is II-removable for $H^p(\Omega \setminus K)$ for all domains $\Omega \supset K$,*

(b) *K is II-removable for $H^p(\Omega \setminus K)$ for some domain $\Omega \supset K$,*

(c) *K is I-removable for $H^p(\Omega \setminus K)$ for all domains $\Omega \supset K$,*

(d) *K is I-removable for $H^p(\Omega \setminus K)$ for some domain $\Omega \supset K$.*

Remarks.

(i) Parts (c) and (d) were not in Hejhal's version, but follow almost trivially from his proof. Hejhal only considered II-removability in his papers.

(ii) A corollary is that I-removability and II-removability are the same for compact sets and do not depend on Ω (as long as $\Omega \supset K$). In general, i.e., for relatively closed subsets, it is not true that I-removability and II-removability are the same.

Finite disjoint unions of compact removable sets are also removable. However, Hejhal [8] gave an example of two sets K_1, K_2 removable for H^1 but with $K_1 \cup K_2$ not removable for H^1. Such examples exist at least for $1 \le p < \infty$, see Björn [1].

In 1978 Morisuke Hasumi [5] proved that different p give different compact sets of removability. The sets he constructed for his proof were of zero Hausdorff dimension. As a consequence, there exists a zero dimensional set which is not removable for any H^p, $p < \infty$.

Definition 7. Let $K \subset \mathbb{C}(=\mathbb{R}^2)$ be a compact set and let $\mu \in \mathcal{M}^+(K)$, the set of positive Borel measures with support on K. We define, for $0 < \alpha < 2$, the *Riesz α-potential*

$$U_\alpha^\mu(z) = \int_K \frac{d\mu(w)}{|z-w|^{2-\alpha}}, \quad z \in \mathbb{C},$$

and the *Riesz α-capacity*

$$\operatorname{cap}_\alpha(K) = \sup\{\mu(K) : \mu \in \mathcal{M}^+(K) \text{ and } U_\alpha^\mu(z) \le 1 \text{ for all } z \in \mathbb{C}\}.$$

Remark. Let $K \subset \mathbb{C}$ be compact. Then

$$\begin{aligned} \dim(K) > 2-\alpha \quad &\Longrightarrow \quad \operatorname{cap}_\alpha(K) > 0, \\ \Lambda_{2-\alpha}(K) < \infty \quad &\Longrightarrow \quad \operatorname{cap}_\alpha(K) = 0, \end{aligned}$$

where Λ_d denotes d-dimensional Hausdorff measure.

Theorem 8. *Let $0 < p \le \infty$, $\alpha = \max(2-p, 1)$, and assume that $\operatorname{cap}_\alpha(K) > 0$ for $K \subset \mathbb{C}$ compact. Then K is not removable for H^p.*

Proof. We start with the case $p \ge 1$. As $\operatorname{cap}_1(K) > 0$, there exists $\mu \in \mathcal{M}^+(K)$, $\mu \ne 0$, with $U_1^\mu(z) \le 1$ for all $z \in \mathbb{C}$. Therefore, the non-constant analytic function

$$f(z) = \int_K \frac{d\mu(w)}{z-w}$$

is bounded on $\mathbb{S} \setminus K$ and we are done.

Assume now instead that $p < 1$. Let $\Omega_1 \subset \Omega_2 \subset \cdots$ be domains with regular boundaries such that $\overline{\Omega}_k \subset \Omega$ and $\mathbb{S} \setminus K = \Omega = \bigcup_{k=1}^\infty \Omega_k$. Let $a \in \Omega_1$ be the norming-point and let $\omega_k = \omega(\,\cdot\,; \Omega_k, a)$ and $g_w(z) = 1/(z-w)$. As $\operatorname{cap}_\alpha(K) > 0$, there exists $\mu \in \mathcal{M}^+(K)$, $\mu \ne 0$, with $U_\alpha^\mu(z) \le 1$ for all $z \in \mathbb{C}$. Then, using (vii) in the remark following Definition 1, the monotone convergence theorem and Fubini's theorem,

$$\begin{aligned} \int_K \|g_w\|_{H^p(\mathbb{S}\setminus K)}^p \, d\mu(w) &= \int_K \lim_{k\to\infty} \int_{\partial\Omega_k} \frac{1}{|z-w|^p} \, d\omega_k(z) \, d\mu(w) \\ &= \lim_{k\to\infty} \int_{\partial\Omega_k} \int_K \frac{1}{|z-w|^p} \, d\mu(w) \, d\omega_k(z) \\ &\le \lim_{k\to\infty} \int_{\partial\Omega_k} d\omega_k(z) = 1. \end{aligned}$$

Hence, almost all, with respect to μ, $g_w \in H^p(\mathbb{S} \setminus K)$ and thus $H^p(\mathbb{S} \setminus K)$ is non-trivial. □

Remark. For $p \geq 1$ this is an old result, but for $p < 1$ it is new.

5. Removable singularities on curves

Definition 9. A *chord-arc curve* is a rectifiable Jordan curve $\Gamma \subset \mathbb{C}$, for which there exists a constant M, such that for any $z_1, z_2 \in \Gamma$, the length of the shorter arc in Γ between z_1 and z_2 is less than $M|z_1 - z_2|$.

Definition 10. A rectifiable Jordan curve Γ is of *bounded rotation* if the forward half-tangent exists at every point and the tangent angle $\tau(s)$, which it makes with a fixed direction, can be defined as a function of bounded variation of the arc length s.

Definition 11. Let f be a continuous function on an arc. Let

$$c(\delta) = c_f(\delta) = \sup_{\text{arc dist}(z,w)<\delta} |f(z) - f(w)|$$

be its modulus of continuity. The function f is called *Dini continuous* if

$$\int_0^\delta \frac{c(t)}{t}\, dt < \infty,$$

for some $\delta > 0$.

Theorem 12. *Let $\Gamma \subset \mathbb{C}$ be a chord-arc curve, with chord-arc constant M. Let $K \subset \Gamma$ be compact and such that $\Lambda_1(K) = 0$. Then the following are true:*
(a) *there exists $p < \infty$, only dependent on M, such that K is removable for H^p,*
(b) *if Γ is of bounded rotation, then K is removable for H^p, for all $p > 1$,*
(c) *if Γ has Dini continuous tangent angles, then K is removable for H^1.*

Remarks.
(i) A chord-arc curve cannot have cusps.
(ii) A $C^{1+\varepsilon}$ curve, $\varepsilon > 0$, satisfies the condition in (c).
(iii) It is known that if K is a subset of a rectifiable curve and $\Lambda_1(K) > 0$, then K is not removable for H^∞, and thus not for any H^p, $p < \infty$.
(iv) Theorem 8 shows that part (c) is sharp in the sense that it is not true for any $p < 1$. Apart from this, it is not known if these conditions are sharp.
(v) For conditions of removability for intersecting curves we refer to Björn [1].

Sketch of the proof. Let $f \in H^p(\mathbb{S} \setminus K)$ and let Ω be the interior of Γ. Let $\varphi : \mathbb{D} \to \Omega$ be a conformal mapping. Let $\Gamma_r = \{\varphi(re^{i\theta}) : 0 \leq \theta \leq 2\pi\}$, $0 < r < 1$, with interior Ω_r. Fix $\zeta \in \Omega_{1/2}$.

Let $1/p + 1/p' = 1$. In (a), we use a result by Mikhail A. Lavrent'ev, Theorem 7 in [9], to show that there exists $p < \infty$ such that $\varphi' \in H^{p'}(\mathbb{D})$. In (b), we let a be the angle at the sharpest corner on Γ, and let $p > \pi/a$, then Stefan E. Warschawski and Glenn E. Schober [14] showed that $\varphi' \in H^{p'}(\mathbb{D})$. In (c), Warschawski in [12], p. 443, and [13], showed that $\varphi' \in C(\overline{\mathbb{D}})$.

By conformal invariance we also know that $f \circ \varphi \in H^p(\mathbb{D})$. Letting $g(z) = z/2\pi(\varphi(z) - \zeta)$, a continuous function on $\left\{z \in \mathbb{C} : \frac{1}{2} \le |z| \le 1\right\}$, and substituting $z = \varphi(re^{i\theta})$, we obtain

$$\begin{aligned}\int_{\Gamma_r} \frac{f(z)}{2\pi i(z-\zeta)}\,dz &= \int_0^{2\pi} \frac{f\circ\varphi(re^{i\theta})}{2\pi i(\varphi(re^{i\theta})-\zeta)} ire^{i\theta}\varphi'(re^{i\theta})\,d\theta \\ &= \int_0^{2\pi} g(re^{i\theta})(f\circ\varphi)(re^{i\theta})\varphi'(re^{i\theta})\,d\theta \\ &\to \int_0^{2\pi} g(e^{i\theta})(f\circ\varphi)(e^{i\theta})(\varphi')^*(e^{i\theta})\,d\theta \\ &= \int_\Gamma \frac{f(z)}{2\pi i(z-\zeta)}\,dz.\end{aligned}$$

In the third integral, $(f \circ \varphi)\varphi' \in H^1(\mathbb{D})$, by Hölder's inequality. The limit is taken as $r \to 1-$, and is obtained using a simple convergence result for functions in $H^1(\mathbb{D})$. The * denotes non-tangential boundary values.

We do the same calculation for the outer region $\mathbb{S} \setminus \overline{\Omega}$. Together with Cauchy's formula and the identity theorem this shows that K is removable.

In (b), the conditions given by Warschawski–Schober [14] only give this result for $p > \pi/a$. To get the result for any given $p > 1$ we study the situation at the finitely many corners with angles $\le \pi/p$ individually.

Using what we have already proved, we know that the only singular points can be at the finitely many corners with angles $\le \pi/p$. It is enough to assume that there is only one such corner.

Assume that the corner is at 0 and has angle η. Using the conformal power mappings $z \mapsto z^{\pi/\eta}$ and $z \mapsto z^{\pi/(2\pi-\eta)}$ together with the results by Warschawski–Schober [14], we are able to show that the corner is also a removable singularity.

This latter technique is also used to study intersection points between curves, as mentioned in remark (v).

For a complete proof see Björn [1].

6. Removability for planar Cantor sets

Let $C_\alpha^{(0)} = [0,1]$ and let $C_\alpha^{(1)}$ be the set remaining after having removed the open middle part of length $(1-2\alpha)$ from the interval. We continue in this way by letting $C_\alpha^{(m+1)}$ be the set remaining after having removed the open middle parts of length

$(1-2\alpha)\alpha^m$ from each of the intervals in $C_\alpha^{(m)}$. Finally we let

$$C_\alpha = \bigcap_{m=0}^{\infty} (C_\alpha^{(m)} \times C_\alpha^{(m)}),$$

which is a self-similar Cantor set in $\mathbb{R}^2 = \mathbb{C}$.

The set $C_\alpha^{(m)} \times C_\alpha^{(m)}$ is composed of 4^m closed squares with sidelength α^m. For us it will be advantageous to work with discs rather than squares. For this reason we here give an alternative definition of C_α.

Let $Q_{\alpha,j}^{(m)}$, $1 \le j \le 4^m$, be the squares constituting $C_\alpha^{(m)} \times C_\alpha^{(m)}$, i.e.,

$$C_\alpha^{(m)} \times C_\alpha^{(m)} = \bigcup_{j=1}^{4^m} Q_{\alpha,j}^{(m)}.$$

Let $D_{\alpha,j}^{(m)}$ be the smallest closed disc containing $Q_{\alpha,j}^{(m)}$, thus having a diameter equal to $\sqrt{2}\alpha^m$. Finally let

$$\widetilde{C}_\alpha^{(m)} = \bigcup_{j=1}^{4^m} D_{\alpha,j}^{(m)} \quad \text{and} \quad \widetilde{C}_\alpha = \bigcap_{m=0}^{\infty} \widetilde{C}_\alpha^{(m)}.$$

Lemma 13. *$C_\alpha = \widetilde{C}_\alpha$ and $\dim C_\alpha = -\log 4/\log\alpha$.*

The set $C_{1/4}$ is perhaps the best known and simplest, but not the first, example of a set with positive generalized length removable for H^∞, see e.g. Garnett [4].

We have compared what happens with the harmonic measure when we move from one generation to the next and have obtained the following result.

Theorem 14. *Let $m \ge 1$ and let $\widetilde{D}_{\alpha,j}^{(m)}$ be the largest open disc with $\widetilde{C}_\alpha^{(m)} \cap \widetilde{D}_{\alpha,j}^{(m)} = D_{\alpha,j}^{(m)}$. Let also $D_{\alpha,k}^{(m+1)} \subset D_{\alpha,j}^{(m)}$. Let $\mathbb{S} = \mathbb{C} \cup \{\infty\}$ be the Riemann sphere. Then*

$$\frac{\omega(\partial D_{\alpha,k}^{(m+1)}; \mathbb{S} \setminus \widetilde{C}_\alpha^{(m+1)}, \infty)}{\omega(\partial D_{\alpha,j}^{(m)}; \mathbb{S} \setminus \widetilde{C}_\alpha^{(m)}, \infty)} \ge \sup_{\Gamma} \inf_{\zeta \in \Gamma} \frac{\omega(\partial D_{\alpha,k}^{(m+1)}; \widetilde{D}_{\alpha,j}^{(m)} \cap (\mathbb{S} \setminus \widetilde{C}_\alpha^{(m+1)}), \zeta)}{\omega(\partial D_{\alpha,j}^{(m)}; \widetilde{D}_{\alpha,j}^{(m)} \cap (\mathbb{S} \setminus \widetilde{C}_\alpha^{(m)}), \zeta)},$$

where the supremum is taken over all circles $\Gamma \subset \widetilde{D}_{\alpha,j}^{(m)} \setminus D_{\alpha,j}^{(m)}$ concentric with $D_{\alpha,j}^{(m)}$. The right hand side is independent of j, k and m, we denote its value by M_α. By the construction we have $0 < M_\alpha \le \frac{1}{4}$. If $p \ge 1$ and

$$p > \frac{\log 1/4M_\alpha}{\log 1/4\alpha},$$

then C_α is removable for H^p.

This result is proved using elementary properties of harmonic measures and harmonic functions, including Harnack's inequalities, together with Cauchy's formula and Hölder's inequality. The restriction $p \geq 1$ is due to the use of Hölder's inequality.

As a corollary of Theorem 14 we get the following result.

Corollary 15. *Let $\alpha < \frac{1}{4}$, then C_α is removable for H^p for some $p < \infty$.*

The denominator in the expression for M_α has a well-known simple expression as the domain considered is an annulus. It is constant for fixed Γ. It is not hard to show that the infimum will increase with the radius of Γ.

In order to give explicit values for p we have used numerical calculations to calculate the harmonic functions in the numerator of M_α, and thus obtaining values for M_α. We used the program PLTMG (Version 6.3), a public domain software using the Finite Element Method. We have reduced the values obtained by a safety factor, covering for numerical errors, and in this way obtained lower bounds for M_α.

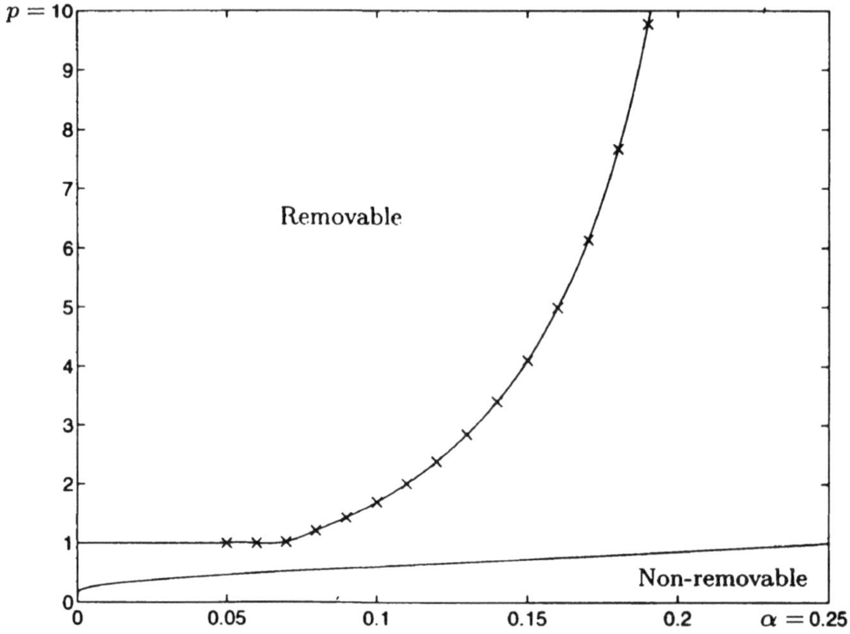

Figure 1. A graph of the removability conditions given for C_α.

The removability conditions obtained, using these lower bounds, are shown in Figure 1. We have done calculations for $\alpha = 0.05, 0.06, \ldots, 0.25$, marked by x in the figure. Below the lower curve Theorem 8 and Lemma 13 show that we do not have removability.

References

[1] Björn, A., Removable Singularities for Hardy Spaces of Analytic Functions. Ph.D. Dissertation, Linköping 1994.

[2] — Some open problems relating removable singularities for Hardy spaces and Hausdorff measures. These Proceedings.

[3] Conway, J. B., Dudziak J. J., Straube, E., Isometrically removable sets for functions in the Hardy space are polar. Michigan Math. J. 34 (1987), 267–273.

[4] Garnett, J. B., Positive length but zero analytic capacity. Proc. Amer. Math. Soc. 21 (1970), 696–699.

[5] Hasumi, M., Hardy classes on plane domains. Ark. Mat. 16 (1978), 213–227; also in [6], Chapter 11, pp. 234–261.

[6] — Hardy Classes on Infinitely Connected Riemann Surfaces. Lecture Notes in Math. 1027, Springer-Verlag, Berlin–Heidelberg 1983.

[7] Hejhal, D. A., Classification theory for Hardy classes of analytic functions. Bull. Amer. Math. Soc. 77 (1971), 767–771.

[8] — Classification theory for Hardy classes of analytic functions. Ann. Acad. Sci. Fenn. Ser. A I Math. 566 (1973), 1–28.

[9] Lavrent'ev, M. A., Boundary problems in the theory of univalent functions. Mat. Sb. (N.S.) 1 (1936), 815–844 (Russian). English transl.: Amer. Math. Soc. Transl. (2) 32 (1963), 1–35.

[10] Parreau, M., Sur certaines classes de fonctions analytiques uniformes sur les surfaces de Riemann. C. R. Acad. Sci. Paris 231 (1950), 751–753.

[11] — Sur les moyennes des fonctions harmoniques et analytiques et la classification des surfaces de Riemann. Ann. Inst. Fourier (Grenoble) 3 (1952), 103–197.

[12] Warschawski, S. E., Über das Randverhalten der Ableitung der Abbildungsfunktion bei konformer Abbildung. Math. Z. 35 (1932), 321–456.

[13] — On differentiability at the boundary in conformal mapping. Proc. Amer. Math. Soc. 12 (1961), 614–620.

[14] Warschawski, S. E., Schober, G. E., On conformal mapping of certain classes of Jordan domains. Arch. Rational Mech. Anal. 22 (1966), 201–209.

Approximation by harmonic functions

Jürgen Bliedtner

Abstract. A simple functional-analytic proof of a version of Stone's approximation theorem and an application to the approximation by harmonic functions will be presented.

1991 Mathematics Subject Classification: 31B05, 31D05, 41A65.

0. Introduction

A (probably well-known) variant of the approximation theorem of M. H. Stone, giving necessary and sufficient conditions, will be proved by using just some notions and some simple properties of the Choquet theory of function spaces. Besides the notion of the relative Choquet boundary introduced in [7] for a treatment of Korovkin's theorem, this paper could have been written after the publication of H. Bauer's article [2]. An application provides a simple proof of a theorem of Khavinson [9] concerning the approximation of continuous functions by harmonic functions.

1. Relative Choquet boundary

Let X be a compact (Hausdorff) space. Denote as usual by $\mathcal{C}(X)$ the vector space of real continuous functions on X with the topology of uniform convergence and by $\mathcal{M}_+^1(X)$ the set of positive Radon measures μ on X such that $\mu(X) = 1$. The Dirac measure in $x \in X$ is denoted by ε_x.

Furthermore, let $\mathcal{H}$ and $\mathcal{K}$ be vector spaces such that

$$1 \in \mathcal{H} \subset \mathcal{K} \subset \mathcal{C}(X).$$

For any $x \in X$, denote by $\mathcal{M}_x(\mathcal{H})$ the set of $\mathcal{H}$*-representing measures* on X, i.e.,

$$\mathcal{M}_x(\mathcal{H}) := \{\mu \in \mathcal{M}_+^1(X) : \int h\, d\mu = h(x) \text{ for all } h \in \mathcal{H}\}.$$

The set

$$\delta(\mathcal{H},\mathcal{K}) := \{x \in X : \mathcal{M}_x(\mathcal{H}) = \mathcal{M}_x(\mathcal{K})\}$$

is called the *relative Choquet boundary* of $\mathcal{H}$ with respect to $\mathcal{K}$.

In the special case $\mathcal{K} = \mathcal{C}(X)$,

$$\delta(\mathcal{H}) ::= \delta(\mathcal{H}, \mathcal{C}(X)) = \{x \in X : \mathcal{M}_x(\mathcal{H}) = \{\varepsilon_x\}\}$$

is called the *Choquet boundary* of $\mathcal{H}$.

Let us finally denote by $A(\mathcal{H})$ the set of all continuous $\mathcal{H}$-*affine* functions on X, i.e.,

$$A(\mathcal{H}) := \{h \in \mathcal{C}(X) : \int h\,d\mu = h(x) \text{ for all } x \in X, \mu \in \mathcal{M}_x(\mathcal{H})\}.$$

$A(\mathcal{H})$ is a closed subspace of $\mathcal{C}(X)$ such that $\mathcal{H} \subset A(\mathcal{H})$.

Remarks 1.1.
(1) $\delta(\mathcal{H}) = \delta(\overline{\mathcal{H}}) = \delta(A(\mathcal{H}))$.
(2) If $x \in X$ and $h \in \mathcal{H}$ such that $0 = h(x) < h(y)$ for all $y \in X, y \neq x$, then $x \in \delta(\mathcal{H})$. Indeed, let $\mu \in \mathcal{M}_x(\mathcal{H})$. Then

$$0 = h(x) = \int h\,d\mu,$$

hence $\mu = \varepsilon_x$.
(3) $\delta(\mathcal{C}(X)) = X$.
(4) $A(\mathcal{C}(X)) = \mathcal{C}(X)$.
(5) $\delta(\mathcal{H}, \mathcal{K}) = X \iff A(\mathcal{H}) = A(\mathcal{K})$.

For any $f \in \mathcal{C}(X)$ let

$$f^{\mathcal{H}} := \inf\{h \in \mathcal{H} : f \leq h\}\ ,\ f_{\mathcal{H}} := -(-f)^{\mathcal{H}}.$$

Lemma 1.2. *Let $f \in \mathcal{C}(X)$. Then the following conditions are equivalent:*
(1) $f \in A(\mathcal{H})$
(2) $f_{\mathcal{H}} = f^{\mathcal{H}} = f$.

Proof. (1) $\Rightarrow$ (2): Let $x \in X$. Then $g \longmapsto g^{\mathcal{H}}(x)$ defines a sublinear functional on $\mathcal{C}(X)$. By the theorem of Hahn–Banach there exists a linear functional $T : \mathcal{C}(X) \longrightarrow \mathbb{R}$ such that

$$T(g) \leq g^{\mathcal{H}}(x) \text{ for all } g \in \mathcal{C}(X) \text{ and } T(f) = f^{\mathcal{H}}(x).$$

If $g \in \mathcal{C}(X), g \leq 0$, then $T(g) \leq g^{\mathcal{H}}(x) \leq 0$, hence T is positive, i.e., T is a measure μ on X. For all $h \in \mathcal{H}$,

$$\int h\,d\mu = T(h) \leq h^{\mathcal{H}}(x) = h(x)$$

and therefore $\int h\,d\mu = h(x)$ since $\mathcal{H}$ is a vector space. Thus $\mu \in \mathcal{M}_x(\mathcal{H})$ and

$$f^{\mathcal{H}}(x) = T(f) = \int f\,d\mu = f(x).$$

Analogously, $f_{\mathcal{H}}(x) = f(x)$.

(2) ⇒ (1): Let $x \in X$ and $\mu \in \mathcal{M}_x(\mathcal{H})$. Then for any $g, h \in \mathcal{H}$, $g \leq f \leq h$,

$$g(x) = \int g\,d\mu \leq \int f\,d\mu \leq \int h\,d\mu = h(x),$$

hence $f(x) = f_{\mathcal{H}}(x) \leq \int f\,d\mu \leq f^{\mathcal{H}}(x) = f(x)$. □

2. Weak Riesz interpolation property

We say that $\mathcal{H}$ has the *weak Riesz interpolation property* (WRIP) with respect to $\mathcal{K}$ if for all $h_1, \ldots, h_4 \in \mathcal{H}$ and $k \in \mathcal{K}$ with $\sup(h_1, h_2) < k < \inf(h_3, h_4)$ there exists $h \in \mathcal{H}$ such that

$$\sup(h_1, h_2) \leq h \leq \inf(h_3, h_4).$$

If $\mathcal{K} = C(X)$, we just say that $\mathcal{H}$ has the property (WRIP).

Remarks 2.1.
(1) If $\mathcal{H}$ is a vector lattice then $\mathcal{H}$ has the property (WRIP).
(2) $\mathcal{H}$ has the property (WRIP) with respect to $\overline{\mathcal{H}}$.

3. Approximation theorem

This section contains the main approximation theorem.

Theorem 3.1. *The following assertions are equivalent:*
(1) $\overline{\mathcal{H}} = \overline{\mathcal{K}}$
(2) (a) $\delta(\mathcal{H}, \mathcal{K}) = X$
(b) $\mathcal{H}$ *has the property (WRIP) with respect to* $\mathcal{K}$.

Proof. (1) ⇒ (2): 1.1 and 2.1.

(2) ⇒ (1): Let $k \in \mathcal{K}$. Then $k \in A(\mathcal{H})$ by 1.1 (5), hence $k^{\mathcal{H}} = k = k_{\mathcal{H}}$ by 1.2. Let $\varepsilon > 0$. By the usual compactness argument, there exist $g_1, \ldots, g_n, h_1, \ldots, h_m \in \mathcal{H}$ such that for $g := \sup(g_1, \ldots, g_n)$ and $h := \inf(h_1, \ldots, h_m)$

$$g < k < h \text{ and } |h - g| < \varepsilon.$$

It follows from (b) that there exists an $h_0 \in \mathcal{H}$ such that $g \leq h_0 \leq h$, hence $|k - h_0| < \varepsilon$ and therefore $\mathcal{H} \subset \mathcal{K} \subset \overline{\mathcal{H}}$. □

For the special case $\mathcal{K} = C(X)$, Theorem 3.1 yields the desired version of the approximation theorem of Stone.

Corollary 3.2. *The following assertions are equivalent:*
(1) $\overline{\mathcal{H}} = \mathcal{C}(X)$
(2) (a) $\delta(\mathcal{H}) = X$
(b) $\mathcal{H}$ *has the property (WRIP).*

Remark 3.3. If moreover $\mathcal{H}$ separates the points of X, then it can be shown that in 3.1 the condition $\delta(\mathcal{H},\mathcal{K}) = X$ is equivalent to $\delta(\mathcal{H}) = \delta(\mathcal{K})$. Then 3.1 is just Alfsen's version of Stone's approximation theorem (see [1]).

4. Theorems of Stone and Stone–Weierstraß

Let $\mathcal{H}$ be a vector space contained in $\mathcal{C}(X)$ such that
(a) $1 \in \mathcal{H}$
(b) $\mathcal{H}$ separates the points of X.

Then the following assertions hold:

(1) **Approximation theorem of Stone:**

If $\mathcal{H}$ is a vector lattice, then $\overline{\mathcal{H}} = \mathcal{C}(X)$.

(2) **Approximation theorem of Stone–Weierstraß:**

If $\mathcal{H}$ is an algebra, then $\overline{\mathcal{H}} = \mathcal{C}(X)$.

Proof of (1): By 3.2, it suffices to show $\delta(\mathcal{H}) = X$. Fix $x \in X$ and $\mu \in \mathcal{M}_x(\mathcal{H})$. Let $h \in \mathcal{H}$ and $f := |h - h(x)|$. Then $f \in \mathcal{H}$ and

$$0 = f(x) = \int f\,d\mu = \int |h - h(x)|d\mu,$$

hence the set $\{y \in X : h(y) = h(x)\}$ contains the support of μ. Since $\mathcal{H}$ separates the points of X one obtains $\mu = \varepsilon_x$, thus $x \in \delta(\mathcal{H})$.

Proof of (2): The assertion follows from (1) since the conditions "$\overline{\mathcal{H}}$ is a vector lattice" and "$\overline{\mathcal{H}}$ is an algebra" are equivalent (see e.g. [6] for a simple proof).

5. Applications

Let $K \subset \mathbb{R}^d$ be a compact set and consider the vector space

$$H_0(K) := \{h \in C(K) : h \text{ harmonic in a neighbourhood of } K\}.$$

For $a \in \mathbb{R}^d$ define the function φ_a by

$$\varphi_a(x) := \|x - a\|^2\ ,\ x \in \mathbb{R}^d.$$

The approximation Theorem 3.2 yields a simple proof of the following theorem of D. Khavinson [9].

Theorem 5.1. *If* $\varphi_{0|K} \in \overline{H_0(K)}$, *then* $\overline{H_0(K)} = \mathcal{C}(K)$.

Proof. (a) $\delta(H_0(K)) = K$: Let $a \in K$. Since for any $x \in K$,

$$\varphi_a(x) = \varphi_0(x) - 2\sum_{i=1}^{d} a_i x_i + \|a\|^2,$$

by assumption $\varphi_{a|K} \in \overline{H_0(K)}$, hence $a \in \delta(\overline{H_0(K)})$ by 1.1 (2) and $a \in \delta(H_0(K))$ by 1.1 (1).

(b) Let $h_1, \dots, h_4 \in H_0(K)$ such that $\sup(h_1, h_2) < \inf(h_3, h_4)$. There exists an open neighbourhood U of K such that $h_1, \dots, h_n$ are continuous on $\overline{U}$ and harmonic on U. The solution h of the generalized Dirichlet problem on U of the boundary function $\sup(h_1, h_2)$ satisfies $\sup(h_1, h_2) \le h \le \inf(h_3, h_4)$ on K. □

Remarks 5.2.

(1) The above Theorem 5.1 could be obtained as an easy consequence of VII.9.2 of [4]. A similar theorem in [8] where the function φ_0 is replaced by a function $\varphi \in \mathcal{C}^2$ such that the set $\{x \in K : \Delta\varphi(x) = 0\}$ has Lebesgue measure zero follows in the same way.

(2) It should be noted that the general assumption "$1 \in \mathcal{H}$" can be replaced by "There exists $h_0 \in H, h_0 > 0$" and a modified version of the notion of representing measures. Furthermore, the approximation Theorem 3.1 remains valid if one assumes that X is only a locally compact space which is countable at infinity. In this case, one has to assume that the space $\mathcal{H}$ is adapted and $\mathcal{C}(X)$ is replaced by $\mathcal{C}_{\mathcal{H}}(X) := \{f \in \mathcal{C}(X) : |f| \le h \text{ for some } h \in \mathcal{H}\}$ such that $\mathcal{H} \subset \mathcal{K} \subset \mathcal{C}_{\mathcal{H}}(X)$. The topology on $\mathcal{C}_{\mathcal{H}}(X)$ has to be the order topology (see [3], [4], [10]).

References

[1] Alfsen, E. M., Un théorème de Weierstraß–Stone pour les sous-espaces vectoriels de $\mathcal{C}_{\mathbb{R}}(X)$. C. R. Acad. Sci. Paris 271 (1970), 725–726.

[2] Bauer, H., Šilovscher Rand and Dirichletsches Problem. Ann. Inst. Fourier 11 (1961), 78–136.

[3] — Theorems of Korovkin type for adapted spaces. Ann. Inst. Fourier 23,4 (1973), 245–260.

[4] Bliedtner, J., Hansen, W., Potential Theory, An Analytic and Probabilistic Approach to Balayage. Springer-Verlag, Berlin 1986.

[5] Choquet, G., Lectures on Analysis III. Benjamin, London 1969.

[6] Dellacherie, C., Un complément au théorème de Weierstraß–Stone. In: Séminaire de Probabilités 1, 52-53. Lecture Notes in Math. 39. Springer-Verlag, Berlin 1969.

[7] Franchetti, G., Convergenza di operatori in sottospazi dello spazio $\mathcal{C}(Q)$. Boll. Un. Mat. Ital. Ser. IV, 3 (1970), 668–675.

[8] Gauthier, P. M., Paramonov, P. V., Approximation by harmonic functions in the $\mathcal{C}^1$-norm and harmonic $\mathcal{C}^1$-content of compact subsets in $\mathbb{R}^n$. Preprint.

[9] Khavinson, D., On uniform approximation by harmonic functions. Mich. Math. J. 34 (1987), 465–473.

[10] Mokobodzki, G., Sibony, D., Cônes adaptés de fonctions continues et théorie du potentiel. In: Sém. Choquet 6 (1968), 5.01–5.35.

Generalized radial limits and the best approach to boundaries

*Jürgen Bliedtner and Peter A. Loeb**

Abstract. The speaker and his coauthor, Jürgen Bliedtner, have recently shown that with a suitable normalization, there is a "best" family of filters for which the Fatou Boundary Limit Theorem holds. The normalization assigns to each positive harmonic function a set of boundary points at which that function must vanish. Known limits, such as those provided by the Lebesgue Differentiation Theorem, or the Fatou Radial Limit Theorem are used to force consistency in this assignment. The zero sets, in turn, are used in constructing the coarsest filters which produce those limits almost everywhere. This general procedure is new even for the case of harmonic functions on the unit disk in the complex plane. In this note, couched in terms of harmonic functions on the unit disk but applicable in quite general settings, "minimal" harmonic functions are used to extend to general domains the use of radial limits in the construction of zero sets.

1991 Mathematics Subject Classification: 31A20, 31B25, 31C35.

1. Introduction

This note outlines a boundary limit theorem for integral representations usually associated with harmonic functions. Details can be found in [6]. First, we review the construction from [4]. There, a procedure is established which gives the best approach to a boundary once a suitable normalization has been fixed. The need for such a normalization follows from a result of J. L. Doob [8]. He showed that for positive harmonic functions on the unit disk, there is no "best", i.e., coarsest, filter for which a Fatou Boundary Limit Theorem holds when the filter is copied by rotation at all points of the unit circle.

The normalization required for our procedure is given by assigning "zero sets". In the last section of this note we will construct such sets for bounded harmonic functions using a sequence of "approximating functions". For the ball, the use of these approximations gives the same zero sets as is obtained by taking radial limits. Finally, employing a theorem from [5], the zero sets for bounded harmonic

* The authors thank Professor Earl Berkson for a helpful conversation. The second author's work was supported by U.S. National Science Foundation grant DMS 92-01494.

functions are used to construct zero sets for those harmonic functions which arise from integrable densities.

2. Best boundary approach filters

Let $(\Delta, \mathcal{B})$ be a measurable space and X a domain of definition for functions obtained by integrating measures on $(\Delta, \mathcal{B})$. Examples are given by letting Δ be a compactifying boundary for a domain X in a Euclidean space or a Riemann surface supplied with ordinary harmonic functions. The space X can also be a domain in an axiomatic, potential theoretic setting (see [9], [7], [1] and [11]). In all these cases, $\mathcal{B}$ would usually be the σ-algebra of Borel subsets of Δ.

Given any finite measure ν on $(\Delta, \mathcal{B})$ and a set $E \in \mathcal{B}$, we write ν_E for the "restriction" of ν to E; i.e., $\nu_E(A) = \nu(A \cap E)$ for each $A \in \mathcal{B}$. Having fixed a finite reference measure σ on $(\Delta, \mathcal{B})$, we will write $d\nu/d\sigma$ for the Radon–Nikodym derivative of the absolutely continuous part of ν with respect to σ.

We assume that a connection exists between $(\Delta, \mathcal{B})$ and X in terms of the following integral representation:

1) There is a convex cone M of finite measures on $(\Delta, \mathcal{B})$, with $\nu_E \in M$ when $\nu \in M$ and $E \in \mathcal{B}$.
2) To each $\nu \in M$, there corresponds a function h_ν on X so that the map $\nu \mapsto h_\nu$ is a positive affine map on M.
3) There is a fixed "reference measure" $\sigma \in M$ with h_σ strictly positive on X.
4) If $\nu \in M$ and $h_\sigma \leq h_\nu$, then $\sigma \leq \nu$; i.e., for each $E \in \mathcal{B}$, $\sigma(E) \leq \nu(E)$.

Examples. For a simple example of such an integral representation, let Δ and X equal the unit circle C and the unit disk D, respectively, in the complex plane. Here, h_ν denotes the Poisson integral of ν for each Borel measure ν. For more general domains, Δ may be the Martin boundary [9], or the boundary constructed in [11]. The space of functions on X may also be quite general. (See [1], [9], and [11].) In these examples, one can let Δ be the set of minimal harmonic functions and use the integral representation given by Choquet theory. The set M may consist of all finite Borel measures, but one can also restrict M to measures generated by densities in $L^p(\sigma)$ for some p with $1 \leq p \leq \infty$.

When $\mathcal{G}$ is a filter base in X and f is a function defined on X, we write $\lim_{\mathcal{G}} f$ to denote the limit of f along the filter generated by $\mathcal{G}$.

Definition. By a *Fatou filter mapping* we mean a function $\mathcal{G} : y \mapsto \mathcal{G}_y$ associating with each point $y \in \Delta$ a filter base $\mathcal{G}_y$ on X so that for each $\nu \in M$,

$$\lim_{\mathcal{G}_y} \frac{h_\nu}{h_\sigma} = \frac{d\nu}{d\sigma}(y) \quad \text{for } \sigma\text{-a.e. } y \in \Delta.$$

Given a filter mapping initially defined only σ-a.e. on Δ, we will assume that the mapping has been extended to the rest of Δ with the filter containing only the set

X. We say that a Fatou filter mapping $\mathcal{F}$ is coarser than a Fatou filter mapping $\mathcal{G}$ if at each $y \in \Delta$ the filter generated by $\mathcal{F}_y$ is contained in the filter generated by $\mathcal{G}_y$; we do not require the containment to be strict. For harmonic functions on the unit disk, the radial approach, nontangential approach, and minimal fine approach are all examples of Fatou filter mappings.

To produce a Fatou filter mapping as in [4], we need to fix for each measure $\nu \in M$ the set of points of Δ where a zero limit is required. That is, we must have in hand a mapping $Z : \nu \mapsto Z_\nu$ from M into the σ-completion of $\mathcal{B}$ satisfying the following: For all ν, $\mu \in M$, all $E \in \mathcal{B}$, and all $\alpha > 0$ in $\mathbb{R}$,

i) $Z_\nu \cap Z_\mu \subseteq Z_{\nu+\mu}$,
ii) $Z_{\alpha\nu} = Z_\nu$,
iii) $Z_\nu = \emptyset$ if $\sigma \leq \nu$,
iv) $Z_0 = \Delta$,
v) if $\nu(E) = 0$, then $\sigma(E \setminus Z_\nu) = 0$.

We will call each Z_ν the *zero set* for ν. One can produce zero sets by using a Fatou filter mapping such as the mapping for radial, nontangential, or minimal fine limits on the disk. Here for each $\nu \in M$ and $y \in \Delta$, one takes the lim sup of the ratio h_ν/h_σ along the filter associated with y; the set Z_ν contains y if the result is 0. One can also use measure differentiation to form zero sets. In general, the selection of zero sets takes into account the singular parts of the corresponding measures.

For each $\nu \in M$, we write $\{h_\nu < h_\sigma\}$ for the set $\{x \in X : h_\nu(x) < h_\sigma(x)\}$. Our approach to the boundary is defined in terms of such level sets.

Theorem. *Given a zero-set mapping Z, for each $y \in \Delta$, set*

$$\mathcal{F}_y := \{\{h_\nu < h_\sigma\} : \nu \in M,\, y \in Z_\nu\} .$$

The mapping $\mathcal{F}$ is the coarsest Fatou filter mapping on Δ with the property that for each $\nu \in M$,

$$Z_\nu \subseteq \{y \in \Delta : \lim\nolimits_{\mathcal{F}_y} \frac{h_\nu}{h_\sigma} = 0\} .$$

Proof. The proof, using our general reduction technique from [3], is given in [4] and also outlined in [5]. □

Remark. Assume that in a potential theoretic setting with M equal to the set of all finite Borel measures, the zero sets Z_ν, $\nu \in M$, are produced by taking limits with respect to a known Fatou-filter map $x \mapsto \mathcal{G}_x$. Also assume that for every potential p on X, the ratio p/h_σ has a limit 0 along $\mathcal{G}_x$ at σ-a.e. $x \in \Delta$. Then for each potential p on X and every $x \in \Delta$ where the limit of p/h_σ is 0, one can add the set $\{p < h_\sigma\}$ to the filter base $\mathcal{F}_x$ and generate a new filter base since $\mathcal{F}_x$ is a subset of the filter generated by $\mathcal{G}_x$ and $\{p < h_\sigma\}$ is an element of that filter. With this change, $\mathcal{F}$ is still a Fatou-filter mapping, and for each potential

p, the ratio p/h_σ will have limit 0 along $\mathcal{F}_x$ at σ-a.e. $x \in X$. In other words, the Ratio Fatou Limit Theorem given in terms of superharmonic functions holds for $\mathcal{F}$ (see [2]).

3. Generalization of radial limits

In this section, we describe preliminary work on a new method for obtaining zero sets. Let us begin by working with ordinary harmonic functions on the unit disk D in the complex plane, so that $X = D$, and Δ is the unit circle C. We will write D_r for the disk of radius r and C_r for the circle of radius r. To start with, M will be the set of all finite Borel measures on C. Again, h_ν denotes the Poisson integral of ν for each $\nu \in M$.

Fix $z_0 = 0$ in D, and for each $r > 0$, let $\mu^r_{z_0}$ denote harmonic measure on C_r with respect to z_0. Let σ be normalized Lebesgue measure $\mu^1_{z_0}$ on C so we have $h_\sigma \equiv 1$.

For each $\nu \in M$, $r < 1$ and $x \in C$, set

$$f^\nu_r(x) := \int_{C_r} P(w,x) \cdot h_\nu(w)\, \mu^r_{z_0}(dw).$$

For points $x, y \in C$,

$$|f^\nu_r(x) - f^\nu_r(y)| \leq \max_{w \in C_r} |P(w,x) - P(w,y)| \cdot |h_\nu(z_o)|\,.$$

Moreover, the mapping $x \mapsto P(\cdot, x)$ from C into the set of positive harmonic functions is continuous with respect to the topology of uniform convergence on compact sets. Therefore, the function $x \mapsto f^\nu_r(x)$ is a continuous real valued function on C.

For each $\nu \in M$, set

$$Z_\nu := \{x \in C : f^\nu_r(x) \to 0 \text{ as } r \to 1\}.$$

Given ν and μ in M and $\alpha > 0$ in $\mathbb{R}$, we have $f^\nu_r + f^\mu_r = f^{\nu+\mu}_r$ and $f^{\alpha\nu}_r = \alpha \cdot f^\nu_r$, and so $Z_\nu \cap Z_\mu \subseteq Z_{\nu+\mu}$ and $Z_{\alpha\nu} = Z_\nu$. If we have $\sigma \leq \nu$, then $h_\nu \geq 1$ on C_r, so $f^\nu_r \geq 1$ on C for each r, whence $Z_\nu = \emptyset$. Since $f^0_r = 0$ for all r, $Z_0 = \Delta$. It follows that the sets Z_ν form zero sets in C if and only if for all $\nu \in M$ and all Borel sets E we have

$$\nu(E) = 0 \Rightarrow f^\nu_r(x) \to 0 \quad \sigma\text{-a.e. on } E.$$

For the unit disk, this implication is a consequence of the Radial Limit Theorem. To see this, fix $\nu \in M$, $x \in C$, and any positive $r < 1$. Note that if w and z

are points on C_r, then $|rw - z| = |w - rz|$. We therefore have the equality

$$\begin{aligned} h_\nu(r^2x) &= \int_{C_r} h_\nu(w) \cdot \frac{r^2 - |r^2x|^2}{|w - r^2x|^2} \cdot \mu^r_{z_0}(dw) \\ &= r^2 \int_{C_r} h_\nu(w) \cdot \frac{1 - r^2}{|w - r(rx)|^2} \cdot \mu^r_{z_0}(dw) \\ &= r^2 \int_{C_r} h_\nu(w) \cdot \frac{1 - r^2}{|rw - rx|^2} \cdot \mu^r_{z_0}(dw) \\ &= \int_{C_r} h_\nu(w) \frac{1 - r^2}{|w - x|^2} \cdot \mu^r_{z_0}(dw) = f^\nu_r(x) \ . \end{aligned}$$

It follows that the sets Z_ν are exactly the sets in C where the radial limit is 0. A similar calculation gives the analogous result for the unit ball in any finite dimensional Euclidean space.

For a general domain X and "boundary" Δ, radial limits are not defined. Nevertheless, in a potential theoretic setting one can still define functions f^ν_r using an exhaustion to replace the family $\{D_r : r < 1\}$, a measure $\sigma \in M$ with $h_\sigma \equiv 1$, "extreme" functions $P(\cdot, x)$, $x \in \Delta$, and a fixed point $z_0 \in X$. We work in our setting with the canonical projection of the Feller compactification of X onto $X \cup \Delta$ and the measure σ_F on the Feller boundary corresponding to the choice of z_0; the projection of σ_F is σ (see [11], [12] or [6]).

Let $\mathcal{H}^+_b$ denote the nonnegative, bounded harmonic functions on X. We will start with the case that M is generated by σ and nonnegative, bounded densities on Δ, so that the corresponding harmonic functions form the class $\mathcal{H}^+_b$. We fix an exhaustion $\{\Omega_n : n \in \mathbb{N}\}$ consisting of regular relatively compact domains in X. We assume that $z_0 \in \Omega_n$ for all $n \in \mathbb{N}$. We then have for each $\nu \in M$, a sequence $\{f^\nu_n : n \in \mathbb{N}\}$ computed in the same was as for the disk.

Lemma. *If $h_\nu \in \mathcal{H}^+_b$, then in the weak* topology on Δ,* $\lim_{n\to\infty} (f^\nu_n \sigma) = \nu$.

Proof. Each $h \in \mathcal{H}^+_b$ has a continuous extension to the boundary of the Feller compactification of X. That is, on the Feller boundary, each $h \in \mathcal{H}^+_b$ corresponds to a continuous function g_h, while on Δ, the correspondence is with a measure ν_h obtained by projecting $g_h \sigma_F$. For any $h \in \mathcal{H}^+_b$, the family $\{f^{\nu_h}_n \sigma : n \in \mathbb{N}\}$ of measures on Δ is uniformly bounded. Given an ultrafilter in the natural numbers, we force a weak* limit for each $h \in \mathcal{H}^+_b$. The corresponding limit on the Feller boundary is $g_h \sigma_F$, and so the limit on Δ represents h. This method of obtaining representing measures on Δ preserves affine combinations in $\mathcal{H}^+_b$. By a corollary of a result of Cartier, Fell and Meyer (see Corollary 4.4 of [11]), for each $h \in \mathcal{H}^+_b$, the resulting measure is the minimal representing measure, and so our lemma follows.□

The lemma gives us weak* convergence, but in general, weak* convergence can not be strengthened. On the other hand, we have the following result if E is a measurable set in Δ and ν is a measure in M with $\nu(E) = 0$.

Proposition. *Fix $h \in \mathcal{H}_b^+$ and the corresponding measure $\nu \in M$. Let $E \subset \Delta$ be a Borel set such that $\nu_h(E) = 0$. Then $f_n^\nu \cdot \chi_E \to 0$ in measure.*

Proof. Fix $\varepsilon > 0$, a compact set K and an open set U with $K \subset E \subset U \subset \Delta$ such that $\sigma(U \setminus K) < \varepsilon/2$ and $\nu(U \setminus E) < \varepsilon^2/4$. Choose a continuous function g on Δ with $0 \leq g \leq 1$, $g = 1$ on K and $g = 0$ on $\Delta \setminus U$. Then $\int g d\nu < \varepsilon^2/4$. By the above lemma, for sufficiently large $n \in \mathbb{N}$, $\int g \cdot f_n^\nu d\sigma < \varepsilon^2/2$, whence $\sigma(\{f_n^\nu \geq \varepsilon\} \cap K) < \varepsilon/2$, and therefore $\sigma(\{f_n^\nu \cdot 1_E \geq \varepsilon\}) < \varepsilon$. □

The result given by the proposition is still not good enough; given the hypotheses, we need to have $f_n^\nu \cdot \chi_E \to 0$ σ -a.e. Let us assume that the Continuum Hypothesis holds, or at least that Martin's Axiom holds (see [10]). By a result of Mokobodzki [13] there is a "rapid ultrafilter" $\mathcal{U}$ in $\mathbb{N}$ along which $f_n^\nu \cdot \chi_E$ converges to 0 σ-a.e. We therefore have the following result for bounded densities and an extension to integrable densities using the technique established in Section 7 of [5].

Theorem. *One can form zero sets Z_ν for the measures associated with $\mathcal{H}_b^+$ by putting $x \in Z_\nu$ if and only if along $\mathcal{U}$, $\lim_{n,\mathcal{U}} f_n^\nu(z) = 0$. For harmonic functions arising from integrable densities on Δ, zero sets can be formed by taking the intersection of the zero sets associated with the dominated, bounded harmonic functions.*

Remark. Consider an appropriate nonstandard model and take a nonstandard $\omega \in {}^*\mathbb{N} \setminus \mathbb{N}$ such that $\omega \in {}^*U$ for all $U \in \mathcal{U}$. The above considerations show that we can form zero sets Z_ν for bounded densities by putting $x \in Z_\nu$ when the following calculation is infinitesimal:

$$ {}^*f_\omega^\nu(x) = \int_{\partial\Omega_\omega} {}^*h_\nu(w)^*P(w,x)\, \mu_{z_0}^{\partial\Omega_\omega}(dw). $$

Note added in proof: It is not necessary to assume in the definition that zero sets are measurable. Measurability will follow, however, from our limit theorem.

References

[1] Bliedtner, J., Hansen, W., Potential Theory. Springer-Verlag, Berlin 1986.

[2] Bliedtner, J., Loeb, P. A., A measure-theoretic boundary limit theorem. Arch. Math. (Basel) 43 (1984), 373–376.

[3] — A reduction technique for limit theorems in analysis and probability theory. Ark. Mat. 30(1992), 25–43.

[4] — Best filters for the general Fatou boundary limit theorem. Proc. Amer. Math. Soc. (to appear).

[5] — The best approach for boundary limits. In: Classical and Modern Potential Theory and Applications (ed. by K. GowriSankaran et al.), 105–112. Kluwer Academic Publishers, Dordrecht 1994.

[6] — Generalized radial limits associated with representing measures. In: Proceedings of the 1994 University of Missouri-Columbia conference on the interaction between functional analysis, harmonic analysis, and probability (to appear).

[7] Constantinescu, C., Cornea, A., Potential Theory on Harmonic Spaces. Springer-Verlag, Berlin 1972.

[8] Doob, J. L., Boundary Approach Filters for Analytic Functions. Ann. Inst. Fourier (Grenoble) 23 (3) (1973), 187–213.

[9] — Classical Potential Theory and its Probabilistic Counterpart. Springer-Verlag, Berlin 1984.

[10] Fremlin, D. H., Consequences of Martin's Axiom. Cambridge University Press, Cambridge 1984.

[11] Loeb, P. A., Applications of nonstandard analysis to ideal boundaries in potential theory. Israel J. Math. 25 (1976), 154–187.

[12] — A regular metrizable boundary for solutions of elliptic and parabolic differential equations. Math. Ann. 251 (1980), 43–50.

[13] Mokobodzki, G., Ultrafiltres rapides sur $\mathbb{N}$, construction d'une densité relative de deux potentiels comparables, Seminaire Brelot–Choquet–Deny (Théorie du Potentiel), 12e année, 1967/68, n° 12, 12-01–12-10.

Dilation operators in excessive structures; existence and uniqueness

Nicu Boboc and Gheorghe Bucur

Abstract. We consider two H-cones S, T such that $S \subset T$ and such that $B \leq B^{\#}$ for any balayage B on T, where $B^{\#}$ is the associated balayage on S given by $B^{\#} = \wedge\{u \in S \mid u \geq Bs\}$. We prove that there exists a map $Q : S \longrightarrow S$ which is a morphism of H-cones such that $s \wedge (Qs + u - Qu + Qf) \in S$ for any $s, u \in S$ and any $f \in (S - S)_+$. In particular, we extend the Shih theorem concerning the subordination of excessive structures in the framework of excessive measures.

1991 Mathematics Subject Classification: 31D05, 60J45, 60J57.

Introduction

Let $\mathcal{V} = (V_\alpha)_{\alpha \geq 0}$ be a proper submarkovian resolvent of kernels on a measurable space $(X, \mathcal{B})$ such that the set $\mathcal{E}_\mathcal{V}$ of all $\mathcal{V}$-excessive functions on X which are finite $\mathcal{V}$-a.s. is min-stable, contains the positive constant functions and generates $\mathcal{B}$. We suppose that X is a Lusin space and that it is semisaturated with respect to $\mathcal{V}$ (i.e., any $\mathcal{V}$-excessive measure on X, dominated by an $\mathcal{V}$-excessive measure on X of the form $\mu \circ V$, is also of the same form). The above conditions are equivalent to the fact that there exists a right process on $(X, \mathcal{B})$ for which $\mathcal{V}$ is the associated resolvent.

In the papers [4] and [5] we considered two submarkovian resolvents $\mathcal{V} = (V_\alpha)_{\alpha \geq 0}$, $\mathcal{W} = (W_\alpha)_{\alpha \geq 0}$ on $(X, \mathcal{B})$ which possess a reference measure and such that the absorbent points with respect to $\mathcal{V}$ and $\mathcal{W}$ are the same. If $\mathcal{E}_\mathcal{W} \subset \mathcal{E}_\mathcal{V}$, if any $s \in \mathcal{E}_\mathcal{V}$ is lower semicontinuous with respect to the fine topology generated by $\mathcal{E}_\mathcal{W}$ and if for any $A \in \mathcal{B}$ and any positive $\mathcal{B}$-measurable function f we have

$$^{\mathcal{V}}B^A f \leq {}^{\mathcal{W}}B^A f$$

(where $\mathcal{V}_{B^A}$ (resp. $\mathcal{W}_{B^A}$) is the balayage on A with respect to $\mathcal{E}_\mathcal{V}$ (resp. $\mathcal{E}_\mathcal{W}$)) then there exists another submarkovian resolvent $\overline{\mathcal{V}} = (\overline{V}_\alpha)_{\alpha \geq 0}$ such that $\mathcal{E}_\mathcal{V} = \mathcal{E}_{\overline{\mathcal{V}}}$ and such that

$$\overline{V}_\alpha f \leq W_\alpha f \quad \forall\, \alpha \geq 0 \text{ and } f \geq 0, \ \mathcal{B}\text{-measurable.}$$

In fact there exists a kernel Q on $(X, \mathcal{B})$ such that $Q(\mathcal{E}_\mathcal{W}) \subset \mathcal{E}_\mathcal{W}$,

$$s \wedge (Qs + t - Qt + Qf) \in \mathcal{E}_\mathcal{W}$$

for any $s, t \in \mathcal{E}_{\mathcal{W}}$ and any positive $\mathcal{B}$-measurable function f on X and such that

$$Wf = \overline{V}f + QWf \quad \forall f \geq 0, \quad \mathcal{B}\text{-measurable}.$$

This type of kernels was first considered by G. Mokobodzki ([11]) in connection with the subordination in excessive structures. Moreover, if Q is such a kernel for which

$$Wf \text{ bounded} \Longrightarrow \inf_n Q^n Wf = 0$$

then for any $u \in \mathcal{E}_{\mathcal{V}}$ we have $u \in \mathcal{E}_{\mathcal{W}} \Longleftrightarrow Qu \preceq_{\mathcal{E}_{\mathcal{V}}} u$. This last problem was considered recently also by R. K. Getoor and M. J. Sharpe ([10]) in the framework of the right processes without reference measure.

In this paper we deal with the above problems in the general framework of H-cones.

We give two H-cones S, T such that S is an H-subcone of T (i.e., $S \subset T$ and for any $M \subset S$ we have $\wedge_S M = \wedge_T M$ and also $\vee_S M = \vee_T M$ if moreover M is increasing and dominated in S). Here S (resp. T) is instead of $\mathcal{E}_{\mathcal{W}}$ (resp. $\mathcal{E}_{\mathcal{V}}$) in the preceding considerations. A map $Q : S \longrightarrow S$ is called an (S, T)-dilation operator on S if

1) $s_1, s_2 \in S,\ s_1 \leq s_2 \Longrightarrow Qs_1, \preceq_T Qs_2 \preceq_T s_2$,

2) $s \in S,\ u \in T,\ Qs \preceq_T u \leq s \Longrightarrow u \in S$.

It is proved that if Q verifies 1) then the property 2) is equivalent to each of the following properties:

3) $s_1 \wedge (Qs_1 + s_2 - Qs_2 + Qf) \in S \quad \forall\, s_1, s_2 \in S,\ f \in (S - S)_+$,

4) $s_1, s_2 \in S,\ s_1 - Qs_1 \leq s_2 - Qs_2 \Longrightarrow s_1 - Qs_1 + Qs_2 \in S$; $s \in S,\ t \in T \Longrightarrow \exists\, s' \in S$ with $(s - Qs) \wedge t = s' - Qs'$.

If B is a balayage on T then the operator $B^{\#}$ on S defined by

$$B^{\#}s = \wedge\{s' \in S \mid s' \geq Bs\}$$

is a balayage on S. If Q is a (S, T)-dilation operator then we have

5) $B^{\#}f - Bf = QB^{\#}f - B(QB^{\#}f) \quad \forall\, f \in (S - S)_+$

and therefore

6) $Bf \leq B^{\#}f \quad \forall\, f \in (S - S)_+$.

Under supplementary conditions (which are quite natural) we prove that if Q verifies 1) then 2) $\Longleftrightarrow$ 5). Moreover if the property 6) holds then there exists a minimal (S, T)-dilation operator P on S (i.e., an (S, T)-dilation operator on S such that any other (S, T)-dilation operator Q verifies the relation $Ps \preceq_T Qs,\ \forall\, s \in S$).

We consider also the problem when there is a unique (S, T)-dilation operator. It is shown, for instance, that the uniqueness holds in one of the following situations:

a) T contains sufficiently many quasi-continuous elements, there exist no absorbent balayages on T and there exists a balayage B on S which is an (S, T)-dilation operator.

b) $S = \mathcal{E}_{\mathcal{W}}$ where $\mathcal{W} = (W_\alpha)_{\alpha > 0}$ is a proper submarkovian resolvent on $(X, \mathcal{B})$ as in the beginning of this introduction such that there are no fine open sets of the form $\{x\}$ and T is the H-cone of all α-excessive functions with respect to $\mathcal{W}$.

Finally, we consider the problem when given an (S,T)-dilation operator Q on S we can extend it to a map $\overline{Q} : D(Q) \longrightarrow T$ defined on a solid subcone $D(Q)$ of T such that $S \subset D(Q)$ and such that we have, for any $u \in D(Q)$, the relation

$$u \in S \Longleftrightarrow \overline{Q}u \preceq_T u.$$

If $\mathcal{V}$, $\mathcal{W}$ are two proper submarkovian resolvents on $(X,\mathcal{B})$ which have all properties from the beginning of this introduction except the existence of reference measure, we can apply the above considerations to the H-cone $\mathrm{Exc}_{\mathcal{V}}$ and $\mathrm{Exc}_{\mathcal{W}}$ of excessive measures on $(X,\mathcal{B})$ associated with $\mathcal{V}$ and $\mathcal{W}$, respectively.

1. Localizable dilation operators in H-cones

Let S be an H-cone. We recall ([4], [5]) that a map $P : S \longrightarrow S$ is called a localizable dilation operator ($l.d$-operator) on S if P is additive, increasing, contractive, continuous in order from below and if for any $s,t \in S$ and any $f \in (S-S)_+$ we have

$$s \wedge (Ps + t - Pt + Pf) \in S.$$

It is known that if P is an $l.d$-operator on S then the convex cone S_P of $S-S$ given by

$$S_P := \{s - Ps \mid s \in S\}$$

endowed with the natural order relation from $S-S$ is also an H-cone. Moreover we have

$$t \in S,\ u \in S_P \Longrightarrow t \wedge u \in S_P,$$
$$f \in (S-S)_+,\ u \in S_P \Longrightarrow Pf \wedge u \in S_P$$

and for any $s \in S$ there exists a unique $s_0 \in S$ such that $s - Ps = s_0 - Ps_0$ and if $t \in S$ is such that $s_0 - Ps_0 \leq t - Pt$ then $s_0 \leq t$. (See [5].)

Concerning the lattice operations on the H-cone S_P we recall that for any subset A of S_P (resp. any upper directed and dominated subset A in S_P) we have

$$\bigwedge_{S_P} A = \bigwedge_{S-S} A \quad (\text{resp. } \bigvee_{S_P} A = \bigvee_{S-S} A).$$

Remark. If $\mathcal{V} = (V_\alpha)_{\alpha>\alpha}$ is a proper submarkovian resolvent on a measurable space $(X,\mathcal{B})$ and there exists a reference measure then the cone $\mathcal{E}_{\mathcal{V}}$ of all $\mathcal{V}$-excessive functions on X which are finite $\mathcal{V}$-a.s. is an H-cone. In this case, supposing that X is semisaturated (i.e., any H-integral on $\mathcal{E}_{\mathcal{V}}$ dominated by a measure on X is also represented as a measure on X) then the above notion of localizable dilation operator is nothing else than a kernel of subordination on $\mathcal{E}_{\mathcal{V}}$ in the sense of Mokobodzki [11].

We recall that for any $f \in (S-S)_+$ the map $B_f : S \longrightarrow S$ given by $B_f^S s = B_f s = \wedge\{t \in S \mid t \geq s \wedge (nf),\ (\forall)\, n \in \mathbb{N}\} = \bigvee_{n\in\mathbb{N}} R(s \wedge (nf))$ is a balayage on S.

Proposition 1.1. *Let P be an l.d-operator on S. The following assertions are equivalent:*

(a) *For any $s, t \in S$ such that*

$$\{u \in S_P \mid u \leq s\} = \{u \in S_P \mid u \leq t\}$$

we have $s = t$.

(b) *For any $s, t \in S$ such that*

$$\{u \in S_P \mid u \leq s\} \subset \{u \in S_P \mid u \leq t\}$$

we have $s \leq t$.

(c) *For any $f \in (S - S)_+$ such that $B_f \leq P$ we have $f = 0$.*

Proof. Obviously (a) $\Longleftrightarrow$ (b).

(b) $\Longrightarrow$ (c): Let $s, t \in S$ be such that $s \leq t$ and let $f = t - s$. If we suppose that $B_f \leq P$ then from the relations

$$B_f u \leq Pu \leq u \qquad (\forall)\, u \in S$$

we get

$$B_f u = B_f B_f u \leq B_f Pu \leq B_f u\,; \quad B_f u = B_f Pu$$

$$B_f u = B_f B_f u \leq PB_f u \leq B_f u\,; \quad B_f u = PB_f u$$

and therefore, using the definition of B_f, we have

$$v \in S,\ v \geq Pu \wedge (nf) \ \ (\forall)\, n \in \mathbb{N} \Longrightarrow v \geq u \wedge (nf) \qquad (\forall)\, n \in \mathbb{N}.$$

In particular, taking $v = Pu$,

$$(Pu) \wedge (nf) = u \wedge (nf) \qquad (\forall)\, n \in \mathbb{N}, \quad (u - Pu) \wedge f = 0,$$

$$(u - Pu) \wedge t = (u - Pu) \wedge [(t - s) + s]$$

$$\leq (u - Pu) \wedge (t - s) + (u - Pu) \wedge s = (u - Pu) \wedge s.$$

Hence if $u - Pu \leq t$ we get $u - Pu \leq s$ and therefore, using the hypothesis, we have $s = t$, $f = 0$.

(c) $\Longrightarrow$ (b): Let $s_0, t_0 \in S$ be such that for any $u \in S$ we have

$$u - Pu \leq s_0 \Longrightarrow u - Pu \leq t_0$$

or equivalently

$$u - Pu \leq s_0 \Longrightarrow u - Pu \leq v_0, \ \text{ where } \ v_0 := s_0 \wedge t_0.$$

Hence, since for any $u \in S$ we have $(u - Pu) \wedge s_0 \in S_P$ we deduce

$$(u - Pu) \wedge s_0 \leq (u - Pu) \wedge v_0, \ (u - Pu) \wedge (s_0 - v_0) = 0$$

for any $u \in S$. Hence

$$0 \leq (u - Pu) \wedge n(s_0 - v_0) \leq n[(u - Pu) \wedge (s_0 - v_0)] = 0,$$

$$u \wedge n(s_0 - v_0) \leq Pu \wedge n(s_0 - v_0) + (u - Pu) \wedge n(s_0 - v_0) = Pu \wedge n(s_0 - v_0)$$

for any $u \in S$ and any $n \in \mathbb{N}$. We get

$$B_{(s_0-v_0)}(u) \leq B_{(s_0-v_0)}(Pu) \leq Pu, \quad B_{(s_0-v_0)} \leq P$$

and using the hypothesis $s_0 - v_0 = 0$, $s_0 \leq t_0$. □

Remark. We recall ([3], [5]) that any balayage B on S is an $l.d$-operator on S and we denote by S_B the set $S_B = \{s - Bs \mid s \in S\}$.

Proposition 1.2. *Let P be an $l.d$-operator on S and let B be a balayage on S such that $B \leq P$. Then the map Q on S_B defined by*

$$Q(s - Ps) = Ps - Bs = Ps - BPs$$

is an $l.d$-operator on S_B and $S_P = (S_B)_Q$.

Proof. Let B be a balayage on S such that $B \leq P$. From

$$s \in S \Longrightarrow BPs \leq Bs = B^2 s \leq BPs; \; B^2 s \leq PBs \leq Bs$$

it follows that

$$PBs = Bs = BPs \quad (\forall)\, s \in S.$$

We have

$$Q(s - Ps) = Ps - Bs \leq s - Bs \quad (\forall)\, s \in S.$$

Obviously Q is additive. If $s_1, s_2 \in S$ are such that $s_1 - Bs_1 \leq s_2 - Bs_2$ we deduce

$$Ps_1 - Bs_1 = Ps_1 - PBs_1 \leq Ps_2 - PBs_2 = Ps_2 - Bs_2,$$

i.e., the map Q is increasing. If the family $(s_i - Bs_i)_{i \in I}$ increases to $s - Bs$, without loss of generality we may suppose that the family $(s_i)_{i \in I}$ increases to s. In this case the family $(Ps_i - Bs_i)_{i \in I}$ is increasing and since $(Ps_i)_i \uparrow Ps$, $(Bs_i)_i \uparrow Bs$ we get

$$\bigvee_{i \in I} Q(s_i - Bs_i) = Q(s - Bs),$$

i.e., Q is continuous in order from below.

Let now $u = s - Bs$, $v = t - Bt$, $f = w_1 - w_2 \in (S_B - S_B)_+$ where $s, t \in S$. We have $f \in (S - S)_+$, $v - Qv = t - Pt$ and

$$u \wedge (Qu + v - Qv + Qf) = s \wedge (Ps + t - Pt + Pf) - Bs.$$

Since P is an $l.d$-operator on S it follows that the element

$$r := s \wedge (Ps + t - Pt + Pf)$$

belongs to S and $Br = Bs$. Hence

$$u \wedge (Qu + v - Qv + Qf) = r - Br \in S_B$$

and therefore Q is an $l.d$-operator on S_B. From the relations

$$s - Ps = (s - Bs) - (Ps - Bs) = (s - Bs) - Q(s - Bs) \quad (\forall)\, s \in S$$

we deduce the equality $S_P = (S_B)_Q$. □

Corollary 1.3. *Let P be an l.d-operator on S and let B be the balayage on S defined by*

$$B = \bigvee\{B_f \mid f \in (S - S)_+, B_f \le P\}.$$

Then the map $Q : S_B \longrightarrow S_B$ given by $Q(s - Bs) = Ps - Bs$ is an l.d-operator on S_B such that there are no functions $g \in (S_B - S_B)_+$, $g \not\equiv 0$ with $B_g^{S_B} \le Q$.

Proof. We consider $g \in (S_B - S_B)_+$ and suppose that $B_g^{S_B} \le Q$. Obviously $g \in (S - S)_+$ and using ([3], [6]) we deduce

$$(B_g^S \vee B)(s) = B_g^{S_B}(s - Bs) + Bs \qquad (\forall)\ s \in S.$$

From the definition of B it follows that $B_g \le B$ and $B_g \vee B = B$. Hence $B_g^{S_B}(s - Bs) = 0$, $(\forall)\ s \in S$. □

Remark. From Proposition 1.2 and Corollary 1.3 we see that for any $l.d$-operator P on S we have $S_P = (S_B)_Q$ where B is a balayage on S and Q is an $l.d$-operator on S_B which satisfies one of the assertions from Proposition 1.1.

In the sequel we suppose that P is an l.d-operator on S which satisfies one of the equivalent assertions a)–c) *from Proposition* 1.1.

We recall ([4]) that if T is an H-cone then a convex subcone S of T is termed an *H-subcone of T* if S, endowed with the natural order relation of T, is an H-cone and for any subset (resp. any upper directed and dominated subset) A of S we have

$$\bigwedge_S A = \bigwedge_T A \quad (\text{resp. } \bigvee_S A = \bigvee_T A).$$

If S is an H-subcone of T then for any balayage B on T the map $B^\# : S \longrightarrow S$ defined by $B^\# s := \wedge\{t \in S \mid t \ge Bs\}$ is a balayage on S and we have $BB^\# s = Bs$ for any $s \in S$ (see [4]).

Definition. An H-cone S is called a *complete H-cone* if any subset A of S such that

$$\bigwedge_{n\in\mathbb{N}^*}\left(\bigvee_{a\in A}\left(p \wedge \left(\frac{a}{n}\right)\right)\right) = 0 \qquad (\forall)\ p \in S$$

is bounded.

As in [8] and [9] we can show that: *for any H-cone S there exists a unique* (up to an isomorphism) *complete H-cone $\overline{S}$ such that S is a solid and increasingly dense convex subcone of $\overline{S}$ which is termed the completion of S.*

Theorem 1.4. *There exists an order preserving embedding* $\theta : S \longrightarrow \overline{S}_P$ *such that*

$$\theta(t) - \theta(Pt) = t - Pt \qquad (\forall)\, t \in S.$$

More precisely for any $s \in S$ *we have* $\theta(s) = \bigvee_{\overline{S}_P}\{u \in S_P \mid u \leq s\}$ *and* $\theta(S)$ *is an* H*-subcone of* $\overline{S}_P$.

Proof. First we remark that for any $s \in S$ the set $A := \{a \in S_P \mid a \leq s\}$ is bounded in $\overline{S}_P$. Indeed, for any $v \in S_P$ we have

$$\bigwedge_{n\in\mathbb{N}^*}\left(\bigvee_{a\in A} v \wedge \left(\frac{a}{n}\right)\right) \leq \bigwedge_{n\in\mathbb{N}^*}\left(v \wedge \frac{s}{n}\right) \leq \bigwedge_{n\in\mathbb{N}^*}\frac{s}{n} = 0.$$

Since S_P is increasingly dense in $\overline{S}_P$ then for any element $w \in \overline{S}_P$ we consider an increasing family $(v_i)_i$ in S_P such that $w = \bigvee_i v_i$.
Let us denote

$$w_0 = \bigwedge_{n\in\mathbb{N}^*}\left(\bigvee_{a\in A}\left(\frac{a}{n} \wedge w\right)\right).$$

The set A is upper directed because for any $t \in S$ we have $(t - Pt) \wedge s \in S_P$.
Hence

$$w_0 \wedge v_i = \bigwedge_{n\in\mathbb{N}^*}\left(\bigvee_{a\in A}\left(\frac{a}{n} \wedge w \wedge v_i\right)\right) = \bigwedge_{n\in\mathbb{N}^*}\left(\bigvee_{a\in A}\left(\frac{a}{n} \wedge v_i\right)\right) = 0$$

and therefore $w_0 = w_0 \wedge w = \bigvee_i (w_0 \wedge v_i) = 0$.

The H-cone $\overline{S}_P$ being complete we deduce that A is bounded. We put, for any $s \in S$,

$$\theta(s) = \bigvee_{\overline{S}_P}\{u \in S_P \mid u \leq s\}.$$

Since, for any $s \in S$, the family $\{u \in S_P \mid u \leq s\}$ is upper directed it follows that for any $s_1, s_2 \in S$ we have $\theta(s_1 + s_2) \geq \theta(s_1) + \theta(s_2)$.

If $u \in S_P$ and $u \leq s_1 + s_2$ then we have

$$u \wedge s_1 \in S_P,\;\; u \wedge s_2 \in S_P,\;\; u \leq u \wedge s_1 + u \wedge s_2,$$
$$u \wedge s_1 \leq \theta(s_1),\;\; u \wedge s_2 \leq \theta(s_2),\;\; u \leq \theta(s_1) + \theta(s_2),$$

i.e., $\theta(s_1 + s_2) \leq \theta(s_1) + \theta(s_2)$. Obviously θ is increasing.

Suppose now that $s_1, s_2 \in S$ are such that $\theta(s_1) \leq \theta(s_2)$. Then for any $u \in S_P$, $u \leq s_1$ we have

$$u \leq \theta(s_1) \leq \theta(s_2) = \bigvee_{\overline{S}_P}\{v \mid v \in S_P,\; v \leq s_2\},$$

$$u = \bigvee_{\overline{S}_P}\{u \wedge v \mid v \in S_P,\; v \leq s_2\} = \bigvee_{S_P}\{u \wedge v \mid v \in S_P,\; v \leq s_2\} =$$

$$\bigvee_S \{u \wedge v \mid v \in S_P,\ v \leq s_2\} \leq s_2.$$

Hence $s_1 \leq s_2$. We show now that for any $s \in S$ we have $\theta(s) - \theta(Ps) = s - Ps$ or equivalently $s - Ps + \theta(Ps) = \theta(s)$. Indeed, if $u \in S_P$ is such that $u \leq Ps$ then we have

$$s - Ps + u \in S_P,\quad s - Ps + u \leq s \text{ and } s - Ps + u \leq \theta(s).$$

Hence u being arbitrary we get $s - Ps + \theta(Ps) \leq \theta(s)$. Let now $u \in S_P$ be such that $u \leq s$. Then we have

$$u \leq s - Ps + Ps,\quad u \leq (s - Ps) + u \wedge Ps.$$

Since $u \wedge Ps$ belongs to S_U we get $u \leq (s - Ps) + \theta(Ps)$ and therefore, u being arbitrary, we obtain

$$\theta(s) \leq s - Ps + \theta(Ps);\quad s - Ps = \theta(s) - \theta(Ps).$$

To prove that $\theta(S)$ is an H-subcone of $\overline{S}_P$ we consider now $A \subset S$ arbitrary. Obviously we have

$$\theta(\bigwedge_S A) \leq \bigwedge_{\overline{S}_P} \theta(A).$$

Conversely, if $u \in S_P$ is such that $u \leq \theta(s)$ for any $s \in A$ then we have $u \leq s$ for all $s \in A$ and therefore

$$u \leq \bigwedge_S A,\quad u \leq \theta(\bigwedge_S A).$$

Since S_P is increasingly dense in $\overline{S}_P$ we deduce

$$\bigwedge_{\overline{S}_P} \theta(A) = \theta(\bigwedge_S A).$$

Let now A be an upper directed and dominated subset of S. Obviously we have

$$\theta(\bigvee_S A) \geq \bigvee_{\overline{S}_P} \theta(A).$$

For the converse inequality we consider an element $u \in S_P$ such that $u \leq \theta(\bigvee_S A)$. We have

$$u \leq \bigvee_S A,\quad u = \bigvee_{S-S} \{u \wedge s \mid s \in A\}.$$

Since $u \wedge s \in S_P$ and $u \wedge s \leq s$ we get

$$u - \bigvee_{S_P}\{s \wedge u \mid s \in A\} \leq \bigvee_{\overline{S}_P}\{\theta(s) \mid s \in A\}.$$

The element $u \in S_P$ being arbitrary and S_P being increasingly dense in $\overline{S}_P$ we get

$$\theta(\bigvee_S A) \leq \bigvee_{\overline{S}_P} \theta(A),\quad \theta(\bigvee_S A) = \bigvee_{\overline{S}_P} \theta(A).$$

□

Remarks. 1. In the sequel we identify S with its image $\theta(S)$ in $\overline{S}_P$. In this way S becomes an H-subcone of $\overline{S}_P$.

2. For any balayage B on $\overline{S}_P$ (or equivalently on S_P) we denote by $B^{\#}$ the balayage on S associated with B by

$$B^{\#}s = \bigwedge\{t \in S \mid t \geq Bs\} \qquad (\forall) s \in S.$$

3. For any $f \in (S-S)_+$ we have $Pf \in \overline{S}_P$. The assertion follows from the fact that $Pf \in (S-S)_+$ and $u \wedge Pf \in S_P$ for any $u \in S_P$.

If $f \in (S-S)_+$ we shall denote by B_f the balayage on $\overline{S}_P$ defined by

$$B_f u = \bigwedge\{v \in \overline{S}_P \mid v \geq v \wedge (nf) \quad (\forall)\ n \in \mathbb{N}\}$$

and we remark that in this case we have

$$B_f^{\#}s = \bigwedge\{t \in S \mid t \geq s \wedge (nf) \quad (\forall)\ n \in \mathbb{N}\}.$$

Theorem 1.5. *For any balayage B on $\overline{S}_P$ and any $s \in S$ we have*

$$B^{\#}s - Bs = PB^{\#}s - BPB^{\#}s.$$

Proof. For any $s \in S$ we put $Ls := Ps + B(s - Ps)$. Since $B(s-Ps) \in S_P$, $B(s-Ps) \leq s - Ps$ we get

$$Ps + B(s-Ps) = s \wedge (Ps + B(s-Ps)) \in S.$$

On the other hand for any $t \in S$ such that $t - Pt \leq s = (s-Ps) + Ps$ there exists $s_1, s_2 \in S$ such that

$$t - Pt = (s_1 - Ps_1) + (s_2 - Ps_2),$$

$$s_1 - Ps_1 \leq s - Ps, \ s_2 - Ps_2 \leq Ps.$$

Hence

$$B(t - Bt) = B(s_1 - Ps_1) + B(s_2 - Ps_2) \leq B(s-Ps) + (s_2 - Ps_2)$$

and therefore

$$B(t-Pt) \leq B(s-Ps) + Ps,$$

$$Bs = \bigvee\{B(u - Pu) \mid u \in S, u - Pu \leq s\} \leq$$

$$\leq B(s-Ps) + Ps = Ls.$$

From the above considerations we have $s \geq Ls \geq B^{\#}s$, $L(B^{\#}s) = B^{\#}s$ and therefore $B^{\#}s - Bs = P(B^{\#}s) - BP(B^{\#}s)$. □

Corollary 1.6. *For any balayage B on $\overline{S}_P$ and any $f \in (S-S)_+$ we have*

$$Bf \leq B^{\#}f.$$

2. (S,T)-dilation operators

In this section we suppose that T is an H-cone and S is an H-subcone of T.

Definition. If $Q : S \longrightarrow S$ is an $(l.d)$-operator on S such that

$$s_1, s_2 \in S,\ s_1 \leq s_2 \Longrightarrow Qs_1 \preceq_T Qs_2 \preceq_T s_2$$

and such that the set $S_Q := \{s - Qs \mid s \in S\}$ is a solid subset of T with respect to the natural order of T then Q will be termed an *(S,T)-dilation operator.*

Remark 1. If S is an H-cone and P is an $l.d$-operator on S which satisfies one of the equivalent properties a)–c) from Proposition 1.1 then P is an (S,T)-dilation operator where T is the completion of the H-cone S_P. Moreover in this case S_P is increasingly dense on T.

Remark 2. If Q is an (S,T)-dilation operator then Q is an (S,T_0)-dilation operator where T_0 is the smallest naturally solid subcone of T such that $S \subset T_0$.

Theorem 2.1. *Let $Q : S \longrightarrow S$ be an additive, increasing an continuous in order from below map such that*

$$s_1, s_2 \in S,\ s_1 \leq s_2 \Longrightarrow Qs_1 \preceq_T Qs_2 \preceq_T s_2.$$

Then the following assertions are equivalent:
(a) *Q is an (S,T)-dilation operator.*
(b) *For any element $u \in T$ such that there exists $s \in S$ for which $Qs \preceq_T u \leq s$ we have $u \in S$.*
(c) *The set S_Q defined by $S_Q := \{s - Qs \mid s \in S\}$ is a solid subcone of T (w.r.t. the natural order of T) and for any $s_1, s_2 \in S$ we have*

$$s_1 - Qs_1 \leq s_2 - Qs_2 \Longrightarrow s_1 - Qs_1 + Qs_2 \in S.$$

(d) *There exists an (S,T)-dilation operator Q_1 such that $Q_1 s \preceq_T Qs,\ (\forall)\, s \in S$.*

Proof. (a) $\Longrightarrow$ (b): If $u \in T$ and there exists $s \in S$ such that $Qs \preceq_T u \leq s$ then the element $u - Qs$ belongs to T and we have $u - Qs \leq s - Qs$. Since S_Q is a solid part of T with respect to the natural order then there exists $s' \in S$ with

$$s' - Qs' = u - Qs \leq s - Qs.$$

Since Q is a localizable dilation operator on S we deduce

$$u = s' - Qs' + Qs = s \wedge (Qs + s' - Qs') \in S.$$

(b) $\Longrightarrow$ (a): Let $s_1, s_2 \in S$ and $f \in (S - S)_+$ be arbitrary and let u be the element of $S - S$ defined by $u := s_2 \wedge (Qs_2 + s_1 - Qs_1 + Qf)$. From hypothesis we have $u \in T$ and $Qs_2 \preceq_T u \leq s_2$ i.e., $u \in S$ and therefore Q is localizable dilation operator on S. It remains to show that the set S_Q is solid in T.

Let $u \in T$ and $s \in S$ be such that $u \le s - Qs$. Let us consider the element $s_0 \in S$ defined by

$$s_0 = \bigwedge \{s' \in S \mid u \le s' - Qs'\}.$$

Obviously we have $u + Qs_0 \le s'$ for any $s' \in S$ such that $u \le s' - Qs'$. Since S is an H-subcone of T we get

$$u + Qs_0 \le \bigwedge \{s' \in S \mid u \le s' - Qs'\} = s_0, \;\; u \le s_0 - Qs_0.$$

Obviously we have $u + Qs_0 \in T$ and $Qs_0 \preceq_T u + Qs_0 \le s_0$. Hence the element $s_0' := u + Qs_0$ belongs to S and we have

$$s_0' \le s_0, \;\; u = s_0' - Qs_0 \le s_0' - Qs_0'.$$

The last inequality implies that $s_0' \ge s_0$ and therefore $s_0' = s_0$. Hence $u = s_0 - Qs_0$.

The implications (a) $\Longrightarrow$ (c) and (a) $\Longrightarrow$ (d) are obvious.

(c) $\Longrightarrow$ (b): Let $u \in T$ and $s \in S$ be such that $Qs \preceq_T u \le s$. The element $t := u - Qs$ belongs to T and $t = u - Qs \le s - Qs$. Hence there exists $s_1 \in S$ such that $t = s_1 - Qs_1$. From the hypothesis and using the inequality $s_1 - Qs_1 \le s - Qs$ we get

$$u = t + Qs = s_1 - Qs_1 + Qs \in S.$$

(d) $\Longrightarrow$ (*b*): Let Q_1 be an (S,T)-dilation operator such that $Q_1 s \preceq_T Qs$ for all $s \in S$ and let $u \in T$, $s' \in S$ be such that $Qs' \preceq_T u \le s'$. Then we have

$$Q_1 s' \preceq_T Qs' \preceq_T u \le s'$$

and therefore $u \in S$.

For any balayage B on T we denote by $B^\#$ the balayage on S given by

$$B^\# s = \bigwedge \{s' \in S \mid s' \ge Bs\} \qquad (\forall)\; s \in S.$$

Obviously we have $Bs \le B^\# s, \;\; B(B^\# s) = Bs, \;\; (\forall)\; s \in S.$ □

Theorem 2.2. *If Q is an (S,T)-dilation operator then for any balayage B on T we have*

$$B^\# s - Bs = QB^\# s - BQB^\# s \qquad (\forall)\; s \in S.$$

Proof. Let $s \in S$ and let B be a balayage on T. We put $u := Bs + QB^\# s - BQB^\# s$. From the relations

$$BB^\# s = Bs, \;\; Bs - BQB^\# s = B(B^\# s - QB^\# s) \le B^\# s - QB^\# s$$

and using the fact that Q is an (S,T)-dilation operator we deduce that $u \in S$ and $Bs \le u \le B^\# s$. Hence $u = B^\# s$ and therefore $B^\# s - Bs = QB^\# s - BQB^\# s$. □

Corollary 2.3. *If there exists an (S,T)-dilation operator Q on S then for any $f \in (S - S)_+$ and any balayage B on T we have $Bf \le B^\# f$.*

Proposition 2.4. *Let Q be an (S,T)-dilation operator and let B be a balayage on T. Then the following assertions are equivalent:*

(a) $B(s - Qs) = 0 \qquad (\forall)\ s \in S.$
(b) $B^{\#}s \le Qs \qquad (\forall)\ s \in S.$
(c) $B^{\#}s = QB^{\#}s \qquad (\forall)\ s \in S.$
(d) $B^{\#}s = B^{\#}Qs \qquad (\forall)\ s \in S.$

Proof. (a) $\Longrightarrow$ (b): If $s \in S$ then the relation (a) implies: $Bs = BQs \le Qs$. Since $Qs \in S$ we deduce $B^{\#}s \le Qs$.

(b) $\Longrightarrow$ (c) and (c) $\Longrightarrow$ (d) follow from the fact that for any $s \in S$ we have

$$B^{\#}s \le Qs \le s \Longrightarrow B^{\#}s = B^{\#}(B^{\#}s) \le Q(B^{\#}s) \le B^{\#}s,$$
$$B^{\#}s = QB^{\#}s \le Qs \Longrightarrow B^{\#}s = B^{\#}(B^{\#}s) \le B^{\#}Qs \le B^{\#}s.$$

The implication (d) $\Longrightarrow$ (a) may be obtained from the fact that for any $s \in S$ we have $Bs = BB^{\#}s$. □

Theorem 2.5. *Let Q be an (S,T)-dilation operator and let B_0 be the greatest balayage on T which vanishes on S_Q. Then the balayage B_0 is absorbent (i.e., $B_0 t \preceq_T t$ for any $t \in T$) and for any $s', s'' \in S$ we have $s' \le s'' \Longrightarrow B_0 s' \preceq_T B_0 s''$. In particular, if there are no absorbent balayages B on T, $B \neq 0$ such that*

$$s', s'' \in S, \ \ s' \le s'' \Longrightarrow Bs' \preceq_T Bs''$$

then S_Q is increasingly dense in T.

Proof. For any $s \in S$ we consider the balayage B_{s-Qs} on T defined by

$$B_{s-Qs}(u) := \bigvee \{u \wedge n(s - Qs) \mid n \in \mathbb{N}\}.$$

Since $s - Qs \in T$, the complement B'_{s-Qs} of the balayage B_{s-Qs} ([3]) is absorbent ([1]). Moreover for any $s', s'' \in S$ we have

$$(s' - Qs') + (s'' - Qs'') = (s' + s'' - Q(s' + s'')),$$

$$B_{s'-Qs'} \le B_{s'+s''-Q(s'+s'')}, \ \ B_{s''-Qs''} \le B_{s'+s''-Q(s'+s'')}$$

and therefore the family $(B'_{s-Qs})_{s\in S}$ of absorbent balayages on T is decreasing. Hence the map $B_0 : T \longrightarrow T$ defined by

$$B_0 u = \bigwedge_{s\in S} B'_{s-Qs}u = \bigwedge_{s\in S} B'_{s-Qs}u$$

is also an absorbent balayage on T and, moreover, we have $B_0(s - Qs) = 0$ for all $s \in S$ and

$$s', s'' \in S, s' \le s'' \Longrightarrow B_0 s' = B_0(Qs') \preceq_T B_0(Qs'') = B_0 s'' \preceq_T s''.$$

The fact that B_0 is the greatest balayage which vanishes on S_Q follows from the fact that for any such a balayage B on T and any $s \in S$ we have

$$B(B_{s-Qs}u) = \bigvee_{n\in\mathbb{N}} B(n(s-Qs)\wedge u) \leq \bigvee_{n\in\mathbb{N}} nB(s-Qs) = 0$$

and therefore $B \wedge B_{s-Qs} = 0$. Hence, using ([1], [3]) we have

$$B = B\wedge I = B\wedge(B_{s-Qs}\vee B'_{s-Qs}) = (B\wedge B_{s-Qs})\vee(B\wedge B'_{s-Qs}) =$$

$$B\wedge B'_{s-Qs} \leq B'_{s-Qs}$$

and therefore, the element $s \in S$ being arbitrary, $B \leq \bigwedge_{s\in S} B'_{s-Qs} = B_0$.

From the preceding considerations we see that B_0 is in fact the complement of the balayage $\bigvee_{s\in S} B_{s-Qs}$. If there are no nontrivial absorbent balayages B on T such that

$$s', s'' \in S,\ \ s' \leq s'' \Longrightarrow Bs' \preceq_T Bs''$$

it follows that $B_0 = 0$ and therefore $I = \bigvee_{s\in S} B_{s-Qs}$ and the convex subcone S_Q is increasingly dense in T. □

Theorem 2.6. *For any balayage B on T which satisfies one of the equivalent properties* a)–d) *from Proposition* 2.4 *for a given (S,T)-dilation operator Q on S we have $s', s'' \in S, s' \leq s'' \Longrightarrow B^{\#}s' \preceq_T B^{\#}s'' \preceq_T s''$. In particular, if there are no nontrivial balayages L on S such that*

$$s', s'' \in S,\ \ s' \leq s'' \Longrightarrow Ls' \preceq_T Ls'' \preceq_T s''$$

then S_Q is increasingly dense in T.

Proof. From Proposition 2.4 we deduce

$$s', s'' \in S, s' \leq s'' \Longrightarrow B^{\#}s' = Q(B^{\#}s') \preceq_T Q(B^{\#}s'') \preceq_T B^{\#}s'',$$
$$s \in S \Longrightarrow B^{\#}s = Q(B^{\#}s) \preceq_T Qs \preceq_T s.$$

The last part of the proof follows from Theorem 2.5. □

3. Existence of (S,T)-dilation operators

In this section S and T are two H-cones such that S is an H-subcone of T. For any balayage B on T we denote by $B^{\#}$ the balayage on S defined by

$$B^{\#}s := \bigwedge\{t \in S \mid t \geq Bs\}.$$

We want to construct, under supplementary conditions, an (S,T)-dilation operator on S.

Theorem 3.1. *The following assertions are equivalent:*
(1) *for any balayage B on T and any $f \in (S-S)_+$ we have $Bf \leq B^{\#}f$;*
(2) *for any balayage B on T and any $f \in (S-S)_+$ there exists an element $t \in T$ such that $B^{\#}f - Bf = t - Bt$;*
(3) *for any balayage B on T, any $f \in (S-S)_+$ and any element $s \in S$ such that $f \leq s$ we have $0 \leq B^{\#}f - Bf \preceq_T s$;*
(4) *for any finite family $(f_i)_{i\in I}$, $f_i \in (S-S)_+$, any finite family $(B_i)_{i\in I}$ of balayages on T and any $s \in S$ such that $\sum_{i\in I} B_i^{\#} f_i \leq s$ we have $0 \leq \sum_{i\in I}(B_i^{\#} f_i - B_i f_i) \preceq_T s$.*

Proof. The implications (4) $\Longrightarrow$ (3) $\Longrightarrow$ (1) and (2) $\Longrightarrow$ (1) are obvious.

(1) $\Longrightarrow$ (2): Let $f \in (S-S)_+$ and let B be a balayage on T. For any balayage M on the H-cone T_B, where $T_B := \{t - Bt \mid t \in T\}$, there exists (see [6]) a balayage B_1 on T, $B_1 \geq B$, such that $M(t-Bt) = B_1 t - Bt$, $(\forall)\, t \in T$.
Since the element $B^{\#}f - Bf$ belongs to $T_B - T_B$ it remains to show that $M(B^{\#}f - Bf) \leq B^{\#}f - Bf$ (see [3]). We have, using the hypothesis, $M(B^{\#}f - Bf) = B_1(B^{\#}f) - B_1Bf \leq B_1^{\#}(B^{\#}f) - B_1Bf = B^{\#}f - Bf$. Hence (1) $\Longleftrightarrow$ (2).

(1) $\Longrightarrow$ (3): Let $f \in (S-S)_+$ and let B be a balayage on T. If $t \in S$ is such that $f \leq t$ we have

$$t - (B^{\#}f - Bf) - Bt = (t - Bt) - (B^{\#}t - Bt) + (B^{\#}(t-f) - B(t-f)),$$

i.e., the element $u := t - (B^{\#}f - Bf) - Bt$ belongs to $T_B - T_B$. We show now that $u \in T_B$. Indeed, if we consider a balayage M on T_B there exists a balayage B_1 on T, $B_1 \geq B$ such that $M(s - Bs) = B_1(s - Bs)$, $(\forall)\, s \in T$ and therefore

$$\begin{aligned} M(u) = B_1(u) &= B_1(t - B^{\#}f) - B_1(Bt - Bf) \\ &\leq B_1^{\#}(t - B^{\#}f) - (Bt - Bf) = B_1^{\#}t - B^{\#}f + Bf - Bt \\ &\leq t - (B^{\#}f - Bf) - Bt, \end{aligned}$$

$$Mu \leq u.$$

Hence $u \in T_B$. On the other hand we have $u \leq t - Bt$ and therefore $u + Bt \in T$, i.e., $t - (B^{\#}f - Bf) \in T$.

(3) $\Longrightarrow$ (4): We proceed inductively and we suppose that for any system $(f_1, f_2, \dots f_n)$ of elements of $(S-S)_+$ and any system $(B_1, B_2, \dots, B_n)$ of balayages on T such that $\sum_{i=1}^{n} B_i^{\#} f_i \leq s$, where $s \in S$, we have $s - \sum_{i=1}^{n}(B_i^{\#} f_i - B_i f_i) \in T$.
Let now $\{f_1, f_2, \dots, f_{n+1}\}$ be a subset of $(S-S)_+$ and $(B_1, B_2, \dots, B_{n+1})$ be a system of balayages on T such that $\sum_{i=1}^{n+1} B_i^{\#} f_i \leq s$ where s is an element of S. We have $s - \sum_{i=2}^{n+1}(B_i^{\#} f_i - B_i f_i) \in T$ and we want to show that $s - \sum_{i=1}^{n+1}(B_i^{\#} f_i - B_i f_i)$

is an element of T. If we denote

$$f = s - \sum_{i=1}^{n+1}(B_i^{\#} f_i - B_i f_i), \quad u = s - \sum_{i=2}^{n+1}(B_i^{\#} f_i - B_i f_i)$$

and by Rf the réduite of f with respect to the H-cone T then we have

$$u \in T, \quad u = f + (B_1^{\#} f_1 - B_1 f_1), \quad u \geq f, \quad u \geq Rf.$$

On the other hand for any $\alpha \in (0,1)$ we have $B_\alpha f \geq B_\alpha(\alpha Rf) = \alpha Rf$, where B_α is the balayage on T defined by

$$t \in T, \quad B_\alpha t = \bigvee_{n \in \mathbb{N}} R(t \wedge ng), \quad g := (f - \alpha Rf)_+$$

(see [2]). On the other hand, we have

$$B_1(B_1^{\#} f_1 - B_1 f_1) = 0, B_1 f = B_1 u \geq B_1 Rf \geq B_1(\alpha Rf).$$

From the relations $B_\alpha f \geq B_\alpha(\alpha Rf) = \alpha Rf$, $B_1 f \geq B_1(\alpha Rf)$ and using the fact that for any $s \in T$ we have $(B_\alpha \vee B_1)s = (B_\alpha s) \vee (B_1 s)$ it follows that

$$(B_\alpha \vee B_1) f \geq (B_\alpha \vee B_1)(\alpha Rf) \geq \alpha Rf.$$

From the relation $0 \leq u - B^{\#} f_1$ we deduce

$$(B_\alpha \vee B_1) f = (B_\alpha \vee B_1)(u - B_1^{\#} f_1) + (B_\alpha \vee B_1)(B_1 f_1)$$
$$\leq (B_\alpha \vee B_1)^{\#}(u - B_1^{\#} f_1) + B_1 f_1 = (B_\alpha \vee B_1)^{\#} u - B_1^{\#} f_1 + B_1 f_1$$
$$\leq u - (B_1^{\#} f_1 - B_1 f_1) = f.$$

Hence for any $\alpha \in (0,1)$ we have $f \geq (B_\alpha \vee B_1) f \geq \alpha Rf$ and therefore $f = Rf \in T$. □

Lemma 3.2. *Suppose that $Bf \leq B^{\#} f$. Denote for any $f \in (S-S)_+$ and any balayage B on T by P_B the map $P_B : S \longrightarrow T$ defined by*

$$P_B s := B^{\#} s - B^{\#} s \wedge_{sp} Bs,$$

where $\wedge_{sp}$ is the infimum in $(T, \preceq_T)$. Then we have
(1) *$P_B(s_1 + s_2) = P_B s_1 + P_B s_2$.*
(2) *$s_1, s_2 \in S$, $s_1 \leq s_2 \Longrightarrow P_B s_1 \preceq_T P_B s_2 \preceq_T s_2$.*
(3) *$P_B(B^{\#} s) = P_B s \quad \forall s \in S$.*
(4) *P_B is continuous in order from below (i.e., in particular, $s_i \uparrow s, s_i,\ s \in S \Longrightarrow P_B s_i \uparrow P_B s$).*

For the proof, see [4].

In the sequel, in this section we suppose that the pair (S, T) satisfies the following two conditions:

a) *Any increasing family of S dominated in T is dominated in S.*

b) *If for any* $t_1, t_2 \in T$ *such that*

$$f \in (S-S)_+,\ \ f \le t_1 \Longrightarrow f \le t_2$$

then $t_1 \le t_2$.

Remarks. 1. If S is complete then the property (a) is satisfied.
2. The property b) is equivalent with the following one: for any $t \in T$ we have

$$t = \vee\{R(f) \mid f \in (S-S)_+, f \le t\}$$

where Rf means the réduite of f in the H-cone T.

Lemma 3.3. *Suppose that* $Bf \le B^{\#} f$ *for any* $f \in (S-S)_+$ *and any balayage* B *on* T. *If* $u \in T$ *is such that*

$$f \in (S-S)_+,\ \ f \le u \Longrightarrow P_B f \preceq_T u$$

for any balayage B *on* T *then* $u \in S$.

Proof. Let $f \in (S-S)_+$, $f \le u$ and let B be a balayage on T. We show that

$$B^{\#} f \le u.$$

Indeed, we have by hypothesis $P_B f \preceq_T u$, and therefore there exists $v \in T$ with

$$u = v + P_B f.$$

Suppose that $f = s - t$ where $s, t \in S$. We have

$$B^{\#}(s-t) + v = B^{\#} s \wedge_{sp} Bs - B^{\#} t \wedge_{sp} Bt + u$$

and therefore

$$B(s-t) + Bv = B^{\#} s \wedge_{sp} Bs - B^{\#} t \wedge_{sp} Bt + Bu.$$

Hence we get

$$u + B(s-t) + Bv = v + B^{\#}(s-t) + Bu.$$

Since $Bv \le v$ and $s - t \le u$ we deduce $B(s-t) \le Bu$ and therefore $u \ge B^{\#}(s-t) = B^{\#} f$.
We show now that $u \ge R^S(f) := \wedge\{s' \in S \mid s' \ge f\}$.
It is known ([2]) that for any $\alpha \in (0,1)$ we have

$$R^S(f) = B_h^S(R^S(f)) \le \frac{1}{\alpha} B_h^S(f)$$

where $h = (f - \alpha R^S f)_+$ and B_h^S is the balayage on S defined by

$$B_h^S s' = \wedge\{s'' \in S \mid s'' \ge s' \wedge nh,\ (\forall)\, n \in \mathbb{N}\}.$$

If B_h denote the following balayage on T

$$B_h t' := \wedge\{t'' \in T \mid t'' \ge t' \wedge nh,\ (\forall)\, n \in \mathbb{N}\}$$

then we have already observed that $B_h^S = (B_h)^{\#}$ and therefore:

$$R^S(f) \leq \frac{1}{\alpha} B_h^S(f) = \frac{1}{\alpha}(B_h)^{\#}(f) \leq \frac{1}{\alpha}u, \ \alpha R^S(f) \leq u.$$

Since $\alpha \in (0,1)$ is arbitrary we get $R^S(f) \leq u$. If we denote, for any $g \in S - S$,

$$R^T(g) := \wedge\{t \in T \mid t \geq g\},$$

obviously we have $R^T(g) \leq R^S g$ and from the preceding considerations we deduce

$$\begin{aligned} u &= \bigvee\{R^T(f) \mid f \in (S-S)_+, \ f \leq u\} \\ &\leq \bigvee\{R^S(f) \mid f \in (S-S)_+, \ f \leq u\} \leq u, \\ u &= \bigvee\{R^S(f) \mid f \in (S-S)_+, \ f \leq u\} \in S. \end{aligned}$$

The above result is an extension of a similar one ([4]) given in the case where S and T are standard H-cones of functions. □

Notation. Suppose that $Bf \leq B^{\#} f$ for any $f \in (S-S)_+$ and any balayage B on T. In the following, for any $s \in S$, we shall denote by Ps the element of T given by

$$Ps := \vee_{sp}\{\sum_{i \in I} P_{B_i} s_i \mid I \text{ finite}, \ s_i \in S, \ \sum_{i \in I} s_i \leq s, \ B_i \text{ balayage on } T\}$$

where $\vee_{sp}$ is the supremum in $(T, \preceq_T)$.

Lemma 3.4. *If $s \in S$ and $u \in T$ are such that $Ps \preceq_T u \leq s$, then $u \in S$.*

Proof. If $f \in (S-S)_+$ is such $f \leq u$ then for any balayage B on T we have

$$P_B f \preceq_T P_B s \preceq_T Ps \preceq_T u, \quad P_B f \preceq u.$$

The assertion follows now from the previous lemma. □

Corollary 3.5. *For any $s \in S$ we have $Ps \in S$ and $Ps \preceq_T s$.*

Proof. Indeed, from Theorem 3.1 we deduce $Ps \preceq_T s$ and from Lemma 3.4 we get $Ps \in S$. □

Theorem 3.6. *The above map $P : S \longrightarrow S$ is an (S,T)-dilation operator on S. Moreover, for any (S,T)-dilation operator Q on S we have $Ps \preceq_T Qs$ $(\forall)$ $s \in S$.*

Proof. By definition and by Corollary 3.5 we have

$$s_1, s_2 \in S, \quad s_1 \leq s_2 \Longrightarrow Ps_1 \preceq_T Ps_2 \preceq_T s_2.$$

From ([4], Theorem 2.8) it follows that P is additive and continuous in order from below. Using now Lemma 3.4 and Theorem 2.1 we deduce that P is (S,T)-dilation operator on S.

Since for any balayage B on S we have

$$B^{\#}s - Bs = QB^{\#}s - BQB^{\#}s \preceq_T QB^{\#}s \preceq_T Qs$$

we deduce, from the definition of P_B, that $P_B s \preceq_T Qs$, and therefore $Ps \preceq_T Qs$. □

Theorem 3.7. *Let $Q : S \longrightarrow S$ be a map which is additive, continuous in order from below and such that*

$$s_1, s_2 \in S,\ s_1 \leq s_2 \Longrightarrow Qs_1 \preceq_T Qs_2 \preceq_T s_2.$$

Then Q is an (S,T)-dilation operator iff for any balayage B on T and any $s \in S$ we have

$$B^{\#}s - Bs = QB^{\#}s - BQB^{\#}s.$$

Proof. If Q is an (S,T)-dilation operator then from Theorem 2.2 we have

$$B^{\#}s - Bs = QB^{\#}s - BQB^{\#}s$$

for any $s \in S$ and any balayage B on T. Conversely suppose that this formula holds for any $s \in S$ and any balayage B on T. We have for any balayage B on T and any $s \in S$,

$$B^{\#}s - Bs = QB^{\#}s - BQB^{\#}s \preceq_T QB^{\#}s \preceq_T Qs$$

and therefore $P_B s \preceq_T Qs$, $Ps \preceq Qs$.

Using Theorem 2.1, the preceding inequality and Theorem 3.6 we deduce that Q is an (S,T)-dilation operator on S. □

4. On the uniqueness of (S,T)-dilation operator

In this section S and T will be H-cones such that S is an H-subcone of T and the following assertions are fulfilled:

a) Any increasing family in S, dominated in T is also dominated in S.

b) For any $t_1, t_2 \in T$ such that

$$\{f \in (S-S)_+ \mid f \leq t_1\} \subset \{f \in (S-S)_+ \mid f \leq t_2\}$$

we have $t_1 \leq t_2$.

c) For any $f \in (S-S)_+$ and any balayage B on T we have $Bf \leq B^{\#}f$.

In the preceding section we have proved that under the above conditions there exist (S,T)-dilation operators on S. Moreover there exists an (S,T)-dilation operator P on S such that for any (S,T)-dilation operator Q on S we have $Ps \leq Qs$ (or, more precisely, $Ps \preceq_T Qs$) for all $s \in S$. This remarkable (S,T)-dilation operator on S will be termed the *minimal (S,T)-dilation operator on S*.

In this section we deal with the uniqueness problem for the family of (S,T)-dilation operators on S.

For simplicity we suppose that the H-cone T contains sufficiently many quasi-continuous elements.

We recall that an element $u \in T$ is termed quasicontinuous if for any increasing family $(u_i)_i$ of T such that $\bigvee_{i \in I} u_i = u$ we have $\bigwedge_{i \in I} R(u - u_i) = 0$ where R means the réduite operator on T. We say that T contains sufficiently many quasi-continuous elements if any element of T is the supremum of the family of its quasi-continuous minorants.

Lemma 4.1. *Let C be an H-cone which contains sufficiently many quasi-continuous elements and let $\varphi : C \longrightarrow C$ be an additive, increasing, continuous in order from below map such that*

$$s_1, s_2 \in C,\ s_1 \leq s_2 \Longrightarrow \varphi(s_1) \preceq \varphi(s_2) \preceq s_2.$$

Then there exists a recurrent balayage ([7]) B on C such that $\varphi(s) \preceq Bs$ for all $s \in C$.

Proof. We denote

$$C_0 := \{s \in C \mid \varphi(s) = 0\},\ C_1 := \{t \in C \mid t \wedge_{sp} s = 0,\ \ (\forall)\ s \in C_0\},$$

where $\wedge_{sp}$ is the infimum in $(C, \preceq)$.

First we show that if $s_1 \in C_1$ and $s_2 \in C$ are such that $s_1 \leq s_2$ then $s_1 \preceq s_2$. Indeed, let us put $u = s_1 \wedge_{sp} s_2$, $s_1' = s_1 - u$, $s_2' = s_2 - u$. We have $s_1' \in C_1$, $s_1' \wedge_{sp} s_2' = 0$, $s_1' \leq s_2'$. Since $\varphi(s_1') \preceq s_1'$, $\varphi(s_1') \preceq \varphi(s_2') \preceq s_2'$ we deduce $\varphi(s_1') = 0$ and therefore $s_1' \in C_0 \cap C_1$, $s_1' = 0$. Hence $s_1 = u \preceq s_2$.

Now, for any quasi-continuous element $s \in C$ we put

$$Bs = \bigvee\{t \mid t \in C_1,\ t \leq s\} = \bigvee\{t \mid t \in C_1 \mid t \preceq s\}.$$

It is easy to see that B is additive, increasing and continuous in order from below. Since $Bs \preceq s$ then Bs is also quasi-continuous and $BBs = Bs$. The map

$$s \longrightarrow \bigvee\{Bt \mid t \leq s,\ t \text{ quasi-continuous}\} = \widetilde{B}s$$

is a balayage on C which extends the above map B and we have

$$s_1, s_2 \in C,\ s_1 \leq s_2 \Longrightarrow \widetilde{B}s_1 \preceq \widetilde{B}s_2 \preceq s_2,$$

and therefore $\widetilde{B}$ is a recurrent balayage on C. Moreover, for any quasi-continuous element s of C we have $s - Bs \in C_0$, $\varphi(s) = \varphi(Bs)$ and therefore $\varphi(s) \preceq Bs = \widetilde{B}s \preceq s$. □

Theorem 4.2. *Suppose that there are no recurrent balayages on T different from zero. Then for any balayage B on S which is an (S, T)-dilation operator we have $B = Q$ for any (S, T)-dilation operator Q in S with $Bs \leq Qs$, $\forall\, s \in S$.*

Proof. Let Q be a (S, T)-dilation operator on S and B be a balayage on S such that $s \in S \Longrightarrow Bs \leq Qs$. Then we have $Bs = B^2 s \leq Q(Bs) \preceq_T Bs$ and therefore $Bs = Q(Bs)$ for any $s \in S$.

We consider now the map $M : T \longrightarrow T$ defined by

$$Mu = \bigvee \{Q(s - Bs) \mid s - Bs \leq u, \ s \in S\}.$$

Since

$$s, t \in S, s - Bs \leq t - Bt \Longrightarrow Q(s - Bs) \preceq_T Q(t - Bt) = Qt - Bt \preceq_T t - Bt$$

we deduce $u, v \in T, u \leq v \Longrightarrow Mu \preceq_T Mv \preceq_T v$. From Lemma 4.1 it follows that there exists a recurrent balayage L on T such that

$$Mu \preceq_T Lu.$$

Using the hypothesis we get $L = 0$ and therefore $M = 0$,

$$s \in S \Longrightarrow Qs = QBs = Bs.$$

□

Remark. If B is a balayage on S such that $S_B := \{s - Bs \mid s \in S\}$ is solid in T then it follows that B is a (S,T)-dilation operator on S. Therefore the preceding theorem shows that if there are no recurrent balayages on T different from zero and B is a balayage on S such that S_B is solid in T then any (S,T)-dilation operator Q on S which dominates B coincides with B.

Corollary 4.3. *If the minimal (S,T)-dilation operator P on S is a balayage on S and there are no recurrent balayages on T then any (S,T)-dilation operator on S equals P.*

The following example shows that the above theorem fails if we drop the supplementary condition about T.

Example 1. We consider an H-cone S which is recurrent (i.e., the natural order coincides with the specific order in S) and for any $\alpha \in [0,1]$ we denote by P_α the map $P_\alpha : S \longrightarrow S$ defined by $P_\alpha s = \alpha \cdot s$. It is easy to see that the pair (S,S) verifies the conditions from the beginning of this section and that P_α is a (S,S)-dilation operator on S for any $\alpha \in [0,1]$. We remark that $P_0 = 0$ is the minimal (S,S)-dilation operator on S and that P_0 and P_1 are balayages on S. In this example $S_{P_\alpha} = S$ for any $\alpha \in [0,1)$ and $S_{P_1} = \{0\}$.

The following example shows that the above corollary holds even if the minimal (S,T)-dilation operator is not a balayage on S.

Example 2. Let S be the H-cone of all positive, increasing and lower semicontinuous real functions on the open interval $(-1,1)$ of $\mathbb{R}$. We consider the map $B^{\{0\}} : S \longrightarrow S$ where for any subset $A \subset (-1,1)$ and any $s \in S$ we have

$$B^A s = \bigwedge \{t \in S \mid t \geq s \text{ on } A\}.$$

It is known that $B^{\{0\}}$ is a localizable dilation operator on S and $S_{B^{\{0\}}}$ is solid and increasingly dense in the set T of all positive, real functions on $(-1,1)$ such that their restrictions to $(-1,0]$ and $(0,1)$ are increasing and lower semicontinuous. We show that any (S,T)-dilation operator on S coincides with $B^{\{0\}}$.

Indeed, we consider the H-cone S_1 of the restrictions to $(-1,0]$ of all $s \in S$ and the map $T : S_1 \longrightarrow S_1$ defined by $Tt = Q(\bar{t})\ |_{(-1,0]}$, where $\bar{t}$ is equal t on $(-1,0]$ and equal $t(0)$ on $(0,1)$. Obviously T verifies the conditions from Lemma 4.1. Since there are no recurrent balayages on S_1 different from zero we get $Tt = 0$ for any $t \in S_1$ and therefore $Q(\bar{t})\ |_{(-1,0]} = 0, \ \forall\, t \in S_1$.

Let now $s \in S$ and let $s_1 := s|_{(-1,0]}$. We have

$$s \le \bar{s}_1 + B^{(0,1)}s, \ Qs \preceq_T Q(\bar{s}_1) + Q(B^{(0,1)}s).$$

Since $Q(\bar{s}_1)|_{(-1,0]} = 0, \ B^{(0,1)}s|_{(-1,0)} = 0$ we deduce $Qs|_{(-1,0)} = 0$. We consider now the H-cone S_2 given by $S_2 := \{s \in S \mid s|_{(-1,0]} = 0\}$. It is easy to see, using the preceding consideration that the map $s \longrightarrow Qs$ satisfies the conditions from Lemma 4.1 with respect to the H-cone S_2 and therefore, since there are no recurrent balayages on S_2 different from zero, we get $Qs = 0, \ \forall\, s \in S_2$.
Because $B^{(-1,0)}s \preceq_S s$ for any $s \in S$ we get $s - B^{(-1,0)}s \in S_2$ and therefore $Q(s) = Q(B^{(-1,0)}s)$.

Hence for any $s,t \in S$ we have $s = t$ on $(-1,0] \Longrightarrow Qs = Qt$ and therefore $s \in S \Longrightarrow Qs = Q(B^{(-1,0)}s) \preceq_T B^{(-1,0)}s$. Since $B^{(-1,0)}s = B^{\{0\}}s$ on $(0,1)$ and $Qs = 0$ on $(-1,0]$ we have $Qs \preceq_T B^{\{0\}}s$.

Let now P be the minimal (S,T)-dilation operator on S. It remains to show that

$$B^{\{0\}}s = Ps \qquad \forall\, s \in S.$$

Indeed, if $s \in S$ is a continuous function then there exists $u \in S$, $u \le s$ such that $s - B^{\{0\}}s = u - Pu$. From $B^{\{0\}}s = 0$ on $(-1,0]$ and $Pu \le B^{\{0\}}u$ it follows that

$$s = u \text{ on } (-1,0]$$

and therefore $Pu = Ps$. Hence $s - u = B^{\{0\}}s - Ps, \ \ Ps \preceq_T B^{\{0\}}s$, and therefore $s - u \in S$. Because s is continuous we deduce that u and $B^{\{0\}}s - Ps$ are also continuous. From the fact that $B^{\{0\}}s - Ps = 0$ on $(-1,0]$ and that $B^{\{0\}}s$ is constant on $(0,1)$ we deduce that $B^{\{0\}}s - Ps$ is also constant on $(0,1)$ and therefore being continuous is equal to zero. Hence $B^{\{0\}}s = Ps$.

Theorem 4.4. *Suppose that there are no absorbent balayages B on T different from zero such that*

$$s_1, s_2 \in S, \ s_1 \le s_2 \Longrightarrow Bs_1 \preceq_T Bs_2$$

and that there exists a balayage B_0 on S which is a (S,T)-dilation operator on S. Then B_0 is the only (S,T)-dilation operator on S.

Proof. From hypothesis it follows that there are no recurrent balayages on T different from zero and therefore, by Theorem 4.2, any (S,T)-dilation operator Q on S such that $Qs \ge B_0 s$ for any $s \in S$, is equal with B_0.

To finish the proof we show that B_0 is the minimal (S,T)-dilation operator on S.

Let P be the minimal (S,T)-dilation operator on S and let s be an arbitrary element of S. We have $s - Bs \leq s - Ps$ and since the set S_B is increasingly dense in T we can choose an increasing family $(s_i - Bs_i)_{i\in I}$ such that $\bigvee_{i\in I}(s_i - Bs_i) = s - Ps$. On the other hand the set S_P being a solid subcone of T we deduce that for any $i \in I$ there exists $t_i \in S$ such that $t_i - Pt_i = s_i - Bs_i$ and moreover taking, for every $i \in I$, the smallest element t_i of S with the above property then we deduce (see [5]) that the family $(t_i)_{i\in I}$ is increasing and dominated by s. If we denote $t = \bigvee_{i\in I} t_i$ we get

$$(Pt_i)_i \uparrow Pt, \quad (t_i - Pt_i)_i \uparrow t - Pt = s - Ps.$$

Using the fact that $s_i - Bs_i = t_i - Pt_i$ we get $Bt_i = BPt_i$ for any $i \in I$ and therefore, passing to the limit $Bt = BPt$ or equivalently $Bs = BPs$. Hence $Bs \leq Ps$, i.e., $Bs = Ps$. □

Let now $\mathcal{V} = (V_\alpha)_{\alpha\geq 0}$ be a submarkovian resolvent on a measurable space $(X, \mathcal{B})$ such that its initial kernel $V_0 = V$ is bounded and absolutely continuous with respect to a finite measure μ. It is known that in this case the convex cone $\mathcal{E} = \mathcal{E}_\mathcal{V}$ of all $\mathcal{V}$-excessive functions on X which are finite $\mathcal{V}$-a.s. is an H-cone. Further we suppose that $\mathcal{E}_\mathcal{V}$ separates the points of X, contains the positive constant functions, is min-stable and generates σ-algebra $\mathcal{B}$. It is known also ([11]) that if $\alpha > 0$ then the kernel $Q := \alpha V_\alpha$ is a $(\mathcal{E}, \mathcal{E}_\alpha)$-dilation operator on $\mathcal{E}$ where $\mathcal{E}_\alpha$ is the H-cone of all α-excessive functions on X with respect to $\mathcal{V}$. In this case we have $Vf - \alpha V_\alpha Vf = V_\alpha f$, and therefore the H-cone $\mathcal{E}_Q = \{s - \alpha V_\alpha s \mid s \in \mathcal{E}\}$ is a solid and increasingly dense subcone of $\mathcal{E}_\alpha$.

Proposition 4.5. *If there are no absorbent points of X with respect to $\mathcal{E}$ then any $(\mathcal{E}, \mathcal{E}_\alpha)$-dilation operator Q on $\mathcal{E}$ such that*

$$\alpha V_\alpha s \preceq_{\mathcal{E}_\alpha} Qs \quad \forall\, s \in \mathcal{E}$$

coincides with αV_α.

Proof. Let Q be a $(\mathcal{E}, \mathcal{E}_\alpha)$-dilation operator on $\mathcal{E}$ such that

$$\alpha V_\alpha s \preceq_{\mathcal{E}_\alpha} Qs \quad \forall\, s \in \mathcal{E}$$

or equivalently

$$\alpha V_\alpha Vf \preceq_{\mathcal{E}_\alpha} QVf \quad \forall\, f \in \mathcal{F}_b.$$

We have

$$Wf := Vf - QVf \prec_{\mathcal{E}_\alpha} Vf - \alpha V_\alpha Vf = V_\alpha f$$

for any $f \in \mathcal{F}_b$ and therefore there exists $g \in \mathcal{F}_b$, $g \leq 1$ such that

$$Wf = V_\alpha(g \cdot f) \quad \forall\, f \in \mathcal{F}_b.$$

Hence

$$QVf = Vf - V_\alpha(g\cdot f) = V_\alpha(f + \alpha Vf - g\cdot f) = V_\alpha((1-g)f + \alpha Vf) = \alpha V_\alpha(\epsilon f + Vf)$$

where $\epsilon = \frac{1-q}{\alpha}$. On the other hand if $f_1, f_2 \in \mathcal{F}_b$ we have

$$Vf_1 \leq Vf_2 \Longrightarrow QVf_1 \preceq_{\mathcal{E}_\alpha} QVf_2$$

or equivalently

$$Vf_1 \leq Vf_2 \Longrightarrow \epsilon f_1 + Vf_1 \leq \epsilon f_2 + Vf_2 \quad \mathcal{V}\text{-a.s.}$$

We want to show that the set $A := [\epsilon \geq r > 0\}$ is $\mathcal{V}$-negligible for any $r > 0$. In the contrary case let $A^* = \{x \in A \mid \lim_{\alpha\to\infty} \alpha V_\alpha(1_A)(x) = 1\}$. It is known that A^* is a sub-basic subset of X and for any $s \in \mathcal{E}$, there exists a sequence $(f_n)_n$ in $\mathcal{F}_b$, $f_n = 0$ on $X \backslash A^*$ such that $(Vf_n)_n$ is increasing and $B^{A^*}s = \sup_n Vf_n$.

We show that any $x \in A^*$ is an absorbent point. We consider $x \in A^*$ and $(U_n)_n$ a decreasing sequence of natural open neighborhoods of x such that $\cap_n U_n = \{x\}$. We take $f_n := 1_{X\backslash U_n \cap A^*}$. Then for any $g \in \mathcal{F}_b$ such that $g = 0$ on $X\backslash U_n \cap A^*$, $Vg \leq Vf_n$ we have $\epsilon g + Vg \leq \epsilon f_n + Vf_n$, and therefore $\epsilon g + Vg \leq Vf_n$ on $U_n \cap A^*$. Let $(g_m)_m$ be a sequence such that $g_m = 0$ on $X\backslash U_m \cap A^*$ and $Vg_m \uparrow B^{U_n \cap A^*}Vf_n$. Since $Vg_m \uparrow Vf_n$ on $U_n \cap A^*$, and since $\epsilon g_m + Vg_m \leq Vf_n$, $\forall m \in \mathbb{N}$ we deduce that

$$g_m \longrightarrow 0, \quad g_m \leq \frac{Vf_n}{r}$$

and therefore $Vg_m \longrightarrow 0$, $Vf_n = 0$ on $U_n \cap A^*$. Hence

$$\{x\} = \left[\left(\sum_n \frac{1}{2^n} Vf_n\right)(x) = 0\right]$$

i.e., $\{x\}$ is absorbent.

From the hypothesis we get that A^* is $\mathcal{V}$-negligible for any $r > 0$ and so $Q = \alpha V_\alpha$. □

Theorem 4.6. *If there are no fine open singleton subsets of X with respect to $\mathcal{E}$ then any $(\mathcal{E}, \mathcal{E}_\alpha)$-dilation operator on $\mathcal{E}$ coincides with αV_α.*

Proof. Let P be the minimal $(\mathcal{E}, \mathcal{E}_\alpha)$-dilation operator on $\mathcal{E}$. We want to show that $P = Q$. We have

$$PVf \preceq_{\mathcal{E}_\alpha} QVf \preceq_{\mathcal{E}_\alpha} Vf \qquad \forall f \in \mathcal{F}_b$$

and

$$Vf = V_\alpha(f + \alpha Vf), \; QVf = \alpha V_\alpha Vf.$$

Since $PVf \in \mathcal{E} \subset \mathcal{E}_\alpha$ there exists $g_f \in \mathcal{F}_b$ such that $PVf = V_\alpha(g_f)$, $g_f \leq \alpha Vf$.

On the other hand, the kernel W on $(X, \mathcal{B})$ given by $Wf = Vf - PVf$, $f \in \mathcal{F}_b$ satisfies the complete maximum principle, $Wf \in \mathcal{E}_\alpha$ and $Wf = V_\alpha(f + \alpha Vf - g_f)$. But the kernel V_α satisfies also the complete maximum principle and

$$W1 = V_\alpha(1 + \alpha V1 - g_1).$$

We deduce that $Wf = V_\alpha((1+\alpha V1 - g_1)\cdot f)\ \forall\ f \in \mathcal{F}_b$ and therefore for any $f \in \mathcal{F}_b$ we have

$$f + \alpha Vf - g_f = (1+\alpha V1 - g_1)\cdot f, \quad \mathcal{V}\text{-a.s.}$$

or equivalently $g_f = \alpha Vf - (\alpha V1 - g_1)\cdot f$, $\mathcal{V}$-a.s. Hence if we put $\epsilon = V1 - \frac{1}{\alpha}g_1$ we get $0 \leq \epsilon$,

$$\epsilon f \leq Vf \quad \mathcal{V}\text{-a.s}, \ \forall\ f \in \mathcal{F}_b,\ PVf = \alpha V_\alpha(Vf - \epsilon f).$$

On the other hand we have, for any $f_1, f_2 \in \mathcal{F}_b$ such that $Vf_1 \leq Vf_2$,

$$PVf_1 \preceq_{\mathcal{E}_\alpha} PVf_2$$

or equivalently

$$\alpha V_\alpha(Vf_1 - \epsilon f_1) \preceq_{\mathcal{E}_\alpha} \alpha V_\alpha(Vf_2 - \epsilon f_2).$$

This relation is equivalent also to the following

$$Vf_1 - \epsilon f_1 \leq Vf_2 - \epsilon f_2 \quad \mathcal{V}\text{-a.s.}$$

Since there exists a reference measure μ on X with respect to $\mathcal{V}$ and since $\mathcal{E}$ generates $\mathcal{B}$ it follows that $\mathcal{B}$ is countably generated and therefore there exists a $\mathcal{V}$-negligible subset M of X, $M \in \mathcal{B}$ for which we have

$$\epsilon f \leq Vf \ \text{ on } X\setminus M, \quad \forall\ f \in \mathcal{F}_b.$$

We want to show that $\epsilon = 0$, $\mathcal{V}$-a.s. Indeed, let $T = \{x \in X\setminus M \mid \epsilon(x) > 0\}$. For any $x \in T$ we have

$$Vf(x) \geq \epsilon(x)f(x) \qquad \forall\ f \in \mathcal{F}_b$$

and therefore $V(\mathbf{1}_{\{x\}}) \geq \epsilon(x)$.

Let now $g \in \mathcal{F}_b$ be such that $Vg \leq V1$ and such that $g(x) = 0$. We have

$$P(Vg) \preceq_{\mathcal{E}_\alpha} P(V\mathbf{1}_x)$$

or equivalently $Vg - \epsilon g \leq V\mathbf{1}_x - \epsilon\mathbf{1}_x$, $\mathcal{V}$-a.s. Hence $\epsilon(x) \leq V\mathbf{1}_x(x) - Vg(x)$. Since

$$B^{X\setminus\{x\}}V\mathbf{1}_x = \sup\{Vg \mid g \in \mathcal{F}_b,\ Vg \leq V\mathbf{1}_{\{x\}},\ g(x) = 0\}$$

it follows

$$\epsilon(x) \leq V\mathbf{1}_x(x) - B^{X\setminus\{x\}}V\mathbf{1}_x(x).$$

But

$$V\mathbf{1}_x = B^{X\setminus\{x\}}V\mathbf{1}_x \ \text{ on } X\setminus\{x\},$$

i.e., $\{x\}$ is finely open with respect to $\mathcal{E}$. From the hypothesis we deduce that $\epsilon = 0$ and therefore $P = \alpha V_\alpha$.

The fact that any $(\mathcal{E}, \mathcal{E}_\alpha)$-dilation operator Q on $\mathcal{E}$ coincides with αV_α follows now from the preceding proposition. □

Remark. The assertion of the above Theorem fails if instead of "there are no fine open singletons in X" we put "there are no absorbent points in X with respect to

$\mathcal{E}$". We consider $X = \{1,2\}$ and

$$Vf(x) = f(x) + \frac{1}{2}(f(x) + f(y)) \qquad \forall\, x, y \in X.$$

The associated resolvent will be $\mathcal{V} = (V_\alpha)_{\alpha \geq 0}$ where

$$V_\alpha f(x) = f(x) + \frac{1}{(1+\alpha)(2\alpha+1)} \left(\frac{f(x) + f(y)}{2} \right).$$

We have

$$\mathcal{E} = \mathcal{E}_{\mathcal{V}} = \{(x_1, x_2) \mid x_1, x_2 \geq 0,\ x_1 \leq 5x_2,\ x_2 \leq 5x_1\},$$
$$\mathcal{E}_1 = \mathcal{E}_{\mathcal{V}_1} = \{(x_1, x_2) \mid x_1, x_2 \geq 0 \mid x_1 \leq 25x_2,\ x_2 \leq 25x_1\}$$

and the map $Q : \mathcal{E} \longrightarrow \mathcal{E}$ defined by

$$QVf(x) = \frac{1}{2} f(x) + \frac{5}{4} \frac{f(x) + f(y)}{2}$$

is a $(\mathcal{E}, \mathcal{E}_1)$-dilation operator on $\mathcal{E}$ such that $QVf \preceq_{\mathcal{E}_{\mathcal{V}_1}} V_1 V f$ and $Q \neq V_1$.

5. The compression operator associated with an (S, T)-dilation operator

In this section S, T are two H-cones as in the preceding sections and Q is a given (S, T)-dilation operator on S. We intend to extend Q to a map $\overline{Q} : D(\overline{Q}) \longrightarrow T$ where $D(\overline{Q})$ a solid convex subcone of T, containing S such that $\overline{Q}$ is additive, increasing, continuous in order from below and such that

$$u, v \in D(\overline{Q}),\ \ u \leq v \Longrightarrow \overline{Q}u \preceq_T \overline{Q}v$$

and to show that for $u \in D(\overline{Q})$ we have $u \in S \Longleftrightarrow \overline{Q}u \preceq_T u$.

Proposition 5.1. *Let us denote by $\overline{Q}$ the operator on $S_Q := \{s - Qs \mid s \in S\}$ defined by $\overline{Q}(s - Qs) = Qs - Q^2 s$. Then $\overline{Q}$ is additive, increasing, continuous in order from below and*

$$s - Qs \leq t - Qt \Longrightarrow \overline{Q}(s - Qs) \preceq_T \overline{Q}(t - Qt).$$

Proof. By definition we get immediately that Q is additive and that

$$s - Qs \leq t - Qt \Longrightarrow \overline{Q}(s - Qs) \preceq_T \overline{Q}(t - Qt).$$

Let now $(s_i - Qs_i)_{i \in I}$ be an increasing family in S_Q such that

$$\bigvee_{i \in I} (s_i - Qs_i) = s - Qs.$$

We denote by s_i' (resp. s') the smallest element in S such that

$$s_i - Qs_i = s_i' - Qs_i', \quad (\text{resp. } s - Qs = s' - Qs').$$

We know that

$$s_i - Qs_i \leq s_j - Qs_j \Longrightarrow s_i' \leq s_j' \leq s'.$$

Denoting $t := \bigvee_{i\in I} s_i'$, we get $t \leq s'$ and

$$\bigvee_{i\in I}(s_i - Qs_i) = t - Qt = s' - Qs'.$$

Hence $s' \leq t,\ s' = t,$

$$\overline{Q}(s - Qs) = \overline{Q}(s' - Qs') = Qs' - Q^2 s',$$
$$Qs_i' \uparrow Qs; \quad Q^2 s_i' \uparrow Q^2 s'.$$

From

$$Q(s_i' - Qs_i') + Q^2 s_i' = Qs_i'$$

we get

$$\bigvee_{i\in I} Q(s_i' - Qs_i') + \bigvee_{i\in I} Q^2 s_i' = \bigvee_{i\in I} Qs_i',$$
$$\bigvee_{i\in I} Q(s_i' - Qs_i') + Q^2 s' = Qs',$$
$$\bigvee_{i\in I} \overline{Q}(s_i - Qs_i) = \bigvee_{i\in I} Q(s_i' - Qs_i') = Q(s' - Qs') = \overline{Q}(s - Qs).$$

□

Corollary 5.2. *If we denote*

$$D(\overline{Q}) := \{u \in T \mid \exists\, v \in T,\ \ s - Qs \leq u \Longrightarrow \overline{Q}(s - Qs) \leq v\}$$

then $D(\overline{Q})$ is a solid convex cone in T, $S_Q \subset D(\overline{Q})$ and

$$u \longrightarrow \bigvee_{s-Qs\leq u} \overline{Q}(s - Qs)$$

is a map from $D(\overline{Q})$ into T which is additive, increasing, continuous in order from below and coincides with $\overline{Q}$ on S_Q. We denote also by $\overline{Q}$ this map and we have
(a) $u_1, u_2 \in D(\overline{Q}),\ \ u_1 \leq u_2 \Longrightarrow \overline{Q}(u_1) \preceq_T \overline{Q}(u_2)$,
(b) $S \subset D(Q)$ *and* $Qs \preceq_T s,\ \ \forall\, s \in S$.

Remark. The map $\overline{Q}$ is a compression operator on T. Hence the set

$$T(\overline{Q}) := \{u \in D(\overline{Q}) \mid \overline{Q}u \preceq_T u\}$$

is an H-subcone of S with $S \subset T(\overline{Q})$ (see [4]).

From now on we suppose that the set

$$S_\Delta := \{s \in S \mid \bigwedge_{\alpha\in\Omega} Q^\alpha s = 0\}$$

is increasingly dense in S where Ω is the first ordinal number which is not countable and where Q^α is defined inductively by $Q^0 s = s$ and

$$Q^\alpha s = Q(\bigwedge_{\beta<\alpha} Q^\beta s).$$

Remark. If S is a standard H-cone then there exists a balayage B on S such that $Bs = \bigwedge_{\alpha\in\Omega} Q^\alpha s$, for any universally continuous element s of S and therefore $Bs = Q(Bs)$ or equivalently $Bs \le Qs$, for any $s \in S$. Hence the fact that S_Δ is increasingly dense in S follows from the fact that there are no balayages B on S, different from zero, dominated by Q.

Theorem 5.3. *If* S_Δ *is increasingly dense in* S *then* $\overline{Q}s = Qs$ *and the set* $\{s - Qs \mid s \in S_\Delta\}$ *is increasingly dense in* T.

Proof. We have inductively, for any $\alpha \in \Omega$ and any $s \in S_\Delta$,

$$s - Q^\alpha s = \sum_{\beta<\alpha} (Q^\beta s - Q(Q^\beta s))$$

and therefore

$$Qs - Q^{\alpha+1} s = \sum_{\beta<\alpha} [Q^{\beta+1} s - Q(Q^{\beta+1} s)].$$

Hence for any $s \in S$ there exists an increasing family $(s_i - Qs_i)_{i\in I}$ in S_Q such that $s_i \in S_\Delta$ and such that

$$\bigvee_{i\in I} (s_i - Qs_i) = s, \ \bigvee_{i\in I} Q(s_i - Qs_i) \ge Qs - Q^{\alpha+1} s \qquad \forall\, \alpha \in \Omega.$$

Since α is arbitrary in Ω we get

$$\bigvee_{i\in I} \overline{Q}(s_i - Qs_i) \ge Qs,$$

$$\overline{Q}(s) > \bigvee_{i\in I} \overline{Q}(s_i - Qs_i) \ge Qs, \ \overline{Q}s = Qs.$$

The equality $\overline{Q}s = Qs$ for any $s \in S$ follows from the fact that S_Δ is increasingly dense in S and from the fact that Q and $\overline{Q}$ are continuous in order from below on S and $D(\overline{Q})$, respectively. We have, for any $u \in T$, $u = \bigvee_{s\in S}(u \wedge s)$. Because S_Q is solid in T it will be sufficient to show that for any $s \in S$ there exists an increasing family $(s_i - Qs_i)_{i\in T}$, where $s_i \in S_\Delta$ such that $\bigvee_{i\in T}(s_i - Qs_i) = s$. From the first part of the proof it follows that this assertion is true if $s \in S_\Delta$. The general assertion follows from the fact that S_Δ is increasingly dense in S. □

Theorem 5.4. *Suppose that $S_\triangle$ is increasingly dense in S. Then for any $u \in D(\overline{Q})$ we have $u \in S \iff \overline{Q}u \preceq_T u$.*

Proof. Let us denote

$$T(\overline{Q}) := \{u \in D(\overline{Q}) \mid \overline{Q}u \preceq_T u\}.$$

Suppose that $u \in T(\overline{Q})$, $v \neq 0$ and there is no $v \in T(\overline{Q})$ such that $u - v \in T(\overline{Q})$ and $\overline{Q}v = v$. Since $u - \overline{Q}u \in T$ then according to Theorem 5.3 there exists an increasing family $(s_i - Qs_i)_{i \in I}$ where $s_i \in S_\triangle$ and

$$\bigvee_{i \in I} (s_i - Qs_i) = u - \overline{Q}u.$$

Since $\bigwedge_{\alpha \in \Omega} Q^\alpha s_i = 0$ it follows that

$$s_i - Qs_i \leq s_j - Qs_j \leq u - \overline{Q}u \Longrightarrow s_i \leq s_j \leq u.$$

If we put $s := \bigvee_{i \in I} s_i$, we get $s \leq u$ and $s - Qs = u - \overline{Q}u$.

Since there is no $v \in T(\overline{Q})$, $v \neq 0$, such that $u - v \in T(\overline{Q})$ and $\overline{Q}v = v$ then we get $u \leq s$ and therefore $u = s$, $u \in S$.

Suppose now that $u \in D(\overline{Q})$ is such that $\overline{Q}u = u$. For any $s \in S_\triangle$ we have $s \in T(\overline{Q})$ and therefore $u \wedge s \in T(\overline{Q})$. Since $\overline{Q}s = Qs$ it follows that

$$\bigwedge_{\alpha \in \Omega} \overline{Q}^\alpha s = 0$$

and therefore there is no $v \in T(\overline{Q})$, $v \neq 0$ with $v \leq u \wedge s$ such that $\overline{Q}v = v$. Hence $u \wedge s$ is as in the first part of the proof and so $u \wedge s \in S$. On the other hand

$$u = \bigvee_{s \in S_\triangle} (u \wedge s)$$

and therefore $u \in S$. □

References

[1] Beznea, L., Boboc N., Absorbent, parabolic, elliptic and quasielliptic balayages in potential theory. Rev. Roum. Math. Pures Appl. 38 (1993), 197–234.

[2] Boboc, N., Cornea A., Cones convexes ordonnées. *H*-cones et adjoints de *H*-cones. C. R. Acad. Sci. Paris 270 (1970), 596–599.

[3] Boboc, N., Bucur Gh., Cornea A., Order and convexity in potential theory: *H*-cones. Lecture Notes in Math. 853. Springer-Verlag, Berlin 1981.

[4] Boboc, N., Bucur Gh., Subordinations in excessive structures. Compressions in *H*-cones. Rev. Roum. Math. Pures Appl. 37 (1992), 7–42.

[5] — Pseudodilations in *H*-cones. Rev. Roum. Math. Pures Appl. 37 (1992), 115–132.

[6] — Compression and dilation operators on excessive structures. Rev. Roum. Math. Pures Appl. 38 (1993), 289–317.

[7] — Recurrence and transience in excessive structures. Rev. Roum. Math. Pures Appl. 38 (1993), 101–117.

[8] Cornea, A., Eriksson S. L., Order continuity of greatest lower bound of two functionals. Analysis 7 (1987), 173–184.

[9] Eriksson, S. L., Completion of H-cones. Ann. Acad. Scientiarum Fennicae Series A I. Mathematica 15 (1990).

[10] Getoor, R. K., Sharpe, M. J., Inverse subordination of excessive functions. Preprint 1993.

[11] Mokobodzki, G., Opérateurs de subordination des resolvantes (1983, unpublished).

On the harmonic morphism for the Kolmogorov type operators

Miroslav Brzezina and Martina Šimůnková*

Abstract. For the Kolmogorov type operator

$$L := \sum_{i,j=1}^{n} a_{ij} \frac{\partial^2}{\partial x_i \partial x_j} + \sum_{i=1}^{l} \sum_{j=1}^{n} b_{ij} x_j \frac{\partial}{\partial y_i} - \frac{\partial}{\partial t}$$

we describe all transformations preserving solutions of the equation $Lu = 0$.

1991 Mathematics Subject Classification: 31B35, 35K10.

0. Introduction

In this note, we consider the Kolmogorov type operators and their harmonic morphisms (see the definition below).

Harmonic morphisms in potential theory have a long history. For the Laplace operator in $\mathbb{R}^n$, the Kelvin transformation is the only one which maps the class of harmonic function into itself (up to compositions with similarities); see [4], [6].

For the heat operator, the Appell transformation plays the same rôle as the Kelvin transformation for the Laplace operator; see [1], [8]. All caloric morphisms (i.e., mappings which map caloric functions into itself) are described by Heinz Leutwiler in [8].

In [2], one of the authors investigated the Appell type transformation for the Kolmogorov operator and described all transformations mapping solutions into solutions.

In what follows, we generalize results from [2] for the Kolmogorov type operators. (For further information on operators of this type, see, e.g., [5], [9], [3].)

In the first section, we give the definition of L-harmonic functions, Appell type transformation and L-harmonic morphism for a special case of the Kolmogorov type operator. In the 2nd section, we derive the explicit form of L-harmonic mor-

* Financial support by the Konferenz der deutschen Akademien der Wissenschaften is gratefully acknowledged.

phisms for this special case. Section 3 contains some auxiliary results. In Section 4, we give the main result for a wide class of Kolmogorov type operators, see Theorem 2.

1. Appell type transformation

For real matrices $A = (a_{ij})_{i,j=1,\ldots,n}$, $B = (b_{ij})_{\substack{i=1,\ldots,l\\ j=1,\ldots,n}}$, we consider the *Kolmogorov type operator*

$$L := \sum_{i,j=1}^{n} a_{ij}\frac{\partial^2}{\partial x_i \partial x_j} + \sum_{i=1}^{l}\sum_{j=1}^{n} b_{ij}x_j\frac{\partial}{\partial y_i} - \frac{\partial}{\partial t}$$

in $\mathbb{R}^{n+l+1}$. For an open set $U \subset \mathbb{R}^{n+l+1}$, let

$$\mathcal{C}^{2,1,1}(U) := \{u : U \to \mathbb{R}; \frac{\partial^2 u}{\partial x_i \partial x_j}, \frac{\partial u}{\partial y_k}, \frac{\partial u}{\partial t} \in \mathcal{C}(U),\ i,j = 1,\ldots,n,\ k = 1,\ldots,l\}.$$

The function $u \in \mathcal{C}^{2,1,1}(U)$ is said to be *L-harmonic* on U, if $Lu = 0$ on U. For any $(x,y,t) \in \mathbb{R}^{n+l+1}, t \neq 0$, let

$$E(x,y,t) := \frac{1}{|t|^{\frac{n+3l}{2}}}\exp\left(-\frac{A^{-1}x\cdot x}{4t} - \frac{3}{t^3}\left(BAB^T\right)^{-1}\left(y + \frac{t}{2}Bx\right)\cdot\left(y + \frac{t}{2}Bx\right)\right).$$

A straightforward calculation shows that the function E is L-harmonic on the set

$$\mathbb{R}_0^{n+l+1} := \{(x,y,t) \in \mathbb{R}^{n+l+1}; t \neq 0\}.$$

(The function E is related to the fundamental solution of the operator L, see, e.g., [3].)

As a special case, we consider the operator

$$\tilde{L} := \sum_{i=1}^{l}\left(\frac{\partial^2}{\partial x_i^2} + x_i\frac{\partial}{\partial y_i}\right) + \sum_{i=1}^{n-l}\frac{\partial^2}{\partial z_i^2} - \frac{\partial}{\partial t}. \tag{1}$$

A straightforward calculation shows that both mappings

$$(\tilde{A}u)(x,y,z,t) := \frac{1}{|t|^{(n+3l)/2}}\exp\left(-\frac{\|x\|^2}{t} - 3\frac{x\cdot y}{t^2} - 3\frac{\|y\|^2}{t^3} - \frac{\|z\|^2}{4t}\right)u\left(\frac{x}{t} + 3\frac{y}{t^2}, \frac{y}{t^3}, \frac{z}{t}, -\frac{1}{t}\right) \tag{2}$$

(we use the symbol $\|\cdot\|$ for the corresponding Euclidean norm) and

$$(\tilde{S}u)(x,y,z,t) := \nu u\left(rRx + a, r^2(rRy - at), rPz + c, r^2t + b\right) \tag{3}$$

(R and P are orthogonal matrices of the order l and $n-l$, respectively, $\nu, b, r \in \mathbb{R}$, $a \in \mathbb{R}^l$, $c \in \mathbb{R}^{n-l}$) map the class of $\tilde{L}$-harmonic functions into itself.

Following H. Leutwiler [8], p. 217, we give the definition of L-harmonic morphism.

Definition. Let $U, V \subset \mathbb{R}^{n+l+1}$ be open sets, $\Psi : U \to V$ a bijection from U onto V, and let $\phi : U \to \mathbb{R}$ be a strictly positive function. The transformation T, defined for $u \in \mathcal{C}^{2,1,1}(V))$ by

$$(Tu)(x,y,t) := \phi(x,y,t)u(\Psi(x,y,t))$$

is said to be an *L-harmonic morphism* on U, provided that Tu is L-harmonic on U whenever u is L-harmonic on V.

Clearly, the mappings $\tilde{A}$, $\tilde{S}$ and their compositions are $\tilde{L}$- harmonic morphisms.

We are going to describe all $\tilde{L}$-harmonic and L-harmonic morphisms on an open set; see Sections 2 and 4. In the proof, we use some ideas from [8], [2].

2. $\tilde{L}$-harmonic morphisms

In this section, we describe all harmonic morphisms for the operator $\tilde{L}$.

Theorem 1. *Let $\alpha, \beta, \gamma, \delta, \nu$ be real numbers, $\alpha\delta - \beta\gamma = 1$, $\gamma^2 + \delta^2 \neq 0$, $\nu > 0$, $\tilde{\gamma}, \tilde{\delta}, \kappa, \omega \in \mathbb{R}^l$, $A, B \in \mathbb{R}^{n-l}$, R, P be orthogonal matrices of the order $l, n-l$, respectively. Let*

$$\chi(x,y,t) = \frac{1}{\delta}Rx - \frac{\tilde{\gamma}}{\delta^3}t^2 - \frac{2\tilde{\delta}}{\delta^3}t - \kappa, \tag{4}$$

$$\xi(y,t) = \frac{1}{\delta^3}Ry + \frac{\tilde{\gamma}}{3\delta^5}t^3 + \frac{\tilde{\delta}}{\delta^5}t^2 + \frac{\kappa}{\delta^2}t + \omega, \tag{5}$$

$$\zeta(z,t) = \frac{1}{\delta}Pz + \frac{1}{\delta}(At + B), \tag{6}$$

$$\tau(t) = \frac{1}{\delta}\left(\frac{1}{\delta}t + \beta\right), \tag{7}$$

$$\varphi(x,y,z,t) = \nu \exp\left(-\frac{1}{\delta^2}(\tilde{\gamma}t + \tilde{\delta}) \cdot Rx + \frac{1}{2}A \cdot Pz - \frac{\tilde{\gamma}}{\delta^2} \cdot Ry \right. \tag{8}$$
$$\left. + \frac{\|\tilde{\gamma}\|^2}{3\delta^4}t^3 + \frac{\tilde{\gamma}\cdot\tilde{\delta}}{\delta^4}t^2 + \frac{\|\tilde{\delta}\|^2}{\delta^4}t + \frac{\|A\|^2}{4}t\right)$$

if $\gamma = 0$, *and*

$$\chi(x,y,t) = \frac{1}{\gamma t+\delta}Rx + \frac{3\gamma}{(\gamma t+\delta)^2}Ry + \frac{\gamma\tilde{\delta}-\tilde{\gamma}\delta}{\gamma^2}\frac{1}{(\gamma t+\delta)^2} + \frac{2\tilde{\gamma}}{\gamma^2}\frac{1}{\gamma t+\delta} + \kappa, \tag{9}$$

$$\xi(y,t) = \frac{1}{(\gamma t+\delta)^3}Ry + \frac{\gamma\tilde{\delta}-\tilde{\gamma}\delta}{3\gamma^3}\frac{1}{(\gamma t+\delta)^3} + \frac{\tilde{\gamma}}{\gamma^3}\frac{1}{(\gamma t+\delta)^2} + \frac{\kappa}{\gamma}\frac{1}{\gamma t+\delta} + \omega, \tag{10}$$

$$\zeta(z,t) = \frac{1}{\gamma t+\delta}Pz + \frac{At+B}{\gamma t+\delta}, \tag{11}$$

$$\tau(t) = \frac{\alpha t+\beta}{\gamma t+\delta}, \tag{12}$$

$$\begin{aligned}\varphi(x,y,z,t) = \frac{\nu}{|\gamma t+\delta|^{(n+3l)/2}}\exp\Bigg(&-\frac{\gamma}{(\gamma t+\delta)}\|x\|^2 - \frac{\gamma}{4(\gamma t+\delta)}\|z\|^2 \\ &- \frac{\tilde{\gamma}t+\tilde{\delta}}{(\gamma t+\delta)^2}\cdot Rx - \frac{3\gamma^2}{(\gamma t+\delta)^2}x\cdot y + \frac{A\delta-B\gamma}{2(\gamma t+\delta)}\cdot Pz \\ &- \frac{3\gamma^3}{(\gamma t+\delta)^3}\|y\|^2 - \frac{3\tilde{\gamma}(\gamma t+\delta)+2(\gamma\tilde{\delta}-\delta\tilde{\gamma})}{(\gamma t+\delta)^3}\cdot Ry \\ &- \frac{\|\gamma\tilde{\delta}-\delta\tilde{\gamma}\|^2}{3\gamma^3}\frac{1}{(\gamma t+\delta)^3} - \frac{\tilde{\gamma}\cdot(\gamma\tilde{\delta}-\delta\tilde{\gamma})}{\gamma^3}\frac{1}{(\gamma t+\delta)^2} \\ &- \frac{\|\tilde{\gamma}\|^2}{\gamma^3}\frac{1}{\gamma t+\delta} - \frac{\|A\delta-B\gamma\|^2}{4\gamma(\gamma t+\delta)}\Bigg),\end{aligned} \tag{13}$$

if $\gamma \neq 0$. *Then*

$$(\mathcal{T}u)(x,y,z,t) := \varphi(x,y,z,t)u(\chi(x,y,t),\xi(y,t),\zeta(z,t),\tau(t)) \tag{14}$$

defines an $\tilde{L}$*-harmonic morphism on* $\mathbb{R}^{n+l+1}$ *if* $\gamma = 0$, *and an* $\tilde{L}$*-harmonic morphism on* $\mathbb{R}^{n+l+1}$ *off the plane* $\gamma t + \delta = 0$, *if* $\gamma \neq 0$.

Conversely, let $\Omega \subset \mathbb{R}^{n+l+1}$ *be a domain and let* $\mathcal{M}$ *be an* $\tilde{L}$*-harmonic morphism on* Ω. *Then there exists a transformation* $\mathcal{T}$ *of the form* (14) *such that* $\mathcal{M} = \mathcal{T}|_\Omega$.

Proof. The proof of the first part of Theorem follows by a straightforward calculation. We omit it. Suppose first that Ω is an open interval U; the general case will be discussed at the end of the proof.

Let $\mathcal{M}$ be an $\tilde{L}$-harmonic morphism on U,

$$(\mathcal{M}u)(x,y,z,t) := \varphi(x,y,z,t)u(\chi(x,y,z,t),\xi(x,y,z,t),\zeta(x,y,z,t),\tau(x,y,z,t)) \tag{15}$$

where χ, ξ or ζ denote l or $n-l$, component vector functions, respectively.

The proof is based on the fact that the following polynomials ($i,j = 1,\dots,l$, $p,q = 1,\dots,n-l$)

$$1,\ x_i,\ x_ix_j + 2t\delta_{ij},\ x_i^3 + 6x_it,\ x_i^4 + 12x_i^2t + 12t^2,$$

$$y_i + x_it,\ x_iy_i + x_i^2t + t^2,\ 3y_i^2 + 6x_iy_it + 3x_i^2t^2 + 2t^3,$$

$$z_p, z_p z_q + 2t\delta_{pq}, x_i z_p$$

are $\tilde{L}$-harmonic which implies equations

$$\tilde{L}(\varphi) = 0, \tag{16}$$

$$\tilde{L}(\varphi\chi_i) = 0, \tag{17}$$

$$\tilde{L}(\varphi(\chi_i^2 + 2\tau)) = 0, \tag{18}$$

$$\tilde{L}(\varphi\chi_i\chi_j) = 0 \text{ for } i \neq j, \tag{19}$$

$$\tilde{L}(\varphi(\chi_i^3 + 6\chi_i\tau)) = 0, \tag{20}$$

$$\tilde{L}(\varphi(\chi_i^4 + 12\chi_i^2\tau + 12\tau^2)) = 0, \tag{21}$$

$$\tilde{L}(\varphi(\xi_i + \chi_i\tau)) = 0, \tag{22}$$

$$\tilde{L}(\varphi(\chi_i\xi_i + \chi_i^2\tau + \tau^2)) = 0, \tag{23}$$

$$\tilde{L}(\varphi(3\xi_i^2 + 6\chi_i\xi_i\tau + 3\chi_i^2\tau^2 + 2\tau^3)) = 0, \tag{24}$$

$$\tilde{L}(\varphi\zeta_p) = 0, \tag{25}$$

$$\tilde{L}(\varphi(\zeta_p^2 + 2\tau)) = 0, \tag{26}$$

$$\tilde{L}(\varphi\zeta_p\zeta_q) = 0 \text{ for } p \neq q, \tag{27}$$

$$\tilde{L}(\varphi\chi_i\zeta_p) = 0 \tag{28}$$

on U.

According to Lemma 1 and Lemma 3, it follows from (16)–(21) that $\tau = \tau(t)$ and

$$\|\nabla_x\chi_i\|^2 + \|\nabla_z\chi_i\|^2 = \tau' \tag{29}$$

for $i = 1, \dots, l$. (Here $'$ means the derivative with respect to t and symbols ∇_x and ∇_z denote gradients with respect to x and z.)

According to Lemma 2, it follows from (16)–(24) that $\xi = \xi(y,t)$ and

$$\sum_{k=1}^{n} x_k \frac{\partial \xi_i}{\partial y_k} - \frac{\partial \xi_i}{\partial t} = \tau'\chi_i \tag{30}$$

for $i = 1, \dots, l$.

Now differentiating in (30) with respect to x_k and z_k we obtain

$$\frac{\partial \xi_i}{\partial y_k} = \tau' \frac{\partial \chi_i}{\partial x_k} \tag{31}$$

and

$$\frac{\partial \chi_i}{\partial z_k} = 0. \tag{32}$$

Relation (31) implies $\tau'(t) \neq 0$ on U; otherwise fixing such t that $\tau'(t) = 0$ and arbitrary $x_1, \dots x_l, z_1, \dots z_{n-l}$ we obtain a constant mapping $y_1, \dots, y_l \to \xi_1, \dots, \xi_l$ contrary to our assumption that Φ is a bijection.

Finally, (19) together with (16), (17), (52) and (32) gives $(i,j=1,\dots,l)$

$$\nabla_x\chi_i\cdot\nabla_x\chi_j=0 \text{ for } i\neq j. \tag{33}$$

From (29), (32), (33) and (31) it follows that

$$\left(\frac{1}{(\tau')^{1/2}}\frac{\partial\chi_i}{\partial x_k}\right)_{i,k=1,\dots,l},\quad \left(\frac{1}{(\tau')^{3/2}}\frac{\partial\xi_i}{\partial y_k}\right)_{i,k=1,\dots,l}$$

are orthogonal matrices.

This implies that fixing t, x and z or t, y and z we obtain conformal mappings $y_1,\dots,y_l\to\xi_1,\dots,\xi_l$, $x_1,\dots,x_l\to\chi_1,\dots,\chi_l$, respectively. Furthermore, the mappings have a constant dilatation equal to $(\tau')^{3/2}$, or to $(\tau')^{1/2}$, respectively.

The only conformal mappings with a constant dilatation are similarities. (In the case $l=2$ this can be easily seen from complex analysis and for $l\geq 3$ according to Liouville theorem the only conformal mappings are Möbius transformations but among them only similarities have a constant dilatation.)

Therefore putting $\tau'(t):=\vartheta^2(t)$ a vector function ξ can be written in the form

$$\xi(y,t)=\vartheta^3(t)Ry+d(t) \tag{34}$$

where $d(t)$ is an integrating constant and R is an orthogonal matrix which may depend on t. It is shown in Lemma 4 that R is in fact independent of t. (Note that U is an open interval.)

Putting (34) into (30) we obtain

$$\chi(x,y,t)=\vartheta(t)Rx-3\vartheta'(t)Ry-\frac{d'(t)}{\vartheta^2(t)}. \tag{35}$$

According to Lemma 5,

$$\zeta(z,t)=\vartheta(t)Pz+a(t) \tag{36}$$

with a constant orthogonal $(n-l)\times(n-l)$ matrix P and $a(t)\in R^{n-l}$.

In the sequel we will assume that R and P are unit matrices, otherwise we will investigate the morphism $\mu'=\mu\circ\mu_1$ with

$$(\mu_1u)(x,y,z,t):=u(R^Tx,R^Ty,P^Tz,t).$$

According to Lemma 6,

$$\vartheta(t)=\frac{1}{\gamma t+\delta}, \tag{37}$$

$$a_p(t)=\frac{A_pt+B_p}{\gamma t+\delta}, \tag{38}$$

and in the case $\gamma=0$:

$$d(t)=\frac{1}{3}\frac{\tilde\gamma}{\delta^5}t^3+\frac{\tilde\delta}{\delta^5}t^2+\frac{\kappa}{\delta^2}t+\omega, \tag{39}$$

$$\varphi(x,y,z,t)=\nu\exp\left(-\frac{1}{\delta^2}(\tilde{\gamma}t+\tilde{\delta})\cdot x+\frac{1}{2}A\cdot z-\frac{\tilde{\gamma}}{\delta^2}\cdot y+\frac{\|\tilde{\gamma}\|^2}{3\delta^4}t^3\right. \tag{40}$$
$$\left.+\frac{\tilde{\gamma}\cdot\tilde{\delta}}{\delta^4}t^2+\frac{\|\tilde{\delta}\|^2}{\delta^4}t+\frac{\|A\|^2}{4}t\right)$$

whereas in the case $\gamma\neq 0$

$$d(t)=\frac{\gamma\tilde{\delta}-\delta\tilde{\gamma}}{3\gamma^3}\frac{1}{(\gamma t+\delta)^3}+\frac{\tilde{\gamma}}{\gamma^3}\frac{1}{(\gamma t+\delta)^2}+\frac{\kappa}{\gamma}\frac{1}{\gamma t+\delta}+\omega, \tag{41}$$

$$\varphi(x,y,z,t)=\frac{\nu}{|\gamma t+\delta|^{(n+3l)/2}}\exp\left(-\frac{\gamma}{(\gamma t+\delta)}\|x\|^2-\frac{\gamma}{4(\gamma t+\delta)}\|z\|^2\right. \tag{42}$$
$$-\frac{\tilde{\gamma}t+\tilde{\delta}}{(\gamma t+\delta)^2}\cdot x-\frac{3\gamma^2}{(\gamma t+\delta)^2}x\cdot y+\frac{A\delta-B\gamma}{2(\gamma t+\delta)}\cdot z-\frac{3\gamma^3}{(\gamma t+\delta)^3}\|y\|^2$$
$$-\frac{3\tilde{\gamma}(\gamma t+\delta)+2(\gamma\tilde{\delta}-\delta\tilde{\gamma})}{(\gamma t+\delta)^3}\cdot y-\frac{\|\gamma\tilde{\delta}-\delta\tilde{\gamma}\|^2}{3\gamma^3}\frac{1}{(\gamma t+\delta)^3}$$
$$-\frac{\tilde{\gamma}\cdot(\gamma\tilde{\delta}-\delta\tilde{\gamma})}{\gamma^3}\frac{1}{(\gamma t+\delta)^2}-\frac{\|\tilde{\gamma}\|^2}{\gamma^3}\frac{1}{\gamma t+\delta}$$
$$\left.-\frac{\|A\delta-B\gamma\|^2}{4\gamma(\gamma t+\delta)}\right),$$

which gives (4)–(13).

The last fact left to prove is generalization from open intervals U to general domains Ω. To every point $A\in\Omega$ there exists an interval $U\subset\Omega$, on which the theorem holds. Divide the domain Ω into sets with the same form of the morphism (i.e., with the same integration constants $\tilde{\gamma},\tilde{\delta},\dots$). These sets cover Ω, they are pairwise disjoint and they are open. Since Ω is connected, there can be at most one nonempty set between them. □

Corollary. *Every $\tilde{L}$-harmonic morphism $\mathcal{M}$ is a composition of morphisms of type $\tilde{\mathcal{A}}$ and $\tilde{\mathcal{S}}$.*

Proof. Relations (9)–(13) describe a morphism $\mathcal{M}$ which can be written as a composition

$$\mathcal{M}=\mathcal{S}_5\circ\mathcal{S}_1^{-1}\circ\mathcal{S}_3^{-1}\circ\mathcal{S}_1\circ\mathcal{S}_3\circ\tilde{\mathcal{A}}\circ\mathcal{S}_4\circ\mathcal{S}_1^{-1}\circ\mathcal{S}_2^{-1}\circ\mathcal{S}_1\circ\mathcal{S}_2 \tag{43}$$

where

$$(\mathcal{S}_1u)(x,y,z,t):=u(x+1,y-t,z,t), \tag{44}$$
$$(\mathcal{S}_1^{-1}u)(x,y,z,t):=u(x-1,y+t,z,t), \tag{45}$$
$$(\mathcal{S}_2u)(x,y,z,t):=u\left(x,y,z,t-\frac{R^T(\gamma\tilde{\delta}+2\tilde{\gamma}\delta)}{3\gamma^3}\right), \tag{46}$$

$$\left(\mathcal{S}_2^{-1}u\right)(x,y,z,t) := u\left(x,y,z,t+\frac{R^T(\gamma\tilde{\delta}+2\tilde{\gamma}\delta)}{3\gamma^3}\right), \tag{47}$$

$$\left(\mathcal{S}_3 u\right)(x,y,z,t) := u\left(x,y,z,t-\omega\right), \tag{48}$$

$$\left(\mathcal{S}_3^{-1}u\right)(x,y,z,t) := u\left(x,y,z,t+\omega\right), \tag{49}$$

$$\left(\mathcal{S}_4 u\right)(x,y,z,t) := u\left(\gamma Rx-\frac{\tilde{\gamma}}{\gamma}, \gamma^3 Ry+\tilde{\gamma}\gamma t, \gamma Pz+B\gamma-A\delta, \gamma^2 t+\delta\gamma\right), \tag{50}$$

$$\left(\mathcal{S}_5 u\right)(x,y,z,t) := \nu|\gamma|^{(n+3l)/2}\, u\left(x+\kappa, y-\kappa t, z+\frac{A}{\gamma}, t+\frac{\alpha}{\gamma}\right) \tag{51}$$

and $\tilde{\mathcal{A}}$ is given by (2).

Relations (4)–(8) describe a morphism $\mathcal{M}$ which can be written in the form:

$$\mathcal{M} = \mathcal{S}_6 \circ \mathcal{S}_1^{-1} \circ \mathcal{S}_3 \circ \mathcal{S}_1 \circ \tilde{\mathcal{A}} \circ \mathcal{M}'$$

where

$$\left(\mathcal{S}_6 u\right)(x,y,z,t) := \nu\, u\left(-x,-y,-z,t+\omega+\frac{\beta}{\delta}\right),$$

$\mathcal{S}_1$, $\mathcal{S}_3$ and $\tilde{\mathcal{A}}$ are given by (35), (37) and (2), respectively and $\mathcal{M}'$ is the morphism of (43) described by parameters:

$$\alpha^0 := 0,\ \beta^0 := -\delta,\ \gamma^0 := \frac{1}{\delta},\ \delta^0 := 0,\ \tilde{\gamma}^0 := \frac{\kappa}{\delta},\ \nu^\circ := \nu\,|\delta|^{(n+3l)/2}\exp\left\{-\frac{\kappa\tilde{\delta}}{\delta}-\frac{AB}{2}\right\},$$

$$\tilde{\delta}^0 := 0,\ \kappa^0 := \frac{\tilde{\delta}}{\delta},\ \omega^0 := \frac{\tilde{\gamma}\delta}{3},\ A^0 := A,\ B^0 := B,\ R^\circ := R,\ P^\circ := P.$$

□

3. Auxiliary assertions

In the sequel we will use the identities

$$\tilde{L}(uv) = u\tilde{L}(v) + v\tilde{L}(u) + 2\left(\nabla_x u \nabla_x v + \nabla_z u \nabla_z v\right), \tag{52}$$

$$\tilde{L}(uvw) = v\tilde{L}(uw) + w\tilde{L}(uv) - vw\tilde{L}(u) + 2u\left(\nabla_x v \nabla_x w + \nabla_z v \nabla_z w\right), \tag{53}$$

$$\tilde{L}(uv^n) - nv^{n-1}\tilde{L}(uv) - (n-1)v^n\tilde{L}(u) + n(n-1)uv^{n-2}\left(\|\nabla_x v\|^2 + \|\nabla_z v\|^2\right) \tag{54}$$

which hold for all sufficiently smooth functions u, v, w. Note that all functions φ, χ_i, ξ_i, ζ_p and τ are C^∞-functions (φ is $\tilde{L}$-harmonic, $\chi_i = \frac{\varphi\chi_i}{\varphi}$ is a ratio of two $\tilde{L}$-harmonic functions with a positive denominator, similarly τ, ξ_i and ζ_p are C^∞-functions by (18), (22) and (25)).

We will assume that U is an open interval in $\mathbb{R}^{n+l+1}$.

Lemma 1. *Let $\varphi, \chi_i, \tau \in C^\infty(U)$ satisfy* (16), (17), (18), (20) *and* (21) *on U and let $\varphi > 0$ on U. Then $\tau = \tau(y,t)$ and*

$$\tilde{L}(\tau) + \|\nabla_x \chi_i\|^2 + \|\nabla_z \chi_i\|^2 = 0 \tag{55}$$

on U for all $i = 1, \ldots l$.

Proof. It follows from (16), (17), (54) that

$$\tilde{L}(\varphi\chi_i^n) = n(n-1)\varphi\chi_i^{n-2}\left(\|\nabla_x\chi_i\|^2 + \|\nabla_z\chi_i\|^2\right). \tag{56}$$

Further, from (56) it follows

$$\tilde{L}(\varphi(\chi_i^2 + 2\tau)) = 2\left\{\varphi\left(\|\nabla_x\chi_i\|^2 + \|\nabla_z\chi_i\|^2\right) + \tilde{L}(\varphi\tau)\right\},$$

from (56), (53), (16) and (17) it follows

$$\begin{aligned}\tilde{L}(\varphi(\chi_i^3 + 6\chi_i\tau)) &= 6\{\varphi\chi_i\left(\|\nabla_x\chi_i\|^2 + \|\nabla_z\chi_i\|^2\right) + \chi_i\tilde{L}(\varphi\tau) \\ &\quad + 2\varphi\left(\nabla_x\chi_i\nabla_x\tau + \nabla_z\chi_i\nabla_z\tau\right)\} \\ &= 6\{\chi_i[\varphi(\|\nabla_x\chi_i\|^2 + \|\nabla_z\chi_i\|^2) + \tilde{L}(\varphi\tau)] \\ &\quad + 2\varphi\left(\nabla_x\chi_i\nabla_x\tau + \nabla_z\chi_i\nabla_z\tau\right)\}.\end{aligned} \tag{57}$$

Equations (18), (20) together with the assumption $\varphi > 0$ imply

$$\nabla_x\chi_i\nabla_x\tau + \nabla_z\chi_i\nabla_z\tau = 0. \tag{58}$$

From (56), (53) and (54) it further follows

$$\begin{aligned}\tilde{L}(\varphi(\chi_i^4 + 12\chi_i^2\tau + 12\tau^2)) &= 12\left\{\varphi\chi_i^2\left(\|\nabla_x\chi_i\|^2 + \|\nabla_z\chi_i\|^2\right) + \chi_i^2\tilde{L}(\varphi\tau)\right. \\ &\quad + 2\varphi\tau\left(\|\nabla_x\chi_i\|^2 + \|\nabla_z\chi_i\|^2\right) + 4\varphi\chi_i\left(\nabla_x\chi_i\nabla_x\tau + \nabla_z\chi_i\nabla_z\tau\right) \\ &\quad + 2\tau\tilde{L}(\varphi\tau) + 2\varphi\left(\|\nabla_x\tau\|^2 + \|\nabla_z\tau\|^2\right)\} \\ &= 12\left\{(\chi_i^2 + 2\tau)\left[\varphi\left(\|\nabla_x\chi_i\|^2 + \|\nabla_z\chi_i\|^2\right) + \tilde{L}(\varphi\tau)\right]\right. \\ &\quad + 4\chi_i\varphi\left(\nabla_x\chi_i\nabla_x\tau + \nabla_z\chi_i\nabla_z\tau\right) + 2\varphi\left(\|\nabla_x\tau\|^2 + \|\nabla_z\tau\|^2\right)\}.\end{aligned}$$

Now (18), (21) and (58) imply

$$\|\nabla_x\tau\|^2 + \|\nabla_z\tau\|^2 = 0. \tag{59}$$

We see that the function τ may depend on y_i and t variables only. Finally, (18) and (58) yield

$$\tilde{L}(\tau) + \|\nabla_x\chi_i\|^2 + \|\nabla_z\chi_i\|^2 = 0 \tag{60}$$

for all $i = 1, \ldots, l$. □

Lemma 2. *Let $\varphi, \chi_i, \xi_i, \tau \in C^\infty(U)$ satisfy* (16)–(24) *on U and let $\varphi(x,y,z,t) > 0$ on U. Then $\xi_i = \xi_i(y,t)$ and*

$$\tilde{L}(\xi_i) + \chi_i\tilde{L}(\tau) = 0 \tag{61}$$

holds on U for all $i = 1, \ldots, l$.

Proof. Using (53), (54) and (59) we can write

$$\begin{aligned}
\tilde{L}(\varphi(\xi_i+\chi_i\tau)) &= \tilde{L}(\varphi\xi_i)+\chi_i\varphi\tilde{L}(\tau),\\
\tilde{L}(\varphi(\chi_i\xi_i+\chi_i^2\tau+\tau^2)) &= \chi_i\tilde{L}(\varphi\xi_i)+2\varphi(\nabla_x\chi_i\nabla_x\xi_i+\nabla_z\chi_i\nabla_z\xi_i)\\
&\quad+\varphi\chi_i^2\tilde{L}(\tau)+2\varphi\tau(\|\nabla_x\chi_i\|^2+\|\nabla_z\chi_i\|^2)+2\varphi\tau\tilde{L}(\tau)\\
&=\chi_i\left(\tilde{L}(\varphi\xi_i)+\chi_i\varphi\tilde{L}(\tau)\right)+2\varphi(\nabla_x\chi_i\nabla_x\xi_i+\nabla_z\chi_i\nabla_z\xi_i)\\
&\quad+2\varphi\tau\left(\tilde{L}(\tau)+\|\nabla_x\chi_i\|^2+\|\nabla_z\chi_i\|^2\right),
\end{aligned}$$

$$\begin{aligned}
\tilde{L}(\varphi\chi_i\xi_i\tau) &= \tau\tilde{L}(\varphi\chi_i\xi_i)+\chi_i\xi_i\tilde{L}(\varphi\tau)\\
&=\tau\chi_i\tilde{L}(\varphi\xi_i)+2\tau\varphi(\nabla_x\chi_i\cdot\nabla_x\xi_i+\nabla_z\chi_i\cdot\nabla_z\xi_i)+\varphi\chi_i\xi_i\tilde{L}(\tau),
\end{aligned}$$

$$\begin{aligned}
&\tilde{L}(\varphi(3\xi_i^2+6\chi_i\xi_i\tau+3\chi_i^2\tau^2+2\tau^3))\\
&=6\Big\{\xi_i\tilde{L}(\varphi\xi_i)+\varphi\left(\|\nabla_x\xi_i\|^2+\|\nabla_z\xi_i\|^2\right)+\tau\chi_i\tilde{L}(\varphi\xi_i)\\
&\quad+2\tau\varphi(\nabla_x\chi_i\cdot\nabla_x\xi_i+\nabla_z\chi_i\cdot\nabla_z\xi_i)+\varphi\chi_i\xi_i\tilde{L}(\tau)+\varphi\tau^2\left(\|\nabla_x\chi_i\|^2+\|\nabla_z\chi_i\|^2\right)\\
&\quad+\varphi\chi_i^2\tau\tilde{L}(\tau)+\varphi\tau^2\tilde{L}(\tau)\Big\}\\
&=6\Big\{(\xi_i+\tau\chi_i)\left(\tilde{L}(\varphi\xi_i)+\chi_i\varphi\tilde{L}(\tau)\right)+\varphi\left(\|\nabla_x\xi_i\|^2+\|\nabla_z\xi_i\|^2\right)\\
&\quad+2\tau\varphi(\nabla_x\chi_i\cdot\nabla_x\xi_i+\nabla_z\chi_i\cdot\nabla_z\xi_i)+\varphi\tau^2\left(\tilde{L}(\tau)+\|\nabla_x\chi_i\|^2+\|\nabla_z\chi_i\|^2\right)\Big\}.
\end{aligned}$$

Now we see that equations (22), (23) and (60) imply

$$\nabla_x\chi_i\cdot\nabla_x\xi_i+\nabla_z\chi_i\cdot\nabla_z\xi_i=0, \tag{62}$$

equations (24), (62), (22) and (60) imply

$$\|\nabla_x\xi_i\|^2+\|\nabla_z\xi_i\|^2=0 \tag{63}$$

and finally (22) together with (63) imply (61). □

Lemma 3. *Let* $\chi_i,\xi_i,\tau\in\mathcal{C}^\infty(U)$ *satisfy* (59), (55), (61), (63) *on* U. *Then*

$$\frac{\partial\tau}{\partial y_i}=0 \tag{64}$$

on U *for all* $i=1,\dots,l$.

Proof. Fix $A:=(x,y,z,t)\in U$, fix i and let first $\tilde{L}(\tau)\neq 0$ at A. Since τ is $\mathcal{C}^\infty$-function there exists a neighbourhood U_A of A where $\tilde{L}(\tau)\neq 0$. Then

$$\chi_i=-\frac{\tilde{L}(\xi_i)}{\tilde{L}(\tau)} \tag{65}$$

on U_A. Differentiating in (65) with respect to x coordinates and putting $\nabla_x \chi_i$ into (55) we get

$$\sum_{j=1}^{l}\left[\frac{\partial \xi_i}{\partial t}\frac{\partial \tau}{\partial y_j}-\frac{\partial \xi_i}{\partial y_j}\frac{\partial \tau}{\partial t}+\sum_{k=1}^{l} x_k\left(\frac{\partial \xi_i}{\partial y_j}\frac{\partial \tau}{\partial y_k}-\frac{\partial \xi_i}{\partial y_k}\frac{\partial \tau}{\partial y_j}\right)\right]^2$$

$$=-\left(\sum_{k=1}^{l} x_k\frac{\partial \tau}{\partial y_k}-\frac{\partial \tau}{\partial t}\right)^5$$

which is an algebraic equation in x_k holding on nonempty open set. This implies that all coefficients at the powers of x_k must be zero on U_A and therefore (64) holds for all $i = 1, \ldots, l$ on U.

Let now $\tilde{L}(\tau) = 0$ at A. Then (55) implies $\nabla_x \chi_i = 0$, $\nabla_z \chi_i = 0$ in A. Now differentiating in (55) with respect to x_k we obtain (64). □

Lemma 4. *Let U_1 be an open interval in $\mathbb{R}^{n+l}$, I be an open interval in $\mathbb{R}$, $U = U_1 \times I$, $\vartheta, d_i, R_{ij} \in C^\infty(I)$ for $i, j = 1, \ldots, l$, $\vartheta > 0$ on I and assume that $R(t)$ is an orthogonal matrix for every $t \in I$. Let $\chi_i(x, y, t) := \vartheta(t)R_{ij}(t)x_j - 3\vartheta'(t)R_{ij}(t)y_j - \vartheta(t)R'_{ij}(t)y_j - d'_i(t)/\vartheta^2(t)$ (the summation convention is used), let $\varphi \in C^\infty(U)$, $\varphi > 0$ on U. Let φ and χ_i satisfy (16), (17) on U. Then $R'_{ij}(t) = 0$ on I for $i, j = 1, \ldots, l$.*

Proof. By differentiating the expression for χ_i we can compute $\frac{\partial \chi_i}{\partial x_k} = \vartheta(t)R_{ik}(t)$. Using this, (52) and (16), we can write (17) in the form

$$\varphi \tilde{L}(\chi_i) + 2\frac{\partial \varphi}{\partial x_j}\vartheta(t)R_{ij}(t) = 0$$

or, using $R_{ij}(t)R_{ik}(t) = \delta_{jk}$, in the form

$$\frac{\partial}{\partial x_k}(\log \varphi) = -\frac{1}{2\vartheta(t)}R_{ik}\tilde{L}(\chi_i). \tag{66}$$

Computing $\tilde{L}(\chi_i)$ and differentiating (66) we obtain

$$\frac{\partial^2}{\partial x_j \partial x_k}(\log \varphi) = 2\cdot\frac{\vartheta'(t)}{\vartheta(t)}\delta_{jk} + \frac{1}{2}R_{ij}(t)R'_{ik}(t).$$

The equality

$$\frac{\partial^2}{\partial x_j \partial x_k}(\log \varphi) = \frac{\partial^2}{\partial x_k \partial x_j}(\log \varphi)$$

implies $R_{ij}(t)R'_{ik}(t) = R_{ik}(t)R'_{ij}(t)$. On the other hand, from $R_{ij}(t)R_{ik}(t) = \delta_{jk}$ it follows $R_{ij}(t)R'_{ik}(t) + R_{ik}(t)R'_{ij}(t) = 0$. This implies $R'_{ij}(t) = 0$. □

Lemma 5. *Let U_1 be an open interval in $\mathbb{R}^{n+l}$, I be an open interval in $\mathbb{R}$, $U = U_1 \times I$. Let χ_i, ξ_i $(i = 1, \ldots, l)$, $\zeta_p (p = 1, \ldots, n-l)$, $\varphi \in C^\infty(U)$, $\tau \in C^\infty(I)$ satisfy (16)–(28) on U. Then there exist a constant orthogonal matrix P of the* *order $n-l$ and functions $\vartheta, a_p \in \mathcal{C}^\infty(I)$, $p=1,\dots,n-l$, such that*

$$\zeta_p(x,y,z,t) = \vartheta(t)P_{pq}z_q + a_p(t). \tag{67}$$

Further, $\vartheta^2(t) = \tau'(t)$.

Proof. It follows from (16), (17), (25), (28) and (53) that

$$\nabla_x\chi_i \cdot \nabla_x\zeta_p + \nabla_z\chi_i \cdot \nabla_z\zeta_p = 0.$$

Since $\nabla_z\chi_i = 0$ and $\nabla_x\chi_i$ form a basis of a vector space, $\nabla_x\zeta_p = 0$ holds for all $p = 1,\dots,n-l$. Now from (16), (25), (26), (27), (53) it follows $\nabla_z\zeta_p \cdot \nabla_z\zeta_q = \tau'\delta_{pq}$ which implies that fixing x,y,t we obtain a conformal mapping $z_1,\dots,z_{n-l} \to \zeta_1,\dots,\zeta_{n-l}$ with a constant dilatation $\sqrt{\frac{\partial\tau}{\partial t}}$, which implies (if we put $\vartheta(t) = \sqrt{\tau'(t)}$)

$$\zeta(y,z,t) = \vartheta(t)P(y,t)z + a(y,t) \tag{68}$$

where $P(y,t)$ is an orthogonal matrix $l \times l$ which may depend on y,t and $a(y,t)$ is an integrating constant.

Now we claim that P in fact doesn't depend on y and t and a doesn't depend on y. Using $P_{pq}P_{pr} = \delta_{qr}$ (summation convention is used) we can write (25) in the form

$$\frac{\partial}{\partial z_p}(\log\varphi) = -\frac{1}{2\vartheta(t)}P_{qp}(y,t)\tilde{L}(\zeta_q) \tag{69}$$

(for details, see the proof of Lemma 4). Putting (68) into (69) and differentiating, we obtain

$$\frac{\partial^2}{\partial z_q\partial z_p}(\log\varphi) = -\frac{1}{2}P_{rp}\frac{\partial P_{rq}}{\partial y_i}x_i + \frac{\vartheta'}{2\vartheta}\delta_{pq} + \frac{1}{2}P_{rp}\frac{\partial P_{rq}}{\partial t}$$

and

$$\frac{\partial^2}{\partial x_i\partial z_p}(\log\varphi) = -\frac{1}{2}P_{qp}\frac{\partial P_{qr}}{\partial y_i}z_r - \frac{1}{2\vartheta}\frac{\partial a_q}{\partial y_i}P_{qp}.$$

From $\frac{\partial^2}{\partial z_p\partial z_q}(\log\varphi) = \frac{\partial^2}{\partial z_q\partial z_p}(\log\varphi)$ we obtain, similarly as in Lemma 4, $\frac{\partial P_{pq}}{\partial y_k} = 0$, $\frac{\partial P_{pq}}{\partial t} = 0$. From (66) it follows that $\frac{\partial^2}{\partial z_p\partial x_j}(\log\varphi) = 0$. Since $\frac{\partial^2}{\partial x_j\partial z_p}(\log\varphi) = \frac{\partial^2}{\partial z_p\partial x_j}(\log\varphi)$ it follows $\frac{\partial a_i}{\partial y_p} = 0$. Therefore P is a constant matrix and a doesn't depend on y. □

Lemma 6. *Let U_1 be an open interval in $\mathbb{R}^{n+l}$, I be an open interval in $\mathbb{R}$, $U = U_1 \times I$, $\vartheta, d_i, a_p \in \mathcal{C}^\infty(I)$ for $i = 1\dots l$, $p = 1\dots n-l$. Let*

$$\chi_i = \vartheta(t)x_i - 3\vartheta'(t)y_i - \frac{d_i'(t)}{\vartheta^2(t)},$$

$$\xi_i = \vartheta^3(t)y_i + d_i(t),$$

$$\zeta_p = \vartheta(t)z_p + a_p(t)$$

and $\varphi \in C^\infty(U)$ satisfy (16), (17) and (25) on U. Then there exists $n-l$ components vectors A, B, l components vectors $\tilde\gamma, \tilde\delta, \kappa, \omega$, constants $\gamma, \delta, \nu > 0$, such that $\gamma^2 + \delta^2 \neq 0$ and $\vartheta(t) = \frac{1}{\gamma t+\delta}$, $a(t) = \frac{At+B}{\gamma t+\delta}$, and in the case $\gamma = 0$

$$\begin{aligned} d(t) &= \frac{1}{3}\frac{\tilde\gamma}{\delta^5}t^3 + \frac{\tilde\delta}{\delta^5}t^2 + \frac{\kappa}{\delta^2}t + \omega, \\ \varphi(x,y,z,t) &= \nu \exp\Bigg(-\frac{1}{\delta^2}(\tilde\gamma t + \tilde\delta)\cdot x + \frac{1}{2}A\cdot z - \frac{\tilde\gamma}{\delta^2}\cdot y \\ &\quad + \frac{\|\tilde\gamma\|^2}{3\delta^4}t^3 + \frac{\tilde\gamma\cdot\tilde\delta}{\delta^4}t^2 + \frac{\|\tilde\delta\|^2}{\delta^4}t + \frac{\|A\|^2}{4}t\Bigg) \end{aligned} \tag{70}$$

while in the case $\gamma \neq 0$

$$\begin{aligned} d(t) &= \frac{\gamma\tilde\delta - \tilde\gamma\delta}{3\gamma^3}\frac{1}{(\gamma t+\delta)^3} + \frac{\tilde\gamma}{\gamma^3}\frac{1}{(\gamma t+\delta)^2} + \frac{\kappa}{\gamma}\frac{1}{\gamma t+\delta} + \omega, \\ \varphi(x,y,z,t) &= \frac{\nu}{|\gamma t+\delta|^{(n+3l)/2}} \exp\Bigg(-\frac{\gamma}{(\gamma t+\delta)}\|x\|^2 - \frac{\gamma}{4(\gamma t+\delta)}\|z\|^2 \\ &\quad - \frac{\tilde\gamma t+\tilde\delta}{(\gamma t+\delta)^2}\cdot x - \frac{3\gamma^2}{(\gamma t+\delta)^2}x\cdot y + \frac{A\delta - B\gamma}{2(\gamma t+\delta)}\cdot z - \frac{3\gamma^3}{(\gamma t+\delta)^3}\|y\|^2 \\ &\quad - \frac{3\tilde\gamma(\gamma t+\delta) + 2(\gamma\tilde\delta - \delta\tilde\gamma)}{(\gamma t+\delta)^3}\cdot y - \frac{\|\gamma\tilde\delta - \delta\tilde\gamma\|^2}{3\gamma^3}\frac{1}{(\gamma t+\delta)^3} \\ &\quad - \frac{\tilde\gamma\cdot(\gamma\tilde\delta - \delta\tilde\gamma)}{\gamma^3}\frac{1}{(\gamma t+\delta)^2} - \frac{\|\tilde\gamma\|^2}{\gamma^3}\frac{1}{\gamma t+\delta} - \frac{\|A\delta - B\gamma\|^2}{4\gamma(\gamma t+\delta)}\Bigg). \end{aligned} \tag{71}$$

Proof. It follows from (16), (17)

$$\frac{\partial}{\partial x_k}(\log\varphi) = -\frac{1}{2\vartheta(t)}\tilde L(\chi_k), \quad k = 1,\dots,l \tag{72}$$

(for details, see the proof of Lemma 4) and, similarly, from (16), (25) it follows

$$\frac{\partial}{\partial z_p}(\log\varphi) = -\frac{1}{2\vartheta(t)}\tilde L(\zeta_p) \quad p = 1,\dots,n-l. \tag{73}$$

From (72), (73) we can calculate

$$\log\varphi(x,y,z,t) = \frac{\vartheta'(t)}{\vartheta(t)}\|x\|^2 + \frac{\vartheta'(t)}{4\vartheta(t)}\|z\|^2 - \frac{3\vartheta''(t)}{2\vartheta(t)}x\cdot y - \Theta(t)\cdot x + b(t)\cdot z + g(y,t), \tag{74}$$

where

$$\Theta_i(t) := \frac{1}{2\vartheta(t)}\left(\frac{d_i'(t)}{\vartheta^2(t)}\right)', \tag{75}$$

$$b_p(t) := \frac{a_p'(t)}{2\vartheta(t)} \tag{76}$$

and $g(y,t)$ is an integrating constant. Putting this into (16) we obtain, after long but simple computation the following equation:

$$\begin{aligned}\Bigg[&\|x\|^2\left\{\left(\frac{2\vartheta'(t)}{\vartheta(t)}\right)^2-\frac{3\vartheta''(t)}{2\vartheta(t)}-\left(\frac{\vartheta'(t)}{\vartheta(t)}\right)'\right\}+\|z\|^2\left\{\left(\frac{\vartheta'(t)}{2\vartheta(t)}\right)^2-\left(\frac{\vartheta'(t)}{4\vartheta(t)}\right)'\right\}\\&+x\left\{-2\frac{2\vartheta'(t)}{\vartheta(t)}\left(\frac{3\vartheta''(t)}{2\vartheta(t)}y+\Theta(t)\right)+\nabla_y g(y,t)+\left(\frac{3\vartheta''(t)}{2\vartheta(t)}\right)'y+\Theta'(t)\right\}\\&+z\left\{2\frac{\vartheta'(t)}{2\vartheta(t)}b(t)-b'(t)\right\}+\left\|\frac{3\vartheta''(t)}{2\vartheta(t)}y+\Theta(t)\right\|^2+\frac{\vartheta'(t)}{2\vartheta(t)}(n+3l)+\|b(t)\|^2-\frac{\partial g}{\partial t}\Bigg]\\&\times\varphi(x,y,z,t)=0.\end{aligned}$$

Since $\varphi>0$ on U and the equation is algebraic in x and z holding on an nonempty open interval, all coefficients at the powers of x and z must be zero. This yields the following equations:

$$\left(\frac{2\vartheta'(t)}{\vartheta(t)}\right)^2-\frac{3\vartheta''(t)}{2\vartheta(t)}-\left(\frac{\vartheta'(t)}{\vartheta(t)}\right)'=0, \tag{77}$$

$$\left(\frac{\vartheta'(t)}{2\vartheta(t)}\right)^2-\left(\frac{\vartheta'(t)}{4\vartheta(t)}\right)'=0, \tag{78}$$

$$-2\frac{2\vartheta'(t)}{\vartheta(t)}\left(\frac{3\vartheta''(t)}{2\vartheta(t)}y+\Theta(t)\right)+\nabla_y g(y,t)+\left(\frac{3\vartheta''(t)}{2\vartheta(t)}\right)'y+\Theta'(t)=0, \tag{79}$$

$$\frac{\vartheta'(t)}{\vartheta(t)}b(t)-b'(t)=0, \tag{80}$$

$$\left\|\frac{3\vartheta''(t)}{2\vartheta(t)}y+\Theta(t)\right\|^2+\frac{\vartheta'(t)}{2\vartheta(t)}(n+3l)+\|b(t)\|^2-\frac{\partial g}{\partial t}=0. \tag{81}$$

Equations (77) and (78) are identical and have a solution: $\vartheta(t)=(\gamma t+\delta)^{-1}$. Equation (80) has a solution $a(t)=(At+B)/(\gamma t+\delta)$, $b(t)=(A\delta-B\gamma)(2(\gamma t+\delta))$. Differentiating in (79) and (81) with respect to t and y and using the equality $\nabla_y\left(\frac{\partial g}{\partial t}\right)=\frac{\partial}{\partial t}(\nabla_y g)$ we obtain

$$\Theta''(t)+\frac{4\gamma}{\gamma t+\delta}\Theta'(t)+\frac{2\gamma^2}{(\gamma t+\delta)^2}\Theta(t)=0$$

which can be simplified to

$$((\gamma t+\delta)^2\Theta(t))''=0$$

which gives $\Theta(t)=(\tilde{\gamma}t+\tilde{\delta})/(\gamma t+\delta)^2$. Now we can calculate d from (75) and g from (79) and (81) and we will get (70) and (71) (after a long but straightforward calculation). □

4. Main result

Theorem 2. *Let $n, l \in \mathbb{N}$, $A = (a_{ij})$ be a positive definite matrix of the order n and $B = (b_{ij})$ be a matrix of the order $l \times n$ such that the matrix BAB^T is regular (we remark that this implies $n \geq l$). For $x \in \mathbb{R}^n$, $y \in \mathbb{R}^l$, $t \in \mathbb{R} \setminus \{0\}$, define*

$$(\mathcal{A}u)(x, y, t) := E(x, y, t) u \left(\frac{x}{t} + \frac{3}{t^2} AB^T \left(BAB^T\right)^{-1} y, \frac{y}{t^3}, -\frac{1}{t} \right). \tag{82}$$

Let $a \in \mathbb{R}^n$, $b, r, \nu \in \mathbb{R}$, $P := AB^T(BAB^T)^{-1}B$ and R be a matrix of the order $n \times n$ satisfying equations

$$RAR^T = A; \qquad PR = RP \tag{83}$$

(e.g., the unit matrix E satisfies these equations). For $x \in \mathbb{R}^n$, $y \in \mathbb{R}^l$, $t \in \mathbb{R}$, define

$$(\mathcal{S}u)(x, y, t) := \nu u \left(rRx + a, r^2(rBRAB^T(BAB^T)^{-1}y - Bat), r^2 t + b \right). \tag{84}$$

Then, for L defined in the introduction, the mappings $\mathcal{A}$ and $\mathcal{S}$ are generators of the group of L-harmonic morphisms.

Proof. The proof will be given in two steps. First we show that in the case $a_{ij} = \delta_{ij}$, $b_{ij} = \delta_{ij}$ the assertion of Theorem 2 is exactly the assertion of Theorem 1. In the second step we prove the following implication: if the assertion holds for an operator L then it holds for the operator L' obtained from L by a regular linear transformation of variables: $x'_i = h_{ij} x_j$, $y'_i = k_{ij} y_j$ (the summation convention is used). Since every operator L can be obtained from $\tilde{L}$ by a regular linear transformation of variables, the proof will be finished.

Let now $a_{ij} = \delta_{ij}$ and $b_{ij} = \delta_{ij}$. Note that from regularity of the matrix BAB^T it follows $n \geq l$. Let for $i = 1, \ldots, n-l$ be $z_i := x_{l+i}$. Then L is equal to $\tilde{L}$. Further $B(x_1, \ldots, x_n)^T = (x_1, \ldots, x_l)^T$, BAB^T is an identity matrix of the order l and $B^T(y_1, \ldots, y_l)^T = (y_1, \ldots, y_l, 0, \ldots, 0)^T$. It can be easily seen that in this case $\mathcal{A}$ is identical with $\tilde{\mathcal{A}}$. In (83) first condition means that R is an orthogonal matrix. The second one means that $R_{ij} = 0$ for $i = 1, \ldots, l$, $j = l+1, \ldots, n$ or $i = l+1, \ldots, n$, $j = 1, \ldots, l$, therefore $R = \begin{pmatrix} R_1 & 0 \\ 0 & P \end{pmatrix}$ for some orthogonal matrices R_1 and P of the order l and $n - l$, respectively. Again, it can be easily seen that in this case $\mathcal{S}$ is the same as $\tilde{\mathcal{S}}$.

Let now L be an arbitrary operator defined in the introduction. Let H and K be regular matrices of the order n and l and denote their elements by h_{ij} and k_{ij}, respectively. Let $T : \mathbb{R}^{n+l+1} \to \mathbb{R}^{n+l+1}$ map the point (x, y, t) to the point $T(x, y, t) := (Hx, Ky, t)$. Let $\Omega \subset \mathbb{R}^{n+l+1}$ be a domain, $v \in C^{2,1,1}(T(\Omega))$ and let

$$L'(v) := L(v \circ T) \circ T^{-1}. \tag{85}$$

Then by differentiation of a composite function we get

$$L' := \sum_{i,j=1}^{n} a'_{ij} \frac{\partial}{\partial x_i \partial x_j} + \sum_{i=1}^{l} \sum_{j=1}^{n} b'_{ij} x_j \frac{\partial}{\partial y_i} - \frac{\partial}{\partial t},$$

where a'_{ij}, b'_{ij} are coefficients of $A' := HA\,H^T$, $B' := KBH^{-1}$, respectively. Let Ω_1, Ω_2 be domains, $\varphi \in C^{2.1.1}(\Omega_1)$, $\Phi : \Omega_1 \to \Omega_2$ such that the morphism $\mathcal{M}$ defined by

$$\mathcal{M}(u) := \varphi\, u \circ \Phi \tag{86}$$

is an L-harmonic morphism on Ω_1. Then $\mathcal{M}'$ defined by

$$\mathcal{M}'(v) := \mathcal{M}(v \circ T) \circ T^{-1} \tag{87}$$

is an L'-harmonic morphism on $T(\Omega_1)$. (Using (85) we obtain $L'(\mathcal{M}'(v)) = L(\mathcal{M}'(v) \circ T) \circ T^{-1} = L(\mathcal{M}(v \circ T)) \circ T^{-1}$.) Furthermore the mapping $\mathcal{M} \to \mathcal{M}'$ given in (87) is a bijection between the group of L-harmonic morphisms and the group of L'-harmonic morphisms.
Define

$$\mathcal{A}'(v) := \mathcal{A}(v \circ T) \circ T^{-1}$$

and

$$\mathcal{S}'(v) := \mathcal{S}(v \circ T) \circ T^{-1},$$

then a straightforward calculation gives

$$(\mathcal{A}'v)(x,y,t) = \frac{1}{|t|^{\frac{n+3l}{2}}} \exp\left(-\frac{A'^{-1}x \cdot x}{t} - \frac{3}{t^3} \left(B'A'B'^T\right)^{-1} \left(y + \frac{t}{2}B'x\right) \cdot \left(y + \frac{t}{2}B'x\right)\right) v\left(\frac{x}{t} + \frac{3}{t^2}A'B'^T \left(B'A'B'^T\right)^{-1} y, \frac{y}{t^3}, -\frac{1}{t}\right)$$

and

$$(\mathcal{S}'v)(x,y,t) := v\left(r'R'x + a', r'^2(r'B'R'A'B'^T(B'A'B'^T)^{-1}y - B'a't), r'^2t + b'\right)$$

with $a' := Ha$, $b' := b$, $r' := r$, $R' := HRH^{-1}$. It is easy to see that matrices A', B', R' fulfil the equations (83). □

References

[1] Appell, P., Sur l'équation $\partial^2 z/\partial x^2 - \partial z/\partial y = 0$ et la théorie du chaleur. J. Math. Pures Appl. (4) 8 (1892), 187–216.

[2] Brzezina, M., Appell type transformation for the Kolmogorov operator. Math. Nachr. 169 (1994), 59–67.

[3] Garofalo, N., Lanconelli, E., Level sets of the fundamental solution and the Harnack inequality for degenerate equations of Kolmogorov type. Trans. Amer. Math. Soc. 321 (1990), 775–792.

[4] Helms, L. L., Introduction to potential theory. Wiley Interscience, New York 1969.

[5] Hoh, W., Jacob, N., On the potential theory of the Kolmogorov equation. Math. Nachr. 154 (1991), 51–66.

[6] Kellogg, O. D., Foundations of potential theory. Springer-Verlag, Berlin 1929 (reissued 1967).

[7] Kuptsov, L. P., Mean value theorem and a maximum principle for Kolmogorov's equation. Math. Notes 15 (1974), 280–286.

[8] Leutwiler, H., On the Appell transformation. In: Potential Theory (ed. J. Král et al.), 215–222. Plenum Press, New York 1987.

[9] Scornazzani, L., Sul problema di Dirichlet per l'operatore di Kolmogorov. Bollettino U. M. I. Anal. Funz. Appl. 18 (1981), 43–62.

Potentiels et attracteurs d'un système dynamique

Mohamed Hmissi

Abstract. Let (X, Φ) be a continuous dynamical system on a locally compact space with countable base. In this note, we characterize the stable attractors of (X, Φ) and we construct the associated Liapunov functions in terms of the potential kernel of (X, Φ).

1991 Mathematics Subject Classification: 31B35, 58F12, 60J35.

0. Introduction

Soit (X, Φ) un système dynamique continu sur un espace X localement compact à base dénombrable, le noyau potentiel V associé est défini par: $f \mapsto Vf := \int_0^\infty f(\Phi(t, .))dt$. L'objet de cette note est l'étude des attracteurs de (X, Φ) en termes du noyau V.

Dans le premier paragraphe, nous rappelons les notions de systèmes dynamiques, d'attracteurs et de fonctions de Liapunov et nous donnons leurs propriétés importantes. Nous déterminons aussi la relation fondamentale entre un attracteur et le noyau potentiel de (X, Φ): Soit M un compact de X, alors les propriétés suivantes sont équivalentes:

(i) M est un attracteur stable de (X, Φ);

(ii) M est invariant et M possède un voisinage ouvert invariant Y tel que le noyau potentiel du système dynamique (Y, Φ) est continu sur $Y \setminus M$.

Dans ce cas, nous associons à M une fonction de Liapunov de la forme Vf telle que f et Vf sont continues et positives et $M = \{Vf = 0\}$.

Dans le second paragraphe, nous montrons que si $\mathcal{A}(M)$ est la région d'attraction d'un attracteur stable M, alors il existe $\alpha > 0$ tel que la fonction

$$\omega(x) = \sup\{s \in \mathbb{R} : Vf(\Phi(s, x)) > \alpha\}$$

est définie et continue sur $\mathcal{A}(M) \setminus M$.

Comme applications, nous montrons d'abord que la fonction $\exp(\omega)$ est une fonction de Liapunov associée à M. Aussi, nous supposons que X est la boule unité de $\mathbb{R}^n$ et que M est la sphère unité et nous déduisons que $X \setminus \mathcal{A}(M)$ est un attracteur stable du système dynamique obtenu à partir de (X, Φ) par retournement du temps.

Ce travail fait suite à nos références [2–4] consacrées à l'étude de la théorie du potentiel sur les systèmes dynamiques.

1. Potentiels et attracteurs

L'objet de ce paragraphe est un rappel des notions et résultats utilisés dans cette note, les références de base sont [1] et [2].

Soit X un espace localement compact et à base dénombrable. Si Y est un sous-espace de X, alors $\mathcal{C}(Y)$ désigne l'espace des fonctions définies, finies et continues sur Y et $\mathcal{K}(Y)$ le sous-espace de $\mathcal{C}(Y)$, des fonctions à support compact. Pour $f, g \in \mathcal{C}(Y)$, nous notons par $\{f \leq g\} := \{x \in X : f(x) \leq g(x)\}$ et par $\operatorname{Supp} f$ le support topologique de f. Pour tout $O \subset X$, soit $\mathcal{U}(O) := \{U \subset X : U$ est un voisinage ouvert de $O\}$. Toutefois, nous utilisons la notation $\mathcal{U}(x) := \mathcal{U}(\{x\})$, pour tout $x \in X$. Notons enfin par $\|.\|$ la norme de la convergence uniforme, par d une métrique associée à l'espace X et par $|.|$ la norme euclidienne.

1.1 Systèmes dynamiques.
1. Une application continue $\Phi : \mathbb{R} \times X \mapsto X$ est appelée *système dynamique* (SD) sur X si $\Phi(0,x) = x$ et $\Phi(s+t,x) = \Phi(s,\Phi(t,x))$, pour tous $x \in X$ et $s,t \in \mathbb{R}$. Un tel système est noté plutôt par le couple (X,Φ).
2. Soit (X,Φ) un système dynamique. Pour $U \subset X$ et $I \subset \mathbb{R}$, nous notons par $\Phi(I,U) := \{\Phi(t,x) \ : \ t \in I \ , \ x \in U \ \}$. Pour tout $x \in X$, nous utilisons souvent les notations $\Gamma(x) := \Phi(\mathbb{R},\{x\})$; $\Gamma^+(x) := \Phi(\mathbb{R}^+,\{x\})$ et $\Gamma^-(x) := \Phi(\mathbb{R}^-,\{x\})$.

Un sous-espace $U \subset X$, est dit *invariant* si $\Gamma(x) \subset U$, pour tout $x \in U$. Dans ce cas la restriction du SD (X,Φ) à U , est aussi un SD noté également par (U,Φ). Les sous-espaces positivement invariants et négativement invariants se définissent d'une manière analogue.
3. Soit (X,Φ) un SD. L'opérateur V défini, pour tout $f \in \mathcal{K}$ $Vf := \int_0^\infty f(\Phi(t,.))dt$ est appelé le *noyau potentiel* de (X,Φ). V est en fait le noyau potentiel du semi-groupe de Feller *déterministe* $\mathbb{H} := (H_t)_{t\geq 0}$ associé à (X,Φ). Ce semi-groupe est défini par $H_t f := f(\Phi(t,.))$, pour tous $t \geq 0$ et $f \in \mathcal{K}$. Nous notons aussi par V le noyau potentiel de toute restriction du SD (X,Φ).

Si M est un compact de X, alors le noyau V est dit *continu sur* $X \setminus M$ si, pour tout $f \in \mathcal{K}(X)$ tel que $\operatorname{Supp} f \subset X \setminus M$, nous avons $Vf \in \mathcal{C}(X)$.
4. Soit (X,Φ) un SD. Une fonction $l \in \mathcal{C}^+(X)$ est dite *de Liapunov* si $l(\Phi(t,x)) \leq l(x)$, pour tous $x \in X$ et $t \geq 0$. Une fonction de Liapunov de (X,Φ) est donc excessive pour le semi-groupe $\mathbb{H}$.

1.2 Attracteurs. Soit (X,Φ) un SD. Pour tout $x \in X$, soient

$$\Lambda^+(x) := \{y \in X : \text{ il existe } (t_n) \to +\infty \text{ tel que } \Phi(t_n,x) \to y\},$$
$$\Lambda^-(x) := \{y \in X : \text{ il existe } (t_n) \to -\infty \text{ tel que } \Phi(t_n,x) \to y\},$$
$$J^+(x) := \{y \in X : \text{ il existe } (t_n) \to +\infty \text{ et } (x_n) \to x \text{ tels que } \Phi(t_n,x_n) \to y\}.$$

Notons que ces trois ensembles sont fermés et invariants dans X.
1. Un compact $M \subset X$ est appelé *attracteur* de (X, Φ) s'il existe $O \in \mathcal{U}(M)$ tel que $\Lambda^+(x) \neq \emptyset$ et $\Lambda^+(x) \subset M$, pour tout $x \in O$.

Dans ce cas, M est invariant et l'ensemble

$$\mathcal{A}(M) := \{x \in X : \Lambda^+(x) \neq \emptyset \text{ et } \Lambda^+(x) \subset M\}$$

est un ouvert invariant contenant M; il est appelé *région d'attraction de M*.
2. Un compact M de X est dit *stable* par rapport à (X, Φ) si pour tout $Y \in \mathcal{U}(M)$, il existe $Z \in \mathcal{U}(M)$ tel que $\Phi(\mathbb{R}^+, Z) \subset Y$.

Soit M un compact de X. Il est connu ([1], Théorème 2.6.13) que si M est un attracteur stable de (X, Φ), alors il existe $O \in \mathcal{U}(M)$ tel que $J^+(x) \neq \emptyset$ et $J^+(x) \subset M$, pour tout $x \in O$. D'autre part, nous avons aussi $M \cap \Lambda^-(x) = \emptyset$, pour tout $x \in \mathcal{A}(M) \setminus M$.

1.3 Remarque. Soit M un attracteur compact de (X, Φ). Puisque X est localement compact et M est compact, alors il existe ([1], Théorème 2.6.15) un compact $D(M)$ qui est un attracteur stable de (X, Φ) et tel que $\mathcal{A}(D(M)) = \mathcal{A}(M)$. Dans la suite, nous ne considérons donc que les attracteurs stables.

1.4 Lemme. *Soient $x \in X$ et $f \in \mathcal{K}(X)$ tels que* $\operatorname{Supp} f \cap J^+(x) = \emptyset$, *alors Vf est finie et continue en x.*

Preuve. Soient $K := \operatorname{Supp} f$ et a_K la fonction définie, pour tout $y \in X$, par

$$a_K(y) := \begin{cases} \sup\{t \geq 0 : \ \Phi(t, y) \in K\} & \text{si } \Gamma^+(y) \cap K \neq \emptyset, \\ 0 & \text{si } \Gamma^+(y) \cap K = \emptyset. \end{cases}$$

Montrons d'abord qu'il existe un voisinage compact L de x tel que

$$a := \sup_{y \in L} a_K(y) < \infty \, .$$

Sinon, il existe une suite $(x_n) \to x$ et une suite $(t_n) \subset \mathbb{R}^+$ telle que $(t_n) \to +\infty$ et $(y_n := \Phi(t_n, x_n))$ est une suite de K. Nous pouvons supposer que $(y_n) \to y \in K$. Nous avons donc $y \in J^+(x) \subset K$, ce qui contredit l'hypothèse.

Maintenant, par définition de V, il est clair que

$$(\forall y \in L) \ : \ Vf(y) = \int_0^a f(\Phi(s, y))\, ds,$$

donc Vf est finie et continue en x. □

1.5 Théorème. *Soit M un compact de X, alors les propriétés suivantes sont équivalentes:*
(i) *M est un attracteur stable de (X, ϕ);*
(ii) *M est invariant et il existe $Y \in \mathcal{U}(M)$ invariant tel que le noyau potentiel du SD (Y, Φ) soit continu sur $Y \setminus M$.*

Preuve. Supposons que M est un attracteur stable de (X, Φ), alors M est invariant et $Y := \mathcal{A}(M)$ est un ouvert invariant contenant M. Soit $f \in \mathcal{K}^+(Y)$ tel que $K := \operatorname{Supp} f \subset Y \setminus M$, alors pour tout $x \in Y$, nous avons $K \cap J^+(x) \subset K \cap M = \emptyset$. Le Lemme 1.4 implique alors que Vf est finie et continue en x et donc $Vf \in \mathcal{C}(Y)$.

Réciproquement, supposons que M est invariant et qu'il possède un voisinage Y ouvert et invariant. Soit (K_n) une exhaustion de $Y \setminus M$ par des compacts et soit (f_n) une suite de fonctions de $\mathcal{K}(Y)$ telle que, pour tout $n \in \mathbb{N}$, nous avons:

$$f_n = 1 \text{ sur } K_n \ ; \ f_n = 0 \text{ sur } Y \setminus (M \cup K_{n+1}) \text{ et } 0 \leq f_n \leq 1.$$

Le principe complet du maximum implique que $c_n := \|Vf_n\| < \infty$, pour tout $n \in \mathbb{N}$. Soit donc la fonction f définie sur Y par

$$f := \sum_{n \in \mathbb{N}.} \frac{f_n}{2^n(1 + c_n)},$$

alors $Vf \in \mathcal{C}^+(Y)$, $M = \{Vf = 0\}$ et $Vf(\Phi(t, x)) < Vf(x)$, pour tous $x \in Y \setminus M$ et $t > 0$.

Nous utilisons maintenant une idée de ([1], page 148). Soient $\varepsilon > 0$ tel que $\{d(., M) \leq \varepsilon\} \subset Y$ et $\alpha < \alpha_0 := \inf\{Vf(x) : d(x, M) = \varepsilon\}$. Alors $P_\alpha := \{d(., M) \leq \varepsilon\} \cap \{Vf \leq \alpha\}$ est un voisinage compact de M. De plus P_α est positivement invariant: En effet, soit $x \in P_\alpha$, alors $\Gamma^+(x) \subset \{Vf \leq \alpha\}$. Si $\Gamma^+(x) \not\subset \{d(., M) \leq \varepsilon\}$, alors il existe $t > 0$ tel que $d(\Phi(t, x), M) = \varepsilon$ et donc

$$Vf(\Phi(t, x)) < Vf(x) \leq \alpha < \alpha_0 \leq Vf(\Phi(t, x)),$$

ce qui est absurde.

Montrons maintenant que M est un attracteur stable: Puisque $\Phi(\mathbb{R}^+, P_\alpha) \subset P_\alpha \subset \{d(., M) \leq \varepsilon\}$, alors M est stable. Soit $x \in P_\alpha$, puisque P_α est compact, alors $\Lambda^+(x) \neq \emptyset$. Soient donc $(t_n) \to +\infty$ et $y = \lim_n \Phi(t_n, x) \in \Lambda^+(x)$, alors $y \in M$ car $M = \{Vf = 0\}$ et

$$Vf(y) = \lim_n Vf(\Phi(t_n, x)) = \lim_n \int_{t_n}^{+\infty} f(\Phi(s, x))\, ds = 0.$$

Donc M est un attracteur de (X, Φ). □

1.6 Remarque. Soit M un compact de X. Il est connu ([1], Paragraphe 2.7) que M est attracteur stable si et seulement s'il existe une fonction de Liapunov l définie sur $\mathcal{A}(M)$ telle que $M = \{l = 0\}$. Cette fonction est définie par $l(x) := \sup\{d(\Phi(t, x), M) : t \geq 0\}$. Dans le Théorème précédent, nous l'avons construite sous la forme $l = Vf$.

2. Potentiels et fonctions de Liapunov

Soient (X, Φ) un SD, M un compact attracteur stable de (X, Φ), $Y := \mathcal{A}(M)$ et V le noyau potentiel du SD (Y, Φ). Soit f une fonction définie sur Y tel que f et Vf sont continues et bornées sur Y, strictement positives sur $Y \setminus M$ et $M = \{f = 0\} = \{Vf = 0\}$. Fixons aussi (comme dans la démonstration du Théorème 1.5 précédent) $\varepsilon > 0$ et $\alpha > 0$ tels que $\alpha < \inf\{Vf(x) : d(x, M) = \varepsilon\}$.

2.1 Lemme. *Pour tout $x \in Y \setminus M$, nous avons $\Gamma(x) \cap \{Vf \geq \alpha\} \neq \emptyset$.*

Preuve. Soit $x \in Y \setminus M$ tel que $\Gamma(x) \subset \{Vf < \alpha\}$. Si $d(x, M) > \varepsilon$, alors il existe $y \in \Gamma(x)$ tel que $d(y, M) = \varepsilon$ et donc $Vf(y) > \alpha$ ce qui est impossible. Si $d(x, M) \leq \varepsilon$, alors $\Gamma(x) \subset \{Vf \leq \alpha\} \cap \{d(., M) \leq \varepsilon\}$ et donc $\Lambda^-(x) \neq \emptyset$. Puisque M est un attracteur stable alors, pour tout $y \in \Lambda^-(x)$, nous avons

$$\emptyset \neq \Lambda^+(y) \subset \Lambda^-(x) \cap M,$$

ce qui est impossible. □

Notation. Pour tout $x \in Y \setminus M$, posons:

$$\omega(x) := \omega_\alpha(x) = \sup\{s \in \mathbb{R} : Vf(\Phi(s, x)) > \alpha\}.$$

2.2 Corollaire. *ω est définie sur $Y \setminus M$ et nous avons*

$$(\forall x \in Y \setminus M) : \alpha = \int_{\omega(x)}^{+\infty} f(\Phi(t, x))\, dt.$$

En particulier

$$(\forall x \in Y \setminus M; s \in \mathbb{R}) \ : \ \omega(\Phi(s, x)) = \omega(x) - s.$$

Preuve. Soit $x \in Y \setminus M$, alors l'ensemble $\{s \in \mathbb{R} : Vf(\Phi(s, x)) > \alpha\}$ est non vide d'après le Lemme 2.1. De plus cet ensemble est majoré dans $\mathbb{R}$ car $\lim_{s \to +\infty} Vf(\Phi(s, x)) = 0$. D'autre part, pour tout $s < \omega(x)$, nous avons

$$\alpha < Vf(\Phi(s, x)) = \int_s^{+\infty} f(\Phi(t, x))\, dt.$$

Par passage à la limite, nous obtenons

$$\alpha \leq \int_{\omega(x)}^{+\infty} f(\Phi(t, x))\, dt.$$

Si l'inégalité précédente est stricte, alors il existe $s > \omega(x)$ tel que $Vf(\Phi(s, x)) > \alpha$, ce qui n'est pas possible en vertu de la définition de ω. Donc

$$(\forall x \in Y \setminus M) : \alpha = \int_{\omega(x)}^{+\infty} f(\Phi(t, x))\, dt.$$

Soient maintenant $x \in Y \setminus M$ et $s \in \mathbb{R}$. En appliquant cette égalité à x et $\Phi(s,x)$, nous aurons

$$\int_{\omega(x)}^{+\infty} f(\Phi(t,x))dt = \alpha = \int_{\omega(\Phi(s,x))}^{+\infty} f(\Phi(t,\Phi(s,x)))dt = \int_{\omega(\Phi(s,x))+s}^{+\infty} f(\Phi(t,x))dt.$$

Donc $\omega(\Phi(s,x)) = \omega(x) - s$, car f est continue et strictement positive sur $Y \setminus M$.□

2.3 Proposition. *La fonction ω est continue sur $Y \setminus M$.*

Preuve. Soient $x \in Y \setminus M$ et $s < \omega(x)$, alors d'après le Corollaire 2.2, nous avons

$$Vf(\Phi(s,x)) = \int_{s}^{+\infty} f(\Phi(t,x))\,dt > \int_{\omega(x)}^{+\infty} f(\Phi(t,x))\,dt = \alpha.$$

Puisque la fonction $x \mapsto Vf(\Phi(s,x))$ est continue, alors il existe $O \in \mathcal{U}(x)$ tel que $Vf(\Phi(s,y)) > \alpha$, pour tout $y \in O$. Ceci implique que $s < \omega(y)$ et donc ω est semi-continue inférieurement.

Soit $s > \omega(x)$, alors il existe $s_0 \in]\omega(x), s[$ tel que $Vf(\Phi(s_0,y)) < \alpha$. Or la fonction $x \mapsto Vf(\Phi(s_0,x))$ est continue, ainsi il existe $O \in \mathcal{U}(x)$ tel que $Vf(\Phi(s_0,y)) < \alpha$, pour tout $y \in O$. Ceci implique que $\omega(y) < s_0 < s$ et donc ω est semi-continue supérieurement. □

2.4 Théorème. *Soient M un compact attracteur stable et f une fonction définie sur $\mathcal{A}(M)$ telle que f et Vf sont continues positives sur $\mathcal{A}(M)$ et $M = \{f = 0\} = \{Vf = 0\}$. Soient $\varepsilon > 0$ et $\alpha > 0$ tels que $\alpha < \inf\{Vf(x) : d(x,M) = \varepsilon\}$. Soit ω tel que*

$$(\forall x \in \mathcal{A}(M) \setminus M) : \alpha = \int_{\omega(x)}^{+\infty} f(\Phi(t,x))\,dt.$$

Alors la fonction l définie sur $\mathcal{A}(M)$ par

$$l(x) := \begin{cases} \exp(\omega(x)) & si\ x \in \mathcal{A}(M) \setminus M, \\ 0 & si\ x \in M \end{cases}$$

est une fonction de Liapunov associée à M.

Preuve. Le Corollaire 2.2 implique que la fonction l est ainsi bien définie et que

$$(\forall x \in \mathcal{A}(M) \setminus M, t > 0) : l(\Phi(t,x)) = l(x)\exp(-t) < l(x).$$

D'autre part, $M = \{l = 0\}$ et l est continue sur $\mathcal{A}(M) \setminus M$, d'après la Proposition 2.3. Pour montrer la continuité de l sur $\mathcal{A}(M)$, nous procédons comme dans ([1], page 152): Soit $(x_n) \subset \mathcal{A}(M) \setminus M$ telle que $(x_n) \to x \in M$ et $(l(x_n))$ ne converge pas vers 0. Alors la suite $((\omega(x_n))$ possède une sous suite notée encore $((\omega(x_n))$, convergente vers $t \leq 0$. Mais alors, d'une part $(\Phi(\omega(x_n), x_n)) \to \Phi(t,x) \in M$ car M est invariant; d'autre part $Vf(\Phi(\omega(x_n), x_n))) = \alpha$, pour tout $n \in \mathbb{N}$ et donc $\Phi(t,x) \in \{Vf = \alpha\}$. Or ceci est contradictoire car $M \cap \{Vf = \alpha\} = \emptyset$. □

2.5 Remarque. L'existence d'une fonction de Liapunov l associée à M et vérifiant $l(\Phi(t,x)) = l(x)\exp(-t)$, est connue ([1], Théorème 2.7.14). Dans le Théorème 2.4, nous avons présenté une autre approche inspirée de la théorie du potentiel.

Définition. Le SD (X, Φ^*) dit *dual* associé à (X, Φ) est défini par

$$(\forall x \in X, t \in \mathbb{R}) \ : \ \Phi^*(t,x) := \Phi(-t,x).$$

2.6 Théorème. *Supposons que $X = \{x \in \mathbb{R}^n : |x| \leq 1\}$ et que $M = \{x \in X : |x| = 1\}$ est un attracteur stable de (X, Φ). Alors $X \setminus \mathcal{A}(M)$ est un attracteur stable du SD dual (X, Φ^*).*

Preuve. Montrons d'abord que $N := X \setminus \mathcal{A}(M) \neq \emptyset$: Puisque $\{Vf \geq \alpha\}$ est un fermé disjoint de la sphère M, il existe une boule fermée B tel que $\{Vf \geq \alpha\} \subset B \subset \{x \in \mathbb{R}^n : |x| < 1\}$. Remarquons aussi que, par définition de ω, nous avons $\Phi(\omega(x), x) \in \{Vf \geq \alpha\}$ si $|x| < 1$. D'autre part, il est clair que $\{Vf \geq \alpha\}$ est négativement invariant. Soit donc $z \in \{Vf \geq \alpha\}$, alors

$$\emptyset \neq \Lambda^-(z) \subset X \setminus \mathcal{A}(M).$$

D'autre part N est un fermé de X car $\mathcal{A}(M)$ est un ouvert. Donc N est un compact non vide de X, de plus $X \setminus M$ est un voisinage ouvert de N. Aussi N est invariant par rapport à (X, Φ) donc invariant par rapport à (X, Φ^*).

Soient maintenant $\mathcal{A}^*(N) := X \setminus M$ et l^* la fonction définie sur $\mathcal{A}^*(N)$ par

$$l^*(x) := \begin{cases} \exp(-\omega(x)) & \text{si } x \in \mathcal{A}^*(N) \setminus N, \\ 0 & \text{si } x \in N. \end{cases}$$

Alors, l^* est continue sur $\mathcal{A}^*(N)$: La démonstration se fait exactement de la même manière que celle de 2.4. D'autre part, il est clair que l^* est une fonction de Liapunov du SD (X, Φ^*) et que $N = \{l^* = 0\}$. Nous concluons alors par ([1], Théorème 2.7.14) que $X \setminus \mathcal{A}(M)$ est un attracteur stable de (X, Φ^*). □

References

[1] Bhatia, N. P., Szegö, G. P., Dynamical Systems: Stability Theory and Applications. Lecture Notes in Math. 35, Springer-Verlag, Berlin 1967.

[2] Hmissi, M., Semi-groupes déterministes. Séminaire de théorie du potentiel (ed. N. Bouleau et al.), 135–144. Lecture Notes in Math. 1393, Springer-Verlag, Berlin 1989.

[3] — Recouvrement parallélisable du plan. Proceedings Euro. Conf. Iteration Theory, Lisboa 91, World Scientific (1992), 136–145.

[4] — Sur les systèmes dynamiques instables. Potential Analysis 3 (1994), 145–152.

Various approaches to the Dirichlet problem for non-local operators satisfying the positive maximum principle

Niels Jacob

Abstract. In our talk "The Dirichlet boundary value problem for pseudo differential operators generating Feller semigroups" delivered at the conference we discussed recent results obtained jointly with W. Hoh and which will be published in [9]. In this note we will elaborate in detail the first part of our considerations, namely the difficulties that arise when looking for a reasonable formulation for the Dirichlet problem for non-local generators of Feller semigroups.

1991 Mathematics Subject Classification: 35S15, 31C99.

Suppose that $A: D(A) \to C_\infty(\mathbb{R}^n)$, $D(A) \subset C_\infty(\mathbb{R}^n)$, is a generator of a Feller semigroup $(T_t)_{t\geq 0}$. Then it follows that A satisfies the *positive maximum principle on* $D(A)$: If $u \in D(A)$ and $u(x_0) = \sup_{x\in R^n} u(x) \geq 0$, then it follows that $Au(x_0) \leq 0$ holds.

In [3] Courrège gave characterizations of those operators satisfying the positive maximum principle. We restrict our attention to the case where $C_0^\infty(\mathbb{R}^n) \subset D(A)$ and that A maps $C_0^\infty(\mathbb{R}^n)$ continuously into $C_b(\mathbb{R}^n)$. In this case A has a representation as an integrodifferential operator

$$Au(x) = -\sum_{i,j=1}^n a_{ij}(x)\frac{\partial^2 u(x)}{\partial x_i \partial x_j} + \sum_{j=1}^n b_j(x)\frac{\partial u(x)}{\partial x_j} + c(x)u(x)$$
$$+ \int_{\mathbb{R}^n} \Big(u(y) - \sigma_x(y)\Big(u(x) - \sum_{j=1}^n \frac{\partial u}{\partial x_j}\cdot (y_j - x_i)\Big)\Big) N(x,dy), \qquad (1)$$

where $a_{ij}, b_j, c: \mathbb{R}^n \to \mathbb{R}$ are continuous and bounded functions, $(a_{ij})_{i,j=1,\dots,n}$ is a positive semidefinite matrix $c \geq 0$ and $N(x,dy)$ is a suitable kernel while $\sigma_x(y)$ is just a cut-off function. Using the Lévy–Khinchin formula Ph. Courrège already gave a representation of A as a pseudo differential operator, namely for

$u \in C_0^\infty(\mathbb{R}^n)$ we have

$$A\, u(x) = -p(x, D)u(x) = -(2\pi)^{-n/2} \int_{\mathbb{R}^n} e^{-ix\xi} p(x,\xi)\hat{u}(\xi)d\xi, \tag{2}$$

where $p: \mathbb{R}^n \times \mathbb{R}^n \to \mathbb{C}$ is a continuous function such that $p(x, \cdot): \mathbb{R}^n \to \mathbb{C}$ is, for each $x \in \mathbb{R}^n$, a negative definite function. Recall that a function $a: \mathbb{R}^n \to \mathbb{C}$ is negative definite if and only if for all $m \in \mathbb{N}_0$ and all $\xi^j \in \mathbb{R}^n$ the matrix $(a(\xi^i) + \overline{a(\xi^j)} - a(\xi^i - \xi^j))_{i,j=1,\ldots,n}$ is positive semidefinite, see [1] as a good reference.

In this note we want to concentrate ourselves on the case of a pure non-local operator, i.e., we assume that in (1) the differential operator part vanishes.

In terms of $p(x, \xi)$ this means that in the Lévy–Khinchin representation of $p(x, \cdot)$ only the integral term contributes, i.e., we have

$$p(x,\xi) = \int_{\mathbb{R}^n} \left(1 - e^{-iy\xi} - \frac{iy \cdot \xi}{1 + |y|^2}\right) \frac{1 + |y|^2}{|y|^2} \mu(x, d\xi), \tag{3}$$

where $\mu(x, d\xi)$ is a suitable kernel. Basic examples for negative definite functions are $\xi \mapsto |\xi|^\alpha, 0 < \alpha < 2$, or $f(a(\xi))$, when $a: \mathbb{R}^n \to \mathbb{R}$ is a continuous negative definite function and f is a Bernstein function, see [1].

In a series of papers [11]–[13] we started with a pseudo differential operator $p(x, D)$ and proved by analytical methods that under suitable conditions on $p(x, \xi)$ the operator $-p(x, D)$ extends to a generator of a Feller semigroup. Moreover, if $p(x, D)$ is symmetric, $-p(x, D)$ also generates a symmetric Dirichlet form, see also [15] for the non-symmetric case. These results had partly been extended by W. Hoh [6]–[8] who combined our analytical approach with the martingale problem approach of D. Stroock and S. Varadhan, see [20] as a reference.

Before discussing the Dirichlet problem for these operators more detailed, we will discuss some generalities about this problem. We start our considerations with a short look onto the classical situation for the Laplacian $-\Delta_n$. One formulation of the Dirichlet problem for the Laplacian is to find $u \in C^2(G) \cap C(\bar{G}), G \subset \mathbb{R}^n$ open and bounded, such that for a given $f \in C(\partial G)$ we have

$$\Delta_n u = 0 \text{ in } G \quad \text{and} \quad u|_{\partial G} = f. \tag{4}$$

Let us suppose that ∂G is smooth. Then the unique solution to (4) is given by

$$u(x) = \begin{cases} \int_{\partial G} \Gamma(x,y) f(y) d\sigma(y), & x \in G, \\ f(x), & x \in \partial G, \end{cases} \tag{5}$$

where Γ is the normal derivative of Green's function for the domain G. We can interpret $\Gamma(x, y)d\sigma(y)$ as a kernel on $G \times \mathcal{B}(\partial G)$ and for fixed $x, \Gamma(x, \cdot)d\sigma(\cdot)$ is the harmonic measure which is supported on ∂G.

The solution of (4) has also another representation using the Feller process generated by $-\Delta_n$. Clearly this process $(B_t)_{t\geq 0}$ is Brownian motion and introducing the stopping time

$$\tau_G := \inf\{t > 0;\ B_t \in G^c\} = \inf\{t > 0;\ B_t \in \partial G\} \tag{6}$$

we find for the solution of (4)

$$u(x) = E^x(f(X_{\tau_G})) \quad x\text{-almost everywhere.} \tag{7}$$

Note that the second equality in (6) and the fact that the harmonic measure is supported on ∂G is caused by the locality of Δ_n, i.e., the property that $\operatorname{supp}\Delta_n u \subset \operatorname{supp} u$ for any u. In particular this locality implies that $(B_t)_{t\geq 0}$ has almost surely continuous paths which implies (6). A generalized solution to (4) is always obtained by (7). However, if ∂G is smooth (regular) we also have $\lim_{G\ni y\mapsto x\in\partial G} E^y(f(X_{\tau_G})) = f(x)$.

A further approach to (4) is given by Hilbert space methods. Introducing the Sobolev space $H^1(G) = \{L^2(G);\ ||u||_1 < \infty\}$, where $||u||_1^2 = (\nabla u, \nabla u)_0 + (u,u)_0$ and defining $H_0^1(G) = \overline{C_0^\infty(G)}^{||\cdot||_1}$, H. Weyl proved the orthogonal decomposition

$$H^1(G) = H_0^1(G) \oplus_B \mathcal{H}(G), \tag{8}$$

where $B(u,v) = (\nabla u, \nabla v)_0$ and

$$\begin{aligned}\mathcal{H}(G) &= \{u \in H^1(G);\ B(u,\varphi) = 0 \text{ for } \varphi \in C_0^\infty(G)\}\\ &= \{u \in H^1(G);\ u \text{ is harmonic in } G\}.\end{aligned} \tag{9}$$

The deep result in (9) is, that $B(u,\varphi) = 0$ for all $\varphi \in C_0^\infty(G)$ already implies $u \in C^\infty(G)$! Denote by $P_{\mathcal{H}}\colon H^1(G) \to \mathcal{H}(G)$ the orthogonal projection and suppose that $f = \tilde{f}|_{\partial G}$, where $\tilde{f} \in H^1(G) \cap C(\bar{G})$. Then we find that

$$u(x) = P_{\mathcal{H}}\tilde{f}(x) \tag{10}$$

solves (4). Let us rewrite the Weyl decomposition result in a little more complicated way. We will work now in the space $H^1(\mathbb{R}^n)$ and define

$$\mathcal{H}_G := \{u \in H^1(\mathbb{R}^n);\ B(u,\varphi) = 0 \text{ for all } \varphi \in C_0^\infty(G)\}, \tag{11}$$

where $C_0^\infty(G)$ is considered as a subspace of $C_0^\infty(\mathbb{R}^n)$. Setting

$$\mathcal{F}_G = \{u \in H^1(\mathbb{R}^n);\ u|_{G^c} = 0 \text{ a.e.}\} \tag{12}$$

we again have the decomposition

$$H^1(\mathbb{R}^n) = \mathcal{F}_G \oplus \mathcal{H}_G \tag{13}$$

and the orthogonal projection $P_{\mathcal{H}_G}\colon H^1(\mathbb{R}^n) \to \mathcal{H}_G$. Given $\tilde{f} \in H^1(\mathbb{R}^n)$ such that $\tilde{f}|_{\partial G} = f$, it follows that $u(x) = P_{\mathcal{H}_G}\tilde{f}(x)$ is a harmonic function in G such that $u(x) - \tilde{f}(x) = P_{\mathcal{H}_G}\tilde{f}(x) - \tilde{f}(x) \in \mathcal{F}_G$, i.e., $P_{\mathcal{H}_G}\tilde{f}|_{G^c} = \tilde{f}|_{G^c}$ a.e., which implies in particular $P_{\mathcal{H}_G}\tilde{f}|_{\partial G} = f$ a.e. Under suitable regularity conditions on $\tilde{f}$ and ∂G we get again a classical solution of (4). Moreover, it can be shown, see [4] (and our discussion below), that we have for suitable $\tilde{f}$

$$P_{\mathcal{H}_G}\tilde{f}(x) = E^x(\tilde{f}(X_{\tau_G})) \quad x\text{-almost everywhere.} \tag{14}$$

Note that by a standard procedure it is possible to transform (4) into the problem

$$-\Delta_n u = g \quad \text{in } G \quad \text{and} \quad u|_{\partial G} = 0, \tag{15}$$

provided f has sufficiently nice regularity properties. Especially the Hilbert space approach to the Dirichlet problem becomes easier in this formulation.

Now let us turn to the most simple, but non-trivial non-local generator of a Feller semigroup, namely a fractional power of $-\Delta_n$, i.e., the operator $(-\Delta_n)^\alpha$, $0 < \alpha < 1$. First note that it is by no means clear how $(-\Delta_n)^\alpha$ is defined. One may consider $(-\Delta)^\alpha$ as an operator on $L^2(\mathbb{R}^n)$ defined by the spectral calculus, as a pseudo differential operator with symbol $|\xi|^{2\alpha}$, as a convolution operator with kernel in $S'(\mathbb{R}^n), \dots$.

The first potential theoretical approach to $(-\Delta_n)^\alpha$ is due to M. Riesz [18] who investigated the (formal) inverse to $(-\Delta)^\alpha$, i.e., the M. Riesz potentials. A rather comprehensive discussion of this theory is given by N. S. Landkof in [16]. Besides other results, M. Riesz succeeded to find the harmonic measure for $(-\Delta_n)^\alpha$ in case of the ball $B_\rho(0)$. Its density is given by

$$\epsilon_{2\alpha}^\rho(x) = \begin{cases} 0, & |x| \leq \rho \\ \dfrac{\Gamma(\frac{n}{2})}{\pi^{n/2+1}} \sin(\pi r)\rho^{2\alpha}(|x|^2 - \rho^2)^{-\alpha}|x|^{-n}, & |x| > \rho. \end{cases} \tag{16}$$

For us it is important to note that $\operatorname{supp}\epsilon_{2\alpha}^\rho = B_\rho^c(0)$ and not $\partial B_\rho(0)$. Thus it is clear that in general no reasonable results could be obtained when looking for solutions to the problem

$$(-\Delta_n)^\alpha u = 0 \quad \text{in } B_\rho(0) \text{ (or } G)) \quad \text{and} \quad u = f \text{ on } \partial B_\rho(0) \text{ (or } \partial G). \tag{17}$$

The same holds true when looking at a probabilistic approach to the Dirichlet problem $(-\Delta_n)^\alpha$. Clearly $(-\Delta_n)^\alpha$ generates a Feller process $(X_t^\alpha)_{t\geq 0}$, namely the α-stable symmetric process. Again we can define the stopping time

$$\tau_G := \inf\{t > 0;\ X_t^\alpha \in G^c\}, \tag{18}$$

where $G \subset \mathbb{R}^n$ is a bounded open set. But now $(X_t^\alpha)_{t\geq 0}$ has càdlàg paths which are not continuous. Thus $\tau_G \neq \inf\{t > 0; X_t^\alpha \in \partial G\}$. For this reason we need a function f defined at least on G^c in order to study

$$u(x) = E^x(f(X_{\tau_G}^\alpha)). \tag{19}$$

In some papers, see [17] or [19], (19) is taken as a definition for a (generalized) 2α-harmonic function in G, since it is possible to prove for $G = B_\rho(0)$ that $(-\Delta_n)^\alpha u|_G = 0$. However, this definition does not link from the very beginning the notion of 2α- harmonicity to the operator $(-\Delta)^\alpha$. As indicated in [10] it is possible to handle the Dirichlet problem for $(-\Delta_n)^\alpha$ in the frame of Hilbert space methods, but this follows also from the following discussion.

Now let us introduce a class of pseudo differential operators generating Feller semigroups, i.e., satisfying the positive maximum principle. The following results are taken from [13]. Let $a^2\colon \mathbb{R}^n \to \mathbb{R}$ be a continuous negative definite function.

We define the anisotropic Sobolev spaces

$$H^{a^2,s}(\mathbb{R}^n) = \{u \in L^2(\mathbb{R}^n);\ ||u||_{a^2,s} < \infty\}, \quad s \geq 0, \tag{20}$$

where

$$||u||^2_{a^2,s} = \int_{\mathbb{R}^n} (1 + a^2(\xi))^{2s} |\hat{u}(\xi)|^2 d\xi. \tag{21}$$

Theorem 1 ([13]). *Let $a^2: \mathbb{R}^n \to \mathbb{R}$ be a continuous negative definite function such that, for some $r > 0$, $a^2(\xi) \geq c|\xi|^r$ holds for all $\xi \in \mathbb{R}^n$, $|\xi|$ sufficiently large. Further let $p: \mathbb{R}^n \times \mathbb{R}^n \to \mathbb{R}$ be a continuous function such that $p(x, \cdot): \mathbb{R}^n \to \mathbb{R}$ is negative definite and such that $p(x,\xi) = p_1(\xi) + p_2(x,\xi)$ with*
(i) *$\gamma_0 a^2(\xi) \leq p_1(\xi) \leq \gamma' a^2(\xi)$, $\xi \in \mathbb{R}^n$ and $|\xi|$ large, $\gamma_0 > 0$;*
(ii) *$p_2(\cdot,\xi) \in C^m(\mathbb{R}^n)$ and $|\partial_x^\alpha p_2(x,\xi)| \leq \varphi_\alpha(x)(1 + a^2(\xi))$ for $\varphi_\alpha \in L^1(\mathbb{R}^n)$ and $|\alpha| \leq m$, m sufficiently large;*
(iii) *$\sum_{|\alpha|\leq m} ||\varphi_\alpha||_{L^1} \leq \lambda\gamma_0$, where λ is small (in a very precise sense, see [13]).*
Then $-p(x,D)$ extends to a generator of a Feller semigroup, where for $u \in C_0^\infty(\mathbb{R}^n)$

$$p(x,D)u(x) = (2\pi)^{-n/2} \int_{\mathbb{R}^n} e^{ix\xi} p(x,\xi)\hat{u}(\xi) d\xi. \tag{22}$$

If $p(x,D)$ is symmetric, then $-p(x,D)$ has a selfadjoint extension on $L^2(\mathbb{R}^n)$ with domain $H^{a^2,1}(\mathbb{R}^n)$ and further the bilinear form $B(u,v) = (p(x,D)u,v)_0$, $u,v \in C_0^\infty(\mathbb{R}^n)$ extends to a Dirichlet form with domain $H^{a^2,1/2}(\mathbb{R}^n)$.

Remark. A. When one is only interested in constructing a Feller semigroup, then condition iii) is not necessary as shown in [8]. But the proof of Theorem 1 relies on several norm estimates which we also need to handle the Dirichlet problem and which are not valid in the more general case.

B. In [15] we extended Theorem 1 to complex-valued symbols, i.e., to functions $p: \mathbb{R}^n \times \mathbb{R}^n \to \mathbb{C}$, but we will not use these results here.

Having in mind the results for $(-\Delta_n)^\alpha$, $0 < \alpha < 1$, we give two formulations of the Dirichlet problem for $-p(x,D)$.

Problem 1. Let $G \subset \mathbb{R}^n$ be an open bounded domain with smooth boundary ∂G. For $f \in C(G^c)$ find $u \in D(-p(x,D)) \subset C_\infty(\mathbb{R}^n)$ such that

$$(p(x,D)u)|_G = 0 \quad \text{and} \quad u|_{G^c} = f|_{G^c} \tag{23}$$

hold.

Problem 2. Let G be as in Problem 1 and $f \in L^2(G^c), L^2(G^c) \subset L^2(\mathbb{R}^n)$. Find $u \in H^{a^2,1}(\mathbb{R}^n)$ such that

$$(p(x,D)u)|_G = 0 \text{ a.e.} \quad \text{and} \quad u|_{G^c} = f|_{G^c} \text{ a.e.} \tag{24}$$

Remark. A. In Problem 1 we mean by $D(-p(x,D))$ the domain of the extension of (22) as a generator of a Feller semigroup.

B. Since the domain of the extension of (22) to a generator of a sub-Markovian semigroup is $H^{a^2,1}(\mathbb{R}^n)$, the condition $u \in H^{a^2,1}(\mathbb{R}^n)$ is analogous to the condition in Problem 1.

C. A solution of Problem 2 will be called a *strong-L^2-solution* to the Dirichlet problem.

It turns out that a direct approach to Problem 1 is rather difficult, while Hilbert space methods can be applied to get at least weak solutions to Problem 2. We assume that $p(x,D)$ satisfies the assumptions of Theorem 1 and is symmetric. For $G \subset \mathbb{R}^n$ an open bounded domain with smooth boundary we consider $C_0^\infty(G)$ as subspace of $C_0^\infty(\mathbb{R}^n)$ and define

$$\mathcal{H}^G_{p(x,D)} := \{u \in H^{a^2,1/2}(\mathbb{R}^n);\ B(u,\varphi) = 0 \text{ for all } \varphi \in C_0^\infty(G)\} \tag{25}$$

and

$$H_0^{a^2,1/2}(G) = \overline{C_0^\infty(G)}^{\|\cdot\|_{a^2,1/2}} \subset H^{a^2,1/2}(\mathbb{R}^n). \tag{26}$$

Using the spectral synthesis theory for general Dirichlet spaces, see [4] or [5], we get the orthogonal decomposition

$$H^{a^2,1/2}(\mathbb{R}^n) = H_0^{a^2,1/2}(G) \oplus_B \mathcal{H}^G_{p(x,D)}. \tag{27}$$

Let $P_\mathcal{H}: H^{a^2,1/2}(\mathbb{R}^n) \to \mathcal{H}^G_{p(x,D)}$ be the corresponding orthogonal projection. If $f \in H^{a^2,1/2}(\mathbb{R}^n)$, then $P_\mathcal{H} f$ is a *weak solution* to Problem 2, i.e.,

$$B(P_\mathcal{H} f, \varphi) = 0 \quad \text{for all } \varphi \in C_0^\infty(G), \tag{28}$$

and

$$P_\mathcal{H} f|_{G^c} = f|_{G^c}. \tag{29}$$

Assuming more regularity for f, we can give another formulation of a weak solution. For $f \in H^{a^2,1}(\mathbb{R}^n)$ and a solution $u \in H^{a^2,1}(\mathbb{R}^n)$ of Problem 2 we set $v = u - f$. Thus $v \in H^{a^2,1}(\mathbb{R}^n)$ and $v|_{G^c} = 0$, which implies that $v \in H_0^{a^2,1/2}(G) \cap H^{a^2,1}(\mathbb{R}^n)$. Furthermore we find for $\varphi \in C_0^\infty(G)$

$$B(v,\varphi) = (p(x,D)v,\varphi)_0 = (p(x,D)u,\varphi)_0 - (p(x,D)f,\varphi)_0$$

or with $g := -p(x,D)f \in L^2(\mathbb{R}^n)$

$$B(v,\varphi) = (g,\varphi)_0 \quad \text{for all } \varphi \in C_0^\infty(G). \tag{30}$$

Thus instead of the projection theorem, we may also use a representation theorem for linear functionals on $H^{a^2,1/2}(G)$ with respect to B to get a weak solution. Using the representation problem (30) we find for $P_\mathcal{H} f$ as defined above the following local regularity result.

Theorem 2 ([14]). *For any $f \in H^{a^2,1/2}(\mathbb{R}^n)$ it follows that $P_{\mathcal{H}}f \in H^{a^2,1}_{\mathrm{loc}}(G)$, i.e., for any $\varphi \in C_0^\infty(G)$ we have $\varphi P_{\mathcal{H}}f \in H^{a^2,1}(\mathbb{R}^n)$.*

Thus we have the interior regularity which is needed in order to prove that the weak solution to Problem 2 is already a strong L^2-solution. However, it seems to us almost impossible to prove in a rather general situation that $P_{\mathcal{H}}f \in H_0^{a^2,1/2}(G) \cap H^{a^2,1}(\mathbb{R}^n)$. The operator $p(x,D)$ is non-local and its "type" is not preserved under a change of coordinates. Thus we have no chance in imitating the classical proof of boundary regularity. However, we can use probability theory to get at least some regularity results for more regular data. By Theorem 1 we know that $-p(x,D)$ generates a Feller semigroup $(T_t)_{t\geq 0}$ and with that Feller semigroup we can associate a Feller process $(X_t^{p(x,D)})_{t\geq 0}$. But since we require $p(x,D)$ to be symmetric, B is a Dirichlet form and therefore there exists a corresponding Hunt process which is equivalent to the Feller process $(X_t^{p(x,D)})_{t\geq 0}$. But then the theory of Dirichlet forms gives us a representation of the projection $P_{\mathcal{H}}$! For this let

$$\tau_G = \inf\{t > 0;\ X_t^{p(x,D)} \in G^c\}. \tag{31}$$

Note that τ_G is a stopping time. Since the paths of $(X_t^{p(x,D)})_{t\geq 0}$ are càdlàg but not continuous, we cannot substitute G^c by ∂G in (31). Now it follows that for $f \in H^{a^2,1/2}(\mathbb{R}^n)$

$$P_{\mathcal{H}}f(x) = E^x(f(X_{\tau_G}^{p(x,D)})) \tag{32}$$

holds for almost every $x \in \mathbb{R}^n$. This representation of the weak solution to Problem 2 enables us to look at the boundary behaviour of $u(x) = E^x(f(X_{\tau_G}^{p(x,D)}))$ from a different point of view, namely to investigate directly

$$\lim_{G \ni z \to x \in \partial G} E^z(f(X_{\tau_G}^{p(x,D)})). \tag{33}$$

Such questions had already been studied very carefully for the case when dealing with a strong Feller semigroup admitting many excessive functions. In this situation J. Bliedtner and W. Hansen, see [2], developed a complete balayage theory, proved that there exists a probabilistic counterpart and gave on an abstract level a discussion of the Dirichlet problem including boundary regularity. Their formulation of the Dirichlet problem looks quite different from Problem 1, but once we proved that $-p(x,D)$ gives rise to a Feller semigroup their theory applies to, it is clear that the two formulations correspond to each other.

Unfortunately, we cannot always prove that $-p(x,D)$ generates a strong Feller semigroup admitting sufficiently many excessive functions. However, under some additional assumptions we succeeded. The following considerations are taken from our joint paper with W. Hoh [9].

Suppose in addition to the assumptions of Theorem 1 that for a continuous function $\tilde{a}^2\colon \mathbb{R}^n \to \mathbb{R}$ such that $\tilde{a}(0) = 0$ and $\tilde{a}^2(\xi) \leq c(1 + |\xi|^2)$ we have

$$p(x,\xi) \leq \tilde{a}^2(\xi). \tag{34}$$

Then it follows that for any $\mu > 0$ the operator $-p_\mu(x, D)$, where $p_\mu(x, D) = p(x, D) + \mu$, generates a strong Feller semigroup admitting sufficiently many excessive functions, or using the terminology of [2], we get a balayage space. Now we can apply the results of [9] to get

Theorem 3 ([9]). *Let $f \in H^{a^2,1}(\mathbb{R}^n) \cap C_\infty(\mathbb{R}^n)$ and suppose that ∂G is smooth. Then for the weak solution $u(x) = P^\lambda_{\mathcal{H}} f(x)$ of Problem 2 posed for $p_\lambda(x, D)$ we have*

$$\lim_{G \ni z \to x \in \partial G} u(z) = \lim_{G \ni z \to x \in \partial G} P^\lambda_{\mathcal{H}} f(z) = \lim_{G \ni z \to x \in \partial G} E^x(f(X^{p_\lambda(x,D)}_{\tau_G})) = f(x) \quad (35)$$

for all $x \in \partial G$.

Remark. A. In [9] we proved that $P^\lambda_{\mathcal{H}} f$ is also a solution of the Dirichlet problem in the sense of the underlying balayage space.

B. After our talk in the conference, W. Hansen pointed out to us that, in his opinion, it should not be very difficult to improve our results in Theorem 3, namely to have Theorem 3 for $\lambda = 0$.

References

[1] Berg, C., Forst G., Potential theory on locally compact Abelian groups. Ergeb. Math. Grenzgeb. (2) 87, Springer-Verlag, Berlin–Heidelberg–New York 1975.

[2] Bliedtner, J., Hansen, W., Potential theory – An analytic and probabilistic approach to balayage. Universitext, Springer-Verlag, Berlin–Heidelberg–New York–Tokyo 1986.

[3] Courrège, Ph., Sur la forme intégro-différentielle des opérateurs de C_K^∞ dans C satisfaisant du principe du maximum. Sém. Théorie du Potentiel 10 (1965/66), Paris, 38p.

[4] Fukushima, M., Dirichlet forms and Markov processes. North Holland Math. Library Vol. 23, North Holland Publ. Comp., Amsterdam–Oxford–New York 1980.

[5] Fukushima, M., Oshima, Y., Takeda, M., Dirichlet forms and symmetric Markov processes. De Gruyter Stud. Math. 19, Walter de Gruyter, Berlin-New York 1994.

[6] Hoh, W., Zum Martingalproblem für eine Klasse von Pseudodifferentialoperatoren. Dissertation, Universität Erlangen-Nürnberg 1992.

[7] Hoh, W., The martingale problem for a class of pseudo differential operators. Math. Ann. 300 (1994), 121–147.

[8] Hoh, W., Pseudo differential operators with negative definite symbols and the martingale problem. Stochastics Stochastics Rep. 55 (1995), 225–252.

[9] Hoh, W., Jacob, N., On the Dirichlet problem for pseudo differential operators generating Feller semigroups. J. Funct. Anal. (to appear).

[10] Jacob, N., Sur les fonctions $2r$-harmoniques de N. S. Landkof. C. R. Acad. Sc. Paris 304 (1987), 169–171.

[11] Jacob, N., Feller semigroups, Dirichlet forms, and pseudo differential operators. Forum Math. 4 (1992), 433–446.

[12] Jacob, N., Further pseudo differential operators generating Feller semigroups and Dirichlet forms. Rev. Mat. Iberoam. 9 (1993), 373–407.

[13] Jacob, N., A class of Feller semigroups generated by pseudo differential operators. Math. Z. 215 (1994), 151–166.

[14] Jacob, N., Local regularity of solutions of a class of pseudo differential equations in Dirichlet spaces. Math. Nachr. 164 (1993), 63–89.

[15] Jacob, N., Non-local (semi-) Dirichlet forms generated by pseudo differential operators. In: Dirichlet forms and stochastic processes, Beijing 1993. Walter de Gruyter, Berlin 1995, 223–233.

[16] Landkof, N. S., Foundation of modern potential theory. Grundlehren Math. Wiss. 180, Springer-Verlag, Berlin–Heidelberg–New York 1972.

[17] Ming, L., The Dirichlet problem of a discontinuous Markov process. Acta Math. Sinica (New Ser.) 5 (1989), 9–15.

[18] Riesz, M., Intégrales de Riemann-Liouville et potentiels. Acta Szeged Sect. Math. 9 (1938), 1–42.

[19] Song, R., Probabilistic approach to the Dirichlet problem of perturbed stable processes. Probab. Theory Relat. Fields 95 (1993), 371–389.

[20] Stroock, D. W., Varadhan, S. R. S., Multidimensional diffusion processes. Grundlehren Math. Wiss. 233, Springer-Verlag, Berlin–Heidelberg–New York 1979.

[21] Weyl, H., The method of orthogonal projection in potential theory. Duke Math. J. 7 (1940), 411-444.

Estimates for eigenvalues of the Laplacian

Pawel Kröger

Abstract. G. Pólya conjectured in 1954 that the first term in Weyl's asymptotic formula is a lower bound for the Dirichlet eigenvalues and an upper bound for the Neumann eigenvalues of the Laplacian on a domain in Euclidean space. Although Pólya succeeded in proving his conjecture for a special class of domains with the property that the Euclidean space can be "paved" with congruent copies of that domain (see [P]), the general case still seems to be open. The aim of this note is to review some estimates for sums of consecutive eigenvalues of the Laplacian which are related to Pólya's conjecture in the sense that the leading term in those estimates is in accordance with the leading term in Weyl's asymptotic formula.

1991 Mathematics Subject Classification: 35P15.

Spectral projections and estimates for eigenvalues

We sketch an elementary method for obtaining bounds for sums of consecutive eigenvalues of the Laplacian on a domain in Euclidean space and on a compact Riemannian manifold.

The first result of this type was the following lower bound for sums of Dirichlet eigenvalues given by P. Li and S. T. Yau in [LY]. We will assume through-out this note that Ω is a bounded (connected) domain with Lipschitz boundary in n-dimensional Euclidean space $\mathbb{R}^n$. The unit ball in $\mathbb{R}^n$ will be denoted by $B_1^{(n)}$. We consider the eigenvalues of the Laplacian $-\Delta$ on Ω under Dirichlet boundary conditions increasingly ordered as follows:

$$0 < \lambda_1 < \lambda_2 \leq \lambda_3 \leq \cdots .$$

Theorem 1 (P. Li and S.-T. Yau). *Let* $\mathrm{vol}(\Omega)$ *denote the volume of* Ω. *Then the following holds:*

$$\sum_{j=1}^{k} \lambda_j \geq (2\pi)^2 \frac{n}{n+2} (\mathrm{vol}(\Omega)\,\mathrm{vol}(B_1^{(n)}))^{-\frac{2}{n}} k^{\frac{n+2}{n}} \quad \textit{for every } k. \tag{1}$$

The proof of the theorem makes use of the following lemma.

Lemma 1 (L. Hörmander, see [LY]). *Let f be a nonnegative measurable function on $\mathbb{R}^n$ with $f \leq M_1$ and $\int_{\mathbb{R}^n} f(z)\,dz \geq M_2$ for positive constants M_1 and M_2. Then $\int_{\mathbb{R}^n} |z|^2 f(z)\,dz \geq \frac{n}{n+2}(M_1 \operatorname{vol}(B_1^{(n)}))^{-\frac{2}{n}} M_2^{\frac{n+2}{n}}$.*

Proof. The assertion of the Lemma follows immediately by comparison with the function $f^{(0)}$ which is given by $f^{(0)}(z) \equiv M_1$ for every z with $|z| \leq r$ and $f^{(0)}(z) \equiv 0$ elsewhere; the positive number r is chosen such that $\int_{\mathbb{R}^n} f^{(0)}(z)\,dz = M_2$. □

The proof of the Theorem given by Li and Yau is based on the consideration of the integral kernel for a spectral projection corresponding to the first k eigenvalues. This integral kernel is extended by zero to the whole Euclidean space. In view of the Dirichlet boundary condition, the new kernel belongs to the form domain for the standard Dirichlet form on the Euclidean space. Thus, we can use the Fourier transform in order to study the total mass and energy of this kernel. On the other hand, we take into account that the Fourier transform of the spectral projection at z is up to a dimension dependent normalization constant equal to the orthogonal projection of the function $x \mapsto \exp(i < z, x >)$ onto the linear span of the corresponding Dirichlet eigenfunctions. This yields a pointwise upper bound for the L^2-norm of the Fourier transform of the spectral projection. The proof is completed by an application of Lemma 1. □

A method which is in some sense dual to that method can be used in order to obtain upper bounds for sums of Neumann eigenvalues (see [K1]). The functions $z \mapsto \exp(i < z, x >)$ for $z \in \mathbb{R}^n$ are eigenfunctions of the Laplacian on $\mathbb{R}^n$. Since we are dealing with Neumann boundary conditions, we obtain test functions for the eigenvalue problem on a bounded domain in $\mathbb{R}^n$ if we simply restrict the above functions to that domain. The Rayleigh–Ritz formula yields a preliminary upper bound for the k-th eigenvalue. In order to apply the Plancherel formula, we have to integrate over all z from suitable balls centered at zero (cf. also the proof of Lemma 1). Finally, Lemma 1 is replaced by an induction argument. □

Before stating the theorem, we recall that the above assumptions on Ω imply that the spectrum of $-\Delta$ on Ω under Neumann boundary conditions is discrete. Hence, the eigenvalues can be increasingly ordered as follows:

$$0 = \mu_1 < \mu_2 \leq \mu_3 \leq \cdots.$$

Theorem 2 (see [K1]). *The following holds:*

$$\sum_{j=1}^{k} \mu_j \leq (2\pi)^2 \frac{n}{n+2} (\operatorname{vol}(\Omega) \operatorname{vol}(B_1^{(n)}))^{-\frac{2}{n}} k^{\frac{n+2}{n}} \quad \textit{for every } k. \tag{2}$$

It turns out that we can modify the above methods for obtaining upper bounds for sums of Neumann eigenvalues and lower bounds for sums of Dirichlet eigenvalues in order to obtain upper bounds for sums of Dirichlet eigenvalues and lower bounds for sums of Neumann eigenvalues, resp. Let $\mathrm{dist}(x, \partial\Omega)$ denote the distance of a point $x \in \Omega$ from the boundary of Ω. We introduce the notation

$$\Omega_\delta \equiv \{x \in \Omega \mid \mathrm{dist}(x, \partial\Omega) < \delta\} \quad \text{for a positive constant } \delta.$$

Theorem 3 (see [K2]). *Suppose that there exists a constant $C_\Omega^{(0)}$ such that* $\mathrm{vol}(\Omega_\delta) \leq C_\Omega^{(0)} \mathrm{vol}(\Omega)^{\frac{n-1}{n}} \delta$ *for every δ. Then the following holds:*

$$\sum_{j=1}^{k} \lambda_j \leq (2\pi)^2 \frac{n}{n+2} (\mathrm{vol}(\Omega)\,\mathrm{vol}(B_1^{(n)}))^{-\frac{2}{n}} (k^{\frac{n+2}{n}} + C_n^{(1)} C_\Omega^{(0)} k^{\frac{n+1}{n}}) \tag{3}$$

for every $k \geq {C_\Omega^{(0)}}^n$ and a constant $C_n^{(1)}$ which depends only on the dimension n.

Example. Suppose that Ω is the disjoint union of l congruent balls with radius $l^{-\frac{1}{n}}$ for n fixed. Obviously, Ω is disconnected but we could easily mend that by joining the balls with tubes of sufficiently small size. Then the optimal constant $C_\Omega^{(0)}$ is proportional to $l^{\frac{1}{n}}$. On the other hand, $\lambda_1 = \cdots = \lambda_l = c_1^{(n)} l^{\frac{2}{n}}$ for some $c_1^{(n)}$ which depends only on n. Hence, $\sum_{j=1}^{k} \lambda_j = c_1^{(n)} l^{\frac{2}{n}} k$ for every $k \leq l$. This cannot be estimated above by $C(k^{\frac{n+2}{n}} + l^{\frac{1}{n}} k^{\frac{n+1}{n}})$ for k arbitrary and a constant C. This explains why the restriction $k \geq {C_\Omega^{(0)}}^n$ is neccessary.

Proof. The proof of Theorem 3 is based on the same method as the proof of Theorem 2. The test functions for the Dirichlet problem are given by $x \mapsto \chi_\delta(x) \exp(i < z, x >)$ for an appropriate cut off function χ_δ. Of course, we could not expect to obtain upper bounds for sums of Dirichlet eigenvalues without additional assumptions on the domain under consideration (i.e., informations in addition to the knowledge of the volume of the domain). The volume of Ω_δ comes into the matter if we estimate the energy integral of the test functions:

$$\int_\Omega |\nabla_x(\chi_\delta(x) \exp(i < z, x >))|^2 \, dx$$
$$\leq |z|^2 \mathrm{vol}(\Omega) + 2|z| \|\chi_\delta\|_\infty \mathrm{vol}(\mathrm{supp}(\nabla\chi_\delta)) + \|\chi_\delta\|_\infty^2 \mathrm{vol}(\mathrm{supp}(\nabla\chi_\delta)).$$

The cut off function χ_δ is chosen in such a way that the closure of Ω_δ is the support of the gradient of χ_δ. The remainder of the proof is similar to the proof of Theorem 2. □

Now suppose that that there exists a bi-Lischitz map f which maps Ω onto an open ball B in $\mathbb{R}^n$. Let C_Ω be a common Lipschitz constant for f and f^{-1}.

Theorem 4 (see [K2]). *Under the above assumptions on Ω the following holds:*

$$\sum_{j=1}^{k} \mu_j \geq (2\pi)^2 \frac{n}{n+2} (\mathrm{vol}(\Omega)\,\mathrm{vol}(B_1^{(n)}))^{-\frac{2}{n}} (k^{\frac{n+2}{n}} - C_n^{(2)} C_\Omega^{2n^2+2n} k^{\frac{n+1}{n}}) \tag{4}$$

for every k and a constant $C_n^{(2)}$ which depends only on the dimension n.

Remark. The assumptions of Theorem 4 are obviously stronger than the assumptions of Theorem 2.

Proof. We extend the products of the Neumannn eigenfunctions of Ω with cut off functions χ_δ by zero to the whole of $\mathbb{R}^n$. We recall that the relation between the Neumann eigenvalues and properties of the domain is more complicated than for the case of Dirichlet eigenvalues, for instance there is no monotonicity property with respect to the domain for the case of Neumann eigenvalues. We can overcome this difficulty if the integral of the trace of the Neumann heat kernel $p^N(\delta^2; x, x)$ over the boundary region Ω_δ can be estimated above. This can be done by a combination of estimates for the behaviour of heat kernels under changes of metric with probabilistic localization technics (cf. [Da] and the proof for the continuity of paths of Brownian motion in [Dy]). The proof is again completed by an application of Lemma 1. □

Theorem 1 in conjunction with Theorem 3 and Theorem 2 in conjunction with Theorem 4 immediately yield the following Corollary:

Corollary. *Suppose that Eqs. (3) and (4) hold. Then we can conclude that there exists a constant $C_n^{(3)}$ which depends only on the dimension n such that the following holds:*

$$|\lambda_k - (2\pi)^2(\operatorname{vol}(\Omega)\operatorname{vol}(B_1^{(n)}))^{-\frac{2}{n}}k^{\frac{2}{n}}| \leq C_n^{(3)}C_\Omega^{(0)}\operatorname{vol}(\Omega)k^{\frac{3}{2n}} \quad \textit{and}$$
$$|\mu_k - (2\pi)^2(\operatorname{vol}(\Omega)\operatorname{vol}(B_1^{(n)}))^{-\frac{2}{n}}k^{\frac{2}{n}}| \leq C_n^{(3)}C_\Omega^{2n^2+2n}\operatorname{vol}(\Omega)k^{\frac{3}{2n}} \quad \textit{for every } k.$$

Remark. 1.) It is easy to see that the order of the second term in k on the right-hand sides of (3) and (4), resp., is correct.

On the other hand, the order of k in the terms on the right-hand sides of the estimates given in the Corollary does not seem to be sharp. It would be a big step towards a proof of Pólya's Conjecture (cf. [P]) to give estimates similar to those given in the Corollary which contain two terms of correct order in k.

2.) We notice that our proofs have no "hard analysis" component. A similar requirement was one of the motivations for the paper [G].

Finally, we derive lower und upper bounds for sums of squares of eigenvalues on a compact Riemannian manifold without boundary. Here we need a lower bound $\mathrm{i}(M)$ for the injectivity radius of the manifold M and an upper bound $|\mathrm{K}(M)|$ for the modulus of the sectional curvature in order to ensure that the following test functions are well-defined. We introduce the test functions $\omega_y^{(\rho,s)}$ on M for every $\rho, s \in \mathbb{R}$ with $\rho \geq \rho_0 \equiv \max\{2/\mathrm{i}(M); |\mathrm{K(M)}|^{1/2}\}$ and $y \in M$ by

$$\omega_y^{(\rho,s)}(\exp_y u) \equiv \exp(is|u|)\chi(\rho|u|) \quad \text{for every } u \in \mathbb{R}^n;$$

here $\exp_y$ stands for the exponential map from T_yM in M and χ is a smooth nonnegative function on $\mathbb{R}$ with $\chi(\lambda) = 0$ for $\lambda \leq 1/2$ and $\lambda \geq 1$. (This construction might look somewhat peculiar but it avoids technical difficulties concerning the smoothness of the test functions. On the other hand, the above construction is equivalent to the construction used in the euclidean case since χ vanishes in a neighbourhood of 0.) The assumption that M is compact without boundary permits us to replace the gradient in the Rayleigh–Ritz formula by the Laplace–Beltrami operator. Thus we are able to give the following estimate for sums of squares of eigenvalues of the Laplacian instead of an estimate for sums of eigenvalues:

$$\left|\sum_{j=1}^{k} \lambda_j^2 - (2\pi)^2 \frac{n}{n+4} (\operatorname{vol}(M)\operatorname{vol}(B_1))^{-\frac{2}{n}} k^{\frac{n+4}{n}}\right| \leq C_n \rho_0 k^{\frac{n+3}{n}} (\operatorname{vol}(M))^{-\frac{2}{n}}$$

for every k and an appropriate constant C_n which depends only on n.

Remark. 1.) We notice that this estimate does not yield a non-trivial lower bound for $\sum_{j=1}^{k} \lambda_j^2$ if $k \leq \tilde{C}_n \frac{\operatorname{vol}(M)}{\rho_0^n}$ for a constant $\tilde{C}_n$. Recall that $\rho_0 \geq \mathrm{i}(M)$.

2.) Corresponding results for the eigenvalues of the Hodge–de Rham Laplacian can be derived in exactly the same way (cf. also [HSU]).

Note added in proof: The estimate for sums of squares of eigenvalues on a compact manifold is only true under the additional assumption that $k \geq \tilde{C}_n \frac{\operatorname{vol}(M)}{\rho_0^n}$ for a constant $\tilde{C}_n$ (cf. the example following Theorem 3 and Remark 1 at the end of this paper).

References

[Da] Davies, E. B., Spectral properties of compact manifolds and changes of metric. Amer. J. Math. 112 (1990), 15–39.

[Dy] Dynkin, E. B., Foundations of the theory of Markov processes. Pergamon Press, Oxford 1960.

[G] Guillemin, V., A new proof of Weyl's formula on the asymptotic distribution of eigenvalues. Adv. in Math. 55 (1985), 131–160.

[HSU] Hess, H., Schrader, R., Uhlenbrock, D. A., Kato's inequality and the spectral distribution of Laplacians on compact Riemannian manifolds. J. Differential Geom. 15 (1980), 27–38.

[K1] Kröger, P., Upper bounds for the Neumann eigenvalues on a bounded domain in Euclidean space. J. Funct. Anal. 106 (1992), 353–357.

[K2] — Estimates for sums of eigenvalues of the Laplacian. J. Funct. Anal. 126 (1994), 217–227.

[LP] Li, P., Yau, S. T., On the Schrödinger equation and the eigenvalue problem. Commun. Math. Phys. 88 (1983), 309–318.

[P] Pólya, G., On the eigenvalues of vibrating membranes. Proc. London Math. Soc. (3) 11 (1961), 419–433.

Fine $\mathcal{A}$-supersolutions and finely $\mathcal{A}$-superharmonic functions

Visa Latvala

Abstract. We consider fine supersolutions and finely superharmonic functions of the equation

$$-\operatorname{div} \mathcal{A}(x, \nabla u) = 0, \tag{1}$$

where $\mathcal{A}(x,h) \cdot h \approx |h|^p$ and $1 < p \leq n$. Equation (1) is similar to the p-Laplacian

$$-\operatorname{div}(|\nabla u|^{p-2}\nabla u) = 0.$$

As the main result we state that each finely $\mathcal{A}$-superharmonic function u in a finely open set $\Omega \subset \mathbb{R}^n$ satisfies

$$u(z) = \text{fine} \liminf_{x \to z} u(x)$$

at each point $z \in \Omega$. This leads, together with results of [3], to a useful characterization of finely $\mathcal{A}$-superharmonic functions by means of fine $\mathcal{A}$-supersolutions.

1991 Mathematics Subject Classification: 31C45.

1. Capacity, fine topology, quasi topology

Throughout this article, $n \geq 2$ is the dimension of the euclidean space $\mathbb{R}^n$ and $1 < p \leq n$ is a fixed real number. We denote by $W^{1,p}(G)$, $W_0^{1,p}(G)$, and $W_{\text{loc}}^{1,p}(G)$ the standard Sobolev spaces on an open set $G \subset \mathbb{R}^n$.

For an arbitrary set $E \subset \mathbb{R}^n$ define

$$\operatorname{cap}_{1,p}(E) = \inf_u \int_{\mathbb{R}^n} (|u|^p + |\nabla u|^p)\, dx,$$

where u runs through the set

$$\mathcal{U}(E) = \{\, u \in W^{1,p}(\mathbb{R}^n) : 0 \leq u \leq 1,\ u = 1 \text{ in an open set containing } E \,\}.$$

If $\mathcal{U}(E) = \emptyset$, we define $\operatorname{cap}_{1,p}(E) = +\infty$. The set function $E \longmapsto \operatorname{cap}_{1,p}(E)$ is called the *Sobolev p-capacity*. For the properties of the Sobolev p-capacity, see, e.g., [2], pp. 48–50. We call a set $E \subset \mathbb{R}^n$ *p-thin at* $x \in \mathbb{R}^n$ if

$$\int_0^1 \left(r^{p-n} \operatorname{cap}_{1,p}(E \cap B(x,r))\right)^{\frac{1}{p-1}} \frac{dr}{r} < \infty.$$

Definition 1.1. A set $E \subset \mathbb{R}^n$ is called *polar* if $\operatorname{cap}_{1,p}(E) = 0$.

Any relation is said to hold *quasi everywhere* (abbreviated *q.e.*) if it holds outside a polar set.

Definition 1.2. A set $U \subset \mathbb{R}^n$ is called *finely open* if $\complement U$ is p-thin at each point of U.

It is not hard to see that finely open sets form a topology. This is called the *fine topology*. In fact, polarity and the fine topology depend on p. However, we prefer to omit the prefix p, since p is assumed to be fixed. We use the word *fine*(*ly*) for the concepts related to the fine topology.

Remark. In nonlinear potential theories, the notion of thinness (resp. polarity) is often defined using the Bessel capacity $B_{1,p}$ (see e.g. [3], p. 373). However (see [3], Remark 1.13), the capacities $B_{1,p}$ and $\operatorname{cap}_{1,p}$ are equivalent, i.e., there are real numbers $\lambda_1 > 0$ and $\lambda_2 > 0$ such that

$$\lambda_1 B_{1,p}(E) \leq \operatorname{cap}_{1,p}(E) \leq \lambda_2 B_{1,p}(E)$$

for all $E \subset \mathbb{R}^n$. This implies that thin sets (resp. polar sets) are the same for the Sobolev p-capacity $\operatorname{cap}_{1,p}$ and the Bessel capacity $B_{1,p}$.

Besides the fine topology we need some quasi topological notions. A set $U \subset \mathbb{R}^n$ is said to be *quasi open* if for every $\varepsilon > 0$ there is an open set $G \subset \mathbb{R}^n$ such that $U \cup G$ is open and $\operatorname{cap}_{1,p}(G) < \varepsilon$. Let $U \subset \mathbb{R}^n$ be quasi open. Then a function $f\colon U \to \mathbb{R} \cup \{\pm\infty\}$ is called *quasi continuous* in U if for every $\varepsilon > 0$ there is an open set $G \subset \mathbb{R}^n$ such that $\operatorname{cap}_{1,p}(G) < \varepsilon$ and the restriction $f_{|U \setminus G}$ is finite and continuous. A family $\mathcal{B}$ of quasi open sets is termed a *quasi covering* of a set $E \subset \mathbb{R}^n$ if the union B of some countable subfamily of $\mathcal{B}$ satisfies $\operatorname{cap}_{1,p}(E \setminus B) = 0$. We write $V \subset\subset \Omega$, if V is bounded (quasi open) and $\bar{V} \subset \Omega$.

2. Fine $\mathcal{A}$-supersolutions

We fix a mapping $\mathcal{A}\colon \mathbb{R}^n \times \mathbb{R}^n \to \mathbb{R}^n$ which satisfies the assumptions A1–A5 for some constants $0 < \alpha \leq \beta < \infty$:

A1. The function $x \longmapsto \mathcal{A}(x,h)$ is measurable for all $h \in \mathbb{R}^n$ and the function $h \longmapsto \mathcal{A}(x,h)$ is continuous for almost every $x \in \mathbb{R}^n$.

For all h, h_1, $h_2 \in \mathbb{R}^n$ and for almost every $x \in \mathbb{R}^n$,

A2. $\mathcal{A}(x,h) \cdot h \geq \alpha |h|^p$,

A3. $|\mathcal{A}(x,h)| \leq \beta |h|^{p-1}$,

A4. $(\mathcal{A}(x,h_1) - \mathcal{A}(x,h_2)) \cdot (h_1 - h_2) > 0$ whenever $h_1 \neq h_2$,

A5. $\mathcal{A}(x,\lambda h) = \lambda |\lambda|^{p-2} \mathcal{A}(x,h)$ for all $\lambda \in \mathbb{R}$, $\lambda \neq 0$.

In order to present Sobolev spaces and test functions on finely open sets, we need the following definition:

Definition 2.1. Let $U \subset \mathbb{R}^n$ be quasi open. Then $V \subset U$ is called a $(1,p)$-*strict* subset of U if V is quasi open and there is a quasi continuous function $u \in W^{1,p}(\mathbb{R}^n)$ such that $0 \leq u \leq 1$, $u = 0$ outside U and $u = 1$ on V.

Remark. By [3], Theorem 1.5, any set $U \subset \mathbb{R}^n$ is quasi open if and only if $U = V \cup A$, where V is finely open and A is polar.

In the sequel, $\Omega \subset \mathbb{R}^n$ is assumed to be finely open. Kilpeläinen and Malý defined in [3] the generalized Sobolev space $W^{1,p}(\Omega)$ as the set of all functions u on Ω that are quasi locally restrictions of ordinary Sobolev functions such that the generalized Sobolev norm is finite. We refer to Section 2 in [3] and Section 3 in [5] for details. The local Sobolev space $W^{1,p}_{\text{loc}}(\Omega)$ consists of all functions u on Ω that belong to $W^{1,p}(V)$ for each $(1,p)$-strict subset $V \subset\subset \Omega$. The Sobolev space $W^{1,p}_0(\Omega)$ is best described by the following result of Kilpeläinen and Malý (see [3], Theorem 2.10):

Lemma 2.2. *Assume that $u \in W^{1,p}(\Omega)$ is quasi continuous in Ω. Let $v = u$ on Ω and $v = 0$ on $\mathbb{R}^n \setminus \Omega$. Then $u \in W^{1,p}_0(\Omega)$ if and only if v is quasi continuous in $\mathbb{R}^n$.*

We denote by $\mathcal{D}(\Omega)$ the collection of all bounded quasi continuous functions from $W^{1,p}(\mathbb{R}^n)$ vanishing outside a $(1,p)$-strict subset V of Ω with $V \subset\subset \Omega$. The set $\mathcal{D}(\Omega)$ has the role of test functions in our theory.

Definition 2.3. A quasi continuous function u on Ω is called a *fine $\mathcal{A}$-supersolution* of

$$-\operatorname{div} \mathcal{A}(x, \nabla u) = 0$$

in Ω if $u \in W^{1,p}_{\text{loc}}(\Omega)$ and

$$\int_\Omega \mathcal{A}(x, \nabla u) \cdot \nabla v \, dx \geq 0$$

for all $v \in \mathcal{D}_+(\Omega) = \{ w \in \mathcal{D}(\Omega) : w \geq 0 \}$.

Fine $\mathcal{A}$-solutions are defined similarly by replacing $\mathcal{D}_+(\Omega)$ with $\mathcal{D}(\Omega)$. In the sequel, we will simply speak about fine $\mathcal{A}$-supersolutions (resp. $\mathcal{A}$-solutions) in Ω instead of fine $\mathcal{A}$-supersolutions (resp. $\mathcal{A}$-solutions) of $-\operatorname{div} \mathcal{A}(x, \nabla u) = 0$ in Ω.

Remarks. (a) If Ω is open, then u is a fine $\mathcal{A}$-supersolution in Ω if and only if u is a quasi continuous ordinary $\mathcal{A}$-supersolution in Ω, see [2], p. 57, and [5], p. 24.

(b) Let $\mathcal{A}(s,t) = |t|^{p-2}t$. Then $\mathcal{A}$ satisfies the assumptions A1–A5 and the equation $-\operatorname{div}\mathcal{A}(x,\nabla u) = 0$ takes the form

$$-\operatorname{div}(|\nabla u|^{p-2}\nabla u) = 0,$$

which is called the *p-Laplace equation.* The classical Laplace equation is included in our theory as the special case $p = 2$.

3. Finely $\mathcal{A}$-superharmonic functions

We remind the reader that $\Omega \subset \mathbb{R}^n$ is always assumed to be finely open. The Dirichlet problem can be extended to bounded finely open sets as follows:

Definition 3.1. Let $\Omega \subset \mathbb{R}^n$ be bounded and suppose that $f \in W^{1,p}(\Omega)$. Then a quasi continuous function $H_f^\Omega \in W^{1,p}(\Omega)$ is called a *solution of the Dirichlet problem in Ω with the boundary condition f* whenever H_f^Ω is a fine $\mathcal{A}$-solution in Ω and $H_f^\Omega - f \in W_0^{1,p}(\Omega)$.

The existence and the uniqueness (as a Sobolev function) of the solution H_f^Ω are proved in [3], p. 389. Assume that $f \in C^1(\mathbb{R}^n)$. Then there is a pointwise unique finely continuous representative $\mu^\Omega(f)$ of the solution H_f^Ω (see [3], p. 398, or [5], p. 25). If $x \in \Omega$, we denote by $\mu_x^\Omega(f)$ the value of $\mu^\Omega(f)$ at the point x. The definition of $\mu^\Omega(f)$ can be extended to the class of all functions $f \in C(\overline{\Omega})$ by using the approximation in the sup-norm, see [3], p. 399, or [5], p. 26, for details.

We are now ready to give the following definition due to Kilpeläinen and Malý (compare with [1], Definition 8.1):

Definition 3.2. A function u on Ω with values in $\mathbb{R} \cup \{+\infty\}$ is called *finely $\mathcal{A}$-superharmonic* in $\Omega \subset \mathbb{R}^n$ if u is finely lower semicontinuous, finite q.e., and there is a basis $\mathcal{B}$ for the fine topology on Ω such that for each $W \in \mathcal{B}$, $x \in W$, and $f \in C(\overline{W})$ the inequality $f \le u$ in $\overline{W} \cap \Omega$ implies that $\mu_x^W(f) \le u(x)$.

Theorem 3.3. *Assume that u is finely $\mathcal{A}$-superharmonic in Ω. Then*

$$u(z) = \operatorname{fine}\liminf_{x \to z} u(x)$$

at each point $z \in \Omega$.

Proof. We will sketch the proof. See [5], pp. 29–31, for details. Let $\mathcal{B}$ be a basis for the fine topology on Ω such that $\mathcal{B}$, together with u, satisfies the comparison condition of Definition 3.2. Assume there is $z \in \Omega$ and $a \in \mathbb{R}$ such that

$$u(z) < a < \operatorname{fine}\liminf_{x \to z} u(x).$$

We are free to assume that $u(z) = 0$. Let $U \subset \Omega$ be a fine neighborhood of z with

$$\inf u(U \setminus \{z\}) > a$$

and let (see [5], Lemma 2.9) $V \in \mathcal{B}$ be a bounded finely open set such that

$$z \in V \subset \overline{V} \subset U.$$

According to [4], Theorem 1.3 (see also [2], Lemmas 7.2 and 9.14), one can find a bounded $\mathcal{A}$-subharmonic function v in an open neighborhood G of $\overline{V}$ with

$$\limsup_{x \to z,\, x \in \complement V} v(x) < u(z) < v(z) < a.$$

Since $u(z) = 0$, we may assume that $v \leq a$. Therefore, it is possible to construct a continuous function g on $\overline{V}$ such that $g \in W^{1,p}(V)$, $v \leq g \leq a$ in $\overline{V} \setminus V$, and $g(z) < u(z)$. Let

$$w = \max(v - H_g^V, 0).$$

One can show with the aid of Lemma 2.2 and [3], Theorem 1.4 that $w \in W_0^{1,p}(V)$. Since v is a fine $\mathcal{A}$-subsolution in V, we may apply [3], Theorem 3.8, and conclude that $v \leq H_g^V$ q.e. in V. Then

$$v(z) \leq \mu_z^V(g)$$

by the fine continuity of v and $\mu^V(g)$. On the other hand, $g \leq u$ in $\overline{V}$. Since $\mathcal{B}$ satisfies Definition 3.2 together with u in Ω, we get

$$u(z) < v(z) \leq \mu_z^V(g) \leq u(z).$$

This contradiction implies the assertion. □

Theorem 3.3 leads together with [3], Theorem 5.6, to the following useful characterization (see [5], Lemma 5.7, for the proof):

Corollary 3.4. *Suppose that u is a function on Ω with values in $\mathbb{R} \cup \{+\infty\}$. Then u is finely $\mathcal{A}$-superharmonic in Ω if and only if $u \in S(\Omega)$ and*

$$u(z) = \text{fine} \liminf_{x \to z} u(x)$$

at each point $z \in \Omega$.

The set $S(\Omega)$ in the previous corollary is defined as follows: A quasi continuous function u on Ω belongs to $S(\Omega)$ if there is a quasi covering $\mathcal{B}$ of Ω by finely open sets such that u is a fine $\mathcal{A}$-supersolution in V for each $V \in \mathcal{B}$.

The characterization of Corollary 3.4 is the key for several potential theoretic results of finely $\mathcal{A}$-superharmonic functions, see [5]. We will only mention here the following removability result of finely $\mathcal{A}$-superharmonic functions:

Corollary 3.5. *Let $E \subset \Omega$ be a polar set. If u is finely $\mathcal{A}$-superharmonic in $\Omega \setminus E$ such that*

$$\operatorname{fine}\liminf_{x \to z} u(x) > -\infty$$

for each $z \in E$, then the function $\tilde{u}\colon \Omega \to \mathbb{R} \cup \{+\infty\}$,

$$\tilde{u}(z) = \operatorname{fine}\liminf_{x \to z} u(x),$$

is finely $\mathcal{A}$-superharmonic in Ω.

Proof. The assertion follows easily from Corollary 3.4 since $U \setminus E$ is finely open for each finely open set $U \subset \mathbb{R}^n$. □

References

[1] Fuglede, B., Finely harmonic functions. Lecture Notes in Math. 289. Springer-Verlag, New York–Berlin–Heidelberg 1972.

[2] Heinonen, J., Kilpeläinen, T., Martio O., Nonlinear potential theory of degenerate elliptic equations. Oxford University Press, Oxford–New York–Tokyo 1993.

[3] Kilpeläinen, T., Malý, J., Supersolutions to degenerate elliptic equations on quasi open sets. Comm. Partial Differential Equations 17 (1992), 371–405.

[4] Kilpeläinen, T., Malý, J., The Wiener test and potential estimates for quasilinear elliptic equations. Acta Math. 172 (1994), 137–161.

[5] Latvala, V., Finely superharmonic functions of degenerate elliptic equations. Ann. Acad. Sci. Fenn. Ser. A I Math. Dissertationes 96 (1994).

Remarks on modified Clifford analysis

Heinz Leutwiler

Abstract. Harmonic 1-forms, considered on the hyperbolic space, lead to the following extension of the Cauchy–Riemann system to $\mathbb{R}^{n+1}$:

$$x_n\left(\frac{\partial u_0}{\partial x_0}-\frac{\partial u_1}{\partial x_1}-\cdots-\frac{\partial u_n}{\partial x_n}\right)+(n-1)u_n=0,$$

$$\frac{\partial u_0}{\partial x_k}=-\frac{\partial u_k}{\partial x_0},\quad \frac{\partial u_i}{\partial x_k}=\frac{\partial u_k}{\partial x_i}\quad (i,k=1,\ldots,n).$$

Embedding $\mathbb{R}^{n+1}$ into the Clifford algebra $\mathcal{C}_n$, one then obtains some generalized function theory, as we showed in [6]. We now look at the results of this article from a potential-theoretic view point.

1991 Mathematics Subject Classification: 30G35, 31C12.

1. Introduction

In [6] an attempt was undertaken to build up a function theory in $\mathbb{R}^{n+1}$, using Clifford numbers instead of the complex ones. Identifying the point $x = (x_0, x_1, \ldots, x_n) \in \mathbb{R}^{n+1}$ with the Clifford number $x = x_0 + x_1e_1 + \cdots + x_ne_n$ from the Clifford algebra $\mathcal{C}_n$(see §2 for its definition), it is easy to see that the powers x^k, $k \in \mathbb{N}$, are again in $\mathbb{R}^{n+1}$. We may thus ask for a generalization of the classical Cauchy–Riemann system, which is satisfied by these powers. As shown in [6], such a system is given by the following *modification* of the Dirac–Fueter system, usually studied in Clifford analysis (see, e.g., [3]): On an arbitrary open subset $\Omega \subset \mathbb{R}^{n+1}$, consider the C^2-solutions $f = \sum_{k=0}^{n} u_ke_k$ of the system

$$x_n\left(\frac{\partial u_0}{\partial x_0}-\frac{\partial u_1}{\partial x_1}-\cdots-\frac{\partial u_n}{\partial x_n}\right)+(n-1)u_n=0, \tag{1.1}$$

$$\frac{\partial u_0}{\partial x_k}=-\frac{\partial u_k}{\partial x_0},\quad \frac{\partial u_i}{\partial x_k}=\frac{\partial u_k}{\partial x_i}\quad (i,k=1,\ldots,n). \tag{1.2}$$

Solutions f of this system are called H_n-solutions in [6], because of their relationship with Hodge's theory of harmonic differential forms. Indeed, considering the 1-form $\omega = \mathrm{Re}(f dx) = u_0dx_0 - u_1dx_1 - \cdots - u_ndx_n$ on the upper half-space $\mathbb{R}^{n+1}_+ = \{(x_0, x_1, \ldots, x_n); x_n > 0\}$, endowed with the hyperbolic metric (see §4), it turns out that ω is harmonic (in the sense of Hodge) if and only if the system

(1.1)–(1.2) is satisfied. This was already remarked by H. Maass in [7]. In [6], in order to keep the theory as elementary as possible, little or no use has been made of this fact. Yet the theory of H_n-solutions becomes much more transparent if one uses this concept, as we shall show in this note.

In Section 2 we recall the concepts from the theory of Clifford algebras relevant to this paper. In Section 3, we briefly look at the harmonic differential forms, defined on an arbitrary, oriented Riemannian manifold. In Sections 4 and 5 we apply the concepts of Section 3 to the case of the hyperbolic space. This provides shorter proofs and new insight to most of the results we obtained in [6].

2. Clifford numbers

We start recalling that by definition, the *Clifford algebra* C_n is the real associative algebra generated by the elements $e_1, e_2, \ldots, e_n$ subject to the relations $e_k^2 = -1$, $k = 1, \ldots, n$, $e_i e_k = -e_k e_i$, $i \neq k$, and no others. Every element q of C_n can uniquely be expressed in the form

$$q = q_0 + \sum_{\nu} q_\nu e_\nu, \tag{2.1}$$

where the sum ranges over all multi-indices $\nu = (\nu_1, \ldots, \nu_p)$ with $1 \leq \nu_1 < \nu_2 < \cdots < \nu_p \leq n$, $e_\nu = e_{\nu_1} e_{\nu_2} \cdots e_{\nu_p}$ and $q_\nu \in \mathbb{R}$. The component q_0 referred to as the *real part*, $\mathrm{Re}(q)$, of q corresponds to the unit element $e_0 = 1$ (often identified with $e_\emptyset$). The term $\sum_\nu q_\nu e_\nu$ may be looked at as the *imaginary part* of q. Note that

$$\mathrm{Re}(ab) = \mathrm{Re}(ba), \tag{2.2}$$

for arbitrary $a, b \in C_n$.

C_n is a real vector space of dimension 2^n and one can identify C_0 with $\mathbb{R}$, C_1 with $\mathbb{C}$ and C_2 with $\mathbb{H}$, the algebra of quaternions. It will also be convenient to identify the subalgebra of C_n, generated by the elements $e_1, e_2, \ldots, e_{n-1}$, with the algebra C_{n-1}. We endow C_n with the Euclidean norm so that if q is of the form (2.1) then

$$|q|^2 = q_0^2 + \sum_{\nu} q_\nu^2. \tag{2.3}$$

There are three basic *involutions* of C_n. The first one, called main involution, is the one obtained by replacing each factor e_{ν_i} in $e_\nu = e_{\nu_1} e_{\nu_2} \ldots e_{\nu_p}$ by $-e_{\nu_i}$. It determines an automorphism $q \to q'$ of C_n, $(a+b)' = a' + b'$ and $(ab)' = a'b'$. The second involution, referred to as reversion, is the anti-automorphism $q \to q^*$, obtained by reversing the order of the factors e_{ν_i} in the product $e_\nu = e_{\nu_1} e_{\nu_2} \ldots e_{\nu_p}$. Consequently, $(a+b)^* = a^* + b^*$ and $(ab)^* = b^* a^*$. The third involution is the composition of the two others and will be denoted by $\overline{q} = (q')^* = (q^*)'$. It is again an anti-automorphism.

Throughout this paper we shall assume that the Euclidean space $\mathbb{R}^{n+1}$ has been identified with the subspace of C_n, spanned by the elements $e_0 = 1, e_1, e_2, \ldots, e_n$. Consequently, we shall call Clifford numbers of the form $x = x_0 + x_1 e_1 + \cdots + x_n e_n$ *vectors*. We thus adapt the terminology of Ahlfors [1] and do not use the term "paravector" as some authors nowadays do. For vectors, we have $x^* = x$ and hence $\overline{x} = x'$. Further, $x\overline{x} = \overline{x}x = |x|^2$, so non-zero vectors are invertible with inverse $x^{-1} = \overline{x}/|x|^2$.

3. Harmonic differential forms

On an arbitrary, oriented Riemannian manifold (M, g) of dimension n ($n \geq 2$), with metric tensor g, one usually considers the harmonic functions h, defined as solutions of the Laplace–Beltrami equation $\Delta_{LB} h = 0$, where

$$\Delta_{LB} = \frac{1}{\sqrt{|g|}} \sum_{i,j=1}^{n} \frac{\partial}{\partial x_i} \left(\sqrt{|g|} g^{ij} \frac{\partial}{\partial x_j} \right). \tag{3.1}$$

Here (g^{kl}) denotes the inverse matrix of $g = (g_{ij})$ and $|g| = det(g_{ij})$.

According to Hodge, the concept of a harmonic function, viewed as a harmonic 0-form, can be extended as follows. Let $\star$ denote the linear operator, called Hodge star, which is well defined by the requirement that for any orthonormal basis $t_1, \ldots, t_n$ of the cotangent space M_m^* at $m \in M$, we have

$$\star (1) = \pm t_1 \wedge t_2 \wedge \cdots \wedge t_n, \quad \star (t_1 \wedge t_2 \wedge \cdots \wedge t_n) = \pm 1,$$
$$\star (t_1 \wedge t_2 \wedge \cdots \wedge t_p) = \pm t_{p+1} \wedge \cdots \wedge t_n,$$

where one takes the sign + if $t_1 \wedge t_2 \wedge \cdots \wedge t_n$ is positively oriented and the sign − if otherwise. Denoting, as usual, the exterior differentiation operator by d, we introduce its *adjoint* δ on p-forms by setting

$$\delta = (-1)^{n(p+1)+1} \star d \star . \tag{3.2}$$

Thus δ maps p-forms to $(p-1)$-forms. On 0-forms δ is simply the zero functional. A p-form ω is now called *harmonic*, if it is closed and coclosed, i.e., if

$$d\omega = 0 \quad and \quad \delta\omega = 0. \tag{3.3}$$

With d and δ at our disposal we may also introduce the operator

$$\Delta_{LB} = \delta d + d\delta, \tag{3.4}$$

called Laplace–Beltrami operator. It maps p-forms into p-forms, and, for $p = 0$, agrees with (3.1), except possibly for a minus sign. Clearly, every harmonic p-form ω satisfies the Laplace–Beltrami equation $\Delta_{LB}\omega = 0$, but in case $p \geq 1$ the converse is not necessarily true. It holds however on compact manifolds ([9], Chap. 6).

Let us also recall that differential forms can be differentiated with respect to any given vector field X on M. The resulting derivative is known as the *Lie derivative* and is defined as follows (see, e.g., [9]): Let X_t denote the local 1-parameter group of transformations associated with X. Follow the integral curve of X through $m \in M$ out to the point $X_t(m)$ and evaluate ω there. Then transfer $\omega(X_t(m))$ back to $\Lambda(M_m^*)$ via the transpose of the differential dX_{-t} of the diffeomorphism X_{-t}. Finally, take the difference with $\omega(m)$, divide by t, and let $t \to 0$. The outcome is the Lie derivative $L_X(\omega)$. In what follows, we shall use the fact that L_X commutes with d, i.e.,

$$L_X(d\omega) = dL_X(\omega). \tag{3.5}$$

A natural question now to ask is the following one: What are the vector fields X on M with the property that their Lie derivative L_X maps the set of harmonic 1-forms into itself. The following result shows that this question can be reduced to an easier one.

Theorem 3.1. *Let X be a (smooth) vector field on M. Then the following statements are equivalent:*
(1) *For all harmonic 1-forms ω on M, the Lie derivative $L_X(\omega)$ is harmonic.*
(2) *For all harmonic functions h on M, $L_X(h)$ is harmonic.*

Proof. (1) $\Rightarrow$ (2): Let h be a harmonic function. Then $\omega = dh$ is a harmonic 1-form, since $d^2h = 0$ and $\delta dh = \Delta_{LB}h = 0$. Hence, by (3.5) and (1), $L_X(\omega) = L_X(dh) = dL_X(h)$ is a harmonic 1-form. Consequently, $0 = \delta dL_X(h) = \Delta_{LB}L_X(h)$.
(2) $\Rightarrow$ (1): Let ω be a harmonic 1-form. Then $d\omega = 0$ and hence, locally, $\omega = dh$ for some real-valued function h. From $\Delta_{LB}h = \delta dh = \delta\omega = 0$, we conclude that h is harmonic. Hence, by (2), so is $L_X(h)$. There results $\delta L_X(\omega) = \delta L_X(dh) = \delta dL_X(h) = \Delta_{LB}L_X(h) = 0$. Since also $dL_X(\omega) = L_X(d\omega) = L_X(0) = 0$, $L_X(\omega)$ must be harmonic. □

4. The H_n-solutions

In this section and in the following one we shall apply the concepts of Section 3 to the case of the hyperbolic space, represented by the upper half-space $\mathbb{R}_+^{n+1}$, endowed with the hyperbolic metric

$$ds^2 = \frac{1}{x_n^2}(dx_0^2 + dx_1^2 + \cdots + dx_n^2). \tag{4.1}$$

An easy calculation shows that the 1-form $\omega = \mathrm{Re}(f dx) = u_0 dx_0 - u_1 dx_1 - \cdots - u_n dx_n$,considered on an open set $\Omega \subset \mathbb{R}_+^{n+1}$, is harmonic with respect to the hyperbolic metric (4.1) if and only if the function $f = \sum_{k=0}^n u_k e_k$ is an H_n-solution, i.e., satisfies the equations (1.1) and (1.2). Note that for this geometrical interpretation of the H_n-solutions the set Ω must be contained in $\mathbb{R}_+^{n+1}$, rather than $\mathbb{R}^{n+1}$. This however is irrelevant for the purpose of this paper, since all the results naturally extend to $\mathbb{R}^{n+1}$.

Introducing the differential operator

$$D = \sum_{k=0}^{n} e_k \frac{\partial}{\partial x_k},$$

called Dirac operator by some authors, Fueter operator by others, the system (1.1)–(1.2) can also be written in the form

$$x_n(Df)(x) + (n-1)u_n(x) = 0, \tag{4.2}$$

where $f = \sum_{k=0}^{n} u_k e_k$ and $x = \sum_{k=0}^{n} x_k e_k$.

As already indicated in the introduction, the main justification for studying this system lies in the fact that the "natural candidate", the power function $f(x) = x^k (k \in \mathbb{N})$, $x = \sum_{k=0}^{n} x_k e_k$, yields an H_n-solution. This has already been established in [6]. In what follows we shall however present a more illuminating proof. It is based on the following adaptation of Theorem 3.6 of [5]:

Lemma 4.1. *Let f be an H_n-solution, defined on some open set $\Omega \subset \mathbb{R}^{n+1}$. Assume that f commutes with the variable x, i.e., satisfies the relation $xf(x) = f(x)x$, for all $x \in \Omega$. Then the function $F : \Omega \to \mathbb{R}^{n+1}$, defined by $F(x) = xf(x)$, also represents an H_n-solution.*

Proof. Let $f = \sum_{k=0}^{n} u_k e_k$ and $x = \sum_{k=0}^{n} x_k e_k$. From $\sum_{k=0}^{n} e_k f e_k = (1-n)\overline{f}$ we conclude that

$$D(fx) = (Df)x + (1-n)\overline{f}. \tag{4.3}$$

Assuming now that $xf = fx$, it is easy to see that

$$x_i u_k = x_k u_i \quad (i, k = 1, \ldots, n) \tag{4.4}$$

and that the component U_n of $F = xf = \sum_{k=0}^{n} U_k e_k$ is given by

$$U_n = x_0 u_n + x_n u_0. \tag{4.5}$$

Consequently, by (4.3)–(4.5), we have

$$\begin{aligned} x_n DF + (n-1)U_n &= x_n\big[(Df)x + (1-n)\overline{f}\,\big] + (n-1)(x_0 u_n + x_n u_0) \\ &= x_n(Df)x + (n-1)\big[x_n(u_0 - \overline{f}) + x_0 u_n\big] \\ &= x_n(Df)x + (n-1)\Big[\sum_{k=1}^{n} x_n u_k e_k + x_0 u_n\Big] \\ &= x_n(Df)x + (n-1)x u_n = \big[x_n Df + (n-1)u_n\big]x. \end{aligned}$$

Assuming therefore that f is an H_n-solution, so is F. □

An inductional argument now shows:

Corollary 4.2. *The power-function* $f(x) = x^k$, $x = x_0 + x_1e_1 + \cdots + x_ne_n$, $k \in \mathbb{N}$, *is an* H_n*-solution.*

Proof. Assuming that, by hypothesis, $f(x) = x^{k-1}$ $(k \in \mathbb{N})$ is an H_n-solution, Lemma 4.1 shows that the same is true for $F(x) = x^k$. □

We now turn to another relation between the hyperbolic space $(\mathbb{R}^{n+1}_+, ds)$ and the Clifford algebra C_n. Consider the conformal maps of the hyperbolic space. It is known (see,e.g., [2]) that they agree with the Möbius transformations γ on $\mathbb{R}^{n+1}$, which map $\mathbb{R}^{n+1}_+$ onto itself. On the other hand, according to an important result of K. T. Vahlen [8], every orientation-preserving Möbius transformation γ, mapping $\mathbb{R}^{n+1}_+$ onto itself, admits the representation

$$\gamma(x) = (ax + b)(cx + d)^{-1}, \quad x = \sum_{k=0}^{n} x_k e_k. \tag{4.6}$$

Here, a,b,c and d are Clifford numbers of the algebra C_{n-1}, satisfying the following conditions (see Ahlfors [1]):

(1) a,b,c and d are representable as finite products of vectors in C_{n-1},

(2) $ad^* - bc^* = 1$,

(3) ac^{-1} and $c^{-1}d$ are in $\mathbb{R}^n$, if $c \neq 0$,

(4) db^{-1} and $b^{-1}a$ are in $\mathbb{R}^n$, if $b \neq 0$.

In order to see how these transformations γ act on the H_n-solutions f we consider the harmonic 1-form $\omega = \mathrm{Re}(f dy)$. Changing coordinates according to $y = \gamma(x)$, taking into account that

$$dy = (xc^* + d^*)^{-1} dx (cx + d)^{-1} \tag{4.7}$$

(see [7]), we obtain from (2.2) that

$$\begin{aligned} \mathrm{Re}(f(y)dy) &= \mathrm{Re}\left[f(\gamma(x))(xc^* + d^*)^{-1} dx (cx + d)^{-1}\right] \\ &= \mathrm{Re}\left[(cx + d)^{-1} f(\gamma(x))(xc^* + d^*)^{-1} dx\right]. \end{aligned}$$

This simple calculation already provides a proof for the following result obtained in [6]:

Theorem 4.3. *Let* Ω *be an open subset of the upper half-space* $\mathbb{R}^{n+1}_+$ *and* γ *a Möbius transformation, mapping* $\mathbb{R}^{n+1}_+$ *onto itself. Furthermore, let* f *be an* H_n*-solution, defined on the image set* $\gamma(\Omega)$*. Assuming that* γ *is represented in the form* (4.6)*, the function* $F : \Omega \to \mathbb{R}^{n+1}$*, defined by*

$$F(x) = (cx + d)^{-1} f(\gamma(x))(xc^* + d^*)^{-1}, \quad x \in \Omega, \tag{4.8}$$

is an H_n*-solution. Hereby* $*$ *denotes the involution introduced in Section* 2.

As an application of this theorem, combined with Corollary 4.2, we get the result that for any $k \in \mathbb{N}$, the power-function $f(x) = x^{-k}$ also represents an H_n-solution.

5. Differential operators

In this section we first answer the following question: What are the vector fields X on the hyperbolic space $(\mathbb{R}^{n+1}_+, ds)$, having the property that their Lie derivative L_X maps the set of harmonic 1-forms into itself. On account of Theorem 3.1 this question is equivalent to the following one: Determine all the vector fields X on $(\mathbb{R}^{n+1}_+, ds)$ with the property that for each harmonic function h the function $L_X h = (dh)(X)$ also defines a harmonic one. This question however has been answered in [6]. There results

Theorem 5.1. *The vector fields* $X = \sum_{k=0}^n a_k \frac{\partial}{\partial x_k}$ *on the hyperbolic space* $(\mathbb{R}^{n+1}_+, ds)$*, whose Lie derivatives* L_X *map harmonic 1-forms into harmonic 1-forms, form a real vector space* $\mathcal{V}_n$ *of dimension* $\frac{1}{2}(n+1)(n+2)$*. A basis for* $\mathcal{V}_n$ *is given by the following vector fields:*

$$X_k = \frac{\partial}{\partial x_k} \quad (k = 0, 1, \ldots, n-1),$$

$$Y_0 = (x_0^2 - x_1^2 - \cdots - x_n^2)\frac{\partial}{\partial x_0} + 2x_0x_1\frac{\partial}{\partial x_1} + \cdots + 2x_0x_n\frac{\partial}{\partial x_n},$$

$$Y_1 = 2x_1x_0\frac{\partial}{\partial x_0} + (-x_0^2 + x_1^2 - x_2^2 - \cdots - x_n^2)\frac{\partial}{\partial x_1} + 2x_1x_2\frac{\partial}{\partial x_2} + \cdots + 2x_1x_n\frac{\partial}{\partial x_n},$$

$$\cdots\cdots$$

$$Y_{n-1} = 2x_{n-1}x_0\frac{\partial}{\partial x_0} + \cdots + 2x_{n-1}x_{n-2}\frac{\partial}{\partial x_{n-2}}$$
$$+ (-x_0^2 - \cdots - x_{n-2}^2 + x_{n-1}^2 - x_n^2)\frac{\partial}{\partial x_{n-1}} + 2x_{n-1}x_n\frac{\partial}{\partial x_n},$$

$$Z = x_0\frac{\partial}{\partial x_0} + x_1\frac{\partial}{\partial x_1} + \cdots + x_n\frac{\partial}{\partial x_n},$$

$$W_{ij} = x_j\frac{\partial}{\partial x_i} - x_i\frac{\partial}{\partial x_j} \quad (0 \le i < j \le n-1).$$

In the language of the H_n-solutions the vector fields, occurring in Theorem 5.1, have the property that their Lie derivatives map H_n-solutions into H_n-solutions. They thus provide differential operators ("rules" for) mapping the set of H_n-solutions into itself.

Corollary 5.2. *Let* f *be an* H_n*-solution. Then so are the functions*

$$\delta_k f = \frac{\partial f}{\partial x_k} \quad (k = 0, 1, \ldots, n-1),$$

$$F_0 f = xf + fx + (x_0^2 - x_1^2 - \cdots - x_n^2)\frac{\partial f}{\partial x_0} + 2x_0x_1\frac{\partial f}{\partial x_1} + \cdots + 2x_0x_n\frac{\partial f}{\partial x_n},$$

$$F_1 f = -e_1xf - fxe_1 + 2x_1x_0\frac{\partial f}{\partial x_0} + (-x_0^2 + x_1^2 - x_2^2 - \cdots - x_n^2)\frac{\partial f}{\partial x_1} +$$

$$+ 2x_1x_2\frac{\partial f}{\partial x_2} + \cdots + 2x_1x_n\frac{\partial f}{\partial x_n},$$

......

$$F_{n-1}f = -e_{n-1}xf - fxe_{n-1} + 2x_{n-1}x_0\frac{\partial f}{\partial x_0} + \cdots + 2x_{n-1}x_{n-2}\frac{\partial f}{\partial x_{n-2}}$$

$$+ (-x_0^2 - \cdots - x_{n-2}^2 + x_{n-1}^2 - x_n^2)\frac{\partial f}{\partial x_{n-1}} + 2x_{n-1}x_n\frac{\partial f}{\partial x_n},$$

$$\mathcal{D}f = x_0\frac{\partial f}{\partial x_0} + x_1\frac{\partial f}{\partial x_1} + \cdots + x_n\frac{\partial f}{\partial x_n},$$

$$G_{0j}f = e_jf + fe_j - 2x_j\frac{\partial f}{\partial x_0} + 2x_0\frac{\partial f}{\partial x_j} \quad (j = 1, \ldots, n-1),$$

$$G_{ij}f = -e_ie_jf - fe_je_i - 2x_j\frac{\partial f}{\partial x_i} + 2x_i\frac{\partial f}{\partial x_j} \quad (0 < i < j \leq n-1).$$

Endowing the real vector space $\mathcal{V}_n$ from Theorem 5.1 with the Lie bracket $[X, Y] = XY - YX$, we infer from the identity (see, e.g., [4])

$$L_{[X,Y]}\omega = L_X(L_Y\omega) - L_Y(L_X\omega) \tag{5.1}$$

that $(\mathcal{V}_n, [,])$ defines a *Lie algebra*. It agrees with the Lie algebra associated with the group of Möbius transformations, mapping $\mathbb{R}^{n+1}_+$ onto itself. The vector fields X_k $(k = 0, 1, \ldots, n-1)$ correspond to the one-parameter subgroups $x \to x + \lambda e_k$ $(\lambda \in \mathbb{R})$ of translations, the vector field Z is associated with the subgroup of dilatations $x \to e^\lambda x (\lambda \in \mathbb{R})$. The vector fields W_{ij} correspond to the rotations, leaving fixed the x_n-axis. For an interpretation of the vector fields Y_k $(k = 0, 1, \ldots, n-1)$ we refer to [6].

References

[1] Ahlfors, L. V., On the fixed points of Möbius transformations in $\mathbb{R}^n$. Ann. Acad. Sci. Fenn. Ser. A. I. Math. 10 (1985), 15–27.

[2] Ahlfors, L. V., Möbius transformations in several dimensions. Lectures in Mathematics. University of Minnesota, Minneapolis 1981.

[3] Brackx, F., Delanghe, R., Sommen, F., Clifford analysis. Pitman, Boston, London, Melbourne 1982.

[4] Helgason, S., Differential geometry and symmetric spaces. Academic Press, New York 1962.

[5] Hengartner, W., Leutwiler, H., Holomorphic functions in $\mathbb{R}^3$. Preprint.

[6] Leutwiler, H., Modified Clifford Analysis. Complex Variables Theory Appl. 17 (1992), 153–171.

[7] Maass, H., Automorphe Funktionen von mehreren Veränderlichen und Dirichletsche Reihen. Abh. Math. Sem. Univ. Hamburg 16 (1949), 72–100.

[8] Vahlen, K. T., Über Bewegungen und complexe Zahlen. Math. Ann. 55 (1902), 585–593.

[9] Warner, F. W, Foundations of differentiable manifolds and Lie groups. Scott, Foresman and Company, Glenview 1971.

Electric fields in domains with complicated structure

Peter V. Malyshev and Dmitry V. Malyshev

Abstract. We consider a problem of finding perturbations of a potential of uniform electric field by an infinite periodic array of dielectric bodies. Solutions are obtained in the form of series consecutively taking into account the influence of multipoles. New formulas are obtained for effective parameters characterizing inhomogeneous media.

1991 Mathematics Subject Classification: 31B20.

In various physical problems, it is necessary to have a detailed information on the behavior of systems of charged particles in the domains with complicated structure, for example, systems that can be regarded as models of membranes of various types. This is quite important, e.g., for the problems of membrane electrolyte separation.

At the same time, many characteristics of these processes, such as distribution functions, selectivity, etc., are determined in terms of potentials of (screened) electric fields that appear in the systems of charged particles near membranes of various types.

Distribution functions of such systems of particles and screened potentials in them were determined in [1] for a membrane in the form of an infinite dielectric layer with cylindrical channels in it. Here, we study perturbations of a uniform electric field by an array of spherical dielectric bodies placed at the points of infinite cubic lattice. This enables us to find effective characteristics of the medium in different situations and apply them in investigating the properties of charged particles near membranes.

1. Assume that an infinite system of identical neutral bodies (inclusions) is placed in a homogeneous medium with dielectric permittivity ε_1. The dielectric permittivity of the inclusions is ε_2 (without loss of generality, we can assume that $\varepsilon_1 > \varepsilon_2$). Also assume that the geometric centers of the bodies coincide with the points of the infinite cubic lattice with period a that lie in a layer $T \subset \mathbb{R}^3$ between two planes perpendicular to one of the edges of the elementary cell of this lattice. Let one of the coordinate axes, say, $\vec{e}_3$ be directed along this edge, and let $\vec{e}_1$ and $\vec{e}_2$ be directed along the other two edges of the elementary cube. Clearly, the number of "layers" of inclusions along the axis $\vec{e}_3$ is bounded; denote it by N. The "layers" are enumerated in the increasing order in the direction chosen as positive along the axis $\vec{e}_3$. We place the origin of coordinates at the geometric center of an inclusion, say, from the m-th "layer" ($1 \le m \le N$). Denote an arbitrary body, e.g., from the j-th "layer" by $F_{\vec{k}^j}$, where $a\vec{k}^j = a(k_1, k_2, j-m)$, $k_1 \in \mathbb{Z}$, $k_2 \in \mathbb{Z}$, $j = \overline{1, N}$ is a radius vector of its center.

Consider a uniform electric field whose potential in the medium without inclusions is

$$\varphi_0(x) = -\frac{\vec{E}\cdot\vec{x}}{\varepsilon_1},$$

where $\vec{E} \parallel \vec{e}_3$ is its intensity in a vacuum. Let us study perturbations of this field by the indicated system of dielectric bodies.

A potential of the resultant electric field in the system under consideration can be found as a solution of the Laplace equation inside and outside the bodies:

$$\Delta\varphi(x) = 0 \tag{1}$$

with the following boundary conditions on the surfaces of the inclusions:

$$\varphi_+(x) = \varphi_-(x), \quad x \in \partial F, \quad \partial F = \bigcup_{j=1}^{N} \bigcup_{k_1,k_2} \partial F_{\vec{k}^j}, \tag{2}$$

$$\varepsilon_1 \left(\frac{\partial\varphi}{\partial\vec{n}}\right)_-(x) = \varepsilon_2 \left(\frac{\partial\varphi}{\partial\vec{n}}\right)_+(x), \quad x \in \partial F, \tag{3}$$

where "+" and "−" denote limiting values of the corresponding functions when approaching the surface from inside and outside, respectively, and $\vec{n}$ is an outer normal to this surface.

We seek solutions of the problem (1)–(3) in the form of a sum of the potential φ_0 and a potential φ_1 of charges induced on the surfaces of the inclusions. Taking into account that, owing to the symmetry of the considered system, the densities of the surface charges are equal for all the inclusions in the same "layer", we get

$$\varphi(x) = -\frac{\vec{E}\cdot\vec{x}}{\varepsilon_1} + \frac{1}{4\pi\varepsilon_0\varepsilon_1}\sum_{j=1}^{N}\int_{\partial F_0}\sum_{k_1,k_2}\frac{\sigma^{(j)}(y)}{\left|\vec{x}-\vec{y}-a\vec{k}^j\right|}dS_y = \varphi_0(x) + \varphi_1(x), \tag{4}$$

where ∂F_0 is the surface of the body located at the origin, $\sum\limits_{k_1,k_2} = \sum\limits_{k_1=-\infty}^{\infty}\sum\limits_{k_2=-\infty}^{\infty}$, and $\sigma^{(j)}$ is a density of charges induced on the surface of an inclusion from the j-th "layer".

The potential φ given by (4) is a solution of the boundary value problem (1)–(3) if the densities of the surface charges $\sigma^{(i)}, i = \overline{1, N}$, satisfy the following system of integral equations [2]:

$$\sigma^{(i)}(x) + \lambda\sum_{j=1}^{N}\int_{\partial F_0}\sigma^{(j)}(y)\sum_{k_1,k_2}\frac{\cos(\vec{x}-\vec{y}-a\vec{k}, \vec{n}_x)}{2\pi\left|\vec{x}-\vec{y}-a\vec{k}\right|^2}dS_y = -2\lambda\varepsilon_0\vec{E}\cdot\vec{n}_x, \tag{5}$$

$$x \in \partial F_0, \quad i = \overline{1, N},$$

where $\lambda = \frac{\varepsilon_1 - \varepsilon_2}{\varepsilon_1 + \varepsilon_2} > 0$ and $\vec{k} = (k_1, k_2, j - i)$, and the condition

$$\int_{\partial F_0} \sigma^{(i)}(x) dS_x = 0, \quad i = \overline{1, N}; \tag{6}$$

reflecting the fact the inclusions are electrically neutral.

If there exists a unique solution of the system (5) satisfying (6), then the following assertion is true:

Proposition 1. *If the surfaces of inclusions are symmetric with respect to the three planes perpendicular to the coordinate axes, then the densities of surface charges $\sigma^{(j)}$ possess the following properties of symmetry:*

$$\sigma^{(j)}(T_i x) = \begin{cases} \sigma^{(j)}(x), & i = 1, 2, \\ -\sigma^{(N+1-j)}(x), & i = 3, \end{cases} \tag{7}$$

where T_i is the operator of inversion of the i-th coordinate axis.

Proof. Let us prove the last equality in (7). For this purpose, we perform in (5) the inversion $\vec{x} \to T_3 \vec{x}$, change $\vec{y} \to T_3 \vec{y}$, $k_1 \to -k_1$, and $k_2 \to -k_2$, and take into account that $\vec{E} \cdot \vec{n}_{T_3 x} = -\vec{E} \cdot \vec{n}_x$. After simple calculation, we obtain the following equation for $\sigma^{(i)}(T_3 x)$:

$$\sigma^{(i)}(T_3 x) + \lambda \sum_{j=1}^{N} \int_{\partial F_0} \sigma^{(j)}(T_3 y) \sum_{k_1, k_2} \frac{\cos(\vec{x} - \vec{y} - a\vec{k}', \vec{n}_x)}{2\pi \left| \vec{x} - \vec{y} - a\vec{k}' \right|^2} dS_y = 2\lambda \varepsilon_0 \vec{E} \cdot \vec{n}_x,$$

where $\vec{k}' = (k_1, k_2, i - j)$. By comparing this equation with the equation for $\sigma^{N+1-i}(x)$ rewritten in the form

$$\sigma^{(N+1-i)}(x) + \lambda \sum_{j=1}^{N} \int_{\partial F_0} \sigma^{(N+1-j)}(y) \sum_{k_1, k_2} \frac{\cos(\vec{x} - \vec{y} - a\vec{k}', \vec{n}_x)}{2\pi \left| \vec{x} - \vec{y} - a\vec{k}' \right|^2} dS_y = -2\lambda \varepsilon_0 \vec{E} \cdot \vec{n}_x,$$

we can easily conclude that if the solution of system (5) is unique, then the following equality holds:

$$\sigma^{(i)}(T_3 x) = -\sigma^{(N+1-i)}(x), \quad i = \overline{1, N}.$$

The other equalities in (7) can be proved similarly. □

Generally speaking, the series in the kernels of integral equations (5) are not absolutely convergent. However, it can be shown that if condition (6) is satisfied, these series converge. For this purpose, we expand each term of the series representing the kernels of equations (5) in a power series in $(x_i - y_i)$, $i = \overline{1, 3}$ (this method is due to Petrina [3] who was the first who applied it in solving the classical

Maxwell problem). If we write these series in the form of the first two terms and the remainder and take relations (6) and (7) into account, then we can rewrite the integral equations (5) as follows [2]:

$$\begin{aligned}\sigma^{(i)}(x) + \frac{\lambda}{2\pi}\int\limits_{\partial F_0} \sigma^{(i)}(y)\frac{\cos(\vec{x}-\vec{y},\vec{n}_x)}{|\vec{x}-\vec{y}|^2}dS_y + \lambda\sum_{j=1}^{N}\int\limits_{\partial F_0}\sigma^{(j)}(y)K_{ij}(x,y)dS_y \\ = -2\lambda\varepsilon_0\vec{E}\cdot\vec{n}_x + \frac{\lambda}{2\pi a^3}\sum_{j=1}^{N}\mu^{(i-j)}p^{(j)}n_{x_3}, \quad x\in\partial F_0, \quad i=\overline{1,N},\end{aligned} \tag{8}$$

where

$$\sum_{k_1,k_2}{}' = \sum_{\substack{k_1,k_2\\ |\vec{k}|\neq 0}}, \qquad \vec{p}^{(j)} = \int\limits_{\partial F_0}\sigma^{(j)}(x)\,\vec{x}\,dS_x = \left(0,0,p^{(j)}\right),$$

$$\mu^{(i-j)} = \mu^{(j-i)} = \sum_{k_1,k_2}{}'\left(|\vec{k}|^3 - 3(i-j)^2|\vec{k}|^5\right),$$

and

$$K_{ij}(x,y) = \frac{1}{4\pi}\sum_{k_1,k_2}{}'\sum_{\alpha,\beta=1}^{3}\left(\frac{\partial^2}{\partial r_\alpha\partial r_\beta}\frac{(\vec{r}-a\vec{k})\cdot\vec{n}_x}{\left|\vec{r}-a\vec{k}\right|^3}\right)\Bigg|_{\vec{r}=\vec{\xi}}(x_\alpha-y_\alpha)(x_\beta-y_\beta),$$

$$0\le|\xi_l|\le|x_l-y_l|, \qquad l=1,2,3.$$

One can easily see that all the series in the integral equations (8) are absolutely convergent.

Let us introduce the Banach space $B(\partial F_0)$ of columns $\chi = \begin{pmatrix}\chi^{(1)}(x)\\ \vdots \\ \chi^{(N)}(x)\end{pmatrix}$, $x\in\partial F_0$, whose elements are functions $\chi^{(i)}(x)$, $i=\overline{1,N}$, integrable over the surface ∂F_0 and satisfying condition (6). A norm in the space $B(\partial F_0)$ is defined as follows:

$$\|\chi\| = \max_{1\le i\le N}\int\limits_{\partial F_0}|\chi^{(i)}(x)|dS_x.$$

In the space $B(\partial F_0)$, the system of integral equations (8) takes the form

$$(I+\lambda A_1)\sigma + \lambda(A_2+A_3)\sigma = \sigma_0, \tag{9}$$

where $(\sigma_0)_i(x) = -2\lambda\varepsilon_0\vec{E}\cdot\vec{n}_x$, $i = \overline{1,N}$, and the operators A_1, A_2, and A_3 are defined as follows:

$$(A_1\chi)_i(x) = \int\limits_{\partial F_0} \chi^{(i)}(y)\frac{\cos(\vec{x}-\vec{y},\vec{n}_x)}{2\pi|\vec{x}-\vec{y}|^2}dS_y, \qquad x\in\partial F_0,\ i=\overline{1,N},$$

$$(A_2\chi)_i(x) = \sum_{j=1}^{N}\int\limits_{\partial F_0} \chi^{(j)}(y)K_{ij}(x,y)dS_y, \qquad x\in\partial F_0,\ i=\overline{1,N},$$

$$(A_3\chi)_i(x) = -\frac{1}{2\pi}n_{x_3}\sum_{j=1}^{N}\mu^{(i-j)}\int\limits_{\partial F_0}\chi^{(j)}(y)y_3dS_y, \qquad x\in\partial F_0,\ i=\overline{1,N}.$$

The following statement is true [2]:

Theorem 1. *Assume that the following inequality holds:*

$$C\frac{|\lambda|v}{1-|\lambda|} < 1,$$

where C is a certain constant and v is a volume density of inclusions in the layer T. Then the solution of the operator equation (9) exists and is unique in the space $B(\partial F_0)$.

If we solve the system of integral equations (8), then we can construct a solution of the problem (1)–(3) by inserting an expression for the densities of surface charges thus obtained in relation (4). However, the series in the definition of the potential $\varphi_1(x)$ are not absolutely convergent. By using the same procedure, as when analyzing system (5), we can prove the following statement for $\varphi_1(x)$ (here, we take into account that, in view of the symmetry of the problem, the potential $\varphi_1(x)$ is a function periodic in the variables x_1 and x_2 with period a).

Theorem 2. *The function $\varphi_1(x)$ is continuous and bounded in the parallelepiped $V_L(\vec{x} : |x_1| \le a/2, |x_2| \le a/2, |x_3| \le L)$, $Na < L < \infty$, to within a constant $\frac{1}{4\pi\varepsilon_0\varepsilon_1}\sum\limits_{j=1}^{N}\sum\limits_{k_1,k_2}{}'' \frac{\vec{p}^{(j)}\cdot\vec{k}^{(j)}}{a^2|\vec{k}^{(j)}|^3}$, where $\sum\limits_{k_1,k_2}{}'' = \sum\limits_{\substack{k_1,k_2\\ |k_1|+|k_2|\ge 1}}$.*

Having found the potential of the electric field in the system, we can determine the effective dielectric permittivity $\varepsilon_{\text{ef}}^{(N)}$ of the system of inclusions. For the system under consideration, the effective dielectric permittivity is given by the relation

$$\varepsilon_{\text{ef}}^{(N)} = \varepsilon_1\frac{\varepsilon_0 E N a^3}{\varepsilon_0 E N a^3 - \sum_{j=1}^{N} p^{(j)}}.$$

We restrict ourselves to the case of spherical inclusions with radius R and assume that the volume density of inclusions $v = \frac{4\pi R^3}{3a^3}$ is sufficiently small so that we can neglect the terms of the order $v^{4/3}$ and higher in system (8). (In other words, we neglect the remainder in the corresponding expansion. Note that, by taking more terms of expansion into account, we can determine effective parameters of the system with any desired degree of accuracy.) Multiplying equations (8) by x_3 and integrating over the surface ∂F_0, we obtain the following system of algebraic equations for the dipole moments $p^{(i)}$:

$$\left(1+\frac{\lambda}{3}-\frac{\lambda v}{2\pi}\mu^{(0)}\right)p^{(i)}-\frac{\lambda v}{2\pi}\sum_{\substack{j=1\\ j\neq i}}^{N}\mu^{(i-j)}p^{(j)}=-2\lambda\varepsilon_0 Eva^3,\quad i=\overline{1,N}. \tag{10}$$

Solving the system (10), we arrive (in a given approximation) at the following relations for the densities of surface charges and the potential of the resultant electric field near inclusions and inside them [2]:

$$\sigma^{(i)}(x)=3\varepsilon_0\frac{\nu-1}{\nu+2}\left(\vec{n}_x\cdot\left(\vec{E}-\frac{1}{4\pi\varepsilon_0}\sum_{j=1}^{N}\mu^{(i-j)}\vec{p}^{(j)}\right)\right),\quad x\in\partial F_0,\ i=\overline{1,N},$$

$$\varphi^{(i)}(x)=\begin{cases}-\frac{3}{\varepsilon_1}\frac{1}{\nu+2}\left(\vec{E}-\frac{1}{4\pi\varepsilon_0}\sum\limits_{j=1}^{N}\mu^{(i-j)}\vec{p}^{(j)}\right)\cdot\vec{x}, & |\vec{x}|\le R,\\ -\frac{1}{\varepsilon_1}\left(\vec{E}-\frac{1}{4\pi\varepsilon_0}\sum\limits_{j=1}^{N}\mu^{(i-j)}\vec{p}^{(j)}\right)\cdot\vec{x}\left(1-\frac{\nu-1}{\nu+2}\frac{R^3}{|\vec{x}|^3}\right), & R\le|\vec{x}|<<\frac{a}{2},\end{cases}$$

where $\nu=\frac{\varepsilon_2}{\varepsilon_1}$; the origin of coordinates in the expression for the potential $\varphi^{(i)}(x)$ is located at the center of the inclusion from the i-th "layer", in the vicinity of which the solution is investigated.

To illustrate these results, we present explicit formulas for the cases $N=1$ and $N=2$.

I. $N=1$. We have

$$p=\frac{3(\nu-1)\varepsilon_0 Eva^3}{2+\nu+\frac{3}{4\pi}(\nu-1)v\mu^{(0)}},$$

$$\sigma(x)=\frac{3(\nu-1)\varepsilon_0\vec{E}\cdot\vec{n}_x}{2+\nu+\frac{3}{4\pi}(\nu-1)v\mu^{(0)}},$$

$$\varphi(x)=\begin{cases}-\frac{1}{\varepsilon_1}\frac{3\vec{E}\cdot\vec{x}}{2+\nu+\frac{3}{4\pi}(\nu-1)v\mu^{(0)}}, & |\vec{x}|\le R,\\ \frac{1}{\varepsilon_1}\frac{\vec{E}\cdot\vec{x}}{2+\nu+\frac{3}{4\pi}(\nu-1)v\mu^{(0)}}\left((\nu-1)\frac{R^3}{|\vec{x}|^3}-2-\nu\right), & R\le|\vec{x}|<<\frac{a}{2},\end{cases}$$

$$\varepsilon_{\text{ef}}^{(1)}=\varepsilon_1\left(1+\frac{3(\nu-1)v}{2+\nu+\frac{3}{4\pi}(\nu-1)v\left(\mu^{(0)}-4\pi\right)}\right).$$

II. $N = 2$. In this case, we have

$$p^{(1)} = p^{(2)} = \frac{3(\nu-1)\varepsilon_0 E v a^3}{2+\nu+\frac{3}{4\pi}(\nu-1)v\left(\mu^{(0)}+\mu^{(1)}\right)},$$

$$\sigma^{(1)}(x) = \sigma^{(2)}(x) = \frac{3(\nu-1)\varepsilon_0 \vec{E}\cdot\vec{n}_x}{2+\nu+\frac{3}{4\pi}(\nu-1)v\left(\mu^{(0)}+\mu^{(1)}\right)},$$

$$\varphi(x) = \begin{cases} -\frac{1}{\varepsilon_1}\frac{3\vec{E}\cdot\vec{x}}{2+\nu+\frac{3}{4\pi}(\nu-1)v\left(\mu^{(0)}+\mu^{(1)}\right)}, & |\vec{x}| \le R, \\ \frac{1}{\varepsilon_1}\frac{\vec{E}\cdot\vec{x}}{2+\nu+\frac{3}{4\pi}(\nu-1)v\left(\mu^{(0)}+\mu^{(1)}\right)}\left((\nu-1)\frac{R^3}{|\vec{x}|^3}-2-\nu\right), & R \le |\vec{x}| << \frac{a}{2}, \end{cases}$$

$$\varepsilon_{\text{ef}}^{(2)} = \varepsilon_1\left(1+\frac{3(\nu-1)v}{2+\nu+\frac{3}{4\pi}(\nu-1)v\left(\mu^{(0)}+\mu^{(1)}-4\pi\right)}\right).$$

Similarly, we can find the parameters of the system for an arbitrary number of "layers". However, our analysis shows that even the values of the parameters obtained in the case of $N = 1$ are close to the solutions of the classical problem for $N = \infty$ (in the case under consideration, this is the Maxwell formula; see Section 2 below). For example, in the case where $\varepsilon_1 = 80$, $\varepsilon_2 = 3$, and $R/a = 1/3$, we have $\varepsilon_{\text{ef}}^{(1)} = 63.4350$ and $\varepsilon_{\text{ef}}^{(\infty)} = 63.6121$.

2. By setting $N = \infty$ in the problem posed in Section 1 (the centers of inclusions are placed at the points of an infinite cubic lattice), we arrive at the classical Maxwell problem. In this case, due to the symmetry of the problem, all the inclusions have the same density of surface charges σ, which satisfies an integral equation

$$\sigma(x) + \lambda\sum_{\vec{k}}\int_{\partial F_0}\sigma(y)\frac{\cos(\vec{x}-\vec{y}-a\vec{k},\vec{n}_x)}{2\pi\left|\vec{x}-\vec{y}-a\vec{k}\right|^2}dS_y = -2\lambda\varepsilon_0\vec{E}\cdot\vec{n}_x, \qquad x\in\partial F_0 \quad (11)$$

and condition (6); here, $\sum_{\vec{k}} = \sum_{k_1=-\infty}^{\infty}\sum_{k_2=-\infty}^{\infty}\sum_{k_3=-\infty}^{\infty}$. In this case, the density of surface charges σ possesses the following symmetry properties:

$$\sigma(T_i x) = \begin{cases} \sigma(x), & i = 1,2, \\ -\sigma(x), & i = 3. \end{cases}$$

By applying the same procedure as in Section 1 to Eq. (11) and keeping only the first two terms of the expansion, we arrive at the following expressions for the density of surface charges, the dipole moment of the inclusions, and the potential of electric field near the bodies:

$$\sigma(x) = \frac{3(\nu-1)\varepsilon_0 \vec{E}\cdot\vec{n}_x}{2+\nu+v(1-\nu)},$$

$$p = \frac{3(\nu-1)\varepsilon_0 E v a^3}{2+\nu+v(1-\nu)},$$

$$\varphi(x) = \begin{cases} -\frac{1}{\varepsilon_1}\frac{3\vec{E}\cdot\vec{x}}{2+\nu+v(1-\nu)}, & |\vec{x}| \le R, \\ \frac{1}{2\varepsilon_1}\frac{\vec{E}\cdot\vec{x}}{2+\nu+v(1-\nu)}\left((\nu-1)\frac{R^3}{|\vec{x}|^3} - 2 - \nu\right), & R \le |\vec{x}| << \frac{a}{2}. \end{cases}$$

In this approximation, we arrive at the classical Maxwell formula for the effective dielectric permittivity,

$$\varepsilon_{\text{ef}}^{(\infty)} = \varepsilon_1\left(1 + \frac{3(\nu-1)v}{2+\nu+(1-\nu)v}\right).$$

Taking the next two terms of expansion into account, we obtain the well-known Rayleigh formula

$$\varepsilon_{\text{ef}}^{(\infty)} = \varepsilon_1\left(1 + \frac{3(\nu-1)v}{2+\nu+(1-\nu)v+1.3091\frac{(1-\nu)^2}{\nu+4/3}v^{10/3}}\right).$$

Taking several more terms of the expansion into account, we get the following expression for the effective dielectric permittivity of this system that generalizes the Maxwell and Rayleigh formulas [4]:

$$\varepsilon_{\text{ef}}^{(\infty)} = \varepsilon_1\left\{3v\left[\frac{\nu+2}{\nu-1} - v + \frac{\frac{49}{9}\frac{1-\nu}{\nu+4/3}\left(\frac{R}{a}\right)^{10}\left(L_1 - \frac{11}{120}\frac{1-\nu}{\nu+6/5}L_2L_3\left(\frac{R}{a}\right)^{11}\right)^2}{1+\frac{1}{18}\frac{1-\nu}{\nu+4/3}L_2\left(\frac{R}{a}\right)^7 - \frac{14399}{16}\frac{(1-\nu)^2}{(\nu+4/3)(\nu+6/5)}L_3^2\left(\frac{R}{a}\right)^{18}}\right.\right.$$
$$\left.\left.+\frac{11}{8100}\frac{1-\nu}{\nu+6/5}L_2^2\left(\frac{R}{a}\right)^{14} + \frac{165}{8}\frac{1-\nu}{\nu+8/7}L_3^2\left(\frac{R}{a}\right)^{18}\right]^{-1} + 1\right\}; \quad (12)$$

here,

$$L_1 = \sum_{\vec{k}}{}' \frac{|\vec{k}|^4 - 15k_1^2k_2^2}{|\vec{k}|^9}, \qquad \sum_{\vec{k}}{}' = \sum_{\substack{\vec{k} \\ |\vec{k}|\neq 0}},$$

$$L_2 = 45\sum_{\vec{k}}{}' \frac{|\vec{k}|^6 + 84k_1^2k_2^2|\vec{k}|^2 - 231k_1^2k_2^4}{|\vec{k}|^{13}},$$

$$L_3 = \sum_{\vec{k}}{}' \frac{|\vec{k}|^8 - 41k_1^2k_2^2|\vec{k}|^4 + 169k_1^2k_2^4|\vec{k}|^2 - 195k_1^2k_2^6}{|\vec{k}|^{17}}.$$

By computing the sums of the convergent series L_1, L_2, and L_3 with an accuracy of 10^{-5}, we get

$$\varepsilon_{\text{ef}}^{(\infty)} = \varepsilon_1 \left\{ 1 + 3v \left[\frac{\nu+2}{\nu-1} - v + \frac{1.3091\frac{1-\nu}{\nu+4/3}v^{10/3}\left(1 - 0.1173\frac{1-\nu}{\nu+6/5}v^{11/3}\right)^2}{1 + 0.4054\frac{1-\nu}{\nu+4/3}v^{7/3} + 6.6568\frac{(1-\nu)^2}{(\nu+4/3)(\nu+6/5)}v^6} \right.\right.$$
$$\left.\left. + 0.0723\frac{1-\nu}{\nu+6/5}v^{14/3} + 0.15256\frac{1-\nu}{\nu+8/7}v^6 \right]^{-1} \right\}.$$

3. Consider the problem posed in Section 2 for the volume-centered cubic lattice (the bodies of the same geometry with dielectric permittivity ε_3 are placed at the points of intersection of the diagonals of elementary cubes). In this case, we have the following expression for the effective dielectric permittivity [5]:

$$\varepsilon_{\text{ef}} = \varepsilon_1 \left(1 - \frac{3\left(\frac{1-\nu_1}{2+\nu_1} + \frac{1-\nu_2}{2+\nu_2}\right)v}{1 + v\left(\frac{1-\nu_1}{2+\nu_1} + \frac{1-\nu_2}{2+\nu_2}\right)} \right),$$

where ν_1 and ν_2 are the ratios of the corresponding dielectric permittivities.

4. All the results presented in this paper also hold for conductive inclusions; one should only pass to the limit as $\varepsilon_2 \to \infty$ ($\lambda \to -1$, $\nu \to \infty$). However, the correctness of this transition should be justified because Theorem 1 does not hold in this case. Consider the same array of bodies as in Section 2.

In this case, a potential is determined as a solution of the Laplace equation (1) with the following boundary conditions on the surfaces of the inclusions:

$$\varphi(x) = \text{const}, \qquad x \in \bigcup_{\vec{k}} \partial F_{\vec{k}}, \tag{13}$$

$$\int_{\partial F_{\vec{k}}} \left(\frac{\partial \varphi}{\partial \vec{n}}\right)(x)\, dS_x = 0. \tag{14}$$

We seek a solution of the problem (1), (13), (14) in the form of a sum of the potential of the external field and a potential of charges induced on the surfaces of the inclusions. Acting as in Section 2, we arrive at the following system of equations:

$$\sigma(x) - \sum_{\vec{k}} \int_{\partial F_0} \sigma(y) \frac{\cos(\vec{x} - \vec{y} - a\vec{k}, \vec{n}_x)}{2\pi \left|\vec{x} - \vec{y} - a\vec{k}\right|^2} dS_y = 2\varepsilon_0 \vec{E} \cdot \vec{n}_x, \qquad x \in \partial F_0 \tag{15}$$

with condition (6). The density of surface charges σ possesses the same properties of symmetry as in Section 2.

Consider the Banach space $B_1(\partial F_0)$ of bounded functions $\chi(x)$ given on the surface ∂F_0 and satisfying condition (6). A norm in the space $B_1(\partial F_0)$ is defined as follows: $\|\chi\| = \sup_{x\in\partial F_0} |\chi(x)|$. In $B_1(\partial F_0)$, equation (15) (transformed as in Section 2) takes the form

$$(I - A)\sigma - \tilde{A}\sigma = \tilde{\sigma}_0, \tag{16}$$

where $\tilde{\sigma}_0 = 2\varepsilon_0 \vec{E}\cdot\vec{n}_x$ and linear operators A and $\tilde{A}$ are defined as follows:

$$(A\chi)(x) = \int\limits_{\partial F_0} \chi(y)\frac{\cos(\vec{x}-\vec{y},\vec{n}_x)}{2\pi|\vec{x}-\vec{y}|^2}\,dS_y,$$

$$(\tilde{A}\chi)(x) = \int\limits_{\partial F_0} \chi(y)K(x,y)\,dS_y + \frac{2}{3a^3}n_{x_3}\int\limits_{\partial F_0}\chi(y)y_3\,dS_y.$$

It is known that [6]

$$\|A\| \le 1 \qquad \text{and} \qquad \|\tilde{A}\| \le C_1 v,$$

where C_1 is a constant.

Theorem 3. *If $C_2 v \le 1$, where C_2 is a constant, then the operator equation* (16) *possesses a unique solution in the space $B_1(\partial F_0)$.*

Proof. Clearly, Theorem 3 holds if the operator $I - A$ has bounded inverse in the space $B_1(\partial F_0)$.

First, let us show that the operator A is completely continuous. Consider a bounded set $D(f\,|\,f \in B_1(\partial F_0),\ \|f\| \le M < \infty)$. According to the definition of the operator A and the Arzelà theorem, it suffices to show that the functions $(Af)(x)$ are uniformly bounded and equicontinuous.

Taking into account the convexity of the surface ∂F_0 and the fact that (see, e.g., [7])

$$\int\limits_{\partial F_0}\frac{\cos(\vec{x}-\vec{y},\vec{n}_x)}{2\pi|\vec{x}-\vec{y}|^2}\,dS_y = \begin{cases} 1, & x\in\partial F_0,\\ 0, & x\in\mathbb{R}^3\backslash(F_0\cup\partial F_0),\end{cases}$$

we get

$$|(Af)(x)| = \left|\int\limits_{\partial F_0} f(y)\,\frac{\cos(\vec{x}-\vec{y},\vec{n}_x)}{2\pi|\vec{x}-\vec{y}|^2}\,dS_y\right| \le \|f\|\int\limits_{\partial F_0}\frac{\cos(\vec{x}-\vec{y},\vec{n}_x)}{2\pi|\vec{x}-\vec{y}|^2}\,dS_y \le M$$

for any $f \in D$ and $x \in \partial F_0$. Therefore, the functions $(Af)(x)$, $f \in D$ are uniformly bounded.

Let us show that the functions $(Af)(x)$, $f \in D$, are equicontinuous, i.e., that, for any $\varepsilon > 0$, one can find $\delta > 0$ (depending only on ε) such that, for an arbitrary function $f \in D$ and any $\vec{x}_1 \in \partial F_0$ and $\vec{x}_2 \in \partial F_0$, satisfying the inequality $|\vec{x}_1 - \vec{x}_2| < \delta$, the following relation holds:

$$|(Af)(x_1) - (Af)(x_2)| < \varepsilon. \tag{17}$$

The kernel $\frac{\cos(\vec{x}-\vec{y},\vec{n}_x)}{2\pi|\vec{x}-\vec{y}|^2}$ is polar. Therefore, for any $\varepsilon > 0$, there exists a degenerate kernel $\mathcal{P}(x,y)$ such that

$$\max_{\vec{x}\in\partial F_0} \int\limits_{\partial F_0} \left| \frac{\cos(\vec{x}-\vec{y},\vec{n}_x)}{2\pi|\vec{x}-\vec{y}|^2} - \mathcal{P}(x,y) \right| dS_y < \frac{\varepsilon}{3M}.$$

Furthermore, since the degenerate kernel $\mathcal{P}(x,y)$ is continuous for any $\varepsilon > 0$, one can find $\delta > 0$ such that the inequality $|\mathcal{P}(x_1,y) - \mathcal{P}(x_2,y)| < \varepsilon/3M$ holds whenever $|\vec{x}_1 - \vec{x}_2| < \delta$.

The inequalities obtained above prove the required estimate (17) and, hence, the equicontinuity of the functions $(Af)(x)$. Thus, we have shown that the operator A is completely continuous.

It remains to prove that the homogeneous equation $(I-A)f = 0$ only possesses the trivial solution $f \equiv 0$ in the space $B_1(\partial F_0)$, i.e., to show that if a continuous bounded function $\chi(x)$, $x \in \partial F_0$, satisfies condition (6), then the homogeneous equation

$$\chi(x) - \int\limits_{\partial F_0} \chi(y) \frac{\cos(\vec{x}-\vec{y},\vec{n}_x)}{2\pi|\vec{x}-\vec{y}|^2}\, dS_y = 0 \tag{18}$$

only possesses a trivial solution $\chi(x) \equiv 0$.

We prove this by contradiction. Assume that a function $\chi_0(x) \not\equiv 0$ is a solution of equation (18).

We construct a function

$$u(x) = \frac{1}{4\pi\varepsilon_0} \int\limits_{\partial F_0} \frac{\chi_0(y)}{|\vec{x}-\vec{y}|}\, dS_y,$$

which has the sense of a potential of a field generated in a vacuum by charges located on the surface ∂F_0 with the density $\chi_0(x)$.

The limiting values of the simple-layer potential $u(x)$ when approaching the surface ∂F_0 from inside and outside are given by

$$\left(\frac{\partial u}{\partial \vec{n}}\right)_+ (x) = \frac{\chi_0(x)}{2\varepsilon_0} - \frac{1}{4\pi\varepsilon_0} \int\limits_{\partial F_0} \chi_0(y) \frac{\cos(\vec{x}-\vec{y},\vec{n}_x)}{|\vec{x}-\vec{y}|^2}\, dS_y, \qquad x \in \partial F_0,$$

$$\left(\frac{\partial u}{\partial \vec{n}}\right)_- (x) = -\frac{\chi_0(x)}{2\varepsilon_0} - \frac{1}{4\pi\varepsilon_0} \int\limits_{\partial F_0} \chi_0(y) \frac{\cos(\vec{x}-\vec{y},\vec{n}_x)}{|\vec{x}-\vec{y}|^2}\, dS_y, \qquad x \in \partial F_0.$$

By virtue of (18), $\left(\frac{\partial u}{\partial \vec{n}}\right)_+ (x) = 0$ and, hence,

$$\left(\frac{\partial u}{\partial \vec{n}}\right)_- (x) = -\frac{\chi_0(x)}{\varepsilon_0}, \qquad x \in \partial F_0.$$

Since the function $u(x)$ is harmonic for $x \in F_0$, we have

$$\int_{F_0} |\nabla u(x)|^2 \, dV_x = \int_{\partial F_0} u(x) \left(\frac{\partial u}{\partial \vec{n}}\right)_+ (x) \, dS_x.$$

Therefore, $u(x) = \text{const}$, $x \in F_0$.

On the other hand, the function $u(x)$ is harmonic in the region $\mathbb{R}^3 \backslash (F_0 \cup \partial F_0)$ and, hence,

$$\int_{\mathbb{R}^3 \backslash (F_0 \cup \partial F_0)} |\nabla u(x)|^2 \, dV_x = -\int_{\partial F_0} u(x) \left(\frac{\partial u}{\partial \vec{n}}\right)_- (x) \, dS_x.$$

Since a simple-layer potential is a continuous function, we have $u(x) = \text{const}$ for $x \in \partial F_0$. Furthermore, taking condition (6) into account, we get $u(x) = \text{const}$ for $x \in \mathbb{R}^3 \backslash (F_0 \cup \partial F_0)$. Consequently, by virtue of continuity of a simple-layer potential, we have $u(x) = \text{const}$ for any $x \in \mathbb{R}^3$. Thus,

$$\chi_0(x) = -\varepsilon_0 \left(\frac{\partial u}{\partial \vec{n}}\right)_- (x) \equiv 0,$$

i.e., we arrive at a contradiction. Theorem 3 is proved. □

5. The method presented in this paper can be applied to the cases of various arrays (both regular and random) of inclusions of various forms. Moreover, it can be used for determining characteristics of porous media. This method is also applicable for describing the processes of diffusion and filtration in various media.

This work was partially supported by the Ukrainian State Committee on Science and Technology.

References

[1] Malyshev, P. V., Gorunovich V. V., Determination of screened potentials and distribution functions for the systems of charged particles near porous membranes. Dokl. Akad. Nauk Ukr. SSR, Ser. A No. 8 (1986), 56–61.

[2] Malyshev, D. V., Malyshev, P. V., Perturbation of a uniform elctrostatic field by a dynamic membrane. Dokl. Akad. Nauk Ukr. SSR, Ser. A No. 2 (1986), 44–48.

[3] Petrina, D. Ya., On solution of a classical electrostatic problem by using a subtraction procedure. Zh. Vychisl. Mat. Mat. Fiz. 59 (1984), 104–116.

[4] Malyshev, D. V., Malyshev, P. V., Effective dielectric permittivity in the periodic system of inclusions. Dokl. Akad. Nauk Ukr. SSR, Ser. A No. 12 (1987), 48–53.

[5] — Determination of the effective characteristics for the systems with complex structure. In: Mathematical Problems in the Theory of Systems of Charged Particles in Inhomogeneous Media, 63–76. Institute of Mathematics, Ukrainian Academy of Sciences, Kiev 1988.

[6] Malyshev, D. V., Perturbations of Electric Fields in the Regions with Conductive Inclusions. Preprint No.40, Institute of Mathematics, Ukrainian Academy of Sciences, Kiev 1992.

[7] Vladimirov, V. S., Equations of Mathematical Physics. Nauka, Moscow 1976.

Renormalization on fractals

Volker Metz

Abstract. The construction of "Laplacians" on nested fractals via Dirichlet forms relies on the solution of a finite dimensional, nonlinear eigenvalue problem which can be interpreted as a renormalization problem. We formulate this problem by a map acting on a cone of Dirichlet forms and fit it into a suitable fixed point theory, namely, Hilbert's projective metric on cones. This improves previous approaches by Brouwer's fixed point theorem, because we obtain information beyond mere existence of solutions. The results and concepts are illustrated by an example, the Vicsek snowflake. In this example we can, in fact, apply Banach's fixed point theorem locally to the Hilbert metric set-up. In another non-nested example the fixed point is not unique.

1991 Mathematics Subject Classification: 31C25.

1. Introduction

The desire to study fractals arose in physics in the early 1980s. Fractals may provide good models for disordered media, although no natural media has a fractal structure on the submolecular scale [5]. One way to study fractals and their physical properties is to consider their "Laplacians", analogously to the investigation of Riemannian manifolds and their Laplace–Beltrami operators.

The only type of fractals for which the existence of "Laplacians" has been proved is the class of nested fractals or generalizations [16, 9]. Therefore, we restrict our attention to nested or closely related fractals. To keep the notation short we consider an example, the Vicsek snowflake. Since fractals are not manifolds, we have to construct their "Laplacians" by hand, that is, for example by the scaled limit of a sequence of discrete Dirichlet forms as invented by Fukushima and Shima [10]. The Dirichlet operator of the limiting form is then our "Laplacian". The basic difficulty is the existence of a suitable scaling, in other words, a renormalization problem. This is the topic of section 2 together with a brief review of the properties of the limiting model.

The problem of uniqueness and approximation of "Laplacians" on nested fractals is still mostly open. This is due to the use of Brouwer's fixed point theorem. In section 3 we will introduce a more appropriate fixed point theory — Hilbert's projective metric on cones of Dirichlet forms.

In section 4 the usefulness of this approach will be illustrated for the Vicsek snowflake and a non-nested variant. In fact, we can apply Banach's fixed point theorem locally in the Vicsek snowflake case. Further research in this direction will eventually lead to stronger existence, uniqueness, and approximation results by the use of Hilbert's metric and Banach's fixed point theorem. On the non-nested variant the fixed point is not unique.

The theorems of this article have very different scopes of application. To avoid lengthy statements I left out those assumptions which are fulfilled for nested fractals. So, all the results in this paper are at least valid for nested fractals except those in section 4. For more precise statements see [19].

2. Dirichlet forms and renormalization

We will sketch the construction of Dirichlet forms on a fractal and define an eigenvalue problem which guarantees the success of this construction.

The abstract definition of *nested fractals* is given in [16]. We restrict ourselves to an example, the Vicsek snowflake. It is a nested fractal constructed by the five similitudes $\psi_1, \dots, \psi_5 : \mathbb{R}^2 \to \mathbb{R}^2$, given by

$$\psi_1(x) := \frac{x}{3} + \frac{2}{3}(1,1), \qquad \psi_2(x) := \frac{x}{3} + \frac{2}{3}(0,1), \qquad \psi_3(x) := \frac{x}{3},$$
$$\psi_4(x) := \frac{x}{3} + \frac{2}{3}(1,0), \qquad \psi_5(x) := \frac{x}{3} + \frac{1}{3}(1,1).$$

They contract the unit square $[0,1]^2$ by $\frac{1}{3}$ and arrange five such small squares in a chessboard pattern. We define $\Psi(A) := \bigcup_{i=1}^{5} \psi_i(A)$ for $A \subset \mathbb{R}^2$. Then the Vicsek snowflake X can be defined by $X := \bigcap_{n\in\mathbb{N}} \Psi^n([0,1]^2)$. Each ψ_i has a unique fixed point x_i. Set $F_0 := \{x_1, \dots, x_4\}$, that is, F_0 consists of the corner points of $[0,1]^2$. Our sequence of "grids" is constructed by $F_n := \Psi^n(F_0)$ for all $n \in \mathbb{N}$. They approximate X because $F_\infty := \bigcup_{n\in\mathbb{N}} F_n$ is dense in X with respect to the Euclidean topology on $\mathbb{R}^2$.

A frequently used construction of a "Laplacian" on X due to Fukushima and Shima is to define a local, regular, and irreducible Dirichlet form $(\mathcal{A}, \mathcal{D}(\mathcal{A}))$ on X as the scaled limit of discrete Dirichlet forms $(\mathcal{E}_n, \mathcal{D}(\mathcal{E}_n))_{n\in\mathbb{N}}$ on $(F_n)_{n\in\mathbb{N}}$ [10]. We call a function $c : F_0^2 \to \mathbb{R}_+$, which is symmetric and 0 on the diagonal, a *conductance* on F_0. Each conductance defines a discrete Dirichlet form $(\mathcal{E}_0, \mathcal{D}(\mathcal{E}_0))$ on F_0 by

$$\begin{aligned} \mathcal{D}(\mathcal{E}_0) &:= \{f \mid f : F_0 \to \mathbb{R}\}, \\ \mathcal{E}_0(f,f) &:= \sum_{x,y \subset F_0} (f(y) - f(x))^2 c(x,y) \,. \end{aligned} \tag{2.1}$$

By (2.1) we can define a map Ξ transforming conductances into Dirichlet forms on F_0 which vanish on constants. It is one-to-one and onto [18: Section 1]. A conductance on F_0 can be visualized by a complete graph with vertices F_0 and weights $c(x,y)$ on the edges $\{x,y\} \subset F_0$. Such a weighted graph is also called an *electric network*. It helps to understand the analogy between $\Psi(F_n) = F_{n+1}$ and the construction of the Dirichlet form $(\mathcal{E}_{n+1}, \mathcal{D}(\mathcal{E}_{n+1}))$ on F_{n+1},

$$\begin{aligned} \mathcal{D}(\mathcal{E}_{n+1}) &:= \{f \mid f : F_{n+1} \to \mathbb{R}\}, \\ \mathcal{E}_{n+1}(f,f) &:= \sum_{i=1}^{5} \mathcal{E}_n(f \circ \psi_i, f \circ \psi_i) \end{aligned} \tag{2.2}$$

for all $n \in \mathbb{N}$. Let $C(X,\mathbb{R})$ denote the set of continuous functions from X to $\mathbb{R}$. The main difficulty is to find a *scaling constant* $\gamma > 0$ such that

$$\begin{aligned} \mathcal{A}(f,f) &:= \lim_{n\to\infty} \gamma^{-n}\mathcal{E}_n(f,f), \\ \mathcal{D}(\mathcal{A}) &:= \{f \in C(X,\mathbb{R}) \mid \mathcal{A}(f,f) \text{ exists}\} \end{aligned} \tag{2.3}$$

defines a local, regular, and irreducible Dirichlet form on $L^2(X,\mu)$, where μ is the normalized $\frac{\ln 5}{\ln 3}$-dimensional Hausdorff measure on X. This is possible if a certain eigenvalue problem can be solved [2: Section 2]. In the remainder of this section we consider this problem.

Like in [2] we define the *decimated form* $\widetilde{\mathcal{E}}_{n+1}$ of $\mathcal{E}_{n+1}$ by

$$\begin{aligned} \mathcal{D}(\widetilde{\mathcal{E}}_{n+1}) &:= \mathcal{D}(\mathcal{E}_n), \\ \widetilde{\mathcal{E}}_{n+1}(f,f) &:= \inf\{\mathcal{E}_{n+1}(g,g) : g \in \mathcal{D}(\mathcal{E}_{n+1}), g|_{F_n} = f\} \end{aligned} \tag{2.4}$$

for all $n \in \mathbb{N}$. This operation is sometimes called "coarse-graining renormalization" because it eliminates vertices in an electric network [11]. It is well known in statistics and the theory of electric networks as "shorting of operators" or in linear algebra as a version of the "Schur complement" [7]. A related concept was studied in algebraic potential theory by Eriksson-Bique and Leutwiler [8]. Note that

$$\widetilde{\mathcal{E}}_{n+1}(f,f) = \mathcal{E}_{n+1}(H_{F_{n+1}\setminus F_n} f, H_{F_{n+1}\setminus F_n} f),$$

where $H_{F_{n+1}\setminus F_n}$ is the harmonic kernel of $\mathcal{E}_{n+1}$ on $F_{n+1} \setminus F_n$. The construction of a scaled limit in (2.3) is straightforward if we are able to arrange that for some $\gamma > 0$ and all $f \in \mathcal{D}(\mathcal{E}_0)$,

$$\widetilde{\mathcal{E}}_{n+1} = \gamma\mathcal{E}_n, \quad H_{F_{n+1}\setminus F_0} f|_{F_n} = H_{F_n\setminus F_0} f \tag{2.5}$$

for all $n \in \mathbb{N}$. Here $H_{F_m\setminus F_0}$ is the harmonic kernel of $\mathcal{E}_m$ on $F_m \setminus F_0$. This problem can be reduced to the case $n = 0$, since X is *self-similar* and *finitely ramified* [15: Section 4]. The last term was coined by physicists and means here that X can be disconnected by removing only finitely many points, namely, points of our "grids" $(F_n)_{n\in\mathbb{N}}$. The set $X \setminus F_m$ for example consists of connected components which are all scaled down copies of $X \setminus F_0$. The invariance of harmonic kernels in (2.5) is a consequence of $\widetilde{\mathcal{E}}_{n+1} = \gamma\mathcal{E}_n$. Therefore, our eigenvalue problem looks as follows: The domain is the cone $\mathbb{D}$ of all discrete Dirichlet forms $\mathcal{E}_0$ on F_0 defined by (2.1) and a conductance on F_0. We define a map Λ by $\Psi(\mathcal{E}_0) := \mathcal{E}_1$, $\Phi(\mathcal{E}_1) := \widetilde{\mathcal{E}}_1$, and $\Lambda(\mathcal{E}_0) := \Phi \circ \Psi(\mathcal{E}_0)$. We immediately see that Ψ is linear and we will see that Φ fails to be additive. Now, the nonlinear, finite dimensional *eigenvalue problem* is

$$\Lambda(\mathcal{F}) = \gamma\mathcal{F} \text{ for a } \gamma > 0 \text{ and an irreducible } \mathcal{F} \in \mathbb{D}\ , \tag{2.6}$$

where *irreducible* means: $\mathcal{F}(f,f) = 0$ if and only if f is constant [cf. 19]. In [16: Theorem V.5], Lindstrøm proved the existence of solutions of (2.6) for nested fractals by Brouwer's fixed point theorem. The questions of the uniqueness and the approximation of solutions are open except for some examples or classes of examples [cf. 3 and the references therein].

We will say that *renormalization works* on $\mathbb{D}$ (for $\gamma^{-1}\Lambda$) if there exists a $\gamma > 0$ such that for all irreducible $\mathcal{E} \in \mathbb{D}$ there exists an irreducible $\mathcal{E}_\infty \in \mathbb{D}$ which satisfies

$$\lim_{n\to\infty} \gamma^{-n}\Lambda^n(\mathcal{E}) = \mathcal{E}_\infty \text{ and } \Lambda(\mathcal{E}_\infty) = \gamma\mathcal{E}_\infty \ . \tag{2.7}$$

When this is true we can use any irreducible $\mathcal{E} \in \mathbb{D}$ to define the scaled limit in (2.3). On the other hand, we could use $\mathcal{E}_\infty$ in (2.3). In both cases we end up in the same Dirichlet form on X. This is the meaning of renormalization in potential theoretic terms [20: Theorem 8.16].

In the case of the Sierpinski gasket, which is also a nested fractal, the properties of the limiting Dirichlet form have been studied extensively. Let μ again denote the appropriate Hausdorff measure on the fractal. The heat kernels don't have Gaussian estimates [2: Section 5], harmonic functions are not differentiable on a dense subset but μ-almost everywhere differentiable with first derivative 0 [17: Satz 7.7 & 7.19], and the energy measures are singular to μ but they are still absolutely continuous with respect to another measure on the fractal constructed by Kusuoka [14].

3. Hilbert's projective metric on cones

We will firstly investigate Λ and the cones it is acting on. Secondly, we apply a fixed point theory to our eigenvalue problem (2.6) which seems to be more appropriate than Brouwer's fixed point theorem [cf. 19]. The new approach is a nonlinear, infinite dimensional generalization of the Perron–Frobenius theory and is mostly termed Hilbert's projective metric on cones.

To simplify the model we now restrict ourselves to "nice" subcones of $\mathbb{D}$. Our fractal X has a non-trivial *symmetry group* $\mathcal{G}$. When we choose our conductances on F_0 to be $\mathcal{G}$-invariant we get a smaller cone $\mathbb{D}_\mathcal{G} \subset \mathbb{D}$ of Dirichlet forms on F_0. We define the cone $\mathbb{P}_\mathcal{G} \supset \mathbb{D}_\mathcal{G}$ of *positive semidefinite forms* by

$$\mathbb{P}_\mathcal{G} := \{\mathcal{E} - \mathcal{F} \mid \mathcal{E},\mathcal{F} \in \mathbb{D}_\mathcal{G},\ \mathcal{E}(f,f) \geq \mathcal{F}(f,f) \text{ for all } f: F_0 \to \mathbb{R}\} \ .$$

This is a subset of the Banach space

$$\mathbb{B}_\mathcal{G} := \mathbb{D}_\mathcal{G} - \mathbb{D}_\mathcal{G} \ ,$$
$$\|\mathcal{E}\| := \sup\{\mathcal{E}(f,f)^{\frac{1}{2}} \mid f: F_0 \to \mathbb{R},\ \|f\|_2 = 1\} \ ,$$

where $\|f\|_2$ is the Euclidean norm of $f \in \mathcal{D}(\mathcal{E}_0)$. Now $\mathcal{E} \in \mathbb{D}_\mathcal{G}$ is irreducible if and only if $\mathcal{E}$ is also an element of the interior $\mathbb{P}^\circ_\mathcal{G}$ of $\mathbb{P}_\mathcal{G}$. The cone $\mathbb{P}_\mathcal{G}$ defines a *partial ordering* $\leq$ on $\mathbb{B}_\mathcal{G}$ by: $\mathcal{E} \leq \mathcal{F}$ if and only if $\mathcal{F} - \mathcal{E} \in \mathbb{P}_\mathcal{G}$.

Proposition 3.1. Λ *has the following properties for all* $\mathcal{E}, \mathcal{F} \in \mathbb{P}_\mathcal{G}$*:*
(a) $\Lambda(\mathbb{P}_\mathcal{G}) \subset \mathbb{P}_\mathcal{G}$, $\Lambda(\mathbb{D}_\mathcal{G}) \subset \mathbb{D}_\mathcal{G}$, $\Lambda(\mathbb{P}^\circ_\mathcal{G}) \subset \mathbb{P}^\circ_\mathcal{G}$, $\Lambda(\mathbb{D}^\circ_\mathcal{G}) \subset \mathbb{D}^\circ_\mathcal{G}$*;*
(b) $\Lambda(\alpha\mathcal{E}) = \alpha\Lambda(\mathcal{E})$ *for all* $\alpha \geq 0$*;*
(c) $\mathcal{E} \leq \mathcal{F}$ *implies* $\Lambda(\mathcal{E}) \leq \Lambda(\mathcal{F})$*;*
(d) $\Lambda(\mathcal{E} + \mathcal{F}) \geq \Lambda(\mathcal{E}) + \Lambda(\mathcal{F})$*;*
(e) $\Lambda \in \mathcal{C}^1(\mathbb{P}^\circ_\mathcal{G}) \cap \mathcal{C}(\mathbb{D}_\mathcal{G})$,
where $\mathcal{C}^1$ *means continuously Fréchet differentiable.*

The results of Proposition 3.1 are mostly deduced from [1]. The complete list can be found in [19: Theorem 2.2]. Some results were also proved in [16] and [11]. In (4.1a) we will see that (3.1c) is in general not true for the semiordering induced by $\mathbb{D}_\mathcal{G}$. The proof of (3.1d) shows that Λ fails to be additive because Φ is superadditive.

Our cones $\mathbb{P}_\mathcal{G}$ and $\mathbb{D}_\mathcal{G}$ have a simple geometry.

Proposition 3.2. $\mathbb{P}_\mathcal{G}$ *and* $\mathbb{D}_\mathcal{G}$ *are polyhedral cones.*

Proof. The cone $\mathbb{D}_\mathcal{G}$ is polyhedral, since it can be characterised by nonnegative conductances in (2.1) and since Ξ is an isomorphism.

We will prove that all forms in $\mathbb{B}_\mathcal{G}$ are simultaneously diagonalizable. This will follow from showing that their operators commute [12: Theorem 1.3.19]. Consequently, $\mathbb{P}_\mathcal{G}$ is polyhedral because $\mathcal{E} \in \mathbb{P}_\mathcal{G}$ if and only if it is in $\mathbb{B}_\mathcal{G}$ and all its eigenvalues are nonnegative.

For $x, y \in F_0$ with $x \neq y$, let $H_{x,y}$ denote the hyperplane of points equidistant from x and y. The reflection in $H_{x,y}$ is denoted by $\rho_{x,y}$. By definition, the symmetry group $\mathcal{G}$ of a nested fractal contains $\{\rho_{x,y} \mid x, y \in F_0;\ x \neq y\}$ [16: Axiom 2].

Each conductance c on F_0 defines a graph $\Gamma(c)$ with vertex set F_0 and edge set $\{\{x,y\} \subset F_0 \mid c(x,y) \neq 0\}$. In this notation, $\Gamma(1)$ is the complete graph with vertices F_0. On the other hand, we can associate to each graph $\Gamma = (F_0, E)$ a conductance c which is 1 on E and 0 elsewhere. This defines a Dirichlet form on F_0 by (2.1).

The symmetry group $\mathcal{G}$ splits the edges of $\Gamma(1)$ into orbits, say $E_1, \ldots, E_l$. Each such orbit defines a graph Γ_i, a conductance c_i, and a Dirichlet form $\mathcal{E}_i$ with Dirichlet operator A_i. The extremal rays of $\mathbb{D}_\mathcal{G}$ are spanned by $\mathcal{E}_1, \ldots, \mathcal{E}_l$. So it suffices to prove that $A_1, \ldots, A_l$ commute in order to conclude that $\mathbb{P}_\mathcal{G}$ is polyhedral. We define

$$d_i := \sum_{y \in F_0} c_i(x,y) \qquad (x \in F_0,\ 1 \leq i \leq l),$$

and remark that d_i is constant because Γ_i is $\mathcal{G}$-invariant and $\mathcal{G}$ acts transitively on F_0. Let I denote the identity map on $\mathcal{D}(\mathcal{E}_0)$ and set $B_i := d_i \cdot I + A_i$ for all $1 \leq i \leq l$. Then the (A_i) commute if and only if the adjacency matrices (B_i) do.

Let $1_{\{z\}}$ denote the indicator function of $z \in F_0$ and $\langle \cdot, \cdot \rangle$ the Euclidean inner product on $\mathcal{D}(\mathcal{E}_0)$. For all $x, y \in F_0$ and $1 \leq i, j \leq l$, we conclude

$$\langle B_i B_j 1_{\{x\}}, 1_{\{y\}} \rangle = \sum_{z \in F_0} \langle B_j 1_{\{x\}}, 1_{\{z\}} \rangle \cdot \langle B_i 1_{\{z\}}, 1_{\{y\}} \rangle.$$

The latter sum can be interpreted as the total amount of paths of length 2 from x to y in $\Gamma(1)$ with the first step along Γ_j and the second along Γ_i. Because of Lindstrøm's symmetry axiom, $\rho_{x,y}$ maps these paths one-to-one onto the paths of length 2 from y to x in $\Gamma(1)$ with the first step along Γ_j and the second along Γ_i. Hence,

$$\langle B_i B_j 1_{\{x\}}, 1_{\{y\}} \rangle = \sum_{z \in F_0} \langle B_j 1_{\{y\}}, 1_{\{z\}} \rangle \cdot \langle B_i 1_{\{z\}}, 1_{\{x\}} \rangle.$$

Since B_i and B_j are symmetric, we infer $B_i B_j = B_j B_i$. □

Proposition 3.2 is a generalization of Lemma 3.1 in [3]. It allows to map $\mathbb{B}_\mathcal{G}$ linearly into $\mathbb{R}^k$, where $\mathbb{P}_\mathcal{G}$ is defined by k halfspaces.

Now we will consider special metric spaces [cf. 21]. Two forms $\mathcal{E}, \mathcal{F} \in \mathbb{P}_\mathcal{G}$ are said to be *comparable* if there exist reals $\alpha, \beta > 0$ such that $\alpha\mathcal{F} \leq \mathcal{E} \leq \beta\mathcal{F}$. This is an equivalence relation which will be written as $\mathcal{E} \sim \mathcal{F}$. The equivalence classes are called *parts*. We are interested in the part $\mathbb{P}^\circ_\mathcal{G}$. For comparable forms $\mathcal{E}, \mathcal{F}$ we define

$$\begin{aligned} m(\mathcal{E}/\mathcal{F}) &:= \sup\{\alpha > 0 \mid \alpha\mathcal{F} \leq \mathcal{E}\}, \\ M(\mathcal{E}/\mathcal{F}) &:= \inf\{\beta > 0 \mid \mathcal{E} \leq \beta\mathcal{F}\}, \\ h(\mathcal{E}, \mathcal{F}) &:= \ln \left(\frac{\mathrm{M}(\mathcal{E}/\mathcal{F})}{\mathrm{m}(\mathcal{E}/\mathcal{F})} \right). \end{aligned}$$

In case $\mathcal{E}, \mathcal{F} \in \mathbb{P}_\mathcal{G}$ are non-comparable we define $h(\mathcal{E}, \mathcal{F}) := \infty$ and, finally, $h(0,0) := 0$. The map $h = h_{\mathbb{P}_\mathcal{G}}$ differs from a metric in the following respects:

$$\begin{aligned} &h(\mathcal{E}, \mathcal{F}) = \infty \text{ if and only if } \mathcal{E} \not\sim \mathcal{F}, \\ &h(\mathcal{E}, \mathcal{F}) = 0 \text{ if and only if } \mathcal{E} = \lambda\mathcal{F} \text{ for some } \lambda > 0, \\ &h(\alpha\mathcal{E}, \beta\mathcal{F}) = h(\mathcal{E}, \mathcal{F}) \text{ for all } \alpha, \beta > 0. \end{aligned}$$

We set $S := \{\mathcal{E} \in \mathbb{P}^\circ_\mathcal{G} : \|\mathcal{E}\| = 1\}$. A theorem of Birkhoff implies that (S, h) is a complete metric space [6]. The metric h is usually called *Hilbert's projective metric* because it is constant on rays of $\mathbb{P}_\mathcal{G}$. A close relative of h is the *part metric* $p = p_{\mathbb{P}_\mathcal{G}}$ defined by

$$p(\mathcal{E}, \mathcal{F}) := \max(\ln M(\mathcal{E}/\mathcal{F}), \ln M(\mathcal{F}/\mathcal{E}))$$

for all $\mathcal{E}, \mathcal{F} \in \mathbb{P}^\circ_\mathcal{G}$ and extended in the same way as h. Results of Thompson or Bauer and Baer imply that $(\mathbb{P}^\circ_\mathcal{G}, p)$ is a complete metric space [23, 4]. We get the following theorem:

Theorem 3.3. *Let $\mathcal{E}, \mathcal{F} \in \mathbb{P}^\circ_\mathcal{G}$ and $\alpha, \beta > 0$. Then*
(a) $h(\mathcal{E}, \mathcal{F}) \geq h(\Lambda(\mathcal{E}), \Lambda(\mathcal{F}))$, $p(\mathcal{E}, \mathcal{F}) \geq p(\Lambda(\mathcal{E}), \Lambda(\mathcal{F}))$.

(b) $\Lambda(\mathcal{E}) = \alpha\mathcal{E}, \Lambda(\mathcal{F}) = \beta\mathcal{F}$ *implies* $\alpha = \beta$.
(c) $\exists\gamma > 0 \exists\mathcal{J} \in \mathbb{D}_{\mathcal{G}}^{\circ} : \Lambda(\mathcal{J}) = \gamma\mathcal{J}$.
(d) $\mathcal{J}$ *is unique (up to a constant multiple) if the Fréchet derivative of* Λ *in* $\mathcal{J}$ *has the property that its* γ*-eigenspace has dimension* 1.
(e) *The fixed point set* $\mathrm{Fix} := \{\mathcal{E} \in \mathbb{P}_{\mathcal{G}}^{\circ} \mid \gamma^{-1}\Lambda(\mathcal{E}) = \mathcal{E}\}$ *is a deformation retract of* $\mathbb{P}_{\mathcal{G}}^{\circ}$. *Especially,* Fix *is simply and arcwise connected.*
(f) *Set* $\widetilde{\Lambda}(\mathcal{E}) := \frac{\Lambda(\mathcal{E})}{\|\Lambda(\mathcal{E})\|}$ *and let* $\mathcal{F} \in S$ *be a fixed point of* $\widetilde{\Lambda}$. *Assume the existence of reals* $\varepsilon > 0$ *and* $0 < \lambda < 1$ *such that*

$$h(\widetilde{\Lambda}(\mathcal{E}), \mathcal{F}) \leq \lambda h(\mathcal{E}, \mathcal{F}) \quad (\mathcal{E} \in B_{\varepsilon}(\mathcal{F})),$$

where $B_{\varepsilon}(\mathcal{F}) = \{\mathcal{E} \in S \mid h(\mathcal{E}, \mathcal{F}) \leq \varepsilon\}$. *Then for every* $\mathcal{E} \in S$, *one has* $\lim_{n\to\infty} \widetilde{\Lambda}^n(\mathcal{E}) = \mathcal{F}$, *and if* M *is any compact subset* M *of* S, *there exist constants* $\delta \geq 0$ *and* $0 < \rho < 1$ *such that*

$$h(\widetilde{\Lambda}^n(\mathcal{E}), \mathcal{F}) \leq \delta\rho^n h(\mathcal{E}, \mathcal{F}) \quad (\mathcal{E} \in M, n \geq 1).$$

(g) *For each* $\mathcal{F} \in \mathbb{P}_{\mathcal{G}}^{\circ}$ *there exists an* $m_{\mathcal{F}} \in \mathbb{N}$ *with* $1 \leq m_{\mathcal{F}} \leq 2^k(k!)$ *and there exists an* $\mathcal{F}_{\infty} \in \mathbb{P}_{\mathcal{G}}^{\circ}$ *such that renormalization works for* $(\gamma^{-1}\Lambda)^{m_{\mathcal{F}}}$ *on* $\mathbb{P}_{\mathcal{G}}^{\circ}$.
(h) *Assume that* Λ *is ray-contracting, that is,* $h(\Lambda^n(\mathcal{E}), \Lambda^n(\mathcal{F})) = h(\mathcal{E}, \mathcal{F})$ *for all* $n \in \mathbb{N}$ *always implies that* $\mathcal{E}$ *and* $\mathcal{F}$ *lie on the same ray. Then renormalization works on* $\mathbb{P}_{\mathcal{G}}^{\circ}$.

Proof. (a), (b), (d), (e) and (f) are general consequences of the theory of Hilbert's metric on cones which can for example be found in [21]. Statement (e) relies among others on Proposition 3.2. The result (b) was proven independently in [11]. A theorem of Lindstrøm which is translated into our setup gives (c) [16: Theorem V.5]. Theorem 3.2 of Krause and Nussbaum in [13] and our Proposition 3.2 imply (g). Statement (h) can be deduced from [22: Theorem 3.5] or [20: Theorem 8.7]. □

We remark that (3.3a) is not in general true for $p_{\mathbb{D}_{\mathcal{G}}}$ by (4.1b). The statements (3.3g & h) do not imply uniqueness of the fixed point. In the case of the non-nested Vicsek snowflake of section 4, (h) was used to prove that renormalization works even on $\mathbb{P}$ and not only on $\mathbb{P}_{\mathcal{G}}$ [20].

4. The Vicsek snowflake

In this section, we analyse the Vicsek snowflake in more detail to illustrate the abstract results in section 3. For comparable results, see [3].

The Vicsek snowflake has only two edge orbits, the sides E_1 and the diagonals E_2 of $[0,1]^2$. Let $\alpha \in \mathbb{R}$ be a weight on E_1 and $\beta \in \mathbb{R}$ a weight on E_2. By the isomorphism Ξ derived from formula (2.1) we get

$$\mathbb{B}_{\mathcal{G}} \cong \mathbb{R}^2,$$
$$\mathbb{D}_{\mathcal{G}} \cong \mathbb{R}_+^2 = \{(\alpha, \beta) \in \mathbb{R}^2 \mid 0 \leq \alpha, 0 \leq \beta\}.$$

A Dirichlet form with conductivity values $\alpha, \beta \geq 0$ has the spectrum 0, 4α, and two times $2(\alpha+\beta)$. Hence,

$$\mathbb{P}_{\mathcal{G}} \cong \mathbb{K} := \{(\alpha,\beta) \in \mathbb{R}^2 \mid 0 \leq \alpha, 0 \leq \alpha+\beta\}.$$

Let $h_{\mathbb{K}}$ denote the Hilbert "metric" on $\mathbb{K}$ and recall that h is the Hilbert "metric" on $\mathbb{P}_{\mathcal{G}}$. Then $(\mathbb{P}_{\mathcal{G}}, h)$ and $(\mathbb{K}, h_{\mathbb{K}})$ are "isometrically" isomorphic. According to [19: Section 5] the formula of $f := \Xi^{-1} \circ \Lambda \circ \Xi : \mathbb{K} \to \mathbb{K}$ is

$$f(\alpha,\beta) = \frac{1}{3} \cdot \frac{\alpha+\beta}{5\alpha+\beta}(3\alpha, 2\alpha+\beta).$$

From this formula the following results were deduced in [19: Section 5]:

Remark 4.1.
(a) f is not $\mathbb{R}^2_+$-monotone on $\mathbb{R}^2_+$. Hence Λ is not $\mathbb{D}_{\mathcal{G}}$-monotone on $\mathbb{D}_{\mathcal{G}}$.
(b) There exist points in $\mathbb{D}^\circ_{\mathcal{G}}$ such that f is strictly $p_{\mathbb{D}_{\mathcal{G}}}$-expansive.
(c) $\mathrm{Fix} = \Xi(\{\alpha(1,1) \mid \alpha > 0\})$.
(d) Reducible eigenforms of Λ are the Ξ-images of $(0,0)$ and $(0,1)$.

Next we normalize f to $\tilde{f} := \frac{f}{\alpha+\beta}$ so that $\tilde{f}$ maps

$$U := \{(\alpha,\beta) \in \mathbb{R}^2 \mid 0 \leq \alpha, \beta \leq 1;\ \alpha+\beta = 1\}$$

into itself. When we calculate $h_{\mathbb{K}}$ via eigenvalues of forms we see that for $0 < \alpha \leq 1$,

$$\frac{h_{\mathbb{K}}\left(f(\alpha, 1-\alpha), f\left(\frac{1}{2},\frac{1}{2}\right)\right)}{h_{\mathbb{K}}\left((\alpha,1-\alpha),\left(\frac{1}{2},\frac{1}{2}\right)\right)} = 1 + \frac{\ln\left(\frac{3}{4\alpha+1}\right)}{\ln(2\alpha)} =: q(\alpha)\ .$$

The function q is strictly decreasing in α and

$$\lim_{\alpha \downarrow 0} q(\alpha) = 1,\ q(1) = 1 + \frac{\ln 3 - \ln 5}{\ln 2} < 0.3 < 1\ .$$

Because of the "isometry" Ξ, this implies

$$h(\Lambda^n(\mathcal{E}), \Lambda^n(\mathcal{F})) \leq 2(q(\alpha))^n h(\mathcal{E},\mathcal{F}) \tag{4.2}$$

for $n \in \mathbb{N}$ and $\mathcal{E},\mathcal{F} \in \Xi\left(\{(\beta, 1-\beta) \in \mathbb{R}^2 \mid 0 < \alpha \leq \beta \leq 1\}\right)$ for some $\alpha > 0$. We see that Banach's fixed point theorem is applicable to Λ as long as we stay away from the boundary of $\mathbb{P}_{\mathcal{G}}$ which is spanned by $\Xi((0,1))$ and $\Xi((1,-1))$.

Finally we switch from $\mathcal{G}$ to a smaller symmetry group $\mathcal{G}'$ [cf. 19: Section 5]. This group corresponds to a *non-nested Vicsek snowflake*. We start its contruction with a rhombus instead of a square and again arrange five copies on a rhombus shaped chessboard. Now we have three edge orbits: One consists of the sides of the rhombus, another one of the short diagonal, and the last one of the long diagonal. Let $\alpha, \beta, \eta \in \mathbb{R}$ be their corresponding weights. This time we get

$$\mathbb{B}_{\mathcal{G}'} \cong \mathbb{R}^3,$$
$$\mathbb{D}_{\mathcal{G}'} \cong \mathbb{R}^3_+,$$
$$\mathbb{P}_{\mathcal{G}'} \cong \mathbb{K}' := \left\{(\alpha,\beta,\eta) \in \mathbb{R}^3 \mid 0 \leq \alpha,\ 0 \leq \alpha+\beta,\ 0 \leq \alpha+\eta\right\}.$$

Now the set of eigenforms in $\mathbb{D}^{\circ}_{\mathcal{G}'}$ is

$$\mathrm{Fix}' = \Xi\left(\left\{(\alpha,\beta,\eta) \in \mathbb{K}' \mid \alpha^2 = \beta\eta\right\}\right).$$

This is a two-dimensional, simply connected manifold and also a cone. Hence, (2.6) has linear independent solutions and we cannot expect contraction in (3.3a) as long as Λ acts on a cone with non-empty interior.

It is the task of ongoing research to find easy conditions distinguishing between cases where Banach's fixed point theorem is locally applicable and those where it is not.

References

[1] Anderson Jr., W. N., Trapp, G. E., Shorted operators II. SIAM J. Appl. Math. 28 (1975), 60–71.

[2] Barlow, M. T., Harmonic analysis on fractal spaces. Séminaire Bourbaki 44 ème année, 1991–92, 75 (1992), 345–368.

[3] — Random walks, electrical resistance and nested fractals. In: Asymptotic Problems in Probability Theory: Stochastic Models and Diffusions on Fractals (ed. by K. D. Elworthy, N. Ikeda), 131–157, Pitman Res. Notes Math. Ser. 283, Longman Scientific & Technical, Harlow 1993.

[4] Bauer, H., Baer, H. S., The part metric in convex sets. Pacific J. Math. 30 (1969), 15–33.

[5] Ben-Avraham, D., Havlin, S., Diffusion in disordered media. Adv. in Phys. 36 (1987), 695–798.

[6] Birkhoff, G., Uniformly semi-primitive multiplicative processes. Trans. Amer. Math. Soc. 104 (1962), 37–51.

[7] Carlson, D., What are Schur-complements, anyway? Linear Algebra Appl. 74 (1986), 257–275.

[8] Eriksson-Bique, S.-L., Leutwiler, H., A generalization of parallel addition. Aequationes Math. 38 (1989), 99–110.

[9] Fitzsimmons, P. J., Hambly, B. M., Kumagai, T., Transition density estimates for Brownian motion on affine nested fractals. Comm. Math. Phys. (to appear).

[10] Fukushima, M., Shima, T., On a spectral analysis for the Sierpinski gasket. Potential Analysis 1 (1992), 1–35.

[11] Hattor, K., Hattori, T., Watanabe, H., Gaussian field theories on general networks and the spectral dimension. Progr. Theoret. Phys. Suppl. 92 (1987), 108–143.

[12] Horn, R. A., Johnson, C. R., Matrix Analysis. Cambridge University Press, 1985.

[13] Krause, U., Nussbaum, R. D., A limit set trichotomy for self-mappings of normal cones in Banach spaces. Nonlinear Anal. 20 (1993), 855–870.

[14] Kusuoka, S., Dirichlet forms on fractals and products of random matrices. Publ. RIMS Kyoto Univ. 25 (1989), 659–680.

[15] — Diffusion processes on nested fractals. In: R. L. Dobrushin, S. Kusuoka, Statistical Mechanics and Fractals, 39–98. Lecture Notes in Math. 1567. Springer-Verlag, Berlin–Heidelberg 1993.

[16] Lindstrøm, T., Brownian Motion on Nested Fractals. Mem. Amer. Math. Soc. 83, No. 420. Amer. Math. Soc., Providence 1990.

[17] Metz, V., Potentialtheorie auf dem Sierpinski gasket. Math. Ann. 289 (1991), 207–237.

[18] — How many diffusions exist on the Vicsek snowflake. Acta Appl. Math. 32 (1993), 227–241.

[19] — Hilbert's projective metric on cones of Dirichlet forms. J. Funct. Anal. 127 (1995), 438–455.

[20] — Renormalization of finitely ramified fractals. Proc. Roy. Soc. Edinburgh Sect. A. 125 (1995), 1085–1104.

[21] Nussbaum, R. D., Hilbert's Projective Metric and Iterated Nonlinear Maps. Mem. Amer. Math. Soc. 75, No. 391. Amer. Math. Soc., Providence 1988.

[22] Takáč, P., Convergence in the part metric for discrete dynamical systems in ordered topological cones. Preprint 1994.

[23] Thompson, A. C., On certain contraction mappings in a partially ordered vector space. Proc. Amer. Math. Soc. 14 (1963), 438–443.

Estimates for the density of the area integral

*Charles N. Moore**

Abstract. We discuss the density of the area integral, a harmonic analysis analogue of local time, and several estimates relating it to the nontangential maximal function and Lusin area function. We borrow techniques from the proofs of these estimates to produce a new proof of a characterization of $L \log L$ due to Brossard and Chevalier.

1991 Mathematics Subject Classification: 42B25, 42A61, 31B05.

1. Introduction: results from probability

Consider a real valued continuous martingale X_t starting at 0. To this are associated two other random variables, $X^* = \sup\{|X_t| : t > 0\}$, the maximal function of the martingale, and $S(X) = \sqrt{\langle X \rangle_\infty}$, the square function of the martingale. Here $\langle X \rangle_t$ is the quadratic variation process of the martingale arising from the Doob–Meyer decomposition of the process X_t^2. Burkholder and Gundy [10] showed that for $0 < p < \infty$, there exist constants c_p and C_p such that $c_p \|X^*\|_p \leq \|S(X)\|_p \leq C_p \|X^*\|$. Barlow and Yor [4], [5] considered a third random variable associated to the martingale, the maximal local time. To define this, use the Riemann-Stieltjes measure $d\langle X \rangle_t$ of the increasing process $\langle X \rangle_t$ on $[0, \infty)$ to define a Borel measure μ on $\mathbb{R}$ by $\mu(E) = d\langle X \rangle_t(\{t : X_t \in E\})$. A theorem of P. Lévy states that μ is absolutely continuous with respect to Lebesgue measure and we write $L(a)$ for its density. We call $L(a)$ the local time of the martingale. Set $L^* = \sup\{L(a) : a \in \mathbb{R}\}$; L^* is called the maximal local time.

Note that by definition, $\int_0^\infty \chi_E(X_t) d\langle X \rangle_t = \int_E L(a)da$ for every Borel set $E \subseteq \mathbb{R}$ and from this it follows readily that for any Borel function f on $\mathbb{R}$,

$$\int_0^\infty f(X_t) d\langle X \rangle_t = \int_{\mathbb{R}} f(a) L(a) da. \tag{1.1}$$

Take $f \equiv 1$ in (1.1) and note that $L(a) = 0$ if $a \in [-X^*, X^*]$ to obtain

$$S(X)^2 = \int_0^\infty d\langle X \rangle_t = \int_{-X^*}^{X^*} L(a) da \leq 2L^* X^*. \tag{1.2}$$

* Partially supported by National Science Foundation grant no. 9204529.

Use Cauchy–Schwarz to deduce that $\|S(X)\|_p \le \sqrt{2}\|L^*\|_p^{1/2}\|X^*\|_p^{1/2}$ and combine with the result of Burkholder and Gundy to conclude that for $0 < p < \infty$, $\|S(X)\|_p \le C_p\|L^*\|_p$. The result of Barlow and Yor [4], [5] is the reverse inequality; that is much more difficult.

The inequality $\|S(X)\|_p \le C_p\|L^*\|_p$, the result of Barlow and Yor, and the Burkholder and Gundy inequalities combine to show that for $0 < p < \infty$, the L^p norms of X^*, $S(X)$, and L^* are all equivalent. In fact, more is true. For each pair of these three random variables there are inequalities relating the distribution functions of the pair. These so called good-λ inequalities in particular imply the equivalence of L^p norms but also imply the equivalence of "norms" formed in a more general way. In the next section we will discuss such good-λ inequalities in the setting of harmonic analysis, and will elucidate these points; we wish to be brief and so will avoid discussing these inequalities in both settings.

2. Harmonic analysis analogues

Let $u(s,t)$ be a harmonic function on $\mathbb{R}^{n+1}_+ = \{(s,t) : s \in \mathbb{R}^n, t > 0\}$. For $x \in \mathbb{R}^n$ and $\alpha > 0$ set $\Gamma_\alpha(x) = \{(s,t) : |x-s| < \alpha t\}$, and set

$$N_\alpha u(x) = \sup\{|u(s,t)| : (s,t) \in \Gamma_\alpha(x)\}$$

$$A_\alpha u(x) = \left(\int_{\Gamma_\alpha(x)} |\nabla u(s,t)|^2 t^{1-n}\,dsdt\right)^{1/2}$$

These are known as the nontangential maximal function of u and the Lusin area function (also called the square function) of u, respectively. $N_\alpha u(x)$ is the harmonic analysis analogue of X^*; $A_\alpha u(x)$ is the harmonic analysis analogue of $S(X)$. Both of these have a long history and play a central rôle in harmonic analysis. Surprisingly, even though L^* had been studied extensively for a number of years, it wasn't until 1983 that R. Gundy [16] proposed a harmonic analysis analogue of L^*. To define this, let $r \in \mathbb{R}$ and consider the subharmonic function $(u(s,t)-r)^+$. Its distributional Laplacian is a nonnegative Radon measure on $\mathbb{R}^{n+1}_+$ which we denote by $\Delta(u(s,t)-r)^+(dsdt)$. We then define

$$D_\alpha u(x;r) = \int_{\Gamma_\alpha(x)} t^{1-n}\Delta(u(s,t)-r)^+(dsdt)$$

and

$$D_\alpha u(x) = \sup_{r\in\mathbb{R}} D_\alpha u(x;r).$$

$D_\alpha u(x;r)$ may also be formed by taking smooth functions of u which approximate $(u-r)^+$, forming approximations using the Laplacian of these smooth approximations and then passing to a limit. See Gundy [17] for details on all the possible ways to define this; see also Brossard [7] for more discussion of this. $D_\alpha u(x;r)$ is called the density of the area integral and $D_\alpha u(x)$ is called the maximal density. The density of the area integral is an analogue of the local time of a martingale and the maximal density is the analogue of the maximal local time. That $D_\alpha u(x;r)$ really does form the proper analogue of local time can easily be seen from a formula of Gundy and Silverstein [18]: If $\psi(s,t)$ is a nonnegative Borel function on $\mathbb{R}_+^{n+1}$ and f is a Borel function on $\mathbb{R}$ then

$$\int\int_{\mathbb{R}_+^{n+1}} \psi(s,t)f(u(s,t))|\nabla u(s,t)|^2 dsdt$$

$$= \int_{\mathbb{R}}\int_{\mathbb{R}_+^{n+1}} \psi(s,t)f(r)\Delta(u(s,t)-r)^+(dsdt)dr. \quad (2.1)$$

Take $\psi(s,t) = t^{1-n}\chi_{\Gamma_\alpha(x)}(s,t)$ in (2.1) and we obtain an analogue of (1.1). If, as before, we also take $f \equiv 1$ then we obtain

$$\int\int_{\Gamma_\alpha(x)} |\nabla u(s,t)|^2 t^{1-n} dsdt = \int_{\mathbb{R}}\int_{\Gamma_\alpha(x)} t^{1-n}\Delta(u(s,t)-r)^+(dsdt)dr.$$

Noting that $\Delta((u(s,t)-r)^+) = 0$ on $\Gamma_\alpha(x)$ unless $|r| \leq N_\alpha u(x)$ and using the definitions of the area function and the density of the area function we can rewrite this last formula as:

$$(A_\alpha u(x))^2 = \int_{-N_\alpha u(x)}^{N_\alpha u(x)} D_\alpha u(x;r)dr$$

and this provides a perfect analogue of (1.2). We can then reason as in the probability case to conclude that $\|A_\alpha u\|_p \leq \|D_\alpha u\|_p$ for $0 < p < \infty$. As before, the reverse inequality is more difficult. This was done by Gundy [16] when $n = 2$ and by Gundy and Silverstein [18] (see also Gundy [17]) in all dimensions.

However, there are more precise estimates. The following appears in Bãnuelos and Moore [1].

Theorem 1. *Suppose $\beta > \alpha$. Then there exist constants K, C_1, C_2 depending only on α, β, n such that if $\lambda > 0$, $0 < \epsilon < 1$,*

(a) $|\{x \in \mathbb{R}^n : N_\alpha u(x) > K\lambda,\ D_\beta u(x) \leq \epsilon\lambda\}|$

$$\leq C_1 \exp\left(\frac{-C_2}{\epsilon}\right) |\{x \in \mathbb{R}^n : N_\alpha u(x) > \lambda\}|,$$

(b) $|\{x \in \mathbb{R}^n : A_\alpha u(x) > K\lambda,\ D_\beta u(x) \leq \epsilon\lambda\}|$

$$\leq C_1 \exp\left(\frac{-C_2}{\epsilon^2}\right) |\{x \in \mathbb{R}^n : A_\alpha u(x) > \lambda\}|.$$

Inequalities such as these hold with the rôles of $N_\alpha u$ and $D_\beta u$, and $A_\alpha u$ and $D_\beta u$ reversed, but the proof of these requires a "smoother" version of $D_\beta u$. Consider a smooth positive radially symmetric function ψ supported on a ball $B(0,\alpha)$ in $\mathbb{R}^n$ such that $\int_{\mathbb{R}^n}\psi(x)dx = 1$. We set $\psi_t(s) = \frac{1}{t^n}\psi\left(\frac{s}{t}\right)$ and for $r \in \mathbb{R}$ we set $Du(x;r) = \int_{\mathbb{R}_+^{n+1}} t\psi_t(x-s)\Delta(u(s,t)-r)^+(dsdt)$, and then set $Du(x) =$

$\sup_{r\in\mathbb{R}} Du(x;r)$. Technically, $Du(x)$ depends on ψ but to simplify the notation we consider ψ fixed and just write $Du(x)$ for the maximal density formed this way. These are comparable to what was defined before; the function $t^n\psi_t(x-s)$ merely provides a smooth approximation to a cone. The following appears in Bãnuelos and Moore [1].

Theorem 2. *There exist constants K, C_1 and C_2 depending only on ψ, β and n such that if $\lambda > 0$, $0 < \epsilon < 1$, then*

(a) $$|\{x \in \mathbb{R}^n : Du(x) > K\lambda,\ N_\beta u(x) \le \epsilon\lambda\}| \le C_1 \exp\left(\frac{-C_2}{\epsilon}\right) |\{x \in \mathbb{R}^n : Du(x) > \lambda\}|,$$

(b) $$|\{x \in \mathbb{R}^n : Du(x) > K\lambda,\ A_\beta u(x) \le \epsilon\lambda\}| \le C_1 \exp\left(\frac{-C_2}{\epsilon^{2/3}}\right) |\{x \in \mathbb{R}^n : Du(x) > \lambda\}|.$$

An immediate consequence of these theorems is well known.

Corollary. *Let u be harmonic on $\mathbb{R}^{n+1}_+$. Let Φ be a nondecreasing function on $\{t : t > 0\}$ with $\Phi(0) = 0$ and which has the property that there exists a constant C_0 such that $\Phi(2\lambda) \le C_0\Phi(\lambda)$. Let $\beta > \alpha$. Then there exists a constant C depending only on α, β, n and C_0 such that $\int_{\mathbb{R}^n} \Phi(N_\alpha u(x))dx \le C \int_{\mathbb{R}^n} \Phi(D_\beta u(x))dx$. Similar results hold for $A_\alpha u$ in place of $N_\alpha u$ and with the roles of $N_\alpha u$ and $D_\beta u$ and $A_\alpha u$ and $D_\beta u$ switched.*

To prove this, note that $\int_{\mathbb{R}^n} \Phi(N_\alpha u(x))dx = \int_{\mathbb{R}^n} |\{x : N_\alpha u(x) > \lambda\}| d\Phi$ and then estimate the integrand using the estimate in Theorem 1(a). (See Lenglart, Lepingle and Pratelli [19] where this type of argument is made explicit.) The similar results follow correspondingly from the remaining parts of Theorem 1 and Theorem 2. We note that since $\int_{\mathbb{R}^n} \Phi(N_\alpha u)dx \approx \int_{\mathbb{R}^n} \Phi(N_\beta u)dx$ (an easy estimate on the distribution functions) for all α and β and since well known inequalities for $A_\alpha u$ and $N_\beta u$ such as those in Theorems 1 and 2 produce an equivalence of these "Φ norms" for $A_\alpha u$ and $N_\beta u$ then one can deduce the equivalence of these Φ norms for $A_\alpha u$, $N_\beta u$, $D_\gamma u$ for any α, β, γ.

The inequalities in Theorems 1 and 2 are called good-λ inequalities. In the setting of probability, exact analogues of these for $S(X)$ and L^*, and X^* and L^* were shown by Bass [6] and independently by Davis [14]. It is possible to formulate versions of $N_\alpha u(x)$, $A_\alpha u(x)$ and $D_\beta u(x)$ on Lipschitz domains and in [1], Theorem 1 is shown in this more general setting. Unfortunately, the proof of Theorem 2 doesn't seem to be adaptable to the case of Lipschitz domains so Theorem 2 is not known to this degree of generality. If in Theorem 2(b) we impose the restriction that α (recall supp $\psi \subseteq B(0,\alpha)$) and β satisfy $\alpha < \frac{\beta}{2048(1+2\sqrt{n})}$ then in Moore [20] it is shown that this estimate holds with the expression $C_1 \exp\left(\frac{-C_2}{\epsilon^2}\right)$ replacing $C_1 \exp\left(\frac{-C_2}{\epsilon^{2/3}}\right)$. This restriction on α and β is probably unnecessary, that is, it seems likely that Theorem 2(b) should hold with the expression $C_1 \exp\left(\frac{-C_2}{\epsilon^2}\right)$ on the right-hand-side and with only the restriction that $\alpha < \beta$ but we have been unable to show this.

In Moore [20], it is shown that Theorem 2(a) as written and Theorem 2(b) with the expression $C_1 \exp\left(\frac{-C_2}{\epsilon^2}\right)$ are in some sense the sharpest attainable in terms of decay as $\epsilon \to 0$, that is, these results would no longer be true with higher powers of ϵ. By comparison with the probability analogue of Theorem 1 which is known to be sharp it is probably the case that Theorem 1 is sharp but we have been unable to show this.

The advantage of Theorems 1 and 2 is that the exponential expressions on the right-hand-side allow us to pass immediately to similar estimates where Lebesgue measure is replaced by any measure which satisfies an A_∞ condition with respect to Lebesgue measure. Another advantage of a "subgaussian" estimate such as Theorem 1(b) is that it leads to laws of the iterated logarithm involving $A_\alpha u$ and $D_\beta u$. (See [1].)

We will not prove Theorems 1 and 2 here; for the proofs see [1] and also [20]. The proofs consist mainly of Green's theorem arguments on certain subdomains of $\mathbb{R}^{n+1}_+$ constructed over cubes $Q \subseteq \mathbb{R}^n$ and in this regard are quite similar to many other proofs of similar good-λ inequalities. These ideas appear in, for example, Burkholder and Gundy [11], Dahlberg [12], Fefferman, Gundy, Silverstein and Stein [15], Murai and Uchiyama [21] or Bañuelos and Moore [2], [3] (to name but a few). These ideas continue to be fruitful. Rather than proving Theorems 1 and 2 whose proofs are recorded in the literature anyway, we will now take the opportunity to illustrate the continuing utility of these techniques by using them to provide a new proof of a result of Brossard and Chevalier.

3. The D function and $L \log L$

In [8], Brossard and Chevalier used the density of the area integral to characterize the class $L \log L$ within $H^1(\mathbb{R}^{n+1}_+)$. In this section, we will show how the techniques used in proving good-λ inequalities can be used to give alternate proofs of their results.

For $\alpha > 0$ we wish to consider a variant of $D_\alpha u(x;0)$ defined by: $D_\alpha u(x;0) = \int_{\Gamma_\alpha(x)} y^{1-n} \Delta |u(x,y)| dx dy$. Here we have considered the subharmonic function $|u(x,y)|$ rather than $u(x,y)^+$ considered previously. However, this $D_\alpha u(x;0)$ is just twice the $D_\alpha u(x;0)$ defined previously and it will be a little easier to work with.

Recall that via the theorem of Burkholder, Gundy and Silverstein we may regard $H^1(\mathbb{R}^{n+1}_+)$ as the space of all harmonic u on $\mathbb{R}^{n+1}_+$ with $\|N_\beta u\|_1 < \infty$. (Here the choice of β is immaterial; these norms are equivalent for different β.) For $u(x,y) \in H^1(\mathbb{R}^{n+1}_+)$ the nontangential limit, call it $u(x)$, exists a.e.; furthermore $u(x,y)$ is just the Poisson integral of $u(x)$. Define $L \log L$ to be the space of all harmonic u on $\mathbb{R}^{n+1}_+$ such that

$$\sup_{y>0} \int_{\mathbb{R}^n} |u(x,y)|(1+\log^+|u(x,y)|)dx < \infty.$$

We consider the Brossard and Chevalier [8] characterization of $L\log L$ as a subspace of $H^1(\mathbb{R}^{n+1}_+)$:

Theorem. *Let $u \in H^1(\mathbb{R}^{n+1}_+)$. The following are equivalent:*
(a) $u \in L\log L$
(b) $\int_{\mathbb{R}^n} D_\alpha u(x;0)\log^+ D_\alpha u(x;0)dx < \infty$.

To show that (a) implies (b), Brossard and Chevalier employ a combination of probabilistic and analytic techniques; in particular, they rely on a probabilistic analogue of this theorem they showed in [9]. They prove the other implication using purely analytic techniques. We will give a purely analytic proof of the implication (a) $\Rightarrow$ (b). Similar ideas work to give an analytic proof of (b) $\Rightarrow$ (a) which is different than the proof of Brossard and Chevalier but we will not do this explicitly and will merely comment on this later. Our proof of (a) $\Rightarrow$ (b) is different than that of Brossard and Chevalier but we will borrow a few of their ideas. However, the central idea of our proof is a technique which is central to many proofs of good-λ inequalities.

We first need some definitions. For $h > 0$ set $\Gamma^h_\alpha(x) = \{(s,t) \in \mathbb{R}^{n+1}_+ : |x-s| < \alpha t, t < h\}$; this is simply a cone truncated at height h. We then define a truncated version $D^h_\alpha u(x;0)$ of $D_\alpha u(x;0)$ by using the same definition but by integrating over $\Gamma^h_\alpha(x)$ instead of $\Gamma_\alpha(x)$.

Fix $\beta > \alpha$. Consider a point $x \in \mathbb{R}^n$ and a cube $Q \subseteq \mathbb{R}^n$ centered at x. Define $\Omega(Q) = \bigcup_{s\in Q} \Gamma^h_\alpha(s)$, where h is chosen so that for every $s \in Q$, $\Gamma_\alpha(s) - \Gamma^h_\alpha(s) \subseteq \Gamma_\beta(x)$. Such an h can be chosen as $h = c\ell(Q)$ where $\ell(Q)$ represents the sidelength of Q and c is a constant which depends only on α, β and n. Define $\tilde{Q}$ to be a cube with the same center as Q but with sidelength $\ell(\tilde{Q}) = k\ell(Q)$ where k is to be chosen momentarily. We then set $\Omega(\tilde{Q}) = \bigcup_{s\in\tilde{Q}} \Gamma^{kh}_\alpha(s)$ where h is as in the definition of $\Omega(Q)$. We choose k large enough so that $k \geq 3$ and for $P \in \Omega(Q)$, $d(P, \partial(\Omega(\tilde{Q}))$ is realized by a point $(s,0) \in \mathbb{R}^n$. Set $P_0 = (x, \frac{kh}{2})$. We can also assume that k is chosen large enough so that $P_0 \in \Gamma_{\alpha/2}(s)$ for every $s \in Q$. Our choice of such a k can be made depending only on α, β, and n, independent of Q. For each cube $Q \subseteq \mathbb{R}^n$ we consider $\Omega(Q)$ and $\Omega(\tilde{Q})$ constructed in this way; since each $\Omega(Q)$ and $\Omega(\tilde{Q})$ can be considered as a translation and a rescaling of a canonical domain, all constants arising in what follows will depend only on α, β, n. For such a cube Q, and $E \subseteq \partial(\Omega(\tilde{Q}))$ we let $\omega(E) = \omega(E, P_0, \partial(\Omega(\tilde{Q})))$ denote the harmonic measure of E relative to $\Omega(\tilde{Q})$. Let $G(P) = G(P, P_0)$, $P \in \Omega(\tilde{Q})$ denote the Green's function so that $\omega(E) = -\int_E \frac{\partial G}{\partial n}(P)d\sigma(P)$ where σ is surface measure on $\partial(\Omega(\tilde{Q}))$ and $\frac{\partial}{\partial n}$ represents the outward normal derivative.

Lemma 1. *There exists a constant $C = C(\alpha, \beta, n)$ so that if $E \subseteq Q$ then*

$$C^{-1}\frac{|E|}{|Q|} \leq \omega(E) \leq C\frac{|E|}{|Q|}.$$

Proof. This follows readily from boundary Harnack comparisons applied to the functions y and $G(P)$. See Dahlberg [13], lemma 3.4.

For $P \in \tilde{\Omega}(Q)$ let $d(P)$ denote the distance from P to $\partial(\Omega(\tilde{Q}))$. Note that $\Omega(\tilde{Q})$ is starlike with respect to the point P_0. For $P \in \Omega(\tilde{Q}) \setminus \{P_0\}$ we let $\hat{P}$ be the point of intersection of $\partial(\Omega(\tilde{Q}))$ with the ray from P_0 through P. For $P \in \partial(\Omega(\tilde{Q}))$ set $A(P, r) = \partial(\Omega(\tilde{Q})) \cap \{P' : |P - P'| < r\}$. The following appears in Dahlberg [12].

Lemma 2. *Fix $\ell > 0$. There is a number $K > 0$ such that for all $P \in \Omega(\tilde{Q}) \setminus \{P_0\}$*

$$K^{-1}d(P)^{n-1}G(P) \leq \omega(A(\hat{P}, \ell d(P))) \leq Kd(P)^{n-1}G(P)$$

The K here depends only on Lipschitz constants of the domain and on ℓ; later we will fix ℓ depending only on α, β, n so that we ultimately have $K = K(\alpha, \beta, n)$.

For $s \in \tilde{Q}$ set $D^T_\alpha u(s; 0) = D_\alpha u(s; 0) - D^h_\alpha u(s; 0)$. This is the contribution to $D_\alpha u(s; 0)$ from the "top" part of the cone $\Gamma_\alpha(s)$. The amount that $D^T_\alpha u(s; 0)$ varies over Q can be controlled by $N_\beta u(x)$, where x is the center of Q. This is possible by our choice of h so that $\Gamma_\alpha(s) - \Gamma^h_\alpha(s) \subseteq \Gamma_\beta(x)$ $\forall s \in Q$. The following is from [1]. (See page 154.)

Lemma 3. *There exists a constant $C = C(\alpha, \beta, n)$ such that if $s \in Q$ then*

$$|D^T_\alpha u(s; 0) - D^T_\alpha u(x; 0)| \leq CN_\beta u(x)$$

The next lemma contains the main idea of the proof.

Lemma 4. *There exists a constant $C = C(\alpha, \beta, n)$ such that for every $x \in \mathbb{R}^n$,*

$$MD_\alpha u(x; 0) \leq C(D_\alpha u(x) + Mu(x)).$$

Here $MD_\alpha u(x; 0)$ is the Hardy–Littlewood maximal function of $D_\alpha u(x; 0)$.

Proof. Fix an $x \in \mathbb{R}^n$ and a cube Q centered at x and consider $\Omega(Q)$ and $\Omega(\tilde{Q})$ as constructed above. We apply Green's theorem to the function $G(P)$ and $|u|$ on $\Omega(\tilde{Q})$. Unfortunately, neither $|u|$ nor $\partial(\Omega(\tilde{Q}))$ is smooth so to do this we must approximate $|u|$ by a smooth function of u, approximate $\Omega(\tilde{Q})$ by a smooth domain, apply Green's theorem and then pass to a limit. These computations can be found explicitly in Gundy and Silverstein [18] and in Gundy [17]. (See also Stein [23], p. 206 for similar computations.) The result appears as if Green's theorem has been formally applied to $|u|$ and $G(P)$ on $\Omega(\tilde{Q})$ and for brevity we will present these computations this way, bearing in mind the necessity for this smoothing procedure.

Green's theorem yields:

$$\int\int_{\Omega(\tilde{Q})} G(P)\Delta|u(P)|dP = -\int_{\partial(\Omega(\tilde{Q}))} \frac{\partial G}{\partial n}(P)|u(P)|dP - |u(P_0)|$$
$$= \int_{\partial\Omega(\tilde{Q})} |u(P)|d\omega(P) - |u(P_0)|.$$

On $\mathbb{R}^{n+1}_+$, $|u(P)|$ is dominated by the Poisson integral of $|u|$, call it v. Then v is harmonic on $\mathbb{R}^{n+1}_+$ and $v(P) \geq |u(P)|$ for $P \in \mathbb{R}^{n+1}_+$; consequently,

$$\int_{\partial(\Omega(\tilde{Q}))} |u(P)|d\omega(P) \leq \int_{\partial(\Omega(\tilde{Q}))} v(P)d\omega(P) = v(P_0)$$
$$\leq N_\beta v(x) = N_\beta|u|(x) \leq C_\beta Mu(x).$$

Thus, $\int\int_{\Omega(\tilde{Q})} G(P)\Delta|u(P)|dP \leq CMu(x)$.

For $P \in \Omega(\tilde{Q})$, $z \in \mathbb{R}^n$ set $\chi(P,z) = 1$ if $P \in \Gamma^h_\alpha(z)$ and 0 otherwise. Note that there exists a constant ℓ, depending only on α, β, and n such that if $P \in \Omega(Q)$ then $\{z \in \tilde{Q} : z \in \Gamma^h_\alpha(P)\} \subseteq A(\hat{P}, \ell d(P))$. Then, using this and Lemma 2 gives

$$\int\int_{\Omega(\tilde{Q})} G(P)\Delta|u(P)|dP \geq C\int\int_{\Omega(\tilde{Q})} d(P)^{1-n}\omega(A(\hat{P},\ell d(P)))\Delta|u(P)|dP$$
$$\geq C\int\int_{\Omega(Q)} \Delta|u(P)|d(P)^{1-n}\int_Q \chi(P,z)d\omega(z)dP$$
$$= C\int_Q\int_{\Gamma^h_\alpha(z)} \Delta|u(P)|d(P)^{1-n}dPd\omega(z)$$
$$\geq \frac{C}{|Q|}\int_Q D^h_\alpha u(z,0)dz$$

where the last inequality follows from Lemma 1 and the fact that for $P = (s,t) \in \Omega(Q)$, $d(P) \approx t$. So we've shown that $\frac{1}{|Q|}\int_Q D^h_\alpha u(z;0)dz \leq CMu(x)$. But then, by Lemma 3,

$$\frac{1}{|Q|}\int_Q D_\alpha u(z;0)dz \leq CMu(x) + D^T_\alpha u(x;0) + CN_\beta u(x)$$
$$\leq C(Mu(x) + D_\alpha u(x)).$$

Taking the supremum over all such Q finishes the proof of Lemma 4.

We now prove (a) implies (b) in the theorem:

$$\int_{\mathbb{R}^n} D_\alpha u(x;0)\log^+ D_\alpha u(x,0)dx = \int_{\{x:D_\alpha u(x;0)>1\}} D_\alpha u(x;0)\log^+ D_\alpha u(x;0)dx$$
$$= \int_1^\infty \left(\frac{1}{\lambda}\int_{\{x:D_\alpha u(x;0)>\lambda\}} D_\alpha u(x;0)dx\right)d\lambda$$
$$\leq C\int_1^\infty |\{x : MD_\alpha u(x;0) > c\lambda\}|d\lambda.$$

This last estimate is just equation (6) in Stein [22]; C, c depend only on n. But by Lemma 4, $|\{x : MD_\alpha u(x;0) > c\lambda\}| \leq |\{x : D_\alpha u(x) > b\lambda\}| + |\{x : Mu(x) > b\lambda\}|$ with $b = \frac{c}{2C_4}$, and where C_4 is the constant appearing on the right in Lemma 4. So using the fact that $\|D_\alpha u\|_1 \leq C\|N_\beta u\|_1$ (which follows from Theorem 2(a) in section 2) and the well known weak type $1-1$ estimate for the Hardy–Littlewood maximal function we obtain:

$$\begin{aligned}\int_{\mathbb{R}^n} D_\alpha u(x;0)\log^+ D_\alpha u(x;0)dx &\leq C\int_1^\infty |\{x : D_\alpha u(x) > b\lambda\}|d\lambda \\ &\quad + C\int_1^\infty |\{x : Mu(x) > b\lambda\}|d\lambda \\ &\leq C\|D_\alpha u\|_1 + \frac{C}{b}\int_b^\infty \frac{1}{\lambda}\int_{\{x:|u(x)|>\frac{\lambda}{2}\}} |u(x)|dxd\lambda \\ &\leq C\|N_\beta u\|_1 + \frac{C}{b}\int_{\mathbb{R}^n} |u(x)|\int_b^{2|u(x)|}\frac{1}{\lambda}d\lambda dx \\ &\leq C\|N_\beta u\|_1 + C\int_{\mathbb{R}^n} |u(x)|(1+\log^+|u(x)|)dx.\end{aligned}$$

So, $\int_{\mathbb{R}^n} D_\alpha u(x;0)\log^+ D_\alpha u(x;0)dx \leq C\|N_\beta u\|_1 + C\int_{\mathbb{R}^n}|u(x)|(1+\log^+|u(x)|)dx$. Since $|u(x,y)|(1+\log^+|u(x,y)|)$ is subharmonic on $\mathbb{R}^{n+1}_+$, the last integral is just $\sup_{y>0}\int_{\mathbb{R}^n}|u(x,y)|(1+\log^+|u(x,y)|)dx$. This completes the proof.

Similar arguments can be used to show that (b) implies (a). The key step in our proof of (a) $\Rightarrow$ (b) was the Green's theorem argument in Lemma 4 and such an argument can be used to show the implication (b) $\Rightarrow$ (a). However, one needs to essentially reverse all the inequalities there and this is where problems arise. Nevertheless, these difficulties can be overcome, but the end result is a proof which is more technical and complicated than the original proof of Brossard and Chevalier.

References

[1] Bañuelos, R., Moore, C. N., Distribution function inequalities for the density of the area integral. Ann. Inst. Fourier (Grenoble) 41, 1 (1991), 137–171.

[2] — Sharp estimates for the nontangential maximal function and the Lusin area function in Lipschitz domains. Trans. Amer. Math. Soc. 312 (1989), 641-662.

[3] — Laws of the iterated logarithm, sharp good-λ inequalities and L^p estimates for caloric and harmonic functions. Indiana Univ. Math. J. 38 (1989), 315-344.

[4] Barlow, M., Yor, M., (Semi) Martingale inequalities and local times. Z. Wahrsch. Verw. Gebiete 55 (1981), 237–354.

[5] — Semimartingale inequalities via the Garsia–Rodemich–Rumsey lemma, and applications to local times. J. Funct. Anal. 49, 2 (1989), 198–229.

[6] Bass, R., L^p-inequalities for Functionals of Brownian Motion. In: Séminare de Probabilitiés XXI, Lecture Notes in Math. 1247, 206–217. Springer-Verlag, New York 1987.

[7] Brossard, J., Densité de l'intégrale d'aire dans $\mathbb{R}_+^{n+1}$ et limites non tangentielles. Invent. Math. 93 (1988), 297–308.

[8] Brossard, J., Chevalier, L., Classe $L \log L$ et densité de l'intégrale d'aire dans $\mathbb{R}_+^{n+1}$. Ann. of Math. 128 (1988), 603–618.

[9] — Classe $L \log L$ et temps local. C.R. Acad. Sci. Sér. I. Math. 305 (1987), 135-137.

[10] Burkholder, D. L., Gundy, R. F., Extrapolation and interpolation of quasilinear operators on martingales. Acta Math. 124 (1970), 249–304.

[11] — Distribution function inequalities for the area integral. Studia Math. 44 (1972), 527–544.

[12] Dahlberg, B. E. J., Weighted norm inequalities for the Lusin area integral and nontangential maximal functions for harmonic functions in Lipschitz domains. Studia Math. 47 (1980), 297-314.

[13] — On the Poisson integral for Lipschitz and C^1 domains. Studia Math 46 (1979), 13-24.

[14] Davis, B., On the Barlow–Yor Inequalities for Local Time. In: Séminaire de Probabilitiés XXI, Lecture Notes in Math. 1247, 218–220. Springer Verlag, New York 1987.

[15] Fefferman, R., Gundy, R. F., Silverstein, M. and Stein, E. M., Inequalities for ratios of functionals of harmonic functions. Proc. Nat. Acad. Sci. U.S.A. 79 (1982), 7958-7960.

[16] Gundy, R. F., The Density of the Area Integral. In: Conference on Harmonic Analysis in honor of A. Zygmund (ed. by W. Beckner, A. P. Calderón, R. Fefferman, and P. Jones), 138–149. Wadsworth, Belmont, California 1983.

[17] — Some Topics in Probability and Analysis (CBMS 70). American Mathematical Society, Providence, Rhode Island 1989.

[18] Gundy, R. F., Silverstein, M. L., The density of the area integral in $\mathbb{R}_+^{n+1}$. Ann. Inst. Fourier (Grenoble) 35, 1 (1985), 215–224.

[19] Lenglart, E., Lepingle, D., and Pratelli, M., Presentation unifiée de certaines inéqualitiés de la théorie des martingales. In: Séminaire de Probabilitiés XIV, Lecture Notes in Math. 784, 26-48. Springer-Verlag, New York 1980.

[20] Moore, C. N., Some Inequalities for the Density of the Area Integral. In: Partial Differential Equations with Minimal Smoothness and Applications, IMA volumes in Math 42 (ed. by B. Dahlberg, E. Fabes, R. Fefferman, D. Jerison, C. Kenig, and J. Pipher), 189–198. Springer-Verlag, New York 1992.

[21] Murai, T., Uchiyama, A., Good-λ inequalities for the area integral and the nontangential maximal function. Studia Math. 83 (1986), 251-262.

[22] Stein, E. M., Note on the class $L \log L$. Studia Math. 32 (1969), 305-310.

[23] — Singular Integrals and Differentiability Properties of Functions. Princeton University Press, Princeton 1970.

Tensor product of standard H-cones and duality

Eugen Popa and Liliana Popa

Abstract. The main result (Theorem 5) asserts that, if there exists M such that, for any $n, k \in \mathbb{N}$:

$$n^{k+2}.V_n^k[v_n(\cdot, y)](x) \leq M$$

then the resolvent $\mathcal{V} = (V_\alpha)_{\alpha>0}$ is associated with an absolutely continuous semi-group, possessing an absolutely continuous adjoint semi-group. This result allows the construction of a tensor product of two standard H-cones $S' = \mathcal{E}_\mathcal{V}$ and S'' (see [10], [11]), as well as that of a tensor product of $S'^* = \mathcal{E}_{\mathcal{V}^*}$ and S''^*, such that $(S' \otimes S'')^* \sim S'^* \otimes S''^*$.

A second theorem establishes the connection with Bertin's construction of a tensor product of Dirichlet spaces (see [1]). Namely, we prove that, under common hypotheses, both constructions lead to the same product H-cone and also to the same Dirichlet product space.

1991 Mathematics Subject Classification: 31D05, 31C25, 60J35.

1. Preliminaries

$(X, \mathcal{X})$ will denote a measurable space. $\mathcal{F}$ will be the set of numerical, positive, measurable functions on $(X, \mathcal{X})$. μ will be a positive, finite measure on $(X, \mathcal{X})$. A kernel V on $(X, \mathcal{X})$ will be called *absolutely continuous* with respect to μ if: $(f \in \mathcal{F}$ and $\mu(f) = 0) \Longrightarrow Vf = 0$. Any resolvent $\mathcal{V} = (V_\alpha)_{\alpha>0}$ or semi-group $\mathcal{P} = (P_t)_{t>0}$ of kernels will be supposed to be sub-markovian. The resolvent $\mathcal{V}$ is associated to the semi-group $\mathcal{P}$ if:

$$V_\alpha f(x) = \int_0^\infty e^{-\alpha t} P_t f(x) dt, \quad \forall f \in \mathcal{F},\ x \in X.$$

The set of $\mathcal{V}$, resp. $\mathcal{P}$-excessive functions will be denoted by $\mathcal{E}_\mathcal{V}$, resp. $\mathcal{E}_\mathcal{P}$. For $\varepsilon > 0$, $\mathcal{V}_\varepsilon$ will denote the resolvent $(V_{\alpha+\varepsilon})_{\alpha>0}$.

2. The connection with Bertin's construction

X will be a locally compact, σ-compact space. $\mathcal{X}$ will denote the Borel σ-algebra of X. μ will be a Radon positive, finite measure on X; H will be a real Hilbert space, together with an imbedding $H \hookrightarrow L^2(\mu)$. H is called a *Dirichlet space* if, for any map $T : \mathbb{R} \to \mathbb{R}$ such that $T(0) = 0$ and $|Tx - Ty| \leq |x - y|$, $\forall x, y \in \mathbb{R}$ (normal contraction), we have:

$$\forall u \in H : \; T \circ u \in H \text{ and } \|T \circ u\| \leq \|u\|.$$

(It is enough that the last property holds only for the fundamental contraction, defined by: $T_1 x = \inf\{|x - t| \mid t \in [0,1]\}$.)

The adjoint of the imbedding $H \hookrightarrow L^2(\mu)$ will be denoted by $f \mapsto u_f$, being thus characterized by:

$$\forall p \in H, \forall f \in L^2(\mu) : \; < p, u_f >_H = \int p f d\mu.$$

P will denote the closure in H of the set $\{u_f \mid f \geq 0\}$.

We consider the following hypothesis:

$$\exists V \text{ bounded kernel on } (X, \mathcal{X}), \text{ absolutely continuous with respect to } \mu, \text{ such that } V1 > 0 \text{ and } \forall f \in L^2(\mu) : \; Vf \in u_f. \qquad \text{(I)}$$

$f \mapsto u_f$ can be extended to a pseudo-kernel, hence there exists a true kernel V (see [4]) such that $Vf \in u_f$. The true content of (I) is thus that the kernel V is absolutely continuous with respect to μ. Of course, if V is a strongly Feller kernel, and supp $\mu = X$, then (I) holds.

It is known (see [5]) that V satisfies the complete maximum principle, hence there exists a sub-markovian resolvent of kernels $\mathcal{V} = (V_\alpha)_{\alpha \geq 0}$ on $(X, \mathcal{X})$, such that $V_0 = V$.

For an absolutely continuous kernel V, we shall also use the notation Vf when f is a class of functions (with respect to the equality a.e.): the result does not depend on the choice of the representative. Also, we shall use the same notation for the class, as well as for any of its representatives: thus Vf may denote a class or a function.

Lemma 1. *The resolvent $\mathcal{V}$ is auto-dual.*

Proof. We have to show that:

$$\forall \alpha \geq 0, \forall f, g \in \mathcal{F} : \; \int f.V_\alpha g d\mu = \int g.V_\alpha f d\mu.$$

By hypothesis: $\forall f, g \in L^2(\mu)$:

$$\int f.V_0 g d\mu = < u_f, V_0 g >_H = < u_f, u_g >_H = < V_0 f, u_g >_H = \int g.V_0 f d\mu.$$

The general case $f, g \in \mathcal{F}$ follows, considering the sequence $f_n = \min(f, n)$. Using the resolvent equation, it remains to prove:

$$\forall \alpha > 0 : \int f.V_0 V_\alpha g d\mu = \int g.V_0 V_\alpha f d\mu.$$

This equality follows from

$$\int f.V_0 V_\alpha g d\mu = \int V_0.V_\alpha g d\mu = \int (V_\alpha f + \alpha V_0 V_\alpha f).V_\alpha g d\mu$$

$$= \int (V_\alpha f.V_\alpha g + \alpha V_\alpha f.V_0 V_\alpha g) d\mu = \int V_\alpha f.V_0 g d\mu = \int g.V_0 V_\alpha f d\mu.$$

□

Lemma 2. *P is canonically imbedded in $\mathcal{E}_V$, as an increasingly dense part.*

Proof. Let $f \in L^2(\mu)$, $f \geq 0$. Since $V_0 f \in u_f$, we deduce that $V_0 f$ is μ-a.e. finite, hence $V_0 f \in \mathcal{E}_V$. Let us define the map $\varphi(u_f) = V_0 f$. It is well defined: from $u_f = u_g$ it follows that $V_0 f = V_0 g$ μ-a.e.; being excessive functions, the equality everywhere follows. φ is clearly additive; it is also injective and increasing. We extend the map φ to the whole P as follows. Let $p \in P$; hence, there exists a sequence $(f_n)_n$ from $L^2(\mu)$ such that $f_n \geq 0$ and $u_{f_n} \to p$ in H. We can find a subsequence, such that $V_0 f_{n_k}$ converges μ-a.e. to an element, denoted by $\hat{p} \in L^2$. We further denote: $\tilde{p} = \bigvee_m \left(\bigwedge_{k \geq m} V_0 f_{n_k} \right)$. Since $\bigwedge_n V_0 f_n = \inf_n V_0 f_n$ μ-a.e., it follows that $\tilde{p}$ and $\hat{p}$ belong to the class of p. Thus, the definition: $\varphi(p) = \tilde{p}$ is correct, each class contains at most one excessive representative. In the same way, we obtain the additivity of the extended map φ. Using the same arguments as above, we obtain that, for all $p, q \in P$, we have: $p \leq q \iff \varphi(p) \leq \varphi(q)$. Finally, the fact that $\varphi(P)$ is increasingly dense in $\mathcal{E}_V$ results from the theorem of Hunt. □

Lemma 3. *$\mathcal{E}_V$ is an auto-dual, standard H-cone, and H is the associated Dirichlet space.*

Proof. It follows from [10, Prop. 6] that $\mathcal{E}_V$ is auto-dual, standard H-cone. By [2], one associates a (symmetric) Dirichlet space. Since P is increasingly dense in $\mathcal{E}_V$, P and $\mathcal{E}_V$ have the same dual H-cone; $P - P$ is total in H, we draw the conclusion.
□

We finally consider the case of the tensor product of Dirichlet spaces. Let H' and H'' be Dirichlet spaces; the tensor product (of linear spaces) $H' \otimes H''$ is endowed with the scalar product:

$$< f' \otimes f'', g' \otimes g'' > = < f', g' > . \int f'' g'' d\mu'' + < f'', g'' > . \int f' g' d\mu'.$$

Let H denote the completion of $H' \otimes H''$. Bertin [1] proved that, under the hypothesis:

$$\{f' \otimes u_{f''} + u_{f'} \otimes f'' \mid f' \in L^2(\mu'), f'' \in L^2(\mu'')\} \text{ is total in } L^2(\mu' \otimes \mu''), \qquad \text{(B)}$$

H is a Dirichlet space. The connection with the product of semi-groups is given by:

Theorem 4. *Let H', H'' be Dirichlet spaces; let V', V'' be absolutely continuous kernels on $(X', \mathcal{X}')$, $(X'', \mathcal{X}'')$ with respect to μ', μ'', such that: $V'1 > 0$, $V''1 > 0$, V' and V'' are bounded. Let us suppose that the kernels are related to the Dirichlet spaces through:*

$$\forall f' \in L^2(\mu'),\ f'' \in L^2(\mu'') \Longrightarrow V'f' \in u_{f'},\quad V''f'' \in u_{f''}.$$

If we denote by $\mathcal{V}'$, $\mathcal{V}''$ the corresponding resolvents, by $\mathcal{P}'$, $\mathcal{P}''$ the associated semi-groups, by $\mathcal{P} = \mathcal{P}' \otimes \mathcal{P}''$ the product semi-group and by $\mathcal{V}$ its associated resolvent, then the following relations hold:

(i) *The cone P of potentials from the product Dirichlet space H is increasingly dense in $\mathcal{E}_{\mathcal{V}}$.*

(ii) *The Dirichlet space associated with the auto-dual (standard) H-cone $\mathcal{E}_{\mathcal{V}}$ is H.*

Proof. As proved by Fukushima [6], $\mathcal{V}$ is associated with an absolutely continuous semi-group. Hence, the product semi-group $\mathcal{P} = \mathcal{P}' \otimes \mathcal{P}''$ is an absolutely continuous semi-group, and the associated resolvent $\mathcal{V}$ is absolutely continuous. It only remains to prove that:

$$\forall f \in L^2(\mu' \otimes \mu'') \Longrightarrow V_0 f \in u_f.$$

We consider first the case $f = f' \otimes u_{f''} + u_{f'} \otimes f''$. We have (see [1]):

$$u_{f'} \otimes u_{f''} = u_{f' \otimes u_{f''} + u_{f'} \otimes f''}.$$

On the other hand, by a formula due to Nicu Boboc (personal communication):

$$V_0 f = V_0(f' \otimes V_0'' f'' + V_0' f' \otimes f'') = V_0' f' \otimes V_0'' f'' \in u_{f'} \otimes u_{f''}.$$

Let us consider the general case $f \in L^2(\mu' \otimes \mu'')$. From the condition (B), we conclude the existence of a sequence $(f_n)_n$ which is convergent in L^2 to f; and, moreover, such that $V_0 f_n \in u_{f_n}, \forall\, n \in \mathbb{N}$. Since V_0 is bounded, we have:

$$f_n \to f \text{ in } L^2 \Longrightarrow f_n \to f \text{ in } L^1 \Longrightarrow V_0 f_n \to V_0 f \text{ in } \mathcal{L}^1.$$

On the other hand:

$$f_n \to f \text{ in } L^2 \Longrightarrow u_{f_n} \to u_f \text{ in } L^2 \text{ and } H \Longrightarrow u_{f_n} \to u_f \text{ in } L^1.$$

Hence, from $V_0 f_n \in u_{f_n}$ it follows that $V_0 f \in u_f$. □

3. A sufficient condition for the existence of an absolutely continuous adjoint semi-group

Let $\mathcal{V} = (V_\alpha)_{\alpha \geq 0}$ and $\mathcal{V}^* = (V^*_\alpha)_{\alpha \geq 0}$ be absolutely continuous resolvents, in duality with respect to μ. We may and do suppose that the cogenerators of these resolvents are bounded. A theorem of Kunita–Watanabe (see [8]) proves the existence of a family $(v_\alpha)_{\alpha \geq 0}$ such that:

- v_α is measurable on $(X \times X, \mathcal{X} \otimes \mathcal{X})$, $\mathcal{V}_\alpha$-excessive in the first variable and $\mathcal{V}^*_\alpha$ -excessive in the second variable;
- $V_\alpha f(x) = \int v_\alpha(x,y) f(y)\mu(dy)$; $V^*_\alpha f(x) = \int v_\alpha(y,x) f(y)\mu(dx)$, $\forall f \in \mathcal{F}$.

These properties define v_α uniquely. Moreover, the following relation holds:

$$v_\alpha(x,y) = v_\beta(x,y) + (\beta - \alpha)\int v_\alpha(x,z) v_\beta(z,y)\mu(dz), \ \forall \alpha < \beta.$$

We denote by $A = \{(x,y) \mid v_0(x,y) = +\infty\}$. The set

$$A_x = \{y \mid (x,y) \in A\} \ \text{ and } \ A_y = \{x \mid (x,y) \in A\}$$

are (co-)polar, $\mathcal{V}$ and μ-negligible.

Since

$$v_\alpha(x,y) = v_{\alpha_0}(x,y) + \sum_{n=1}^{\infty}(-1)^n(\alpha - \alpha_0)^n a_n(x,y),$$

where

$$\begin{aligned} a_n(x,y) &= \underbrace{\int \cdots \int}_{n \text{ times}} v_{\alpha_0}(x,y_1)\ldots v_{\alpha_0}(y_n,y)\mu(dy_1)\ldots\mu(dy_n) \\ &= V^n_{\alpha_0}[v_{\alpha_0}(\cdot,y)](x) = V^{*n}_{\alpha_0}[v_{\alpha_0}(x,\cdot)](y), \end{aligned}$$

we deduce that the derivatives of the function $f(\alpha) = v_\alpha(x,y)$ are:

$$f^{(k)}(\alpha) = (-1)^k.k! V^k_\alpha[v_\alpha(\cdot,y)](x).$$

Thus:

$$\begin{aligned} V^k_n[v_n(\cdot,y)](x) &= \frac{(-1)^k}{k!} f^{(k)}(n) = \frac{1}{k!}|f^{(k)}(n)| \\ &= \underbrace{\int \cdots \int}_{k \text{ times}} v_n(x,y_1)\ldots v_n(y_k,y)\mu(dy_1)\ldots\mu(dy_k). \end{aligned}$$

We recall (see [7, p. 221]) that, with each function $f \in \mathcal{C}^\infty((0,+\infty))$ one associates the Phillips' operator:

$$G_n(f)(t) = e^{-nt}\sum_{k=0}^{\infty}\frac{(-1)^k(n^2.t)^{k+1}}{k!(k+1)!} f^{(k)}(n).$$

In our special case:

$$f_n(t) = e^{-nt} \sum_{k=0}^{\infty} \frac{(n^2t)^{k+1}}{(k+1)!} V_n^k[v_n(\cdot, y)](x).$$

Since

$$V_n^k v_n(\cdot, y)(x) \leq V_n^k[v_0(\cdot, y)](x) \leq \frac{1}{n^k} v_0(x, y),$$

it follows that, for $(x, y) \notin A$, each function f_n is entire. Moreover, for $t \in [0, +\infty)$ $0 \leq f_n(t) \leq n.v_0(x, y)$. From the analyticity of the function f we get:

$$\int_0^{\infty} e^{-\alpha t} f_n(t)dt = \sum_{k=0}^{\infty} \frac{(-1)^k n^{2(k+1)} f^{(k)}(n)}{k!(k+1)!} \int_0^{\infty} e^{-(\alpha+n)t} t^{k+1} dt$$

$$= \sum_{k=0}^{\infty} \frac{(-1)^k n^{2(k+1)} f^{(k)}(n)}{k!(\alpha+n)^{k+2}} = \left(\frac{n}{\alpha+n}\right)^2 \sum_{k=0}^{\infty} \frac{f^{(k)}(n)}{k!} \left(\frac{\alpha n}{\alpha+n} - n\right)^k$$

$$= \left(\frac{n}{\alpha+n}\right)^2 f\left(\frac{\alpha n}{\alpha+n}\right), \ \forall \alpha \geq 0.$$

Hence

$$\lim_{n \to +\infty} \int_0^{\infty} e^{-\alpha t} f_n(t)dt = f(\alpha), \ \forall \alpha \in [0, +\infty),$$

the convergence being uniform on compacts in $(0, +\infty)$. The relation also holds for complex α, such that $\operatorname{Re}\alpha \geq 0$, $\alpha \neq 0$ and the convergence is uniform on compacts in the half-plane $[\operatorname{Re}\alpha > 0]$. In particular,

$$\int_0^{\infty} f_n(t)dt = v_0(x, y).$$

Moreover,

$$f_n'(t) = e^{-nt} \sum_{k=0}^{\infty} \frac{(-1)^k n^{2k+1} t^k}{k!k!} [k f^{(k-1)}(n) + n f^{(k)}(n)].$$

Theorem 5.
(a) *Let us suppose that, for each $(x, y) \notin A$, there exists $M \geq 0$ such that:*

$$\forall n, k \in \mathbb{N}: \quad n^{k+1} \underbrace{\int \cdots \int}_{k \text{ times}} v_n(x, y_1) \ldots v_n(y_k, y)\mu(dy_1)\ldots\mu(dy_k) \leq M.c_k$$

where $c_k \geq 0$, $\forall k$ and $\sum_{k=0}^{\infty} c_k < +\infty$. Then there exists a function

$$(t, (x, y)) \mapsto p_t(x, y)$$

defined and measurable on $(0, +\infty) \times X \times X$ such that, for each $(x, y) \notin A$, the function $t \mapsto p_t(x, y)$ is bounded and with bounded variation on $(0, +\infty)$

and satisfies the equality:

$$\forall \alpha \geq 0: \quad v_\alpha(x,y) = \int_0^\infty e^{-\alpha t} p_t(x,y)dt.$$

(b) *Let us suppose that, for each $(x,y) \notin A$, there exists $M \geq 0$ such that:*

$$\forall n, k \in \mathbb{N}: \quad n^{k+2} \underbrace{\int \ldots \int}_{k \text{ times}} v_n(x, y_1) \ldots v_n(y_k, y)\mu(dy_1) \ldots \mu(dy_k) \leq M.$$

Then there exists a function

$$(t,(x,y)) \mapsto p_t(x,y)$$

defined and measurable on $(0,+\infty) \times X \times X$ such that, for each $(x,y) \notin A$, the function $t \mapsto p_t(x,y)$ is continuous on $(0,+\infty)$.

Proof. a) Directly from the definition of f_n we obtain:

$$\forall t \in (0,+\infty): \quad 0 \leq f_n(t) \leq M.$$

Hence [14, §16], there exists a subsequence $(f_{n_k})_k$, weakly convergent to a bounded function, that is, for each $(x,y) \notin A$, there exists $p_t(x,y)$ such that, for any $g \in L^1((0,+\infty))$,

$$\int_0^\infty f_{n_k}(t)g(t)dt \to \int_0^\infty p_t(x,y)g(t)dt.$$

Choosing $g(t) = e^{-\alpha t}$, for $\alpha > 0$, we obtain:

$$v_\alpha(x,y) = \int_0^\infty e^{-\alpha t} p_t(x,y)dt.$$

Moreover, the same hypothesis ensures the uniform bounded variation of our sequence:

$$\bigvee_0^\infty f_n = \int_0^\infty |f_n'(t)|dt \leq \sum_{k=0}^\infty \frac{n^{2k+1}}{k!k!} \left[k|f^{(k-1)}(n)| + n|f^{(k)}(n)|\right] \int_0^\infty t^k e^{-nt} dt$$
$$\leq 2M. \sum_{k=0}^\infty c_k.$$

From [14, Th. 16.3], it follows the existence of a pointwise convergent subsequence to a function $p_t(x,y)$, which is bounded and of bounded variation. The measurability is clear.

b) From the hypothesis we obtain: $\forall\, t \in [0,+\infty), \forall\, n \geq 0: \quad |f_n'(t)| \leq 2.M$. Using Ascoli–Arzelà's theorem, we obtain a subsequence uniformly convergent on compact sets in $(0,+\infty)$ to a continuous function $t \mapsto p_t(x,y)$. □

Let us consider now the complex case.

Proposition 6.

(a) *If for each* $(x,y) \notin A$ *there exists* $\alpha \in (0,1)$ *and* $M > 0$ *such that:*

$$\forall n,k \in \mathbb{N}: \quad n^{k+1} \underbrace{\int \dots \int}_{k \text{ times}} v_n(x,y_1)\dots v_n(y_k,y)\mu(dy_1)\dots\mu(dy_k) \leq \alpha^{k+1} M,$$

then, for each $(x,y) \notin A$, *the function* $t \mapsto p_t(x,y)$ *is analytic in the angle*

$$D_\alpha = \left\{ t = t_1 + i.t_2 \;\middle|\; |t_2| \leq t_1.\sqrt{\frac{1}{\alpha^2} - 1}, \; t_1 > 0 \right\}.$$

(b) *If, for each* $(x,y) \notin A$, *the function* $\beta \mapsto v_{i\beta}(x,y)$ *is absolutely integrable on* $\mathbb{R}$, *then the function* $t \mapsto p_t(x,y)$ *is continuous and bounded on* $(0,+\infty)$.

Proof. a) Indeed, for any $t \in D_\alpha$ we have

$$|f_n(t)| \leq e^{-n.\operatorname{Re} t} \sum_{k=0}^{\infty} \frac{n^{k+2}|t|^{k+1}}{(k+1)!} \frac{\alpha^{k+1}}{n^{k+1}} M \leq e^{-n(\operatorname{Re} t - \alpha|t|)} M.$$

Since for any $t \in D_\alpha$: $\operatorname{Re} t - \alpha|t| \geq 0$, the sequence $(f_n)_n$ is uniformly bounded in D_α. By Montel's theorem, there exists a subsequence uniformly convergent on compacts in D_α, to an analytic function in D_α.

b) It is well known that the function is given by:

$$p_t(x,y) = \frac{1}{2\pi i} \int_{-\infty}^{+\infty} e^{it\beta} v_{i\beta}(x,y) d\beta.$$

□

Remark. The two preceding results hold also if we suppose that the inequalities are valid only for sufficiently large k.

The imposed conditions are "almost" necessary since for a bounded function, resp. with bounded first derivative, resp. with continuous second order derivative, its Laplace transform F satisfies:

$$\forall \alpha > 0: \; \alpha^{k+1}|F^{(k)}(\alpha)| \leq M.k!,$$

resp.

$$\forall \alpha > 0: \; \alpha^{k+2}|F^{(k)}(\alpha)| \leq M.k!,$$

resp.

$$\beta \mapsto F(i\beta) \text{ is absolutely integrable on } \mathbb{R}$$

In [7], this result is applied to generators of (C_0)-semi-groups of operators (and *not* to the case of absolutely continuous resolvents of kernels).

Proposition 7. *If the function*

$$(t,(x,y)) \mapsto p_t(x,y)$$

is defined and measurable on $(0, +\infty) \times (X \times X \setminus A)$, *and, for each* $(x, y) \notin A$, *the function is right continuous in* $t \in (0, +\infty)$, *then:*

$$P_t f(x) = \int p_t(x, y) f(y) \mu(dy); \; P_t^* f(x) = \int p_t(y, x) f(y) \mu(dy)$$

define absolutely continuous semi-groups in duality.

Proof. (See [10].) Since, for any $x \in X$, $p_t(x, \cdot)$ (resp. $p_t(\cdot, x)$) is defined a.e. and measurable, the definition is correct. We obtain the kernels P_t, P_t^*, which are absolutely continuous and in duality. Let us prove that $(P_t)_{t>0}$ and $(P_t^*)_{t>0}$ are semi-groups. We have to show that:

$$\forall (x, y) \notin A \; , \forall s, t > 0 : p_{s+t}(x, y) = \int p_t(x, z) p_s(z, y) \mu(dz).$$

It is enough to verify the equality of the Laplace transforms (in the variable t), since both members depend right continuously on t. This amounts to:

$$\int_0^\infty e^{-\alpha t} p_{t+s}(x, y) dt = \int v_\alpha(x, z) p_s(z, y) \mu(dz).$$

Since the right continuity in the variable s is preserved, we further verify the equality of the Laplace transforms in the variable s:

$$\int_0^\infty e^{-\beta s} ds \int_0^\infty e^{-\alpha t} p_{t+s}(x, y) dt = \int v_\alpha(x, z) v_\beta(z, y) \mu(dz).$$

But the first member can be transformed, when $\beta < \alpha$:

$$\begin{aligned}
&\int_0^\infty e^{-(\alpha-\beta)t} dt \left(\int_t^\infty e^{-\beta u} p_u(x, y) du \right) \\
&\quad = \int_0^\infty e^{-(\alpha-\beta)t} \left[v_\beta(x, y) - \int_0^t e^{-\beta u} p_u(x, y) du \right] dt \\
&\quad = \frac{1}{\alpha - \beta} v_\beta(x, y) - \int_0^\infty \int_u^\infty e^{-(\alpha-\beta)t} e^{-\beta u} p_u(x, y) dt du \\
&\quad = \frac{1}{\alpha - \beta} [v_\beta(x, y) - v_\alpha(x, y)] = \int v_\alpha(x, z) v_\beta(z, y) \mu(dz).
\end{aligned}$$

□

Remark. The tensor product of various harmonic structures (harmonic spaces, standard spaces of balayage, standard H-cones) is most easily obtained through the product of the attached semi-groups (see: R. Cairoli [3] for the product of semi-groups, L. Stoica [13], U. Schirmeier [12], E. Popa [9] for the product of harmonic structures).

We are interested in those semi-groups having an absolutely continuous associated resolvent: in this case $\mathcal{E}_{\mathcal{P}}$ is a standard H-cone. However, it may happen that two semi-groups have the associated resolvents absolutely continuous, yet the resolvent associated to the product semi-group need *not be* absolutely continuous. In [11] we have proved that if, moreover, one of the semi-groups is absolutely continuous, then the product resolvent is absolutely continuous. This is the reason why we are interested in situations which ensure the absolute continuity of a semi-group.

Since Theorem 5 also guarantees the absolute continuity of the dual semi-group, it allows the construction of both tensor products $\mathcal{E}_{\mathcal{P}} \otimes \mathcal{E}_{\mathcal{Q}}$ and $\mathcal{E}_{\mathcal{P}^*} \otimes \mathcal{E}_{\mathcal{Q}^*}$. Moreover (see [10]), there exists a canonical identification, through the energy form:

$$(\mathcal{E}_{\mathcal{P}} \otimes \mathcal{E}_{\mathcal{Q}})^* \sim \mathcal{E}_{\mathcal{P}^*} \otimes \mathcal{E}_{\mathcal{Q}^*} = \mathcal{E}_{\mathcal{P}^* \otimes \mathcal{Q}^*}.$$

References

[1] Bertin, E. M. J., Un produit d'espaces de Dirichlet. Comm. Math. Rijk. Univ. Utrecht 12 (1980), 26–41.

[2] Boboc, N., Bucur, Gh., Cornea, A., Order and Convexity in Potential Theory: H-Cones. Lecture Notes in Math. 853, Springer-Verlag, Berlin–Heidelberg–New York 1981.

[3] Cairoli, R., Semigroupes de transition et fonctions excessives. In: Séminaire de Probabilités I, 18–33. Lecture Notes in Math. 39, Springer-Verlag, Berlin 1967.

[4] Dellacherie, C., Meyer, P.-A., Probabilités et Potentiel. Act. Sci. Ind. Hermann, Paris 1987.

[5] Deny, J., Hilbertian Methods in Potential Theory. In: Potential Theory, CIME Stresa 1969, Ed. Cremonese, Roma 1970.

[6] Fukushima, M., Dirichlet Forms and Markov Processes. North Holland, Amsterdam 1980.

[7] Hille, E., Phillips, R. S., Functional Analysis and Semi-groups. A. M. S. Coll. Publ. 31, Providence, Rhode Island 1981.

[8] Kunita, H., Watanabe, T., Markov Processes and Martin Boundaries. Illinois J. Math. 9 (1965), 485–526.

[9] Popa, E., Localisation and Product of H-cones. Rev. Roum. Math. Pures Appl. 28(8) (1983), 408–430.

[10] — Tensor Product of Standard H-cones. An. Ştiinţ. Univ. Iaşi (to appear).

[11] — Products of absolutely continuous semi-groups. C. R. Acad. Sci. Paris, 317 (1993), 1119–1122.

[12] Schirmeier, U., Produkte harmonischer Räume. Bayerische Akad.Wiss. München (1978), 5–22.

[13] Stoica, L., Local Operators and Markov Processes. Lecture Notes in Math. 816. Springer-Verlag, Berlin–Heidelberg–New York 1980.

[14] Widder, D. V., The Laplace Transform, Princeton Univ. Press, Princeton 1946.

An example of packing related small sets

Pavel Pyrih *

Abstract. We give an example of disappearing "Champagne bubbles" obtained from packings in the plane. The Lebesgue measure and the Hausdorff capacity of these sets approach zero, but the logarithmic capacity equals one.

1991 Mathematics Subject Classification: 30G12.

1. Introduction

We begin with some basic definitions

Definition 1.1. Let D be an open unit disc in the plane. By a *packing* P of D we will mean a sequence $\{D_n\}_{n=1}^{\infty}$, of open disjoint subdiscs of D, such that the plane Lebesgue measures of D and $\bigcup_{n=1}^{\infty} D_n$ are the same. Let $D_n = D(z_n, r_n)$ ($D(z,r)$ denotes the disc with centre z and radius r). There exists a uniquely defined number $e(P)$, such that $1 \le e(P) \le 2$ and

$$\sum_{n=1}^{\infty} r_n^a$$

converges for $a > e(P)$ and diverges for $a < e(P)$. (See [5].) The number $e(P)$ will be called the *exponent of the packing* P. By an *osculatory packing* of D we understand a collection $\{D_n\}_{n=1}^{\infty}$ of disjoint open subdiscs of D constructed in such a way, that D_1 is internally tangent to D and every disc from the second one onward is the largest possible.

The following theorem was proved by Melzak in [5, Theorem 3].

Theorem 1.2. *An osculatory packing P of D is a packing and its exponent satisfies*

$$1.035 < e(P) < 1.999971\,.$$

* Research supported by the grant No.201/93/2174 of Czech Grant Agency and by the grant No. 354 of Charles University.

We recall that the *fine topology* on $\mathbb{R}^2$ is the smallest topology on $\mathbb{R}^2$ for which all superharmonic functions are continuous in the extended sense. In connection with this topology we will use terms such as finely open, fine interior, ... A set $E \subset \mathbb{R}^2$ is *thin* at x if x is not a fine limit point of E. The *Wiener test* relates thinness of E to the *logarithmic capacity* (cap) of certain subsets of E.

Using the fine topology *finely holomorphic functions* were defined in the 80s. We state one of the equivalent definitions (we identify $\mathbb{R}^2$ with $\mathbb{C}$)

Definition 1.3. A function $f : U \to \mathbb{C}$ (U finely open in $\mathbb{C}$) is said to be *finely holomorphic* in U if every point of U has a $p > 1$ and a fine neighbourhood $V \subset U$ in which

$$f(z) = \int_{\mathbb{C}} \frac{\varphi(\zeta)}{z-\zeta} d\zeta, \quad z \in V,$$

where $\varphi \in L^p_c(\mathbb{C})$ and $\varphi = 0$ almost everywhere (a.e.) with respect to Lebesgue measure in V.

The unique continuation property of finely holomorphic functions is expressed by

Theorem 1.4. *A finely holomorphic function f on a fine domain has at most countably many zeros (unless $f = 0$).*

(See [4].)

2. An example

Theorem 2.1. *Let $P = \{D(z_m, \rho_m)\}_{m=1}^{\infty}$ be an osculatory packing in the unit disc D. Fix $t \geq 1$. For $0 < s < 1$ we set*

$$E_s = \bigcup_{m=1}^{\infty} D(z_m, s.\rho_m^t).$$

Then the fine closure $\widetilde{E}_s^f$ of E_s contains the unit circle T for $0 < s < 1$.

Proof. Suppose there exist $0 < s < 1$ and $x \in T$ such that $x \notin \widetilde{E}_s^f$. Define $U := \mathbb{C} \setminus \widetilde{E}_s^f$. Then $x \in U$. Let V be the fine component of U containing x. The discs $D(z_m, s.\rho_m^t)$ are disjoint for all $m \in \mathbb{N}$ due to the choice of t and s. We define

$$\varphi(\zeta) = \begin{cases} \left(\frac{\rho_m}{s.\rho_m^t}\right)^2, & \text{for } \zeta \in D(z_m, s.\rho_m^t),\ m \in \mathbb{N}, \\ 0, & \text{elsewhere.} \end{cases}$$

Choose $p > 1$ such that $2t + 2p - 2pt > e(P)$ (we use $1 < e(P) < 2$ from Theorem 1.2). Thus

$$\|\varphi\|_p^p = \sum_{m=1}^{\infty} \pi.(s.\rho_m^t)^2.((\frac{\rho_m}{s.\rho_m^t})^2)^p = \sum_{n=1}^{\infty} \pi.s^{2-2.p}.\rho_m^{2t+2p-2pt} < \infty$$

due to the choice of t and p. Hence $\varphi \in \mathbf{L}^p(C)$ with $\varphi = 0$ on U. We define

$$f(z) = \int_C \frac{\varphi(\zeta)}{\zeta - z} d\zeta$$

for $z \in U$. Then f is finely holomorphic on U. For $|z| > 1$ we have

$$f(z) = \sum_{m=1}^{\infty} \int_{B(z_m, s.\rho_m^t)} \frac{\rho_m^2}{s^2.\rho_m^{2.t}} \cdot \frac{1}{\zeta - z} d\zeta = \sum_{m=1}^{\infty} \int_{B(z_m, \rho_m)} \frac{1}{\zeta - z} d\zeta$$
$$= \int_{B(0,1)} \frac{1}{\zeta - z} d\zeta = -\frac{\pi}{z}$$

due to the mean value property and Theorem 1.2 (an osculatory packing has full measure). Hence f has a finely holomorphic extension from V to $\mathbb{C} \setminus \{0\}$. Denote this extension by h. Then $h(z) = -\pi/z$ for $z \in \mathbb{C} \setminus \{0\}$. Now we construct another finely holomorphic extension of f, which leads to the contradiction.

We observe that $\rho_1 + \rho_2 = 1$ due to the construction of the packing P. The point 0 belongs either to $D(z_1, \rho_1)$ (iff $\rho_1 > \frac{1}{2}$), or to $D(z_2, \rho_2)$ (iff $\rho_2 > \frac{1}{2}$), or does not belong to any disc $D(z_m, \rho_m)$ (iff $\rho_1 = \rho_2 = \frac{1}{2}$). Thus $z_m \neq 0$ for $m \in \mathbb{N}$. Define

$$M := \bigcup_{m=1}^{\infty} D(z_m, \rho_m),$$
$$e_n := \{z \in \mathbb{C}, \frac{1}{2^n} \leq |x - z| < \frac{2}{2^n}\},$$
$$M_n := M \cap e_n$$

for $n \in \mathbb{N}$. Then $\lambda(M_n) \geq \pi.(\frac{1}{2^{n+1}})^2$ (since $x \in T$ and the packing has full measure in the unit disc). We use the classical inequality

$$\lambda(A) \leq \pi.(\mathrm{cap}\,(A))^2$$

(for a bounded Borel set A; see [4]) and we estimate

$$\mathrm{cap}\, M_n \geq \frac{1}{2^{n+1}}, \quad n \in \mathbb{N}.$$

Due to the Wiener test M is not thin at x. There exist $m_0 \in \mathbb{N}$ and $y_0 \in D(z_{m_0}, \rho_{m_0}) \cap V$. We have ($z \in V$)

$$f(z) = \sum_{m=1}^{\infty} \int_{D(z_m, s.\rho_m^t)} \frac{\varphi(\zeta)}{\zeta - z} d\zeta = \int_{\mathbb{C} \setminus D(z_{m_0}, \rho_{m_0})} \frac{\varphi(\zeta)}{\zeta - z} d\zeta + \frac{\pi.\rho_{m_0}^2}{z_{m_0} - z}$$

by the mean value theorem. The function written as the last integral is finely holomorphic in the finely open and finely connected set

$$V_0 := V \cup D(z_{m_0}, \rho_{m_0}),$$

and is therefore finely continuous at z_{m_0}. Thus f has finely holomorphic extension from V to the fine domain $V_0 \setminus \{z_{m_0}\}$. Let g denote this extension. Therefore (for $z \in V_0 \setminus \{z_{m_0}\}$)

$$g(z) = \int_{\mathbb{C}\setminus D(z_{m_0},\rho_{m_0})} \frac{\varphi(\zeta)}{\zeta - z} d\zeta + \frac{\pi.\rho_{m_0}^2}{z_{m_0} - z}.$$

The finely holomorphic extensions g and h of the function f from V to

$$W := D(z_{m_0}, \rho_{m_0}) \setminus \{z_{m_0}, 0\}$$

should be equal on W due to Theorem 1.4. Therefore $h = g$ on W and (since we know $z_{m_0} \neq 0$)

$$\lim_{\substack{z \to z_{m_0} \\ z \neq z_{m_0}}} h(z) = h(z_{m_0}).$$

But

$$\lim_{\substack{z \to z_{m_0} \\ z \neq z_{m_0}}} g(z)$$

does not exist (because of the term $(z_{m_0} - z)^{-1}$). This is a contradiction. □

Definition 2.2. For a set $E \subset \mathbb{R}^2$ we define the *Hausdorff capacity* $H^d_\infty(E)$ of E of dimension d $(0 < d \leq 2)$ by

$$H^d_\infty(E) := \inf \sum_j r_j^d$$

where the infimum is over all countable open covers of E by balls $D(z_j, r_j)$.

This H^d_∞ is a monotone, countably subadditive set function on the class of all subsets in $\mathbb{R}^2$ which vanishes on the empty set. This means that H^d_∞ is a capacity in the sense of N. Meyers ([6, p. 257]). (See [1].)

Remark 2.3. We denote by $E(s,t)$ the set E_s from Theorem 2.1 for $t \geq 1$, $0 < s < 1$. The Lebesgue measure of $E(s,t)$ approaches zero as $s \to 0$, and the Hausdorff capacity $H^d_\infty(E(s,t))$ approaches zero as $s \to 0$ for $d > e(P)/t$, but the logarithmic capacity $\operatorname{cap}(E(s,t)) = 1$ because the logarithmic capacity is a fine capacity, i.e., $\operatorname{cap}(E) = \operatorname{cap}(\widetilde{E}^f)$ (see [2, p. 66, Theorem VIII, 12]).

References

[1] David, R. A., A note on Choquet integrals with respect to Hausdorff capacity. Lecture Notes in Math. 1302, 115–124. Springer-Verlag, New York–Berlin–Heidelberg 1988.

[2] Brelot, M., On Topologies and Boundaries in Potential Theory. Lecture Notes in Math. 175, Springer-Verlag, New York–Berlin–Heidelberg 1968.

[3] Fuglede, B., Fine Topology and Finely Holomorphic Functions. Proc. 18-th Scand. Congr. Math. Aarhus 1980, 22–38.

[4] — Sur les fonctions finement holomorphes. Ann. Inst. Fourier (Grenoble) 31.4 (1981), 57–88.

[5] Melzak, Z. A., Infinite packings of discs. Canad. J. Math. 18 (1966), 838–852.

[6] Meyers, N., A theory of capacities for potentials of functions in Lebesgue classes. Math. Scand. 26 (1970), 255–292.

On the domain of the generator of a subordinate semigroup

René L. Schilling

Abstract. For a large subclass of Bernstein functions f we prove a new integral representation formula for the generator A^f of a subordinate (C_0)-semigroup on a Banach space. We show that $A^f = -f(-A)$ in the sense of Dunford's operational calculus. If A, B are two self-adjoint generators, order relations for the associated quadratic forms are preserved, and if $D(A) = D(B)$ we then have $D(A^f) = D(B^f)$. As an application, we can describe the domain of certain pseudo-differential operators generating Feller semigroups.

1991 Mathematics Subject Classification: 47D03, 47D07, 47G30, 47B44.

1. Introduction

Subordination of (C_0)-semigroups of operators is a well-known technique due to Bochner [3] to obtain new (C_0)-semigroups $\{T_t^f\}_{t\geq 0}$ from a given one $\{T_t\}_{t\geq 0}$—the superscript f refers to the Bernstein function that characterizes the underlying subordinator. When we consider the semigroups' generators A and A^f, respectively, subordination turns out to give a functional calculus for a certain class of not necessarily bounded operators, cf. [1], [15] and, in a somewhat different direction, [7]. In [15] an explicit formula for A^f in terms of the original semigroup $\{T_t\}_{t\geq 0}$ is given. In this general setting little more than $D(A) \subset D(A^f)$ can be said on the domains of the generators. In fact, $D(A)$ is a core for A^f (Lemma 2.4) and $D(A) = D(A^f)$ if and only if f grows at least linearly or A is a bounded operator (Thm. 5.1).

More information can be obtained if we restrict ourselves to *complete* Bernstein functions $f \in \mathcal{CBF}$. This class of functions is—with different definitions and names—well-known and has been investigated by various authors, e.g., [1], [6], [13], [16]. That their definitions coincide with ours (Def. 3.1) is, therefore, shown in Section 3. The main feature about complete Bernstein functions is that they allow a nicer integral representation (4.2) of A^f involving A and its resolvents only. In this setting we show that, in the sense of the Dunford–Taylor operational calculus, $A^f = -f(-A)$ holds for all $f \in \mathcal{CBF}$.

Applying the representation formula (4.2) to Hilbert space operators we are able to recover Heinz' result [6, Satz 2] that complete Bernstein functions preserve the monotonicity of operators, i. e., that in the quadratic form sense $A \leq B$ implies

$A^f \le B^f$. Furthermore, we prove a new Heinz–Kato type theorem (Thm. 5.3, Corollary 5.6) stating that $D(A^f) \subset D(B^f)$ whenever A, B generate self-adjoint (C_0)-semigroups and satisfy $\|Au\| \le c(\|Bu\|+\|u\|)$. In particular, $D(A^f) = D(B^f)$ holds if $\|Au\| + \|u\| \sim \|Bu\| + \|u\|$.

This situation is, however, quite familiar when considering stochastic (Lévy-type) processes which are generated by a pseudo-differential operator $-p(x, D)$. In [9], [10], [11], see also Example 5.7, sufficient conditions on the symbol $p(x,\xi)$ are given such that $-p(x, D)$ generates a stochastic (Feller) process. Moreover, the technique employed there shows $\|Au\| + \|u\| \sim \|p(x, D)u\| + \|u\|$ where A denotes the generator of a fixed Lévy process. Having in mind that the domains $D(A)$ and $D(A^f)$ can be neatly described using a scale of anisotropic Sobolev spaces $H^{1,a}$, resp., $H^{1,f\circ a}$, see Example 5.7 and [2, §12], [9], [10, [11], we immediately find $H^{1,f\circ a} = D(-p(x, D))$. This, in turn, has an interesting consequence: although in the above situation the classical symbolic calculus for pseudo-differential operators is not available (the symbols lack to be differentiable), the equality $H^{1,f\circ a} = D(-p(x, D))$ indicates that the pseudo-differential operator with symbol $f(p(x,\xi))$ should be a reasonable first-order approximation for $-p(x, D)^f = -f(-p(x, D))$, as it were natural in the classical setting.

2. Notations and preliminary results

Let us, first of all, recall some basic notions from semigroup theory, see, e. g., [4]. A (C_0)-semigroup on a Banach space $\mathcal{X}$ is a one-parameter family of operators $\{T_t\}_{t\ge0}$ on $\mathcal{X}$ satisfying $T_{t+s}u = T_tT_su$ and $\lim_{t\downarrow0} \|T_tu - u\|$ for all $u \in \mathcal{X}$ and $s, t \ge 0$. The semigroup is called *contractive* if $\|T_t\| \le 1$ holds for all $t \ge 0$. The *infinitesimal generator* A of a (C_0)-semigroup is a densely defined closed operator on $\mathcal{X}$ given by

$$Au = \lim_{t\downarrow0} \frac{T_tu - u}{t} = \frac{d}{dt}T_tu\Big|_{t=0} \quad \text{on} \quad D(A) = \{u \in X : \lim \text{ exists in } \mathcal{X}\}.$$

Since $\{T_t\}_{t\ge0}$ is uniquely determined by A (and vice versa), we can always write e^{tA} for T_t. The number $\omega_0 = \omega_0(A) \in [-\infty, \infty)$ defined by

$$\omega_0 = \inf_{t>0} \frac{\log \|e^{tA}\|}{t} = \lim_{t\to\infty} \frac{\log \|e^{tA}\|}{t}$$

is called the semigroup's *type*. For any $\omega > \omega_0$ there is a constant $M = M_\omega \ge 1$ such that

$$e^{\omega_0 t} \le \|e^{tA}\| \le M\, e^{\omega t}, \qquad t \ge 0, \tag{2.1}$$

holds. If $\{e^{tA}\}_{t\ge0}$ is equi-bounded we can always choose $\omega = 0$. The *resolvent set* $\rho(A)$ of A contains the half-plane $\{z \in \mathbb{C} : \Re z > \omega_0\}$ and we have the following resolvent inequalities

$$\|(t-A)^{-1}\| \le \frac{M}{(\Re t - \omega)} \quad \text{and} \quad \|A(t-A)^{-1}\| \le \frac{M|t|}{(\Re t - \omega)} + 1. \tag{2.2}$$

A *subordinator* is a vaguely continuous convolution semigroup of sub-probability measures $\{\mu_t\}_{t\ge0}$ on $[0,\infty)$. A well-known theorem due to Bernstein establishes the following one-one correspondence

$$\int_0^\infty e^{-sx}\,\mu_t(ds) = e^{-tf(x)}, \qquad t, x \ge 0,$$

between $\{\mu_t\}_{t\ge0}$ and the *Bernstein functions* $\mathcal{BF}$, i.e., C^∞-functions with representation

$$f(x) = a + bx + \int_0^\infty (1-e^{-sx})\,\mu(ds), \qquad x \ge 0, \tag{2.3}$$

with constants $a, b \ge 0$ and a measure μ on $(0,\infty)$ satisfying $\int_0^\infty \frac{s}{s+1}\,\mu(ds) < \infty$. A good reference for Bernstein functions is the book [2, §9].

We will now discuss how to obtain new (C_0)-semigroups from a given one $\{e^{tA}\}_{t\ge0}$. If the subordinator $\{\mu_t\}_{t\ge0}$ satisfies $\int_0^\infty \|e^{sA}\|\,\mu_t(ds) < \infty$ for all $t \ge 0$, the (Bochner) integral

$$(e^{tA})^f u \equiv (e^{tA^f})u \equiv \int_0^\infty e^{sA}u\,\mu_t(ds), \qquad u \in \mathcal{X}, \tag{2.4}$$

makes sense and defines again a (C_0)-semigroup.

Definition 2.1. Let $\{\mu_t\}_{t\ge0}$ be a subordinator and $\{e^{tA}\}_{t\ge0}$ be a (C_0)-semigroup on $\mathcal{X}$.

(i) $\{\mu_t\}_{t\ge0}$ is called *admissible* (w.r.t. $\{e^{tA}\}_{t\ge0}$) if $\int_0^\infty \|e^{sA}\|\,\mu_t(ds) < \infty$ for all $t \ge 0$.

(ii) If $\{\mu_t\}_{t\ge0}$ is admissible, the semigroup $\{e^{tA^f}\}_{t\ge0}$ defined by (2.4) is called *subordinate* to $\{e^{tA}\}_{t\ge0}$ (in the sense of Bochner [3, chap. 4.4]).

It is often convenient to assume that an admissible $\{\mu_t\}_{t\ge0}$ satisfies the stronger condition $\int_0^\infty e^{s\omega}\,\mu_t(ds) < \infty$ with ω as in (2.1). Such subordinators will be called *cute*. A useful characterization of admissible, resp., cute subordinators is contained in the following Proposition.

Proposition 2.2 ([17, Satz 2.13]). *Let $\{\mu_t\}_{t\ge0}$ be a subordinator with $f \in \mathcal{BF}$ as in (2.3). For all $\beta \ge 0$ the following assertions are equivalent:*

(i) $\int_0^\infty e^{s\beta}\,\mu_t(ds) < \infty$ *for all* $t \ge 0$.

(ii) f *has a C^∞-extension onto $(-\beta,\infty)$ which is continuous up to $-\beta$.*

(iii) f *has an analytic extension onto $\{z \in \mathbb{C} : \Re z > -\beta\}$ which is continuous up to the boundary.*

(iv) $\int_1^\infty e^{s\beta}\,\mu(ds) < \infty$.

The infinitesimal generator A^f of a subordinate semigroup was explicitly calculated by Phillips ([15, Thms. 4.3, 4.4]):

Theorem 2.3. *Let $\{e^{tA}\}_{t\geq 0}$ be a (C_0)-semigroup and $\{\mu_t\}_{t\geq 0}$ be an admissible subordinator with $f \in \mathcal{BF}$ as in (2.3). Then $D(A)$ is contained in the domain $D(A^f)$ of the generator A^f of $\{e^{tA^f}\}_{t\geq 0}$ and*

$$A^f u = au + bAu + \int_0^\infty (e^{sA}u - u)\,\mu(ds), \qquad u \in D(A), \tag{2.5}$$

is valid. The spectrum of A^f satisfies $\sigma(A^f) \supset -f(-\sigma(A))$.

In general, the domain $D(A^f)$ cannot be explicitly calculated. We have, however, $\overline{A^f|D(A)} = A^f$ as the following lemma shows.

Lemma 2.4. *Let $\{e^{tA}\}_{t\geq 0}$ be a (C_0)-semigroup and $\{\mu_t\}_{t\geq 0}$ be an admissible subordinator. Then $D(A^k)$ is for any $k \in \mathbb{N}$ a core of $(A^f, D(A^f))$.*

Proof. It is sufficient to show that e^{tA^f} maps $D(A^k)$ into itself, cf. [4, Thm. 1.9]. To this end, fix $t \geq 0$, choose any $u \in D(A^k)$, and note that the sequence $\left\{\int_0^n e^{sA}u\,\mu_t(ds)\right\}_{n\in\mathbb{N}}$ converges to $e^{tA^f}u$. By the closedness of the operator A,

$$\left\|A^k \int_m^n e^{sA}u\,\mu_t(ds)\right\| = \left\|\int_m^n e^{sA}A^k u\,\mu_t(ds)\right\| \leq \int_m^n \|e^{sA}\|\,\mu_t(ds)\,\|A^k u\|,$$

that is, $\left\{A^k \int_0^n e^{sA}u\,\mu_t(ds)\right\}_{n\in\mathbb{N}}$ is a Cauchy sequence. Hence, an induction argument using the closedness of A implies $e^{tA^f}u \in D(A^k)$. □

3. Complete Bernstein functions

We will now introduce a subclass of $\mathcal{BF}$ that was repeatedly investigated by various authors, e. g., [1], [6], [13], [16], for a more recent bibliography consult [1].

Definition 3.1. A function $f : (0,\infty) \to \mathbb{R}$ is called *complete Bernstein function*, $\mathcal{CBF}$ for short, if it has the following representation

$$f(x) = x^2 \int_0^\infty \phi(s)e^{-sx}\,ds, \qquad x > 0, \tag{3.1}$$

with $\phi \in \mathcal{BF}$.

Example 3.2. The functions $x \mapsto c$, $x \mapsto cx$, $x \mapsto \frac{x}{x+c}$, $(c \geq 0)$, $x \mapsto x^\alpha$, $(0 \leq \alpha \leq 1)$, $x \mapsto \sqrt{x}\arctan\frac{c}{x}$, $(c \geq 0)$, or $x \mapsto \log(1+x)$ are complete Bernstein functions, whereas $x \mapsto 1 - e^{-cx}$ $(c > 0)$ is in $\mathcal{BF}$ but not of class $\mathcal{CBF}$.

One can show that every Bernstein function f has a representation similar to (3.1) with ϕ from a larger class of functions than $\mathcal{BF}$, see [16, chap. I.4.1]. Some authors call complete Bernstein functions *operator monotone* for these are the only functions that preserve the positivity of Hilbert space operators. In finite-dimensional spaces this is due to Löwner [13], the necessity part of his proof also applies to infinite dimensional settings. That $\mathcal{CBF}$ are operator monotone in any Hilbert space is due to Heinz [6], a new proof is given in Theorem 5.3 below.

The next theorem shows the equivalence of the class $\mathcal{CBF}$ with the various different definitions used by the above mentioned authors.

Theorem 3.3. *Each of the following five properties of a function* $f : (0,\infty) \to \mathbb{R}$ *implies the other four:*

(i) $f \in \mathcal{CBF}$

(ii) *The function* $x \mapsto \frac{f(x)}{x}$ *is a Stieltjes transform, i. e.* $\frac{f(x)}{x} = \frac{a}{x} + b + \int_{0+}^{\infty} \frac{\sigma(dt)}{t+x}$ *with positive constants* $a, b \geq 0$ *and the measure* σ *on* $(0,\infty)$.

(iii) $f : (0,\infty) \to [0,\infty)$ *extends analytically onto* $\mathbb{C}\setminus(-\infty,0]$ *s. t.* $f(\bar{z}) = \overline{f(z)}$ *and* $\Im z \cdot \Im f(z) \geq 0$.

(iv) $f : \mathbb{C}\setminus(-\infty,0) \to \mathbb{C}$ *has a unique representation*

$$f(z) = \alpha + \beta z + \int_{0-}^{\infty} \frac{tz-1}{t+z}\,\tau(dt), \qquad z \in \mathbb{C}\setminus(-\infty,0),$$

where $\int_0^\infty \frac{1+t}{t}\,\tau(dt) < \infty$, $\alpha \geq \int_0^\infty \frac{1}{t}\,\tau(dt)$, $\beta \geq 0$.

(v) $f \in \mathcal{BF}$ *with representation*

$$f(x) = a + bx + \int_{0+}^{\infty} (1-e^{-sx})\,m(s)\,ds, \qquad x > 0,$$

where $a, b \geq 0$, $m(s) = \int_{0+}^{\infty} e^{-ts}\,\rho(dt)$, *and* $\int_0^\infty \frac{\rho(dt)}{t(1+t)} < \infty$.

The implication (iii) $\Rightarrow$ (iv) can be found in [6], the equivalence of (ii) and (iii) in [1]. A detailed proof of the theorem is given in [17, Satz 5.6].

Remark 3.4. The representation formula (3.1) indicates that complete Bernstein functions enjoy very much the same properties as do completely monotone functions (see [2, §9] for the definition). In particular, $\mathcal{CBF}$ is a convex cone, closed under pointwise limits and composition of functions.

Since the map $z \mapsto \frac{z}{s+z}$, $s > 0$, contracts angels, $\left|\arg\left(\frac{z}{s+z}\right)\right| \leq |\arg(z)|$, so does $f \in \mathcal{CBF}$ and we get $f(S_{\pm\theta}) \subset S_{\pm\theta}$ where $S_{\pm\theta} = \{z \in \mathbb{C} : z = re^{\pm i\psi}, 0 \leq \psi \leq |\theta|\}$, $\theta \in [0,\pi)$. Therefore, the function $z \mapsto \left(f(z^\alpha)\right)^{1/\alpha}$ is well-defined for all $\alpha \in [-1,1]$ and—by Theorem 3.3(iii)—itself of class $\mathcal{CBF}$. Further details are given in [1], [17].

4. A representation formula for A^f

Starting with formula (2.5) for the generator A^f of a subordinate (C_0)-semigroup we get for *complete* Bernstein functions f a somewhat nicer integral representation. For contraction semigroups almost the same expression can be found in [1]. The proof presented here includes the non-contractive case and might even be a bit more elementary.

Theorem 4.1. *Let $\{e^{tA}\}_{t\geq 0}$ be a (C_0)-semigroup of type ω_0 and $\{\mu_t\}_{t\geq 0}$ be a cute subordinator with $f \in \mathcal{CBF}$ as in 3.3(v). Then the representing measure ρ satisfies*

$$\rho\big|_{(0,\omega]} \equiv 0 \quad \text{and} \quad \int_{\omega}^{\omega+1} \frac{\rho(dt)}{(t-\omega)} + \int_{\omega+1}^{\infty} \frac{\rho(dt)}{(t-\omega)^2} < \infty \tag{4.1}$$

(here, ω is the 'cuteness' constant from (2.1)) and the generator of the subordinate semigroup is given by

$$A^f u = au + bAu + \int_{\omega+}^{\infty} \frac{1}{t} A(t-A)^{-1} u\, \rho(dt), \qquad u \in D(A). \tag{4.2}$$

Proof. Since $\{\mu_t\}_{t\geq 0}$ is cute, we see from Proposition 2.2 and Theorem 3.3(v)

$$\begin{aligned} \infty > \int_1^{\infty} e^{s\omega}\, m(s)\, ds &= \int_1^{\infty}\int_0^{\infty} e^{s(\omega-t)}\, \rho(dt)\, ds \\ &\geq \Big(\int_0^{\omega+} + \int_{\omega+}^{\omega+1}\Big)\Big[\frac{1}{\omega-t}\, e^{s(\omega-t)}\Big]_{s=1}^{s=\infty} \rho(dt), \end{aligned}$$

which shows that $\rho\big|_{(0,\omega]} \equiv 0$ and $\int_{\omega+}^{\omega+1} \frac{\rho(dt)}{t-\omega} < \infty$. If $t \geq \omega+1$, the inequality $\frac{1}{t-\omega} \leq \frac{\omega+1}{t}$ holds and we find

$$\int_{\omega+1}^{\infty} \frac{\rho(dt)}{(t-\omega)^2} \leq (\omega+1)^2 \int_{\omega+1}^{\infty} \frac{\rho(dt)}{t^2} < \infty.$$

Together with the resolvent estimates (2.2) this yields for $u \in D(A)$

$$\begin{aligned} &\Big\| \int_{\omega}^{\infty} \frac{1}{t} A(t-A)^{-1} u\, \rho(dt) \Big\| \\ &\qquad \leq \int_{\omega}^{\omega+1} \frac{M+1}{t-\omega}\, \rho(dt)\, \|u\| + \int_{\omega+1}^{\infty} \frac{M+1}{(t-\omega)^2}\, \rho(dt)\, \|Au\|. \end{aligned} \tag{4.3}$$

Hence, the right-hand side in (4.2) is well-defined. In order to see the asserted equality we use Phillip's formula (2.5) and the fact that $f \in \mathcal{CBF}$:

$$A^f u = au + bAu + \int_0^{\infty}\int_{\omega}^{\infty} (e^{sA}u - u) e^{-st}\, \rho(dt)\, ds.$$

Since for $s \geq 0$

$$\|e^{sA}u - u\| = \left\| \int_0^s e^{tA} Au\, dt \right\| \leq M \int_0^s e^{t\omega}\, dt\, \|Au\| \leq M\omega s \|Au\|$$

is valid, essentially the same calculation as in (4.3) gives

$$\int_0^\infty \int_\omega^\infty \|e^{sA}u - u\| e^{-ts}\, \rho(dt)\, ds \leq c_1\|u\| + c_2\|Au\|$$

with c_1, c_2 as in (4.3). Thus we may change the order of integration and obtain

$$\begin{aligned} A^f u &= au + bAu + \int_\omega^\infty \int_0^\infty (e^{sA}u - u)e^{-st}\, ds\, \rho(dt) \\ &= au + bAu + \int_\omega^\infty ((t-A)^{-1} - t^{-1})u\, \rho(dt) \end{aligned}$$

which is just (4.2). □

Example 4.2. Let $\{e^{tA}\}_{t\geq 0}$ be a contractive (C_0)-semigroup and $\{\mu_t\}_{t\geq 0}$ be a subordinator.

(i) The fractional powers x^α, $0 < \alpha < 1$, can be written as

$$f_\alpha(x) := x^\alpha = \frac{\sin \alpha\pi}{\pi} \int_0^\infty \frac{x}{t+x}\, t^{\alpha-1}\, dt.$$

Therefore, we get on $D(A)$

$$A^{f_\alpha} u = \frac{\sin \alpha\pi}{\pi} \int_0^\infty t^{\alpha-1} A(t-A)^{-1} u\, dt, \qquad u \in D(A),$$

which is the well-known integral representation for fractional powers of positive operators, compare [18, chap. IX.11(4)].

(ii) For the logarithm $f(x) = \log(1+x)$ we have

$$f(x) = \log(1+x) = \int_1^\infty \frac{x}{t(t+x)}\, dt.$$

Hence,

$$A^f u = \int_1^\infty \frac{1}{t} A(t-A)^{-1} u\, dt, \qquad u \in D(A).$$

Using the integral formula for logarithms of positive operators in Banach space (see [14, Satz 1]) we find $A^f|_{D(A)} = -\log(1-A)|_{D(A)}$.

(iii) With f as in (ii) and $0 < \gamma < 1$ we see $(A+\gamma)^f u = -\log((1-\gamma) - A)$. Letting $\gamma \to 1$—this can be justified by (4.6) below—we obtain another expression for the logarithm of positive operators in Banach space.

A routine calculation shows that

$$\Big\|(\omega+1-A)^{-1}\Big(a+bA+\int_\omega^\infty \frac{1}{t}A(t-A)^{-1}\,\rho(dt)\Big)u\Big\|$$
$$\le \Big(a+bM+(\omega+2)2M^2\int_\omega^{\omega+1}\frac{\rho(dt)}{t-\omega}+M^2(\omega+1)\int_{\omega+1}^\infty\frac{\rho(dt)}{(t-\omega)^2}\Big)\|u\|.$$

This and the injectivity of the resolvent operator at $\omega+1$ immediately imply that the operator given by the left-hand side above is closable. Since $D(A)$ is a core for A^f, we have in the above examples $A^f = \overline{A^f|_{D(A)}} = \overline{-f(-A)|_{D(A)}}$. That, in general, $A^f = -f(-A)$—with $-f(-A)$ defined by the Dunford operational calculus—is the essence of the following theorem.

Theorem 4.3. *Let $\{e^{tA}\}_{t\ge0}$ be a (C_0)-semigroup of type ω_0 and $\{\mu_t\}_{t\ge0}$ be a cute subordinator with $f\in\mathcal{CBF}$. Then the resolvent of $(A-\epsilon)^f$, $\epsilon>0$, is given by*

$$(t-(A-\epsilon)^f)^{-1}u=(t+f(\epsilon-A))^{-1}u,\qquad t>-f(-\omega),\tag{4.4}$$

where the right-hand side is defined via the Dunford–Taylor integral. Both sides in (4.4) converge in $\mathcal{X}$ as $\epsilon\to0$.

If the 'cuteness' constant satisfies $\omega>\omega_0$, (4.4) holds for $\epsilon=0$.

Proof. In 3.4 we remarked that $f\in\mathcal{CBF}$ implies that $z\mapsto\frac{1}{f(1/z)}$ is again a complete Bernstein function. Thus, $\{z\in\mathbb{C}:\Re z<\omega\}\ni z\mapsto\frac{1}{t+f(-z)}$ is analytic at infinity. Using the Dunford–Taylor integral—see, e.g., [5, chap. XV.1]—we find for all $\epsilon>0$ and $u\in D(A^2)$

$$(t+f(\epsilon-A))^{-1}u=(t+f(\epsilon-A))^{-1}(A-\omega-2\epsilon)^{-2}(A-\omega-2\epsilon)^2u,$$

and putting $v=(A-\omega-2\epsilon)^2u$, $\omega'=\frac12(\omega_0+\omega)$, and $\omega''=\frac14(3\omega_0+\omega)$

$$\begin{aligned}Ru&\equiv(t+f(\epsilon-A))^{-1}u\\&=\frac{1}{2\pi i}\int_{\omega'-i\infty}^{\omega'+i\infty}\frac{1}{t+f(-z)}\,\frac{1}{(z-\epsilon-\omega)^2}\,(z+\epsilon-A)^{-1}v\,dz\\&=\frac{1}{2\pi i}\int_{\omega'-i\infty}^{\omega'+i\infty}\int_0^\infty e^{-s(t+f(-z))}\,\frac{1}{(z-\epsilon-\omega)^2}\,(z+\epsilon-A)^{-1}v\,ds\,dz\\&=\frac{1}{2\pi i}\int_{\omega'-i\infty}^{\omega'+i\infty}\int_0^\infty\int_0^\infty e^{-st}e^{rz}\,\frac{1}{(z-\epsilon-\omega)^2}\,(z+\epsilon-A)^{-1}v\,\mu_s(dr)\,ds\,dz.\end{aligned}$$

Here, we used that $f(z)$ is defined on the half-plane $\Re z>-\omega$ and that $e^{-sf(z)}=\int_0^\infty e^{-rz}\,\mu_s(dr)$. The above integrand can be estimated by

$$e^{-st}e^{r\omega'}\frac{1}{(\omega'-\omega-\epsilon)^2+y^2}\,\frac{M\omega''}{\omega'-\omega''+\epsilon},\quad y\equiv\Im z,$$

which is integrable w. r. t. $\mu_s(dr)\,ds\,dy$. Hence, we may change the order of integration,

$$Ru = \frac{1}{2\pi i}\int_0^\infty\int_0^\infty\int_{\omega'-i\infty}^{\omega'+i\infty} e^{rz}\frac{1}{(z-\epsilon-\omega)^2}(z+\epsilon-A)^{-1}v\,dz\,e^{-st}\mu_s(dr)\,ds$$

and obtain, applying the Dunford–Taylor calculus to the inner integral

$$\begin{aligned} Ru &= \int_0^\infty\int_0^\infty e^{r(A-\epsilon)}(A-\omega-2\epsilon)^{-2}v\,e^{-st}\mu_s(dr)\,ds \\ &= \int_0^\infty\int_0^\infty e^{r(A-\epsilon)}u\,e^{-st}\mu_s(dr)\,ds \qquad (4.5)\\ &= \int_0^\infty e^{s(A-\epsilon)^f}ue^{-st}\,ds \\ &= (t-(A-\epsilon)^f)^{-1}u. \end{aligned}$$

Since the integrand in (4.5) can be estimated by $Me^{-st}e^{r\omega}$, Lebesgue's dominated convergence theorem yields $\lim_{\epsilon\to 0}(t+f(\epsilon-A))^{-1}u = (t-A^f)^{-1}u$, and we are done.

If we may choose $\omega > \omega_0$, the above calculations remain valid for $\epsilon = 0$. □

In the above situation, we even have for every $\epsilon > 0$

$$\begin{aligned} \|A^f u-(A-\epsilon)^f u\| &\le b\epsilon\|u\| + \int_\omega^\infty \frac{1}{t}\|A(t-A)^{-1}u-(A-\epsilon)(t+\epsilon-A)^{-1}u\|\,\rho(dt) \\ &= b\epsilon\|u\| + \int_\omega^\infty \|\epsilon(t-A)^{-1}(t+\epsilon-A)^{-1}u\|\,\rho(dt) \\ &\le b\epsilon\|u\| + M^2\int_\omega^\infty\left(\frac{\epsilon-\omega}{t(t+\epsilon-\omega)}-\frac{-\omega}{t(t-\omega)}\right)\rho(dt)\,\|u\| \\ &\le M^2(f(\epsilon-\omega)-f(-\omega))\|u\|. \qquad (4.6) \end{aligned}$$

5. On the domain $D(A^f)$ of A^f

In this section we will employ the above results on the cone $\mathcal{CBF}$ and the integral formula for A^f to obtain more information on the domain of $D(A^f)$. If we consider a Hilbert space $\mathcal{H}$ we are able to prove some Heinz–Kato type theorem (Theorem 5.3) for complete Bernstein functions.

Theorem 5.1. *Let $\{e^{tA}\}_{t\ge 0}$ be a (C_0)-semigroup of type ω_0 on $\mathcal{X}$ and $\{\mu_t\}_{t\ge 0}$ be a cute subordinator with $f \in \mathcal{BF}$ as in (2.3). Then the operator A^f on $D(A)$ is closed if and only if A is bounded or if $b > 0$, that is, if f has a linear component.*

Proof. If A is bounded, $D(A) = D(A^f) = \mathcal{X}$ and nothing is to prove. Let us therefore assume *that A is an unbounded operator.*

The integral in Phillip's formula (2.5) can be estimated in the following way:

$$\begin{aligned}\Big\| \int_0^\infty (e^{tA}u - u)\,\mu(dt)\Big\| &\le \int_0^\epsilon \|e^{tA}u - u\|\,\mu(dt) + \int_\epsilon^\infty \|e^{tA}u - u\|\,\mu(dt)\\ &\le \int_0^\epsilon \Big\| \int_0^t e^{sA}Au\,ds\Big\|\,\mu(dt) + \int_\epsilon^\infty (\|e^{tA}\| + 1)\,\mu(dt)\,\|u\|\\ &\le M\int_0^\epsilon t\,\mu(dt)\,\|Au\| + M\int_\epsilon^\infty (e^{t\omega}+1)\,\mu(dt)\,\|u\|\\ &\equiv c_\epsilon\|Au\| + d_\epsilon\|u\| \qquad (5.1)\end{aligned}$$

where $c_\epsilon + d_\epsilon < \infty$ for $\{\mu_t\}_{t\ge 0}$ was assumed to be cute.

Suppose that $(A^f, D(A))$ is a closed operator but $b = 0$. Since $(A, D(A))$ is closed, too, we have by a theorem due to Hörmander (see [18, chap. II.6 Thm. 2])

$$\|Au\| \le c(\|A^f u\| + \|u\|), \qquad u \in D(A),$$

with a suitable constant $c > 0$. Choosing ϵ in (5.1) so small that $c_\epsilon < \frac{1}{2c}$ we obtain

$$\|Au\| \le c(\|A^f u\| + \|u\|) \le \frac{1}{2}\|Au\| + c(d_\epsilon + a)\|u\|, \qquad u \in D(A).$$

This, however, contradicts our assumption that A is unbounded, hence $b > 0$.

Conversely, if $b > 0$, we get from (2.5) using the estimate (5.1)

$$\|A^f u\| \ge (b - d_\epsilon)\|Au\| - (a + c_\epsilon)\|u\|, \qquad u \in D(A).$$

Pick $\epsilon > 0$ such that $b - d_\epsilon > 0$ and observe that then

$$\|Au\| \le c\|A^f u\| + c'\|u\|, \qquad u \in D(A), \qquad (5.2)$$

holds. Since $(A, D(A))$ is closed, (5.2) immediately implies the closedness of the operator $(A^f, D(A))$. □

In the remaining part of this section we will *only* consider *self-adjoint* (C_0)-semigroups $\{e^{tA}\}_{t\ge 0}$ in a Hilbert space $\mathcal{H}$ with inner product $(\cdot,\cdot)$. Note, that in this setting $\|e^{tA}\| = e^{t\omega_0}$ and that cute and admissible subordinators coincide. Having in mind Theorem 5.1 we may *restrict ourselves to* $f \in \mathcal{CBF}$ *with* $a = b = 0$.

The following lemma compiles some standard results on closed, (in the quadratic form sense) positive operators on $\mathcal{H}$. A proof can be found in [4, chap. 4].

Lemma 5.2. *Let $\{e^{tA}\}_{t\ge 0}$ be a self-adjoint (C_0)-semigroup on $\mathcal{H}$ and $B : D(B) \subset \mathcal{H} \to \mathcal{H}$ a further self-adjoint operator satisfying $D(B) \subset D(A)$. Then we have in the quadratic form sense*

(i) $-B \ge -A \ge 0$ *if and only if* $-B(t-B)^{-1} \ge -A(t-A)^{-1} \ge 0$ *for all* $t \in \rho(A) \cap \rho(B) \cap (0,\infty)$;

(ii) *if $\{\mu_t\}_{t\ge 0}$ is an admissible subordinator, then $A \le \omega_0$, $e^{tA} \le e^{t\omega_0}$, and A^f is self-adjoint satisfying $A^f \le -f(-\omega_0)$.*

Combining Theorem 4.1 with the above lemma, we recover a result first proved by Heinz [6, Satz 2].

Theorem 5.3. *Let $\{e^{tA}\}_{t\geq 0}$ and $\{e^{tB}\}_{t\geq 0}$ be two self-adjoint (C_0)-semigroups on $\mathcal{H}$ of type $\omega_0(A)$ and $\omega_0(B)$, respectively, $D(B) \subset D(A)$, and $\{\mu_t\}_{t\geq 0}$ be an admissible subordinator with $f \in \mathcal{CBF}$. Then*

$$B \leq A \leq \omega_0(A) \quad \textit{implies} \quad B^f \leq A^f \leq -f(-\omega_0(A)).$$

Proof. Lemma 5.1(i) shows $B(t-B)^{-1} \leq A(t-A)^{-1} \leq \omega_0(A)(t-\omega_0(A))^{-1}$ for all $t > \omega_0(A)$. Using the integral representation (4.2) we find for $u \in D(B)$

$$\begin{aligned}(B^f u, u) &= \int_{\omega_0(A)}^{\infty} \frac{1}{t}(B(t-B)^{-1}u, u)\,\rho(dt) \\ &\leq \int_{\omega_0(A)}^{\infty} \frac{1}{t}(A(t-A)^{-1}u, u)\,\rho(dt) \\ &= (A^f u, u).\end{aligned}$$

Since, by Lemma 2.4, $D(B)$ is a core for $(B^f, D(B^f))$, a standard density argument shows $B^f \leq A^f$ on $D(B^f)$. □

The spectral calculus for self-adjoint operators shows that under the above assumptions $\{e^{-tA^2}\}_{t\geq 0}$ and $\{e^{-tB^2}\}_{t\geq 0}$ are self-adjoint contraction semigroups.

Theorem 5.4. *Let A, B, $\{\mu_t\}_{t\geq 0}$ and f be as in Theorem 5.3. Then*

$$-B^2 \leq -A^2 \leq -(\omega_0(A))^2 \quad \textit{implies} \quad -(B^f)^2 \leq -(A^f)^2 \leq -f^2(|\omega_0(A)|).$$

In particular, $D(B^f) \subset D(A^f)$.

Proof. We remarked in 3.4 that $f \in \mathcal{CBF}$ implies $g \in \mathcal{CBF}$, $g(x) = f^2(\sqrt{x})$. On $D(B^2)$ Theorem 5.3 therefore shows

$$(-B^2)^g \leq (-A^2)^g \leq -g(\omega_0(A)^2) = -f^2(|\omega_0(A)|).$$

It is easy to see from the very definition that subordination is a transitive operation. In particular, $A^{f\circ\phi} = (A^\phi)^f$ holds for all $f, \phi \in \mathcal{CBF}$. This and the fact that $A^f = -f(-A)$ allows us to write

$$-f^2(\sqrt{B^2}) \leq -f^2(\sqrt{A^2}) \leq -f^2(|\omega_0(A)|)$$

or, equivalently,

$$-(B^f)^2 \leq -(A^f)^2 \leq -f^2(|\omega_0(A)|).$$

Again by Lemma 2.4, $D(B^2)$ is a core of $(B^f, D(B^f)$ which gives $\|A^f u\| \leq \|B^f u\|$ for all $u \in D(B^f)$. □

Corollary 5.5. *Let $\{e^{tA}\}_{t\geq 0}$ be a self-adjoint (C_0)-semigroup on $\mathcal{H}$ and $\{\mu_t\}_{t\geq 0}$ an admissible subordinator with $f \in \mathcal{CBF}$. Then $D(A^f) = D((cA)^f) = D(A^{f(c\cdot)})$ for all $c \in (0,1)$.*

Proof. The inclusion $D(A^f) \subset D((cA)^f)$ follows from Theorem 5.4. Conversely, put $B = -A^2$ and observe, that by the spectral calculus applied to the operator B

$$c^2 B(t - c^2 B)^{-1} \le c^2 B(t - B)^{-1}, \qquad t > 0, c \in (0,1)$$

holds. The same reasoning as in Theorem 5.3 and Theorem 5.4 finally shows $c\|A^f u\| \le \|(cA)^f u\|$ for all $u \in D((cA)^f)$ and, in particular, $D((cA)^f) \subset D(A^f)$.□

Corollary 5.6. *Let A, B, $\{\mu_t\}_{t\ge 0}$ and f be as in Theorem 5.3. Then*

$$\|Au\| \le c(\|Bu\| + \|u\|), \qquad u \in D(B)$$

implies $D(B^f) \subset D(A^f)$.

Proof. For $\epsilon > \omega_0(B)$ the operator $(\epsilon - B)$ is closed with domain $D(B)$ and satisfies $\|(\epsilon - B)u\| \ge (\epsilon - \omega_0(B))\|u\|$. Applying Hörmander's comparison theorem for closed operators—see [18, chap. II.6 Thm. 2]—to $(\epsilon - B)$ and A, $D(\epsilon - B) \subset D(A)$, we find

$$\|Au\| \le c(\|(\epsilon - B)u\| + \|u\|) \le \gamma\|(\epsilon - B)u\|$$

with a suitable constant $\gamma = \gamma(\epsilon, \omega_0) > 1$. Hence, Corollary 5.5 and Theorem 5.4 show

$$D((B - \epsilon)^f) \subset D((\gamma^{-1}A)^f) \subset D(A^f).$$

By (4.6) we get $D(B^f) = D((B - \epsilon)^f)$ and the proof is finished. □

As outlined in the introduction we may apply the above results to pseudo-differential operators that occur quite naturally as generators of (Fellerian) stochastic processes.

Example 5.7. In the following examples we need a scale of anisotropic Sobolev spaces $H^{r,a}$ in terms of a continuous negative definite function $a : \mathbb{R}^d \to \mathbb{R}$ (see [2, §7] and also [12], these Proceedings):

$$H^{r,a} = \{u \in L^2(\mathbb{R}^d) : \|u\|_{r,a}^2 \equiv \int_{\mathbb{R}^d} |1 + a(\xi)|^{2r} |\hat{u}(\xi)|^2 \, d\xi < \infty\}, \qquad r \ge 0$$

($\hat{u}$ denotes the Fourier transform of u). Note that for $a(\xi) = |\xi|^2$ the relation $H^{r,|\cdot|^2} = H^{2r}$ with the usual L^2-Sobolev spaces holds.

(i) (see [9], and for a similar example [8], [10]) Let $b_j : \mathbb{R}^d \to \mathbb{R}$, $1 \le j \le d$, be functions of class C_b^∞, bounded below by $b_j \ge$ const. > 0 and satisfying $\frac{\partial}{\partial x_j} b_j \equiv 0$ for all j. For $0 < r \le 1$ we put $a_j^2(\xi) = |\xi_j|^{2r}$ and $a^2(\xi) = \sum_{j=1}^d a_j^2(\xi) \sim |\xi|^{2r}$. Then $p(x,\xi) = \sum_{j=1}^d b_j(x) a_j^2(\xi)$ is the symbol of the pseudo-differential operator

$$p(x, D)u(x) = (2\pi)^{-d/2} \int_{\mathbb{R}^d} p(x,\xi)\hat{u}(\xi) e^{i(x,\xi)} \, d\xi, \qquad u \in C_c^\infty(\mathbb{R}^d). \tag{5.3}$$

It is shown in [9, Thm. 3.1] that $-p(x, D)$ extends to a self-adjoint generator of a (C_0)-semigroup on L^2. If the coefficient functions b_j satisfy $\|b_j - b_j(x_0)\|_\infty \le c$

for a fixed $x_0 \in \mathbb{R}^d$ and a suitable constant $c > 0$, the following estimates

$$\|u\|_{1,a^2} = \|u\|_{H^{2r}} \le c(\|p(x,D)u + \mu u\|_{L^2} + \|u\|_{L^2}), \qquad u \in H^{2r} \tag{5.4}$$

$$\|p(x,D)u + \mu u\|_{L^2} \le c'\|u\|_{H^{2r}} = c'\|u\|_{H^{1,a^2}}, \qquad u \in D(p(x,D)) \tag{5.5}$$

hold true for any $\mu \ge 0$ as is proved in [12, Thms. 5.1, 5.2]. In particular, (5.4) and (5.5) imply $H^{1,a^2} = D(p(x,D))$. Using Corollary 5.6 we therefore get for all $\mu \ge 0$ and every cute subordinator with $f \in \mathcal{CBF}$

$$D((p(x,D)+\mu)^f) = D(f \circ a^2(D)) = H^{1,f\circ a^2} = H^{1,f(|\cdot|^{2r})}.$$

In particular, $f(x) = x^\alpha$ for $0 < \alpha \le 2$ gives $D((p(x,D)+\mu)^\alpha) = H^{\alpha r}$.

(ii) (see [11]) By $p(x,D)$ we denote again the pseudo-differential operator (5.3); its symbol $p(x,\xi)$ is now assumed to be real-valued, continuous and—for fixed $x \in \mathbb{R}^d$—to be negative definite in $\xi \in \mathbb{R}^d$. For an arbitrary x_0 we put

$$p(x,\xi) = p(x_0,\xi) + (p(x,\xi) - p(x_0,\xi)) \equiv p_1(\xi) + p_2(x,\xi).$$

If for a fixed continuous negative definite function $a^2(\xi)$ the conditions

$$|p_1(\xi)| \le \gamma_1(1 + a^2(\xi)), \quad \xi \in \mathbb{R}^d \quad \text{and} \quad p_1(\xi) \ge \gamma_0 a^2(\xi), \quad \xi \in \mathbb{R}^d,\ |\xi| \text{ large}$$

and

$$p_2(\cdot,\xi) \in C^q \qquad \text{for a } q > d+1,$$

$$\left|\frac{\partial^{|\alpha|}}{\partial x^\alpha} p_2(x,\xi)\right| \le \phi_\alpha(x)(1 + a^2(\xi)) \qquad \text{for all } \alpha \in \mathbb{N}_0^d,\ |\alpha| \le q$$

are satisfied with functions $\phi_\alpha \in L^1$ and $\sum_{|\alpha|\le q} \|\phi_\alpha\|_{L^1} \ll \gamma_0$, we have (cf. [11, Lemma 3.1, Thm. 4.2])

$$\|p(x,D)u\|_{L^2} \le c\|u\|_{1,a^2}, \qquad u \in H^{1,a^2} \tag{5.6}$$

and

$$\|u\|_{1,a^2} \le c'(\|p(x,D)u\|_{L^2} + \|u\|_{L^2}), \qquad u \in D(p(x,D)). \tag{5.7}$$

Under the assumption that $p(x,D)$ is symmetric it is shown in [11, Thm. 3.1] that $p(x,D)$ extends to a self-adjoint generator of a (C_0)-semigroup. Again, by Corollary 5.6, we obtain

$$H^{1,f\circ a^2} = D(p(x,D))$$

for every $f \in \mathcal{CBF}$ of a cute subordinator.

References

[1] Berg, Chr., Boyadzhiev, Kh., deLaubenfels, R., Generation of generators of holomorphic semigroups. J. Austral. Math. Soc. (Ser. A) 55 (1993), 246–269.

[2] Berg, Chr., Forst, G., Potential Theory on Locally Compact Abelian Groups. Ergebnisse Math. Grenzgeb (2) 87. Springer-Verlag, Berlin 1975.

[3] Bochner, S., Harmonic Analysis and the Theory of Probability. California Monographs in Math. Sci., University of California Press, Berkeley, CA 1955.

[4] Davies, E. B., One-Parameter Semigroups. L. M. S. Monographs 15. Academic Press, London 1980.

[5] Gohberg, I., Goldberg, S. and M. A. Kaashoek, Classes of Linear Operators Vol. 1. Operator Theory: Advances and Applications 49. Birkhäuser, Basel 1990.

[6] Heinz, E., Beiträge zur Störungstheorie der Spektralzerlegung. Math. Ann. 123 (1951), 415–438.

[7] Hirsch, F., Intégrales de résolvantes et calcul symbolique. Ann. Inst. Fourier (Grenoble) 22 (1972), 239–264.

[8] Hoh, W., Jacob, N., Some Dirichlet forms generated by pseudo differential operators. Bull. Sci. Math. (2) 116 (1992), 383–398.

[9] Jacob, N., A class of elliptic pseudo differential operators generating symmetric Dirichlet forms. Potential Analysis 1 (1992) 221–232.

[10] — Further pseudodifferential operators generating symmetric Feller semigroups and Dirichlet forms. Revista Mat. Iberoamericana 9 (1993), 373–407.

[11] — A class of Feller semigroups generated by pseudo differential operators. Math. Z. 215 (1994), 151–166.

[12] — Various approaches to the Dirichlet problem for non-local operators satisfying the positive maximum principle. These Proceedings.

[13] Löwner, K., Über monotone Matrixfunktionen. Math. Z. 38 (1934), 177–216.

[14] Nollau, V., Über den Logarithmus abgeschlossener Operatoren in Banachschen Räumen. Acta Sci. Math. 28 (1967), 107–121.

[15] Phillips, R. S., On the generation of semigroups of linear operators. Pacific J. Math. 2 (1952), 343–369.

[16] Prüß, J., Evolutionary Integral Equations and Applications. Birkhäuser Monographs in Math. 87. Birkhäuser, Basel 1993.

[17] Schilling, R. L., Zum Pfadverhalten von Markovschen Prozessen, die mit Lévy-Prozessen vergleichbar sind. Dissertation, Erlangen 1994.

[18] Yosida, K., Functional Analysis (6th ed.). Grundlehren Math. Wiss. 123. Springer-Verlag, Berlin 1980.

The double layer potentials for a bounded domain with fractal boundary

Hisako Watanabe

Dedicated to Professor F. Maeda on the occasion of his 60th birthday

Abstract. For a bounded domain D with fractal boundary we define the double layer potentials of α-Hölder continuous functions on ∂D and study the boundary behavior of the double layer potentials.

1991 Mathematics Subject Classification: 31B25, 31B99.

1. Introduction

Let D be a bounded domain in $\mathbb{R}^d$ with fractal boundary. We say that a domain D has a fractal boundary if the Hausdorff dimension β of ∂D is greater than $d-1$. There are many Jordan domains which have fractal boundaries. A typical example is the von Koch snowflake in $\mathbb{R}^2$. In $\mathbb{R}^d$ ($d \geq 3$) we can also construct many domains with fractal boundary using finite similitudes (cf. [Hu]).

We consider double layer potentials for these domains D. A double layer potential is an useful concept mathematically as well as physically. For example, it is well-known that some boundary value problems for the Laplacian in a smooth domain can be solved by using double layer potentials. Let D be a bounded $C^{1,\alpha}$-domain in $\mathbb{R}^d$. Recall that the double layer potential Φg of $g \in L^p(\partial D)$ is defined by

$$\Phi g(x) = -\int_{\partial D} \langle \nabla_y N(x-y), n_y \rangle g(y) d\sigma(y), \tag{1.1}$$

where $N(x-y)$ is the Newton kernel and n_y is the unit outer normal to ∂D. Furthermore if g is a C^1-function with compact support, then we see by the Green formula that

$$\Phi g(x) = \int_{\mathbb{R}^d \setminus \overline{D}} \langle \nabla_y g(y), \nabla_y N(x-y) \rangle dy \tag{1.2}$$

for $x \in D$ and

$$\Phi g(x) = -\int_{D} \langle \nabla_y g(y), \nabla_y N(x-y) \rangle dy \tag{1.3}$$

for $x \in \mathbb{R}^d \setminus \overline{D}$.

If D is a domain with non-smoouth boundary, then the integral in (1.1) cannot be considered, in general, but the integrals in (1.2) and (1.3) may be defined for sufficiently smooth functions g on $\mathbb{R}^d$.

J. Harrison and A. Norton introduced an abstract line integral over a fractal curve in the plane in [HN1] and a surface integral over a fractal surface in $\mathbb{R}^d$ in [HN2]. In their theory the box dimension of the boundary plays an important role.

In this paper we suppose D is a bounded domain in $\mathbb{R}^d$. Moreover we assume that ∂D has the following property (u).

(u) There are a positive Radon measure μ on ∂D and positive real numbers β, γ, r_0, b_1, b_2 such that $d-1 \le \gamma \le \beta < d$ and

$$b_1 r^\beta \le \mu(B(z,r) \cap \partial D) \le b_2 r^\gamma$$

for all $z \in \partial D$ and all $r \le r_0$, where $B(z,r)$ stands for the open ball with center z and radius r in $BbbR^d$.

We note that, if D is a bounded Lipschitz domain, then the surface measure μ has the property (u) for $\beta = \gamma = d-1$. Further if the boundary of D consists of finite self-similar sets, which satisfy the open set condition and whose similarity dimension is β, then the β-dimensional Hausdorff measure μ has the property (u) for $\gamma = \beta$ (cf. [Hu]).

Under these conditions we will define the double layer potential of a function defined on ∂D. To do so, let $0 < \alpha \le 1$ and F be a closed subset of $\mathbb{R}^d$.

We denote by $\Lambda_\alpha(F)$ the Banach space of all bounded α-Hölder continuous real-valued functions on F with norm

$$\|f\|_{\Lambda_\alpha(F)} = \sup\{|f(z)| : z \in F\} + \sup\{\frac{|f(z)-f(w)|}{|z-w|^\alpha} : z,\ w \in F, z \ne w\}.$$

It is known that there exists a linear bounded extension operator $\mathcal{E}_0$ from $\Lambda_\alpha(F)$ to $\Lambda_\alpha(\mathbb{R}^d)$ (cf. [St, Theorem 3 on p. 174]). Multiplying $\mathcal{E}_0(f)$ by a fixed smooth function ϕ such that $\phi = 1$ on $B(0,R)$ and supp $\phi \subset B(0,2R)$ we have

Theorem A. *Let $0 < \alpha \le 1$ and F be a compact subset of $\mathbb{R}^d$ satisfying $F \subset B(0, \frac{R}{2})$. Then there exists a bounded linear operator $\mathcal{E}$ from $\Lambda_\alpha(F)$ to $\Lambda_\alpha(\mathbb{R}^d)$ such that supp $\mathcal{E}(f) \subset B(0,2R)$, $\mathcal{E}(f) = f$ on F and the restriction of $\mathcal{E}(f)$ to the complement of F is a C^∞-function satisfying*

$$|\nabla \mathcal{E}(f)(x)| \le c d(x,F)^{\alpha-1} \|f\|_{\Lambda_\alpha(F)} \text{ for all } x \in \mathbb{R}^d \setminus F,$$

where c is a constant independent of x and f and $d(x,F)$ denotes the distance of x from F.

Let $d \ge 2$ and $0 < \beta - (d-1) < \alpha < 1$. We define *the double layer potential* of $f \in \Lambda_\alpha(\partial D)$ by

$$\Phi f(x) = \int_{\mathbb{R}^d \setminus \overline{D}} \langle \nabla_y \mathcal{E}(f)(y), \nabla_y N(x-y) \rangle dy \tag{1.4}$$

for $x \in D$ and

$$\Phi f(x) = -\int_D \langle \nabla_y \mathcal{E}(f)(y), \nabla_y N(x-y)\rangle dy \tag{1.5}$$

for $x \in \mathbb{R}^d \setminus \overline{D}$, where

$$N(x-y) = \begin{cases} \frac{1}{\omega_d (d-2)|x-y|^{d-2}} & \text{if } d \geq 3, \\ -\frac{3R}{2\pi} \log \frac{|x-y|}{3R} & \text{if } d = 2 \end{cases}$$

and ω_d stands for the surface area of the unit ball in $\mathbb{R}^d$. Then we have

Theorem. *Suppose D is a bounded domain in $\mathbb{R}^d$ $(d \geq 2)$ such that ∂D satisfies the condition (u). Furthermore, assume that $\beta - (d-1) < \alpha < 1$. Then, for every $f \in \Lambda_\alpha(\partial D)$, Φf is harmonic in $\mathbb{R}^d \setminus \partial D$ and for every $z \in \partial D$*

$$\lim_{x \to z, x \in D} \Phi f(x) = \int_{\mathbb{R}^d \setminus \overline{D}} \langle \nabla_y \mathcal{E}(f)(y), \nabla_y N(z-y)\rangle dy$$

and

$$\lim_{x \to z, x \in \mathbb{R}^d \setminus \overline{D}} \Phi f(x) = -\int_D \langle \nabla_y \mathcal{E}(f)(y), \nabla_y N(z-y)\rangle dy.$$

Moreover Φf can be extended to two continuous functions on $\overline{D}$ and on $\mathbb{R}^d \setminus D$, respectively.

Remark. Recently I learned that J. Král and D. Medková have considered double layer potentials on general sets under different assumptions and investigated the existence of angular limits of them (cf. [KM]).

2. Fundamental lemmas

In this section let D be a bounded domain in $\mathbb{R}^d$ such that ∂D satisfies the condition (u) in §1.

Fix a positive real number R satisfying $\overline{D} \subset B(0, R/2)$ and let r_0 be the same number as in the condition (u). The following lemma is fundamental.

Lemma 2.1. *Let $\epsilon < r_0/6$, $r > 0$ and $z \in \partial D$. Then*

$$|\{x \in \mathbb{R}^d : d(x, \partial D) < \epsilon,\ |x-z| < r\}| \leq c\epsilon^{d-\beta} r^\gamma, \tag{2.1}$$

where $|E|$ is the d-dimensional Lebesgue measure of a set E and c is a constant independent of ϵ, r and z.

Proof. First we assume that $3r < r_0$ and $4\epsilon < r$. Denote by E the set

$$\{x \in \mathbb{R}^d : d(x, \partial D) < \epsilon\} \cap B(z, r)$$

and choose δ satisfying $\epsilon < \delta < r_0$ and $\delta < 2\epsilon$. Further, cover $\overline{E}$ by finitely many balls $B(z_j, \delta)$, where $z_j \in \partial D \cap B(z, 2r)$. By the covering lemma we can select a subcovering $\{B(z_{j_k}, \delta)\}_k^m$ such that each two of them are disjoint and

$$\cup B(z_j, \delta) \subset \cup_{k=1}^m B(z_{j_k}, 5\delta).$$

Noting that $\cup_{k=1}^m B(z_{j_k}, \delta) \subset B(z, 3r)$, we have, by (u),

$$\begin{aligned}|E| &\leq \omega_d \sum_{k=1}^{m} 5^d \delta^d \leq 5^d 2^{d-\beta} \omega_d b_1^{-1} \epsilon^{d-\beta} \sum_{k=1}^{m} \mu(B(z_{j_k}, \delta)) \\ &\leq 5^d 2^{d-\beta} \omega_d b_1^{-1} \epsilon^{d-\beta} \mu\left(\bigcup_{k=1}^{m} B(z_{j_k}, \delta)\right) \\ &\leq 5^d 2^{d-\beta} \omega_d b_1^{-1} b_2 \epsilon^{d-\beta} (3r)^\gamma \leq c_1 \epsilon^{d-\beta} r^\gamma.\end{aligned}$$

Next, if $3r < r_0$ and $4\epsilon \geq r$, we have

$$|E| \leq \omega_d r^d \leq \omega_d (4\epsilon)^{d-\beta} r^\beta \leq c_2 \epsilon^{d-\beta} r^\gamma.$$

Finally assume that $3r \geq r_0$. Cover ∂D by the finitely many balls $\{B(z_j, \frac{r_0}{6})\}_{j=1}^l$ with $z_j \in \partial D$ and set $r_1 = r_0/6 + \epsilon$. Then $3r_1 < r_0$. From

$$E \subset \cup_{j=1}^l B(z_j, r_1) \cap \{x \in \mathbb{R}^d : d(x, \partial D) < \epsilon\}$$

and the above consideration we deduce

$$|E| \leq c_3 l \epsilon^{d-\beta} r_1^\gamma \leq c_4 l \epsilon^{d-\beta} r^\gamma.$$

Thus we have the estimate (2.1). □

Lemma 2.2. *Let $\beta - (d-1) < \alpha < 1$. Then, for every fixed $R_0 > 0$, there exists a constant c such that*

$$\int_{B(z,r)} d(x, \partial D)^{\alpha - 1} dx \leq c r^{d-1+\alpha-\beta+\gamma}$$

for all $z \in \partial D$ and all positive real numbers $r < R_0$.

Proof. Set

$$E_n = \{x \in \mathbb{R}^d : d(x, \partial D)^{\alpha-1} > 2^n\}$$

for each integer n and

$$F = \{x \in \mathbb{R}^d : d(x, \partial D) < \frac{r_0}{6 \times 2^{1/(1-\alpha)}}\}.$$

If $x \in E_n$, then $d(x, \partial D) < 2^{-n/(1-\alpha)}$. Put $r_n = 2^{-n/(1-\alpha)}$. Pick the integers m, l satisfying $r_m \geq r > r_{m+1}$ and $r_l \geq r_0/(6 \times 2^{1/(1-\alpha)}) > r_{l+1}$, respectively and set

$t = \max\{m, l\}$. On account of Lemma 2.1 we obtain

$$\int_{B(z,r)\cap F} d(x,\partial D)^{\alpha-1}dx \le c_1 \sum_{n=t}^{\infty} 2^n \int_{B(z,r)} \chi_{E_n}(x)dx$$
$$\le c_2 \sum_{n=m}^{\infty} 2^n r_n^{d-\beta} r^{\gamma} \le c_3 r^{\gamma} \sum_{n=m}^{\infty} \left(2^{1-(d-\beta)/(1-\alpha)}\right)^n.$$

Noting that $\frac{d-\beta}{1-\alpha} > 1$, we have

$$\int_{B(z,r)\cap F} d(x,\partial D)^{\alpha-1}dx \le c_4 r^{\gamma} \left(2^{1-(d-\beta)/(1-\alpha)}\right)^m$$
$$\le c_5 r^{\gamma} r^{d-\beta-1+\alpha} = c_5 r^{d-1+\alpha-\beta+\gamma}.$$

On the other hand we also obtain

$$\int_{B(z,r)\cap(\mathbb{R}^d\setminus F)} d(x,\partial D)^{\alpha-1}dx \le c_6 \left(\frac{r_0}{6\times 2^{1/(1-\alpha)}}\right)^{\alpha-1} \int_{B(z,r)} dx$$
$$\le c_7 r^d \le c_8 r^{d-1+\alpha-\beta+\gamma}.$$

Thus we have the desired estimate. □

Lemma 2.3. *Let $\beta - (d-1) < \alpha < 1$, $d - 1 + \alpha - \beta + \gamma > k$, $0 < r < R_0$ with a fixed $R_0 > 0$, and $z \in \partial D$. Then*

$$\int_{B(z,r)} d(x,\partial D)^{\alpha-1}|x-z|^{-k}dx \le c r^{d-1+\alpha-\beta+\gamma-k},$$

where c is a constant independent of z and r.

Proof. Set

$$F_n = \{x \in B(z,r) : |x-z|^{-k} > 2^n\}$$

and $r_n = 2^{-n/k}$. On account of Lemma 2.2 we have

$$\int_{B(z,r)} d(x,\partial D)^{\alpha-1}|x-z|^{-k}dx \le \sum_{n=m}^{\infty} 2^n \int_{B(z,r_n)} d(x,\partial D)^{\alpha-1}dx$$
$$\le c_1 \sum_{n=m}^{\infty} 2^n r_n^{d-1+\alpha-\beta+\gamma} \le c_1 \sum_{n=m}^{\infty} 2^{((k-d+1-\alpha+\beta-\gamma)/k)n}, \tag{2.2}$$

where m is the integer satisfying $r_m \ge r > r_{m+1}$. Since $d - 1 + \alpha - \beta + \gamma > k$, we have the conclusion by (2.2). □

Lemma 2.4. *Let $r > 0$, $\beta - (d-1) < \alpha < 1$ and $d - 1 + \alpha - \beta + \gamma > k$. Define*

$$g(x) = \int_{B(0,r)} d(y,\partial D)^{\alpha-1}|x-y|^{-k}dy.$$

Then g is bounded on $\mathbb{R}^d$.

Proof. Let $x \in \mathbb{R}^d$ and choose $z \in \partial D$ such that $d(x, \partial D) = |x - z|$.
Split $g(x)$ into an integral extended over

$$B(0,r) \cap \{y : |y - z| \leq 3|y - x|\},$$

to be denoted by $g_1(x)$, and into an integral extended over

$$B(0,r) \cap \{y : |y - z| > 3|y - x|\},$$

to be denoted by $g_2(x)$. Then

$$g_1(x) \leq c_1 \int_{B(z,r+R)} d(y, \partial D)^{\alpha-1}|z - y|^{-k} dy \leq c_2(r + R)^{d-1+\alpha-\beta+\gamma-k},$$

where we used Lemma 2.3 to obtain the last inequality.

Note that $|y-z| > 3|y-x|$ implies $|x-z| > 2|x-y|$, whence $d(y, \partial D) > |x-y|$; moreover, if $y \in B(0,r)$, then $|y - z| < r + R$ and $|y - x| < (r + R)/3$.

Therefore we obtain

$$g_2(x) \leq c_3 \int_{B(x,(r+R)/3)} |x - y|^{\alpha-1-k} dy \leq c_4(r + R)^{d-1-k+\alpha}.$$

Thus we see that g is bounded on $\mathbb{R}^d$. □

3. Properties of double layer potentials

Hearafter we assume that D is a bounded domain such that ∂D satisfies the condition (u). We first show that for $f \in \Lambda_\alpha(\partial D)$ the double layer potential Φf is defined in $\mathbb{R}^d \setminus \partial D$.

Lemma 3.1. *Let* $1 > \alpha > \beta - (d - 1)$ *and* $f \in \Lambda_\alpha(\partial D)$. *Then the double layer potential* Φf *defined by* (1.4) *and* (1.5) *is harmonic in* $\mathbb{R}^d \setminus \partial D$.

Proof. Let K be a compact subset of $\mathbb{R}^d \setminus \overline{D}$.

Since

$$|\langle \nabla_y \mathcal{E}(f)(y), \nabla_y N(x - y)\rangle| \leq c_1 \|f\|_{\Lambda_\alpha(\partial D)} d(y, \partial D)^{\alpha-1} d(K, \overline{D})^{1-d}$$

for $x \in K$ and $y \in D$, we see by Lemma 2.2 that

$$\int_D \langle \nabla_y \mathcal{E}(f)(y), \nabla_y N(x - y)\rangle dy$$

converges for $x \in \mathbb{R}^d \setminus \overline{D}$ and Φf is harmonic in $\mathbb{R}^d \setminus \overline{D}$.

Analogously we can show that Φf is harmonic in D. □

Lemma 3.2. *Let* $\beta - (d-1) < \alpha < 1$ *and* $f \in \Lambda_\alpha(\partial D)$. *Then the functions defined by*

$$g(x) = \int_{\mathbb{R}^d \setminus \overline{D}} \langle \nabla_y \mathcal{E}(f)(y), \nabla_y N(x - y)\rangle dy$$

and

$$h(x) = \int_D \langle \nabla_y \mathcal{E}(f)(y), \nabla_y N(x-y) \rangle dy$$

are continuous on $\mathbb{R}^d$.

Proof. Since supp $\mathcal{E}(f) \subset B(0,2R)$, it is clear that g is continuous on $\mathbb{R}^d \setminus \overline{B(0,2R)}$. To show that g is continuous at $z \in B(0,3R)$, let $x \in B(z,R/2)$ and choose t satisfying $\alpha - \beta + d - 1 > t > 0$. Noting that

$$\begin{aligned}&|\langle \nabla_y \mathcal{E}(f)(y), \nabla_y N(z-y) - \nabla_y N(x-y) \rangle| \\ &\leq c_1 \|f\|_{\Lambda_\alpha(\partial D)} d(y,\partial D)^{\alpha-1} |x-z|^t (|x-y|^{1-d-t} + |z-y|^{1-d-t}),\end{aligned}$$

we have, by Lemma 2.4,

$$|g(z) - g(x)| \leq c_2 \|f\|_{\Lambda_\alpha(\partial D)} |x-z|^t,$$

which shows that g is continuous at z.

Similarly we see that h is also continuous on $\mathbb{R}^d$. □

Proof of Theorem. Theorem is an easy consequence of Lemmas 3.1 and 3.2. □

We next show that the definition of Φf is independent of the extention of f. To approximate D by a sequence of open sets with better boundaries, we use the Whitney decomposition.

For an open set G we denote by $\mathcal{W}(G)$, $\mathcal{W}_k(G)$ the Whitney decomposition of G and the family of all cubes with side length 2^{-k} in $\mathcal{W}(G)$, respectively (cf.[St], [HN1], [HN2]).

Lemma 3.3. *Let m_k, n_k be the numbers of all cubes in $\mathcal{W}_k(D)$ and in $\mathcal{W}_k(\mathbb{R}^d \setminus \overline{D})$, respectively. Then*

$$m_k \leq c 2^{k\beta} \quad \text{and} \quad n_k \leq c 2^{k\beta},$$

where c is a constant independent of k.

Proof. Choose finitely many boundary points z_j $(j = 1, \ldots, l)$ such that

$$\cup_{Q \in \mathcal{W}_k(D)} Q \subset \cup_{j=1}^{l} B(z_j, 2^{-k}a),$$

where $a = 6\sqrt{d}$. By the covering lemma there exists a subfamily $\{B(z_{j_i}, 2^{-k}a)\}_{i=1}^n$ of $\{B(z_j, 2^{-k}a)\}$ such that each two of them are disjoint and

$$\cup_{Q \in \mathcal{W}_k(D)} Q \subset \cup_{i=1}^{n} B(z_{j_i}, 5 \times 2^{-k}a).$$

From the condition (u) it follows that

$$\begin{aligned} m_k 2^{-kd} &\leq |\cup_{i=1}^n B(z_{j_i}, 5 \times 2^{-k}a)| \leq c_1 n (5 \times 2^{-k}a)^d \\ &= c_1 5^d (2^{-k}a)^{d-\beta} n (2^{-k}a)^\beta \\ &\leq c_2 2^{-k(d-\beta)} b_1^{-1} \mu(\cup_{i=1}^n B(z_{j_i}, 2^{-k}a)) \\ &\leq c_3 2^{-k(d-\beta)} \mu(\partial D), \end{aligned}$$

whence $m_k \le c_4 2^{k\beta}$.

Similarly we can show the second inequality. □

The following proposition shows that the definition of the double layer potential of $f \in \Lambda_\alpha(\partial D)$ is independent of the extention of f if it has some properties.

Proposition 3.4. *Let $\beta - (d-1) < \alpha < 1$ and $g \in \Lambda_\alpha(\mathbb{R}^d)$. If g is a C^1-function in $\mathbb{R}^d \setminus \partial D$, with compact support, and $g = 0$ on ∂D, then*

$$\int_D \langle \nabla_y g(y), \nabla_y N(x-y)\rangle dy = 0 \tag{3.1}$$

for every $x \in \mathbb{R}^d \setminus \overline{D}$ for which the integral in (3.1) exists and

$$\int_{\mathbb{R}^d \setminus \overline{D}} \langle \nabla_y g(y), \nabla_y N(x-y)\rangle dy = 0 \tag{3.2}$$

for every $x \in D$ for which the integral in (3.2) exists.

Moreover, if g satisfies

$$|\nabla_y g(y)| \le c\, d(y, \partial D)^{\alpha - 1} \quad \text{for all } y \in \mathbb{R}^d \setminus \partial D, \tag{3.3}$$

then the integrals in (3.1), (3.2) exist and (3.1), (3.2) also hold on ∂D.

Proof. Let $y \in D$ and $y \in Q \in \mathcal{W}_k(D)$ and choose $z \in \partial D$ satisfying $d(y, \partial D) = |y - z|$. Since $g(z) = 0$ and $|y - z| \le c_1 2^{-k}$, we have

$$|g(y)| = |g(y) - g(z)| \le c_2 2^{-k\alpha}.$$

Denote by A_n the interior of $\cup_{k=-\infty}^{n} \cup_{Q \in \mathcal{W}_k(D)} Q$. We note that $\mathcal{W}_k(D) = \emptyset$ for every $k \le k_0$.

On account of Lemma 3.3 we obtain, for $x \in \mathbb{R}^d \setminus \overline{D}$,

$$\begin{aligned} \left|\int_{A_n} \langle \nabla g(y), \nabla_y N(x-y)\rangle dy\right| &= \left|\int_{\partial A_n} g(y) \langle \nabla_y N(x-y), n_y\rangle d\sigma(y)\right| \\ &\le c_3 2^{-n\alpha} d(x, \partial D)^{1-d} 2^{n\beta} (2^{-n})^{d-1} \\ &= c_4 2^{-n(\alpha - \beta + d - 1)} d(x, \partial D)^{1-d}. \end{aligned}$$

Noting that $\alpha - \beta + d - 1 > 0$, we have

$$\int_D \langle \nabla g(y), \nabla_y N(x-y)\rangle dy = \lim_{n\to\infty} \int_{A_n} \langle \nabla g(y), \nabla_y N(x-y)\rangle dy = 0$$

for $x \in \mathbb{R}^d \setminus \overline{D}$. Thus we have (3.1) for $x \in \mathbb{R}^d \setminus \overline{D}$.

Moreover, if (3.3) holds, then we can see that the function defined by the left-hand side of (3.1) is continuous at every boundary point by the same method as in the proof of Lemma 3.2. Therefore (3.1) holds on ∂D.

Similarly we can show (3.2). □

4. Uniformly elliptic operators

Let $A(x)$ be a $C^{1,\lambda}$, symmetric, $d \times d$ matrix-valued fuction in $\mathbb{R}^d$ such that $A(x)$ defines the uniformly elliptic differential operator div$(A(x))$grad (cf. [Mr, §2]). Further let $L(x,y)$ be a Levy function for the differential operator (cf. [Mr, §8]) and suppose that $\partial L/\partial y_j$, $\partial^2 L/\partial x_k \partial y_j$ exist and are continuous outside the diagonal set. Moreover, suppose there exists $\lambda > 0$ such that

$$|\frac{\partial(L-H)}{\partial y_j}(x,y)| \leq c|x-y|^{\lambda+1-d}, \qquad |\frac{\partial^2(L-H)}{\partial x_k \partial y_j}(x,y)| \leq c|x-y|^{\lambda-d}$$

for all x, $y \in B(0,4R)$, $x \neq y$, where H is the function defined by (8.1) in [Mr]. Then

$$|\frac{\partial L}{\partial y_j}(x,y) - \frac{\partial L}{\partial y_j}(z,y)| \leq c|x-z|^t \left(|x-y|^{1-d-t} + |z-y|^{1-d-t}\right)$$

for all x, y, $z \in B(0,2R)$, $x \neq y$, $z \neq y$ and all t, $0 \leq t \leq 1$.

Let D be a bounded domain with $D \subset B(0,R/2)$, satisfying (u) and suppose $\alpha > \beta - (d-1)$. Furthermore denote by A the linear operator on $\mathbb{R}^d$ defined by $A(x)$. We define, for $f \in \Lambda_\alpha(\partial D)$, *the generalized double layer potential* Φf of f by

$$\Phi f(x) = \int_{\mathbb{R}^d \setminus \overline{D}} \langle \nabla \mathcal{E}(f)(y), A(\nabla_y L(x,y)) \rangle dy$$

if $x \in D$ and

$$\Phi f(x) = -\int_D \langle \nabla \mathcal{E}(f)(y), A(\nabla_y L(x,y)) \rangle dy$$

if $x \in \mathbb{R}^d \setminus \overline{D}$.

Then we can also prove that div$(A(x))$grad$(\Phi f) = 0$ in $\mathbb{R}^d \setminus \partial D$ and Φf extends to two continuous functions on $\overline{D}$ and $\mathbb{R}^d \setminus D$, respectively.

References

[HN1] Harrison, J., Norton, A., Geometric integration on fractal curves in the plane. Indiana Univ. Math. J. 40 (1991), 567–594.

[HN2] — The Gauss–Green theorem for fractal boundaries. Duke Math. J. 67 (1992), 575–588.

[Hu] Hutchinson, J. E., Fractals and selfsimilarity. Indiana Univ. Math. J. 30 (1981), 713–747.

[KM] Král, J., Medková, D., Angular limits of double layer potentials. Akademie Věd České Republiky, preprint 83 (1993), 1–26.

[Mr] Miranda, C., Partial differential equations of elliptic type. 2nd edition, Springer-Verlag, Berlin–Heidelberg–New York, 1970.

[St] Stein, E. M., Singular integrals and differentiability properties of functions. Princeton University Press, Princeton- New Jersey, 1970.

PROBLEMS

The Pompeiu property for spherical polygons

David H. Armitage

Let σ denote area measure on the unit sphere $\mathbb{S}^2$. A closed set K in $\mathbb{S}^2$ is said to have the *Pompeiu property* (PP) if the only function $f \in C(\mathbb{S}^2)$ satisfying

$$\int_{A(K)} f d\sigma = 0$$

for every element A of the rotation group SO(3) is $f \equiv 0$. By an n-gon ($n = 2, 3, 4, \dots$) we mean an n-sided regular spherical polygon, bounded by arcs of great circles, strictly contained in some hemisphere of $\mathbb{S}^2$.

Ungar [3] showed that, according to its radius, a spherical cap in $\mathbb{S}^2$ may or may not have (PP) and deduced that for some n there exist n-gons which do not have (PP). A refinement of his argument shows that such n-gons exist for every $n \geq 4$. On the other hand, recent work of Laquer [1] and Ullrich [2] shows that 2-gons always have (PP). The question naturally arises: do all 3-gons have (PP)?

The question seems to be related to the theory of harmonic polynomials, for it is implicit in Ungar's work that a spherical cap C in $\mathbb{S}^2$ lacks (PP) if and only if there exists a positive integer m such that

$$\int_C h d\sigma = 0$$

for every homogeneous harmonic polynomial of degree m in R^3.

For a survey of related work, see Zalcman's article [4].

References

[1] Laquer, H. T., The Pompeiu problem. Amer. Math. Monthly 100 (1993), 461–467.

[2] Ullrich, D. C., More on the Pompeiu problem. Amer. Math. Monthly 101 (1994), 164–168.

[3] Ungar, P., Freak theorem about functions on a sphere. J. London Math. Soc. (2) 29 (1954), 100–103.

[4] Zalcman, L., A bibliographic survey of the Pompeiu problem. In: Approximation by Solutions of Partial Differential Equations (ed. by B. Fuglede et al.), 185–194. NATO ASI Series C, vol. 365. Kluwer Academic Publishers, Dordrecht 1992.

Some open problems relating removable singularities for Hardy spaces and Hausdorff measures

Anders Björn

We refer to [1] for notation.

Problem 1. Does there exist a set K, with $\Lambda_1(K) > 0$, and a $p < \infty$, such that K is removable for H^p?

In the case of $p = \infty$ it is well-known that there exist such sets, and also that there cannot exist removable sets with dimension greater than 1. A consequence of Theorem 8 in [1] is that we must have $p \geq 1$.

For $p \geq 1$ we know that there exist removable sets with dimension 1, see, e.g., Theorem 12 in [1]. As remarked in [1], just before Definition 7, there exist zero-dimensional sets not removable for any $p < \infty$.

If $0 < p < 1$ we have fewer examples. We do, in fact, not know if there are sets of positive dimension removable for H^p. We do, however, have Theorem 8 in [1] which implies that if $\dim(K) > p$, then K is not removable for H^p. This leads us to propose the following problems.

Problem 2a. Let $0 < p < 1$. Does there exist a set K removable for H^p, with $\Lambda_p(K) > 0$?

Problem 2b. Let $0 < p < 1$. Does there exist a set K removable for H^p, with $\dim(K) = p$?

Problem 2c. Let $0 < p < 1$. Does there exist a set K removable for H^p, with $\dim(K) > 0$?

References

[1] Björn, A., Removable singularities for Hardy spaces of analytic functions. These Proceedings.

Convergence on the Martin boundary

Jürgen Bliedtner and Peter A. Loeb

In an appropriate potential theoretic setting (e.g., a Brelot harmonic space), let X be a domain with Martin boundary Δ, and K be the associated Martin kernel with respect to a fixed point $x_0 \in X$. We assume that 1 is harmonic. Fix an increasing exhaustion $\{\Omega_n : n \in \mathbb{N}\}$ consisting of regular, relatively compact domains in X with $x_0 \in \Omega_n$ for all $n \in \mathbb{N}$. For each $n \in \mathbb{N}$, let $\mu^n_{x_0}$ and μ_{x_0} denote harmonic measure with respect to x_0 on $\partial\Omega_n$ and on Δ, respectively. Of course, μ_{x_0} is the representing measure of 1. For each finite Borel measure ν on Δ let h_ν denote the harmonic extension, and for each $n \in \mathbb{N}$, set

$$f^\nu_n(z) := \int_{\partial\Omega_n} K(y,z) \cdot h_\nu(y)\, \mu^n_{x_0}(dy).$$

This is a continuous function of z on Δ. Show that for any Borel set $E \subseteq \Delta$,

$$\nu(E) = 0 \quad \Rightarrow \quad f^\nu_n(z) \to 0 \ \ \mu_{x_0}\text{-a.e. on } \ E.$$

For the unit disk with $x_0 = 0$ and an exhaustion by concentric closed disks of radii r_n about 0, this implication follows from the Radial Limit Theorem since

$$f^\nu_n(z) = h_\nu(r^2_n z).$$

The motivation for this problem can be found in our article in these Proceedings. We also show there that if $\nu(E) = 0$, then by a result of Mokobodzki, there is a "rapid ultrafilter" $\mathcal{U}$ in $\mathbb{N}$ along which $f^\nu_n \cdot \chi_E$ converges to 0 μ_{x_0}-a.e.

Mean value property of L-harmonic functions

Jürgen Bliedtner

For a large class of linear partial differential operators L on open subsets of $\mathbb{R}^d$, the solutions of the equation $Lu = 0$ satisfy a mean value property. The usual proofs of the converse statement, i.e., the continuous functions u which enjoy the mean value property are solutions of the differential equation $Lu = 0$ make extensive use of some deeper results, e.g., existence of a base of regular or resolutive sets. Question: Is it possible to prove this converse statement in an elementary way, i.e., using just calculus?

An open question about convergence in D-law

Nicolas Bouleau

Anyone who taught a course in probability theory knows the slight difficulty existing with the notion of *convergence in law* or *in distribution*: The random variables X_n converge in law means that the laws of the X_n's converge weakly, and the question arises naturally whether there exists a random variable X defined on the same probability space such that

$$\mathbb{E}f(X_n) \to \mathbb{E}f(X) \qquad \forall f \text{ bounded and continuous.}$$

In the context of probability theory the answer is positive:

Theorem. *Let $(\Omega, \mathcal{A}, \mathbb{P})$ be a probability space and let (X_n) be a sequence of random variables defined on Ω with values in $\mathbb{R}^d$ converging in law. Then there exists a random variable X defined on Ω with values in $\mathbb{R}^d$ whose law is the limit one.*

This can be proved by splitting Ω into its atomic part Ω_1 and its part without atoms Ω_2. A sequence of random variables defined on Ω_1 always possesses a subsequence converging a.s. and for every probability law μ defined on Ω_2 there exists a random variable defined on Ω_2 with law μ. That gives easily the result. □

Now the same question arises in the context of Dirichlet structures:
Let

$$(\Omega,\ \mathcal{F},\ m,\ \mathbb{D},\ \mathcal{E})$$

be a Dirichlet structure as defined in [B]. Let (X_n)be a sequence in $\mathbb{D}^d$ converging in D-law, that is (cf. [BH]):

There exists a Dirichlet structure $(\mathbb{R}^d, \mathcal{B}(\mathbb{R}^d), \nu, \hat{\mathbb{D}}, \hat{\mathcal{E}})$ such that

a) $X_{n*}\mathbb{P} \to \nu$ *weakly*

b) $\hat{\mathbb{D}} \supset C_c^1(\mathbb{R}^d)$,

$$\lim_{n\to\infty} \mathcal{E}(f \circ X_n) = \hat{\mathcal{E}}(f) \qquad \forall f \in C_c^1(\mathbb{R}^d).$$

Is it true that there is an $X \in \mathbb{D}^d$ such that the image of $(\Omega,\ \mathcal{F},\ m,\ \mathbb{D},\ \mathcal{E})$ by X is $(\mathbb{R}^d, \mathcal{B}(\mathbb{R}^d), \nu, \hat{\mathbb{D}}, \hat{\mathcal{E}})$?

References

[B] Bouleau, N., Construction of Dirichlet structures. These Proceedings.

[BH] Bouleau, N., Hirsch, F., Dirichlet forms and analysis on Wiener space. De Gruyter Stud. Math. 14. Walter de Gruyter, Berlin 1991.

Brelot spaces

Gustave Choquet

Does there exist a Brelot axiomatics on every locally compact topological space E which is connected, locally connected and with a denumerable base of open sets?

Inverse potential theory

Gustave Choquet

In "Modèles finis en théorie du potentiel" (Choquet–Deny, Journal d'Analyse mathématique 1956–1957, pp. 77–135), "inverse principles" have been defined; these notions can be extended to more general locally compact groups or spaces, and lead to an "inverse potential theory".

Some work in that direction has been done in the sixties by Japanese mathematicians, but inverse potential theory seems worthy of deeper investigations, for instance concerning its links with probability theory.

Convolution kernels and balayage

Gustave Choquet

In "Noyaux de convolution et balayage sur tout ouvert" (Choquet–Deny, Springer Lecture Notes 404, 1974) the class $\mathcal{B}_g$ of convolution kernels on an Abelian locally compact group G for which balayage on **every** open set is possible has been characterized.

However, we have not been able to characterize the class $\mathcal{B}$ of all convolution kernels for which balayage is possible only on **bounded** open sets of G.

This is challenging; Masayuki Itô has disproved a conjecture of mine concerning $\mathcal{B}$, but has not gone further.

Successive averages

Gustave Choquet

Let μ_r be the normalized Lebesgue measure on the ball $B(0,r)$ of $\mathbb{R}^n$. For any $f \in C(\mathbb{R}^n, \mathbb{R})$, put $f_r = f * \mu_r$; and more generally: $f_{r_1,r_2,\ldots,r_n} = (f_{r_1,r_2,\ldots,r_{n-1}}) * \mu_{r_n}$ where $(r_1, r_2, \ldots)$ is an infinite sequence of positive numbers. Under what conditions on f and the sequence $(r_1, r_2, \ldots)$ is it true that $(f_{r_1,r_2,\ldots,r_n})$ converges to a harmonic function (resp. belongs to some quasi-analytic class) ?

Analytic capacity of Cantor sets

Vladimir Ya. Eiderman

Let E_n be a set which is obtained at stage n of the construction of the plane Cantor set: $E_0 = [0,1] \times [0,1]$, E_n consists of 4^n squares with sides of length $\sigma_n = 4^{-n}$. Let $\gamma(E)$ be the analytic capacity of a compact set $E \subset \mathbb{R}^2$, let $C(E)$ be the "classical" capacity corresponding to the kernel $K(r) = 1/r$. T. Murai [1] has proved that $\gamma(E_n) < c_1/\log n$. On the other hand, it is easy to see that $\gamma(E_n) > c_2/n$.

Problem 1. *Give the asymptotic estimate of* $\gamma(E_n)$.

In particular, prove (or disprove) the following

Conjecture 1. *There exists a sequence* $\{p_n\}$ *such that* $p_n \to \infty$ *as* $n \to \infty$ *and* $\gamma(E_n) = p_n/n$.

These problems are connected with the following Conjecture 2 of J.Garnett.

Conjecture 2 [2]. *Let E be a plane Cantor set, let σ_n be the side of squares at stage n of the construction of E (now, generally speaking, $\sigma_n \neq 4^{-n}$). Then $\gamma(E) > 0$ if and only if $\sum_n 4^{-n}/\sigma_n < \infty$.*

(This is equivalent to $\gamma(E) = 0 \iff C(E) = 0$.)

There is a proof of this statement [2, p. 88, Theorem 2.2] under the assumption that $\{\sigma_n/\sigma_{n-1}\}$ is a non-increasing sequence, but this proof contains a gap.

References

[1] T. Murai, Construction of H^1 functions concerning the estimate of analytic capacity. Bull. London Math. Soc. 19 (1987), 154–160.

[2] J. Garnett, Analytic capacity and measure. Lecture Notes in Math. 297. Springer-Verlag, Berlin 1972.

Boundary layers

Matts Essén

Let Ω be an open connected subset of the unit disc U. We assume that Ω is regular with respect to the Dirichlet problem. The complement $E = U \setminus \Omega$ is relatively closed in U. In [2], Volberg defines Ω to be a boundary layer if there exists a positive constant c such that

$$\omega(0, I) \geq c|I|$$

for all arcs I on the unit circle T. Here $\omega(0, I)$ is the harmonic measure of the arc I in the domain Ω evaluated at the origin and $|I|$ is the Lebesgue measure of I.

Can we characterize boundary layers in terms of concepts from potential theory? In [1], it is proved that if Ω is a boundary layer, then E will be minimally thin at all points of T. The converse statement is false: there exists a set E which is relatively closed in U and minimally thin at all points of T but $\Omega = U \setminus E$ is not a boundary layer.

More details on boundary layers can be found in [2]. The picture is far from complete. It would be of great interest to understand more about the interplay between different concepts from potential theory and boundary layers.

Note added in proof: There are some recent results concerning this problem in Aikawa H. and Lundh T., On boundary layers. U.U.D.M. Report 1995:1, ISSN 1101-3591.

References

[1] Essén, M., On minimal thinness, reduced functions and Green potentials. Proc. Edinburgh Math. Soc. 36 (1992), 87–106.

[2] Volberg, A. L., A criterion on a subdomain of the disc for its harmonic measure to be comparable with Lebesgue measure. Proc. Amer. Math. Soc. 112 (1991), 153–162.

Integrals with respect to a gaussian measure

Denis Feyel

Let Ω be a Fréchet space endowed with a centered gaussian measure μ, for example $\mathbb{R}^\infty$ with the canonical gaussian, or the Wiener space with the Wiener measure. Let $f : \Omega \to \mathrm{R}^n$. What can be said about

$$J_t = t^{-n/2} \iint \exp[-(f(x) - f(y))^2/2t]\, d\mu(x)d\mu(y) \ ?$$

This is also

$$J_t = \int \exp(-t.u^2/2)\hat{\xi}(u)^2\, du$$

up to a constant multiplicative factor where $\hat{\xi}$ is the Fourier transform of the image $\xi = f(\mu)$. More precisely, what can be said about the limit J_0 of J_t as t tends to 0?

On Carleman approximation by harmonic functions

Stephen J. Gardiner

Let Ω be a connected open set in $\mathbb{R}^n (n \geq 2)$ and E be a relatively closed proper subset of Ω. If $A \subseteq \mathbb{R}^n$, let $\mathcal{H}(A)$ denote the collection of all functions which are harmonic on an open set containing A. Gardiner and Goldstein [3] (or see [2, Theorem 10] in these Proceedings) have characterized those pairs (Ω, E) which have the following property:

(a) for each u in $C(E) \cap \mathcal{H}(E^\circ)$ and each continuous function $\epsilon : E \to (0, 1]$, there exists v in $\mathcal{H}(\Omega)$ such that $|v - u| < \epsilon$ on E.

Problem. Characterize those pairs (Ω, E) which have the property:

(b) for each u in $\mathcal{H}(E)$ and each continuous function $\epsilon : E \to (0, 1]$, there exists v in $\mathcal{H}(\Omega)$ such that $|v - u| < \epsilon$ on E.

Clearly pairs (Ω, E) which have property (a) also have property (b). It can be deduced from the results of [1] that the converse to this statement is false, at least when $n \geq 3$.

References

[1] Gardiner, S. J., Tangential harmonic approximation on relatively closed sets. Illinois J. Math. 39 (1995), 143–157.

[2] — Harmonic approximation. These Proceedings.

[3] Gardiner, S. J., Goldstein, M., Carleman approximation by harmonic functions. Amer. J. Math. 117 (1995), 245–255.

One sphere problem

Wolfhard Hansen

Let f be a continuous function on the open unit ball U in $\mathbb{R}^d$, $d \geq 3$, such that for every $x \in U$ the value $f(x)$ is the average of f on the boundary of some closed ball centered at x and contained in U. Is f harmonic?

The answer to the corresponding question for $d \leq 2$ is known to be negative (R. Courant, D. Hilbert, *Methoden der mathematischen Physik II*, Grundlehren Math. Wiss. 48, Springer-Verlag 1937; W. Hansen, N. Nadirashvili, *Littlewood's one circle problem*, J. London Math. Soc. 50 (1994), 349–360).

On the modulus of a harmonic vector field

*Viktor P. Havin**

Let h be a strictly positive function defined in the ball $\mathbb{B}^n = \{x \in \mathbb{R}^n : |x| < 1\}$. We call h *admissible* if there is a function u harmonic in $\mathbb{B}^n$ such that

$$|\nabla u| = h. \tag{1}$$

(The term "harmonic vector field" in the title refers to the field ∇u.) There are no formal obstacles to drop the assumption of the *strict* positivity of h, but this would complicate the problems to be discussed below and change their character essentially.

* This work was supported by the RFFI grant No. 94-01-01628.

For $n = 2$ the admissibility of h means that h is the modulus of a non-vanishing function f complex analytic in the unit disc $\mathbb{B}^2$ ($f = \frac{\partial u}{\partial x} - i\frac{\partial u}{\partial y}$). Thus for $n = 2$, h is admissible iff $h > 0$ and $\log h$ is harmonic in $\mathbb{B}^2$. The equation $\Delta \log h = 0$ can be given in the following form:

$$h\Delta h - |\nabla h|^2 = 0 .$$

The problem we propose in this short note is *to write down a differential equation (a system of equations?) characterizing admissible functions in* $\mathbb{R}^n$, $n \geq 3$.

This equation could be useful in solving the following problem: *given a function* H *on the sphere* $\mathbb{S}^{n-1}$ (= *the boundary of* $\mathbb{B}^n$), *find a function* u *harmonic in* $\mathbb{B}^n$ *and satisfying the following boundary condition:*

$$|\nabla u| \restriction \mathbb{S}^{n-1} = H. \tag{2}$$

This condition needs an interpretation. If H is smooth (say, $H \in C^1(\mathbb{S}^{n-1})$) and strictly positive, then it is natural to look for a harmonic u whose gradient is continuous up to $\mathbb{S}^{n-1}$ and satisfies (2). For $n = 2$ a solution can be given explicitly (just take $\frac{\partial u}{\partial x} - i\frac{\partial u}{\partial y}$ as the outer function whose modulus is H on $\mathbb{S}^{n-1}$, see Chapter 5 of [6]; *all* solutions can be easily expressed in terms of this particular u and some inner factors). For $n \geq 3$ only a partial answer is known. Namely, Aleksandrov proved in [5] that for any continuous and strictly positive H there is a function u harmonic in $\mathbb{B}^n$ with bounded gradient satisfying (2) in the following sense: $H(x) = \lim_{r\uparrow 1} |\nabla u(rx)|$ for almost all $x \in \mathbb{S}^{n-1}$. It is unclear whether Aleksandrov's solution u belongs to $C^1(\mathbb{B}^n \cup \mathbb{S}^{n-1})$ even for a strictly positive C^∞-function H.

A differential equation satisfied by all admissible functions is shown to exist in [1]. It is also proved in [1] that the order of this equation is four (if $n = 3$) and not less. These conclusions are based on indirect (and quite elementary) dimension computations and yield no explicit differential equation.

Admissible functions and related questions appear more or less explicitly in applied potential theoretic works on hydrodynamics (see [2]) and geophysics (see [3], [4]).

References

[1] Giesecke, M., Havin, V. P., On the length of a harmonic vector. Vestnik LGU (series 1) 15 (3) (1987), 33–38. Russian.

[2] Geffen, N., A nonstandard nonlinear boundary-value problem for harmonic functions. Quart. Appl. Math. 41 (3) (1983), 289–300.

[3] Backus, G., Application of a non-linear boundary value for Laplace's equation to gravity and geomagnetic intensity surveys. Quarterly J. Mech. Appl. Math. 21 (2) (1968), 196–220.

[4] — Non-uniqueness of the External Geomagnetic Field Determined by Surface Intensity Measurement. J. of Geophys. Research, Space Physics 75 (3)1 (1970), 6339–6341.

[5] Aleksandrov, A. B., Inner functions on spaces of homogeneous type. Zapiski Nauchn. Sem. LOMI 126 (1983), 7–14. Russian.

[6] Hoffman, K., Banach Spaces of Analytic Function. Prentice Hall, Inc., Englewood Cliffs 1962.

Double layer potentials

Josef Král

Let $D \subset \mathbb{R}^m$ ($m \geq 2$) be a bounded domain such that each point $x = (x_1, \ldots, x_m)$ in the boundary ∂D has a neigbourhood in ∂D which is homeomorphic with $\mathbb{R}^{m-1}$. We shall suppose that ∂D can be covered by finitely many hyperplanes; if each of these hyperplanes can be chosen parallel to one of the coordinate hyperplanes $\{x_k = 0\}$ ($k = 1, \ldots, m$), then D will be termed rectangular. Boundary value problems on such domains and domains obtained from them by smooth deformations have attracted considerable attention (cf. [3], [4], [8], [10] where additional references are given). In connection with methods of integral equations an important tool is represented by double layer potentials (cf. [2], [5], [7]).

Let $\mathcal{H}^k$ stand for the k-dimensional Hausdorff measure with the appropriate normalization (so that $\mathcal{H}^k([0,1]^k) = 1$). Put

$$B(x,r) := \{y \in \mathbb{R}^m; |y - x| < r\}, \qquad A := \mathcal{H}^{m-1}(\partial B(x,1))$$

and define the density of D at x by

$$d_D(x) := \lim_{r \downarrow 0} \frac{\mathcal{H}^m(B(x,r) \cap D)}{\mathcal{H}^m(B(x,r))}.$$

Let $n(z)$ denote the unit exterior normal of D at $z \in \partial D$, which is uniquely determined for $\mathcal{H}^{m-1}$-a.e. $z \in \partial D$. Let $\mathcal{C}(\partial D)$ be the space of all continuous real-valued functions on ∂D equipped with the maximum norm

$$\mu(f) := \sup\{|f(z)|; z \in \partial D\}, \qquad f \in \mathcal{C}(\partial D).$$

Given $f \in \mathcal{C}(\partial D)$, define $T^D f$ on ∂D by

$$T^D f(z) := [1 - 2d_D(z)]\, f(z) - \frac{2}{A} \int_{\partial D} f(y) n(y) \frac{z - y}{|z - y|^m}\, d\mathcal{H}^{m-1}(y);$$

it is easily seen (cf. §2 in [5]) that $T^D f \in \mathcal{C}(\partial D)$.

Representation of the solution of the Dirichlet problem for D with the prescribed boundary condition $g \in \mathcal{C}(\partial D)$ by means of the double layer potential with an unknown momentum density $f \in \mathcal{C}(\partial D)$ reduces to the equation

$$(I + T^D)\, f = 2g, \tag{1}$$

where I is the identity operator on $\mathcal{C}(\partial D)$. Let $\mathfrak{A}$ be the space of all compact linear operators acting on $\mathcal{C}(\partial D)$. Given a norm π inducing the topology of uniform convergence in $\mathcal{C}(\partial D)$ we denote by

$$\omega_\pi(T^D) := \inf\{\pi(T^D - V);\ V \in \mathfrak{A}\}$$

the corresponding essential norm of T^D. Because of the presence of points z with $d_D(z) \neq \frac{1}{2}$ in ∂D, the operator T^D is not compact and validity of the estimate

$$\omega_\pi(T^D) < 1 \tag{2}$$

becomes useful (though not indispensable) in connection with the equation (1) (compare [9]). If $D \subset \mathbb{R}^2$ is rectangular, then $\omega_\mu(T^D) = \frac{1}{2}$. On the other hand, simple examples are known of rectangular domains D in $\mathbb{R}^3$ with $\omega_\mu(T^D) > 1$ (cf. [6]). Nevertheless, as shown in [1], [6], for any rectangular domain $D \subset \mathbb{R}^3$ an explicit geometric construction of a norm π in $\mathcal{C}(\partial D)$ satisfying (2) is possible so that π is comparable with the maximum norm μ in the sense that

$$c^{-1}\mu \leq \pi \leq c\mu$$

for suitable constant $c > 0$. The question is, whether such a construction is possible for all dimensions.

Problem. Given any rectangular domain $D \subset \mathbb{R}^m$ with $m > 3$, construct a norm π in $\mathcal{C}(\partial D)$ satisfying (2) which is comparable with the maximum norm μ.

References

[1] Angell, T., Kleinman, R., Král, J., Layer potentials on boundaries with corners and edges. Časopis Pěst. Mat. 113 (1988), 387–402.

[2] Atkinson, K. E., A survey of boundary integral equation methods for the numerical solution of Laplace's equation in three dimensions. In: Numerical Solutions of Integral Equations (ed. Golberg, M.) 1–34, Plenum Press, New York 1990.

[3] Elschner, J., The double layer potential operator over polyhedral domains I. Solvability in weighted Sobolev spaces. Appl. Anal. 45 (1992), 117–134.

[4] Grachev, N. V., Representations and estimates for inverse operators of the potential theory integral equations in a polyhedron. In: Potential Theory (ed. Kishi, M.) 203–206, Walter de Gruyter, Berlin 1991.

[5] Král, J., Integral operators in potential theory. Lecture Notes in Math. 823, Springer-Verlag, Berlin 1980.

[6] Král, J., Wendland, W., Some examples concerning applicability of the Fredholm–Radon method in potential theory. Appl. Mat. 31 (1986), 293–308.

[7] Maz'ya, V. G., Boundary integral equations. Encyclopaedia of Mathematical Sciences (ed. Gamkrelidze, R. V.), Analysis IV (eds. Maz'ya, V. G., Nikolskij, S. M.), Springer-Verlag, Berlin 1991.

[8] Medková, D., The third boundary value problem in potential theory for domains with a piecewise smooth boundary. Czechoslovak Math. J. (to appear).

[9] Radon, J., Über die Randwertaufgaben beim logarithmischen Potential. Collected works vol. 1, 228–272, Austrian Acad. of Sci., Wien 1987.

[10] Rathsfeld, A. The invertibility of the double layer potential operator in the space of continuous functions defined on a polyhedron: the panel method. Appl. Anal. 45 (1992), 135–177.

Wave singularities

Josef Král

Let $G \subset \mathbb{R}^2$ be an open set. If $0 < \alpha \leq 1$ we denote by $\mathcal{K}_\alpha(G)$ the class of all functions $u : G \to \mathbb{R}$ satisfying locally in G the Hölder condition with exponent α which means that, for each compact $Q \subset G$, there exists a constant $k_Q \in [0, \infty[$ such that

$$z_1, z_2 \in Q \Rightarrow |u(z_1) - u(z_2)| \leq k_Q |z_1 - z_2|^\alpha.$$

$\mathcal{C}_\alpha(G)$ will stand for the subclass of all Hölder continuous functions $v \in \mathcal{K}_\alpha(G)$ for which

$$\sup\{|v(z_1) - v(z_2)|;\ z_1, z_2 \in Q, |z_1 - z_2| \leq \delta\} = o(\delta^\alpha) \quad \text{as } \delta \downarrow 0$$

for every compact $Q \subset G$; $\mathcal{C}_0(G)$ denotes the class of all continuous real-valued functions in G. Consider the wave operator

$$\Box := \partial_1^2 - \partial_2^2$$

(where ∂_j is the partial derivative w.r. to the j-th variable) acting on distributions in G. Given a class $\mathcal{K}(G)$ of distributions in G, then a relatively closed set $F \subset G$ will be termed removable for $\mathcal{K}(G)$ provided

$$(u \in \mathcal{K}(G), \Box u = 0 \text{ in } G \setminus F) \Rightarrow \Box u = 0 \text{ in } G$$

(cf. [4]). $\mathcal{V}(G)$ is the class of all $v \in \mathcal{C}_0(G)$ for which $\Box v$ is representable by a signed Radon measure in G. If $M \subset \mathbb{R}^2$ and $\alpha \in [0, \infty[$, then $\mathcal{H}_\alpha(M)$ will denote the usual outer α-dimensional Hausdorff measure of M (cf. [6]). Let

$$L^+ = \{(x, y) \in \mathbb{R}^2; x + y = 0\} \quad \text{and}$$
$$L^- = \{(x, y) \in \mathbb{R}^2; x - y = 0\}$$

be the characteristic lines of $\Box$; consider the corresponding orthogonal projections

$$\pi^+ : \mathbb{R}^2 \to L^+ \qquad \text{and} \qquad \pi^- : \mathbb{R}^2 \to L^-$$

of $\mathbb{R}^2$ onto L^+ and L^-, respectively. The following theorem follows from more general results in [2].

Theorem. *Let* $0 < \alpha \leq 1$. *A relatively closed* $F \subset G$ *is removable for* $\mathcal{K}_\alpha(G) \cap \mathcal{V}(G)$ *iff there are subsets* $F_1, F_2 \subset F$ *such that*

(1) $F = F_1 \cup F_2$,
(2) $\mathcal{H}_\alpha(\pi^+ F_1) = 0 = \mathcal{H}_\alpha(\pi^- F_2)$.

Remark 1. If the sets F_1, F_2 occurring in (1) are of the type F_σ, then (2) suffices for $F = F_1 \cup F_2$ to be removable for $\mathcal{K}_\alpha(G)$. Other subspaces of $\mathcal{K}_\alpha(G)$ (besides $\mathcal{K}_\alpha(G) \cap \mathcal{V}(G)$) for which removable sets F are characterized by (1) and (2) are specified in [2]. It is of interest to compare the above theorem with related results concerning the Laplace operator (cf. [1]) and other partial differential operators (cf. [5]).

Question 1. Is it true for a general relatively closed $F \subset G$ that the existence of subsets F_1, F_2 satisfying (1) and (2) is sufficient for removability of F for $\mathcal{K}_\alpha(G)$?

Remark 2. It follows from [2] that, for $0 \leq \alpha < 1$, any relatively closed $F \subset G$ is removable for $\mathcal{C}_\alpha(G) \cap \mathcal{V}(G)$ provided F can be expressed in the form (1) in such a way that

$$\pi^+ F_1 \quad \text{and} \quad \pi^- F_2 \quad \text{have } \sigma\text{-finite } \mathcal{H}_\alpha\text{-measure;} \tag{3}$$

in the special case $\alpha = 0$ (when $\mathcal{H}_0$ reduces to the counting measure) the condition (1) and (3) is necessary and sufficient for F to be removable for $\mathcal{C}_0(G)$ (cf. [3], [7]).

Question 2. Is it true for a general relatively closed $F \subset G$ that the existence of subsets F_1, F_2 satisfying (1) and (3) is necessary and sufficient for removability of F for $\mathcal{C}_\alpha(G)$ whenever $0 < \alpha < 1$?

References

[1] Carleson, L., Selected problems on exceptional sets. D. Van Nostrand Co., New York–London 1967.

[2] Chlebík, M., Král, J., Removability of the wave singularities in the plane. Aequationes Math. 47 (1994), 123–142.

[3] Dont, M., Sets of removable singularities of an equation. Acta Univ. Carolin. Math. Phys. 14 (1973), 23–30.

[4] Harvey, R., Polking, J., Removable singularities of solutions of linear partial differential equations. Acta Math. 125 (1970), 39–56.

[5] Král, J., Removable singularities of solutions of partial differential equations. In: 17th Seminar in Partial Differential Equations, Cheb 1992, 1–68. University of West Bohemia in Pilsen, Pilsen 1992.

[6] Rogers, C. A., Hausdorff measures. Cambridge University Press, Cambridge 1970.

[7] Schulze, B.-W., Potentiale bei der Wellengleichung im $\mathbb{R}^2$ und Charakterisierung der Mengen der Kapazität Null. In: Elliptische Differentialgleichungen I (Kolloq. Berlin 1969, G. Anger ed.), 137–157. Akademie-Verlag, Berlin 1970.

An eigenvalue extremal problem on convex surfaces

Pawel Kröger

Let M denote the surface of a convex body with smooth surface in 3-dimensional Euclidean space. We consider the set of all Lipschitz continuous functions ϕ on M with $\int_M \phi(x)\,d\omega(x) = 0$ where ω stands for the surface measure on M. Let ϕ_1 be a function from that set such that the quotient

$$\frac{\int_M |\nabla\phi_1(x)|^2\,d\omega(x)}{\int_M |\phi_1(x)|^2\,d\omega(x)}$$

is minimal. In other words, ϕ_1 is an eigenfunction for the first non-zero eigenvalue of the Laplace-Beltrami operator on M.

Our problem is to determine the following supremum:

$$\sup_M \sup_{\phi_1} \frac{\max\phi_1}{-\min\phi_1};$$

the first supremum is taken over all surfaces M of the above type.

A lower bound for that quantity is the ratio of the first maximum value (at zero) of the zero order Bessel function and the modulus of the first minimum of that function. We conjecture that this ratio is the exact value of the above supremum.

On the other hand, we obtain from $2n(n+4) \geq (\mathrm{diam}\, M)^2\lambda_1 \geq \pi^2 + (\ln \frac{\max\phi_1}{-\min\phi_1})^2$ where $n = 2$ is the dimension of M, diam M the diameter of M, and λ_1 the first nontrivial eigenvalue of the Laplace–Beltrami operator that the above supremum is equal or less than $\exp(\sqrt{24-\pi^2})$ (see Cheng, S. Y., Math. Z. 143 (1975), 289–297 for the first part and Kröger, P., J. Differential Geom. 36 (1992), 315–330, Cor. 3 for the second part of the above inequality).

We can conclude that 43.0 is a numerical upper bound and 2.48 is a lower bound for the above quantity.

List of lectures

Invited lectures

H. BAUER
Recollections, reflections, results — Introductory thoughts to the ICPT 94

N. BOULEAU
Construction of Dirichlet structures

G. CHOQUET
Les processus mentaux de la création

S. J. GARDINER
Harmonic approximation

W. HANSEN
Restricted mean value property and harmonic functions

V. P. HAVIN
Approximation properties of harmonic vector fields and differential forms

J. MALÝ
Nonlinear potentials and quasilinear PDE's

P. MATTILA
Removable sets for bounded analytic and Lipschitz harmonic functions and their relations to geometric measure theory and singular integrals

J. SICIAK
Bernstein-Walsh type theorems for pluriharmonic functions

G. VERCHOTA
Potentials for the Dirichlet problem in Lipschitz domains

W. WOESS
Dirichlet problem at infinity for harmonic functions on graphs

Communications

H. AIKAWA
Quasiadditivity and measure property of capacity and the tangential boundary behaviour of harmonic functions

D. ARMITAGE
Inequalities for derivatives of harmonic functions

Z. BEN NAHIA
Poisson kernel associated with the Weinstein operator

N. BEN SALEM
Mean value property and Poisson semigroup associated with the Weinstein operator

L. BEZNEA
Once more about semipolar sets and regular excessive functions

M. BEZZARGA
Duality for semidynamical systems

A. BJÖRN
Removable singularities for Hardy spaces of analytic functions

J. BJÖRN
On the regularity of the point at infinity for the Zaremba problem in the half-cylinder

J. BLIEDTNER
Approximation by harmonic functions

N. BOBOC
Dilation operators in excessive structures; Existence and uniqueness

M. BRZEZINA
On harmonic morphism for the Kolmogorov operator

A. DE LA PRADELLE
On Gaussian sheets

M. DENZER
A classification of balayage spaces by properties of Green functions

V. Ya. EIDERMAN
On exceptional sets for superharmonic functions in a halfspace: an inverse problem

M. ESSÉN

Harmonic majorization of $|x_1|$ *in subsets of* $\mathbb{R}^N$, $N \geq 2$

D. FEYEL

Hausdorff measure on the Wiener space

M. GOLDSTEIN

The solution of the Pompeiu problem

K. GOWRISANKARAN

Fatou-Doob limits and the best filters

F. HMISSI

Some remarks about subtractivity and u-processes

M. HMISSI

On potential theory of attractors of dynamical systems

A. IBRAGIMOV

Neumann problem in unbounded domains

N. JACOB

The Dirichlet boundary value problem for pseudodifferential operators generating Feller semigroups

K. JANSSEN

Some properties of the representing measure of a delta-excessive function

T. KILPELÄINEN

The Lavrentiev phenomenon and superharmonic functions

P. KRÖGER

Estimates of sums of eigenvalues of the Laplacian

V. LATVALA

Finely superharmonic functions of quasilinear elliptic equations

H. LEUTWILER

Modified Clifford analysis

P. A. LOEB

Generalized radial limits and the best approach to boundaries

T. LUNDH

Discontinuous groups and thin sets

H. MAAGLI

Perturbation of triangular resolvents and balayage of excessive pairs of functions

P. V. MALYSHEV

Systems of charged particles in the domains with complicated structure. Electric fields

V. METZ

Renormalization on fractals

Ch. MOORE

Sharp inequalities for the density of the area integral

E. POPA

Tensor product and duality

A. SAUER

The Nevanlinna-order of Hadamard-products of meromorphic functions

R. L. SCHILLING

Some properties of the sample paths of Markov processes obtained by subordination

M. SELMI

Riesz's semigroup on the half space

H. WATANABE

Double layer potentials for a bounded domain with fractal boundary

List of participants

Hiroaki AIKAWA
Department of Mathematics, Faculty of Science, Kumamoto University, 860 Kumamoto, Japan haikawa@sci.eecs.kumamoto-u.ac.jp

Kaouther AMMAR
Département de Mathématiques, Faculté des Sciences de Tunis, Campus Universitaire, 1002 Tunis, Tunisia

David ARMITAGE
Department of Pure Mathematics, Queen's University, Belfast BT7 1NN, Nothern Ireland, United Kingdom
d.armitage@queens-belfast.ac.uk

Vítěslav BABICKÝ
Faculty of Mathematics and Physics, KMA, Charles University, Sokolovská 83, 18600 Praha 8, Czech Republic

Heinz BAUER
Mathematisches Institut, Universität Erlangen-Nürnberg, Bismarckstraße 1a, 91054 Erlangen, Germany bauerh@mi.uni-erlangen.de

Jürgen BAUER
Fachbereich Mathematik, Universität Frankfurt, Robert-Mayer-Straße 6-8, Postfach 11 19 32, 60054 Frankfurt am Main, Germany
jbauer@math.uni-frankfurt.de

Zouhir BEN NAHIA
Département de Mathématiques, Faculté des Sciences de Tunis, Campus Universitaire, 1060 Tunis, Tunisia

Chedlia BEN SALAH
Département de Mathématiques, Faculté des Sciences de Tunis, Campus Universitaire, 1002 Tunis, Tunisia

Nejib BEN SALEM
Département de Mathématiques, Faculté des Sciences de Tunis, Campus Universitaire, 1060 Tunis, Tunisia

Caroline BERTIN
Mathematical Institute, Budapestlaan 6, Postbus 80.010, 3508 TA Utrecht, The Netherlands

Lucian BEZNEA
Institute of Mathematics, Romanian Academy, P.O. Box 1-764, 70700 Bucharest, Romania beznea@imar.ro

Mounir BEZZARGA
Département de Mathématiques, Institut Préparatoire aux Etudes d'Ingénieurs de Monastir, Route de Kairouan, 5000 Monastir, Tunisia

Anders BJÖRN
Matematiska Institutionen, Linköpings Universitet, 58183 Linköping, Sweden anbjo@math.liu.se

Jana BJÖRN
Matematiska Institutionen, Linköpings Universitet, 58183 Linköping, Sweden jabjo@math.liu.se

Jürgen BLIEDTNER
Fachbereich Mathematik, Universität Frankfurt, Robert-Mayer-Straße 6-8, Postfach 11 19 32, 60054 Frankfurt am Main, Germany
bliedtne@math.uni-frankfurt.de

Nicu BOBOC
Faculty of Mathematics, University of Bucharest, Str. Academiei 14, 70109 Bucharest, Romania

Nicolas BOULEAU
Laboratoire de Math. Appliquées, U.A.CNRS 1502, École National des Ponts et Chaussées, La Courtine, 93167 Noisy Le Grand, France
nb@enpc.enpc.fr

Miroslav BRZEZINA
Department of Numerical and Applied Mathematics, Technical University, Hálkova 6, 46117 Liberec, Czech Republic
miroslav.brzezina@vslib.cz

Fioralba CAKONI
Department of Mathematics, Faculty of Natural Sciences, University of Tirana, Tirana, Albania

Eric A. CATOR
Mathematical Institute, Budapestlaan 6, Postbus 80.010, 3508 TA Utrecht, The Netherlands cator@math.ruu.nl

Gustave CHOQUET
16, Avenue d'Alembert, 92160 Antony, France

Jaroslav ČERNÝ
Faculty of Mathematics and Physics, KMA, Charles University, Sokolovská 83, 18600 Praha 8, Czech Republic

Arnaud DE LA PRADELLE
30, Avenue de l'observatoire, 75014 Paris 5, France

Markus DENZER
Fachbereich Mathematik, Universität Frankfurt, Robert-Mayer-Straße 6-8, Postfach 11 19 32, 60054 Frankfurt am Main, Germany
denzer@math.uni-frankfurt.de

Miroslav DONT
Department of Mathematics, FEL, Czech Technical University, Suchbátarova 2, 16627 Praha 6, Czech Republic
dont@math.feld.cvut.cz

Eva DONTOVÁ
Department of Mathematics, FJFI, Czech Technical University, Trojanova 13, 12000 Praha 2, Czech Republic

Vladimir Ya. EIDERMAN
Department of Higher Mathematics, Moscow State Civil Engineering University, Yaroslavskoe shosse 26, 129337 Moscow, Russia

Matts ESSÉN
Department of Mathematics, University of Uppsala, Box 480, 75106 Uppsala, Sweden, matts.essen@math.uu.se, matts@math.uu.se

Eva FAŠANGOVÁ
Faculty of Mathematics and Physics, KMA, Charles University, Sokolovská 83, 18600 Praha, Czech Republic fasanga@karlin.mff.cuni.cz

Denis FEYEL
Département de Mathématiques, Université Evry-Essone, Boulevard des Coquibus, 91025 Evry, France

Jiřina FROLÍKOVÁ
Faculty of Mathematics and Physics, Charles University, Sokolovská 83, 18600 Praha 8, Czech Republic frolik@karlin.mff.cuni.cz

Jaroslav FUKA
Mathematical Institute, Czech Academy of Sciences, Žitná 25, 11567 Praha 1, Czech Republic

Stephen J. GARDINER
Department of Mathematics, University College Dublin, Dublin 4, Ireland
gardiner@irlearn.ucd.ie

Myron GOLDSTEIN
Department of Mathematics, Arizona State University, Tempe, Arizona 85287, USA

Kohur GOWRISANKARAN
Department of Mathematics, McGill University, 805 Sherbrooke W, Montreal H3A 2K6, Canada gowri@math.mcgill.ca

Wolfhard HANSEN
Fakultät für Mathematik, Universität Bielefeld, Postfach 10 01 31, 33501 Bielefeld, Germany hansen@mathematik.uni-bielefeld.de

Victor P. HAVIN
Department of Mathematics, St. Petersburg State University, Bibliotechnaja ploshchad 2, St. Peterhof, 198904 St. Petersburg, Russia
havin@math.lgu.spb.su

Farida HMISSI
Département de Mathématiques, Faculté des Sciences de Tunis, Campus Universitaire, 1060 Tunis, Tunisia

Mohamed HMISSI
Département de Mathématiques, Faculté des Sciences de Tunis, Campus Universitaire, 1060 Tunis, Tunisia mhmissi@stardent.cck.tn

Akif IBRAGIMOV
Profsojuznaya 18/1, Apt. 129, 117292 Moscow, Russia

Niels JACOB
Institut für angewandte Mathematik und Statistik, Technische Universität München, Postfach 20 10 32, 80010 München Germany,
n.jacob@appl-math.tu-muenchen.de

Klaus JANSSEN
Mathematisches Institut, Universität Düsseldorf, Universitätsstraße 1, Gebäude 25.13.01, 40225 Düsseldorf 1, Germany
janssenk@clio.rz.uni-duesseldorf.de

Tero KILPELÄINEN
Department of Mathematics, University of Jyväskylä, P.O. Box 35, 40351 Jyväskylä, Finland terok@math.jyu.fi

Pawel KRÖGER
Mathematisches Institut, Universität Erlangen-Nürnberg, Bismarckstraße 1a, 91054 Erlangen, Germany kroeger@mi.uni-erlangen.de

Josef KRÁL
Mathematical Institute, Czech Academy of Sciences, Žitná 25, 11567 Praha 1, Czech Republic

Peter KUBACH
LG Numerische Mathematik, Fernuniversität Hagen, Lützowstraße 125, 58084 Hagen, Germany kubach@numerik.fernuni-hagen.de

Visa LATVALA
Department of Mathematics, University of Joensuu, P.O. Box 111, 80101 Joensuu, Finland latvala@joyl.joensuu.fi

Heinz LEUTWILER
Mathematisches Institut, Universität Erlangen-Nürnberg, Bismarckstraße 1a, 91054 Erlangen, Germany leutwil@mi.uni-erlangen.de

Peter A. LOEB
Department of Mathematics, University of Illinois, 1409 West Green Street, Urbana, IL 61801, USA loeb@math.uiuc.edu

Jaroslav LUKEŠ
Faculty of Mathematics and Physics, Charles University, Sokolovská 83, 18600 Praha 8, Czech Republic lukes@karlin.mff.cuni.cz

Torbjörn LUNDH
Department of Mathematics, University of Uppsala, Box 480, 75106 Uppsala, Sweden tobbe@math.uu.se

Habib MAAGLI
Département de Mathématiques, Faculté des Sciences de Tunis, Campus Universitaire, 1060 Tunis, Tunisia

Jan MALÝ
Faculty of Mathematics and Physics, Charles University, Sokolovská 83, 18600 Praha 8, Czech Republic maly@karlin.mff.cuni.cz

Peter V. MALYSHEV
Institute of Mathematics, ul. Repina 3, 252601 Kiev, Ukraine
malysh@sovam.com

Pertti MATTILA
Department of Mathematics, University of Jyväskylä, Seminaarink 15, 40100 Jyväskylä, Finland pmattila@jylk.jyu.fi

Dagmar MEDKOVÁ
Mathematical Institute, Czech Academy of Sciences, Žitná 25, 11567 Praha 1, Czech Republic

Volker METZ
Fakultät für Mathematik, Universität Bielefeld, Postfach 10 01 31, 33501 Bielefeld, Germany metz@mathematik.uni-bielefeld.de

Charles MOORE
Department of Mathematics, Kansas State University, Manhattan, KS 66506, USA cnmoore@ksuvm.ksu.edu

Ivan NETUKA
Faculty of Mathematics and Physics, Charles University, Sokolovská 83, 18600 Praha 8, Czech Republic netuka@karlin.mff.cuni.cz

Iva OBDRŽÁLKOVÁ
Faculty of Mathematics and Physics, KMA, Charles University, Sokolovská 83, 18600 Praha 8, Czech Republic obdrzal@karlin.mff.cuni.cz

Eugen POPA
University of Iaşi, Aleea Copou Nr.1, 6600 Iaşi, Romania
epopa@uaic.ro

Pavel PYRIH
Faculty of Mathematics and Physics, Charles University, Sokolovská 83, 18600 Praha 8, Czech Republic pyrih@karlin.mff.cuni.cz

Murali RAO
Department of Mathematics, University of Florida, Gainesville, FL 32611, USA rao@math.ufl.edu

Andreas SAUER
Fachbereich Mathematik, Gerhard-Mercator-Universität Duisburg, Lotharstraße 65, 47048 Duisburg, Germany
sauer@math.uni-duisburg.de

René L. SCHILLING
Mathematisches Institut, Universität Erlangen-Nürnberg, Bismarckstraße 1a, 91054 Erlangen, Germany schilli@mi.uni-erlangen.de

Mohamed SELMI
Département de Mathématiques, Faculté des Sciences de Tunis, Campus Universitaire, 1060 Tunis, Tunisia

Jozef SICIAK
Institute of Mathematics, Ul. Reymonta 4, 30-059 Kraków, Poland
siciak@im.uj.edu.pl

Martina ŠIMŮNKOVÁ
Department of Numerical and Applied, Mathematics of VŠST, Hálkova 6, 46117 Liberec, Czech Republic martina.simunkova@vslib.cz

Frans VAN GOOL,
Magnoliastraat 7, 2651 TD Berkel en Rodenrijs, Netherlands
frans@west.nl

Gregory VERCHOTA
Department of Mathematics, Syracuse University, 215 Carnegie Hall, Syracuse, NY 13244-150, USA mathdept@mailbox.syr.edu

Jiří VESELÝ
Faculty of Mathematics and Physics, Charles University, Sokolovská 83, 18600 Praha 8, Czech Republic jvesely@karlin.mff.cuni.cz

Hisako WATANABE
Department of Mathematics, Ochanomizu University, 2-1-1, Ohtsuka, Bunkyo-ku, 112 Tokyo, Japan watanabe@math.ocha.ac.jp

Wolfgang WOESS
Dipartimento di Matematica, Università di Milano, "F. Enriques", Via C.Saldini 50, 20133 Milano, Italy woess@vmimat.mat.unimi.it

Bei Fragen zur Produktsicherheit wenden Sie sich bitte an:
If you have any questions regarding product safety,
please contact:

Walter de Gruyter GmbH
Genthiner Straße 13
10785 Berlin
productsafety@degruyterbrill.com